Fifth Edition

The North American City

Maurice Yeates

QUEEN'S UNIVERSITY, KINGSTON
and
Centre for the Study of Commercial Activity,
RYERSON POLYTECHNIC UNIVERSITY, TORONTO

 LONGMAN

An imprint of Addison Wesley Longman, Inc.

New York • Reading, Massachusetts • Menlo Park, California • Harlow, England
Don Mills, Ontario • Sydney • Mexico City • Madrid • Amsterdam

EDITOR IN CHIEF: PRISCILLA MCGEEHON

EXECUTIVE EDITOR: ALAN MCCLARE

PROJECT COORDINATION AND TEXT DESIGN: RUTTLE, SHAW & WETHERILL, INC.

COVER DESIGNER: CHRIS HIEBERT

COVER PHOTO: ©1997 PHOTODISC

ELECTRONIC PRODUCTION MANAGER: CHRISTINE PEARSON

MANUFACTURING MANAGER: WILLE LANE

PRINTER AND BINDER: THE MAPLE-VAIL BOOK MANUFACTURING GROUP

COVER PRINTER: CORAL GRAPHIC SERVICES, INC.

Library of Congress Cataloging-in-Publication Data

Yeates, Maurice.
    The North American city / Maurice Yeates. — 5th ed.
      p.   cm.
    Includes bibliographical references and index.
    ISBN 0-321-01364-6
    1. Cities and towns—North America.   I. Title
HT122.Y4 1997   97-2378
307.76'097—dc21   CIP

ISBN 0-321-01364-6

12345678910—MA—00999897

Acknowledgments

Since nearly all of the material cited below is also listed alphabetically by author in the reference section at the back of the book, the reader may find it easier to consult that section for further information about particular source materials used in the text.

The author and publisher wish to express their thanks for permission to reprint, redraw, and/or modify material from copyrighted works:

Academic Press, Inc., New York, and the authors for Fig. 1, p. 375, from the *Journal of Urban Economics*, vol. 2 (paper by M. Edel and E. Sclar).

The American Academy of Political and Social Science for Figs. 1-4 and 5, pp. 8-9 and 11, from *The Annals of the American Academy of Political Science*, vol. 242, 1945 (paper by C. D. Harris and E. L. Ullman).

The publisher of *Architecture* for a map on p. 87 (redrawn and simplified), April issue, 1995 (article by B. McKee).

The Association of American Geographers, Washington, D.C., for the following from the *Annals of the Association of American Geographers*: data on pp. 558-559, vol. 69, 1979 (with special thanks to the author of the paper, S.H. Olsen); Figure 9, p. 106, vol. 24, 2, 1977 (paper by M. P. Conzen); and Table 1, p. 261, vol. 81, 1991 (paper by B. OhUallachain and N. Reid). From *The Professional Geographer*, Figure 2 on p. 37, vol. 45, redrawn (paper by D. Z. Sui and J. O. Wheeler). Also, a map of housing in Los Angeles on p. 22 of *A Comparative Atlas of America's Great Cities: Twenty Metropolitan Regions* (R. Abler, and J. S. Adams (eds.), 1976).

The Advisory Commission on Intergovernmental Relations for the following from *Fiscal Disparities Between Central Cities and Suburbs*, 1984: extracts from Tables A-1 and A-8.

The Ballinger Publishing Company, Cambridge, Mass., for the diagram on p. 91 from J. S. Adams (ed.), *Urban Policy-Making and Metropolitan Dynamics: A Comparative Geographical Analysis*, 1976 (paper by M. J. Dear).

The Brookings Institution, Washington, D.C., for Table 16, p. 243 (to which more recent information has been added) from *The Metropolitan Transportation Problem*, by W. Owen, 1966 (revised edition).

Irwin Clark, Publishers, for maps of land use (modified and redrawn) on Toronto on pp. 31 and 34 in *Toronto: An Urban Study*, by R. P. Baine and A. L. McMurry, 1977.

Wayne K. D. Davies for data from Table 3.1 in *The Changing Social Geography of Canadian Cities* edited by L. S. Bourne and D. Ley, 1993 (chapter by W. K. D. Davies and R. A. Murdie).

The editor and authors of the following from *Economic Geography*: Figure on p. 223, vol. 60, 1984, redrawn and modified (paper by C. L. Moore and M. Jacobson); Figures 1 and 2, vol. 41, 1965, redrawn and modified (paper by M. Yeates); Figures on pp. 357, 359, 361, vol 59, 1983, recompiled and redrawn (paper by A. J. Scott); Figure 2 on p. 48, vol. 67, 1991, redrawn (paper by Chatterjee and Abuttasnath).

The publisher of *The Economist* for the diagrams (redrawn and modified slightly) in the issues of June 22, 1991, on p. 26, May 2, 1992, and Aug. 24, 1985, p. 68; and, tabular material from the insert "The Endless Road" (Tables 5.4, 5.5, and 5.6 in text) of Oct 17, 1992.

Edward Arnold Ltd., London, for table on p. 80 from vol. 10, 1986 *International Journal of Urban and Regional Research* (paper by R. Harris).

Gordon Fielding for Figure 12.1 (p. 289), and 12.3 (p. 298), both redrawn and modified, from *The Geography of Urban Transportation* edited by S. Hanson, 1995.

Irene Fabbro for permission to redraw maps 4, 5, and 6, from her research paper "The Social Geography of Metropolitan Toronto: A Factorial Ecology Approach," 1986, Department of Geography, York University.

The editor of *Geographical Analysis* and Ohio State University Press for Table 1, p. 7, vol 25, 1993 (paper by P. B. Waddell, B. J. L. Berry, and I. Hoch).

Peter Gordon and Pion, London, for Table 1, p. 224, in vol. 21, 1989 in *Environment and Planning A* (paper by E. Heikkila, D. Dale-Johnson, P. Gordon, R. B. Peiser, and H.W. Richardson).

David L. A. Gordon for Table 12.4 in text.

Susan Hanson and Geraldine Pratt, and Routledge (publishing company), for Figure 4.1, p. 96, from *Gender, Work, and Space.*

Harvard University Press, Cambridge, Mass., for extracts from the table on p. 131 of B. Chinitz *Freight and the Metropolis*, 1960.

Graham Humphrys for Figure 4.9, p. 322, modified and redrawn, from "Power and Industrial Structure," in J. W. House (ed.), *The U.K. Space: Resources, Environment and the Future*, 1986 (paper by G. Humphrys).

D. L. Huff for the diagrammatic model of factors affecting consumer space preferences and behavior on p. 165, vol. 6 of *Papers and Proceedings of the Regional Science Association*, 1960 (paper by D. L. Huff).

The editors of the *International Journal of Urban and Regional Research* for Figure 1, p. 543, vol. 16, 1992 (paper by E. G. Goetz).

Ken Jones, Director, Centre for the Study of Commercial Activity, Ryerson Polytechnic University, for Figure 1, and data from Tables 3 and 7 from *International Comparisons of Commercial Structure* (J. W. Simmons, 1995).

The editor of the *Journal of Planning Education and Research* for extracts from Tables 1 (p. 151) and 2 (p. 156) from vol. 12, 1993 (paper by A. Black).

The editor of the *Journal of Regional Science* for the diagram on p. 418 from vol. 25, 1985 (paper by J. E. Anderson), and Table 2, p. 230 from vol. 27, 1987 (paper by J. C. Stabler).

Kendall-Hunt Publishing Co., Dubuque, Iowa, and the Geographical Society of Chicago, to redraw and update Fig. 10, p. 143 from *Chicago: Metropolis of the*

Mid-Continent, by I. Cutler, 1982; and, Kendall-Hunt and the author for the diagram on p. 201 in *The Los Angeles Metropolis*, by H. J. Nelson, 1983.

The Longman Group Ltd., Harlow, England, for the following from *Urban Studies*: Table 1 and extracts from Table 6, pp. 209-217, vol. 22, 1985 (paper by B. Edmonston, M. A. Goldberg, and J. Mercer); and Figure 1 (redrawn), p. 356, vol. 14, 1977 (paper by C.E. Reid).

James Lorimer and Co., Ltd., Publishers, Toronto, for Figure 4 (modified and redrawn) from *A Citizen's Guide to City Politics*, by J. Lorimer, 1972.

McGill-Queens University Press, Montreal and Kingston, to modify and redraw from *The Changing Social Geography of Canadian Cities* edited by L. Bourne and D. Ley, 1993: Figure 7.1, p, 139 (chapter by S. H. Olsen and A. L. Kobayashi); and, Figure 16.1, p. 301 (chapter by M. J. Dear and J. Wolch).

McGraw-Hill Book Co., New York, for Fig. 2, p. 34, from *The Black Ghetto: A Spatial Behavioral Perspective*, by H. M. Rose, 1971 (which has been updated for 1981 and 1991).

The National League of Cities for Figure 1, p. 4, redrawn from *All in it Together: Cities, Suburbs, and Local Economic Regions* (by L. C. Ledebur and W. R. Barnes).

John G. Nelson for table on p. 17 of *Toward a Sense of Civics*, 1993.

The Population Reference Bureau Inc., Washington, D.C., for Figure 6 from "Americans on the Move" *Population Bulletin*, vol. 48 (paper by P. Gober).

R. B. Potter for Fig. 5.9, p. 159, from *The Urban Retailing System*, printed by the Gower Publishing Co., Aldershot (Eng.), 1982.

Praeger Publications and N. Mager for Figure 2.1, redrawn and modified, from *The Kondratieff Long Waves,* by N. H. Mager, 1987.

The Queen's Printer for Ontario, for the following from the *Report of the Greater Toronto Area Task Force* (listed as GTA, 1996): Fig. 2.1, p. 41; Table 2.10, p. 51; Table 4.1, p. 112; and, The Task Force for an unpublished map (Figure 10.10 in the text).

The Regional Science Association for: extracts from Table 2, pp. 55-80, vol. 37, 1976 (paper by K. F. McCarthy); and Table 1, p. 34, vol. 72, 1993 (paper by D. S. West); both in *Papers of the Regional Science Association*.

E. Relph for illustrations on pp. 78 and 79 from *The Toronto Guide, The City, The Metro, The Region*, 1990, redrawn. Revised and reprinted as Major Report 35, Centre for Urban and Community Studies, University of Toronto, 1997.

Sage Publications Inc., Beverly Hills, Calif., for Fig. 1, p. 68, from *Urbanization, Planning and National Development*, by J. Friedmann, 1973; and for material used by the author for the compilation of Table 15.8 from pp. 182-194 from *Small and Large Together: Governing the Metropolis*, by H. W. Hallman, 1977. Also, extracts from Table 4, p. 476, in vol. 29, 1994 (paper by S. K. Ward); Table 3, p. 34, in vol. 30, 1994 (paper by S. Negrey and M. D. Zickel); and Table 4, p. 198, vol. 30, 1994 (paper by A. DiGaetano), all in *Urban Affairs Quarterly*.

Andrew Sancton for Table 2, p. 16 in *Governing Canada's City-Regions*, 1994.

James W. Simmons for extracts from the figure on p. 64 from *Systems of Cities: Readings on Structure, Growth and Policy*, edited by L. S. Bourne and J. W. Simmons, 1978, and published by the Oxford University Press, New York; and the figure on p. 118, redrawn and updated, from *Canadian Cities in Transition*, edited by T. Bunting and P. Filion (eds.), 1991, and published by Oxford University Press, Toronto.

Robert Sinclair for permission to modify and redraw a figure on p. 216 from *Managing and Marketing of Urban Development and Urban Life* edited by G. O. Braun, 1994.

R. Sivitanidou and Pion, London, for extracts from Table 3, p. 17, 1994, from "W & D Facilities and Community Attributes: An Empirical Study", subsequently published in *Environment and Planning, A* (1996), 1261–1278.

Stanford University Press, Stanford, Calif., and the Board of Trustees of the Leland Stanford Junior University, for Figure V.14, pp. 42-43, and Table II.1, p. 4, from *Social Area Analysis: Theory, Illustrative Applications and Computational Procedures*, by E. Shevky and W. Bell, 1955.

The Minister of Supply and Services, Ottawa, Canada, for Figure 2a, p. 7 from *Land in Canada's Urban Heartland*, by M. Yeates, 1985.

The editor for Fig. 2 (redrawn with modifications), p. 301, vol. 67, from *Tijdschrift voor Economische en Sociale Geografie* (paper by H. Popp).

The University of Chicago Press for charts I and II (redrawn) from *The City*, by R. E. Park, E. W. Burgess, and R. D. McKenzie, 1925.

V. H. Winston and Sons, Inc., for the following from *Urban Geography*: the figure on p. 135 (redrawn), vol. 6, 1985 (paper by T. C. Archer and E. R. White); modification of the table on pp. 109-110 in vol. 8, 1987 (paper by R. C. Morrill); and Fig. 1 on p. 5 in vol. 13, 1992 (paper by A. Schwartz). Also, Fig. 4.9 (modified), p. 89, from *The Geography of Housing*, by L. S. Bourne, 1981.

Contents

Preface　　　　　　　　　　　　　　　　　　　　　　　**xv**

CHAPTER 1　　*Urban North America: A Spatial View*　　**1**

Population Clusters and Aspatial Processes　　**2**

Defining Metropolitan Areas　　3
Aspatial Processes Influencing Spatial Patterns　　8
The Effect of Accumulated Urban Infrastructure　　17

The Components of Urban Analysis　　**18**

Urban Patterns　　19
Philosophical Approaches　　20
Theoretical Bases　　22
Scales of Analysis　　25
Techniques for Analysis　　27

Conclusion　　**30**

Part 1　　*North American Urban Systems*　　**31**

CHAPTER 2　　*The Processes of Urbanization*　　**33**

Urbanization　　**33**

North American Urbanization　　35
Causes of Urbanization　　38
The Location of Urban Areas　　39

The Spread and Growth of Urban Areas　　**42**

The Spread of Innovation　　42
Migration　　43
Capital Flows　　44
A Model of Urban System Growth　　48

The Urban Hierarchy　　**51**

The Importance of the Urban Hierarchy　　52

The Economic Bases of Urban Areas　　**52**

The Urban Multiplier　　53
The Productivity of Urban Areas　　57

Conclusion　　**59**

CHAPTER 3   *Waves of Urban Development:
The Evolution of the United States
Urban System*   *60*

Economic Cycles and Urban Development   **62**

Long Waves and Major Eras of Urban Development   64

Urban System Development in the United States   **67**

The Frontier Mercantile Era, 1790–1845   67
Early Industrial Capitalism, 1845–1895   70
National Industrial Capitalism, 1895–1945   76
Mature Industrial Capitalism, 1945–Date   80

Conclusion   **93**

CHAPTER 4   *The Evolution of the Canadian Urban
System*   *94*

Urban Development in Mercantile and Staples Economies   **94**

The Urban System in the Mercantile Era   95
The Urban System in the Era of Early Industrial Capitalism,
    1845–1895   96

**Urban Development in the Era of Industrial Capitalism,
1895–1945**   **98**

A Latent Core-Periphery Framework   101

**Urban Development in the Era of Mature Industrial
Capitalism, 1945–Date**   **104**

Bases of Urban Growth   104
The Changing Spatial Organization of the Urban System   108

Conclusion: Urbanism in Canada and the United States   **120**

CHAPTER 5   *Urban Areas as Centers of Manufacturing*   *123*

The Importance and Nature of Manufacturing   **123**

Examining Manufacturing   124

Regional Changes in the Location of Manufacturing   **128**

Snow Belt–Sun Belt Shifts   131

Least-Cost Theory and the Location of Manufacturing   **132**

The Assumed Nature of the Economy   133
Least-Cost Theory   134

Industrial Restructuring   **142**

Manufacturing's New Era   143

The Location of Multiplant Manufacturing Firms   151
The Effect of Restructuring on the Location of Manufacturing   154
Conclusion   **160**

CHAPTER 6 *Urban Areas as Centers of Service Activities*   *161*

Types of Tertiary Activities   **162**
Employment Change in the Tertiary Sector   163
The Interurban Location of Consumer Services   **165**
The Central Place Model   166
Concepts Concerning Central Places   166
Changes in the Central Place Hierarchy   177
The Interurban Location of Producer Services   **179**
Factors Influencing the Concentration of Producer Services   180
The "World City" Hypothesis   181
The Concentration of Producer Services   183
Trade in Producer Services   185
**The Interurban Location of Governmental and Public
Services**   **186**
The Production and Consumption of Governmental and
    Public Services   187
Conclusion   **190**

Part 2   *Internal Structure of Urban Areas*   *193*

CHAPTER 7   *Internal Structure and Deconcentration*   *195*

Major Periods of Accumulation of Urban Infrastructure   **196**
Urban Areas During the Mercantile Era   198
Urban Areas During the Early Industrial Era   199
Metropolitanism, 1895–1945   202
**The Current Era: The Suburban Reinvention of Urban
North America**   **211**
Forces Accelerating Deconcentration   211
Edge Cities   223
Characteristics of Late-Twentieth-Century Suburbs   224
Urban Structure in the Suburban Era   226
Conclusion   **230**

CHAPTER 8   *Urban Land Use and Population Change*   **232**

Land Use, Economic Rent, and Land Values   **232**

Land Use Theory   233
Land Use Zones in the Urban Environment   236
Empirical Examinations of the Changing Patterns of
    Economic Rent   238
Sources of Economic Rent   245

Models of Urban Population Redistribution   **247**

Population Densities in Monocentric Urban Areas   247
Population Densities in a Multicentered Urban Area   252

Public Policy and Land Use Change   **255**

The Increasing Consumption of Urban Land   255
Zoning   258
The Property Tax and Land Development   261

Conclusion   **263**

CHAPTER 9   *The Intraurban Location of Manufacturing*   **264**

**The Accumulation of Manufacturing Infrastructure
Within Urban Areas**   **266**

The Middle-Industrial-Revolution Railway Era   267
The Late-Industrial-Revolution Highway Era   269

A Typology of Intraurban Manufacturing Locations   **273**

Central Clusters   273
Decentralized Clusters   274
Random Spread   277
Peripheral Patterns   277

**Restructuring Processes and the Changing Location of
Manufacturing**   **278**

Product Life Cycles   278
Capital-for-Labor Substitution   280

**Consequences of Manufacturing Restructuring in an
Industrial Metropolis**   **284**

Growth of Detroit   284
Economic and Social Impacts of Restructuring   285
Spatial Consequences of Restructuring   286

Conclusion   **288**

CHAPTER 10   *The Intraurban Location of Employment
in Service Activities*   **290**

The Intraurban Location of Consumer Services   **293**

Urban Retail Location   294
Spatial Aspects of Consumer Behavior and Cognition   308

The Intraurban Location of Producer Services   **310**

The Location of Wholesale and Distribution Activities   311
Intraurban Office Location   312

The Intraurban Location of Public Services   **321**

Characteristics of Public Facilities and Location   321
Public Facility Location and Public Policy   325

Conclusion   **326**

CHAPTER 11   *Social Spaces Within Urban Areas*   *327*

The Human Ecology Approach and Urban Social Space   **327**

The Human Ecology Approach   328
Defining Social Space   330

Behavioral Processes, Societal Structures, and Urban Social Space   **342**

Behavioral Perspectives on Change in Social Space   342
Ethnicity and Race   348
Structural Perspectives on the Formation of Social Space   358

Conclusion   **365**

CHAPTER 12   *Residential Location and Urban Revitalization*   *367*

Macro Influences on the Demand for and Supply of Housing   **367**

The Demand for Housing   368
The Supply of Housing   370
Filtering and Housing Policy   373

Market-Oriented Perspectives on Residential Location   **375**

The Residential Location Model   375

Behavioral Perspectives on Residential Location   **379**

The Brown-Moore Model   379
A Sovereign Consumer?   382

Housing Submarkets and Residential Location   **383**

The Property Industry   383
The Role of Financial and Governmental Institutions   386
The Role of the Real Estate Industry   389
Landlords and Housing Submarkets   390
The Abandoned Housing Submarket   391
Homelessness   395

Approaches to Urban Revitalization   **399**

Redevelopment   400

Gentrification 404

Conclusion 406

CHAPTER 13　*Women and Urban Form and Structure* 408

Change in Home and Work 409

Women in the Paid Workplace 409
Families and Households 412

Women in Men's Cities 417

From Sex Roles to Urban Structure 420
Urban Aspects of the Feminization of Poverty 427

Improving Urban Living 430

Symmetrical Families 432
Child Care and Urban Environments 434
Women-Friendly Urban Areas 439

Conclusion 440

Part 3　*Improving Livability in Urban North America* 443

CHAPTER 14　*Intra- and Interurban Transportation* 445

Interurban Transportation 446

Intraurban Activity Patterns 450

Movement Characteristics 452
Modeling Movements Related to Activity Patterns 455

Urban Spatial Structure, the Journey to Work, and Modal Choice 460

Transportation in Monocentric and Polycentric Urban Structures 460
Outer- and Inner-Urban-Area Contrasts in Modal Choice 462

Rapid Transit Approaches to Metropolitan Spatial Restructuring 468

Experience in the United States with New Rapid Transit Systems 469
Subsidizing Public and Private Transport 473

Conclusion 476

CHAPTER 15 *Managing Urban Areas in a Global Economy* *477*

The Changing Role of Local Government **478**

Local Government at the End of the Nineteenth Century 478
Local Government in the Middle of the Twentieth Century 480
Local Government in the Second Half of the Twentieth
   Century 482

**Local Government Fragmentation and Metropolitan
Finance** **483**

The Proliferation of Local Governments 484
Sources of Local Government Revenue 486

Issues Facing Local Government **494**

Fiscal Problems 494
Organizational and Representational Problems 496

Toward Solutions **497**

Structural Reforms 500
Fiscal Reform 503

Conclusion **505**

CHAPTER 16 *Future Growth and Sustainable Urban Development* *507*

**From Resource Management to Sustainable
Development** **508**

Sustainable Development 508
North America and Sustainable Living 509
Industry and Sustainability 514
Transportation and Sustainability 521

Urban Form and Sustainability **525**

The Costs of Dispersion 525

Future Urban Patterns **528**

Macro-Urban Regions 528
Future Population Growth 532

Conclusion **535**

Bibliography **537**

Index **567**

Preface

The fifth edition of *The North American City*, like its predecessors, is intended as a textbook for undergraduates, principally at colleges and universities in the United States and Canada. The aim is to provide an overview of the broad scope and increasingly diversified nature of the geographer's interests in, and contributions to, the study of urban areas. This overview, therefore, lays the foundations for more advanced study in the field relating to the book's three main themes: the organization of the North American urban system, the internal spatial structure of urban areas, and public policy relating to urban areas.

In the twenty-five years since the first edition of this book appeared, coauthored, as were the second and third editions, with Barry Garner, urban geography has evolved as an important field within the discipline. For example, the urban geography specialty group has the largest number of members of any of the specialty groups in the Association of American Geographers. The number of courses in urban geography has proliferated, and many specialties (or subfields), such as urban social geography, urban systems, retail structure and location, women in the city, urban industrial geography, urban transportation, and sustainable urban development, have emerged as significant bodies of knowledge. Furthermore, theoretical approaches to urban analysis have become quite diverse, often giving rise to quite different approaches to the study of urban issues.

FIFTH EDITION REVISIONS

The fifth edition of *The North American City* has been expanded to take into account the ever-widening range of urban interests and variety of fruitful theoretical approaches that are being applied to the study of urban questions. It is perhaps impossible for any one person to address adequately the variety of interests and diversity of perspectives in contemporary urban geography. Nevertheless, an attempt has been made to develop a distinct and cohesive approach that reflects the range of current interests.

Following the advice of many colleagues, particularly detailed commentaries on the fourth edition provided by five anonymous publisher's reviewers, the fifth edition provides a thorough revision. An underlying theme relating urban growth and change to major eras of economic development fosters cohesiveness and provides a framework that embraces different theoretical perspectives. This theme also permits the text to focus on some of the abiding postindustrial issues of the time relating to the complexities and challenges of perceived transitions to new types of economies and social/cultural relations.

The section on urban systems has been expanded from two to five chapters to include discussions of: processes of urbanization; the evolution of the United States urban system; the evolution of the Canadian urban system; urban areas as centers of manufacturing; and urban areas as centers of service activities. The chapters on the Canadian urban system and service activities are essentially new. In the latter case the chapter highlights the importance of consumer, producer, and governmental and public services in modern urban development, and in the establishment of a layer of "world cities."

The section on the internal structure of urban areas has been expanded from six to eight chapters to include discussions of: deconcentration and the suburban reinvention of urban North America; land use theory, rural-to-urban land conversion, and taxation; the intraurban location of manufacturing; the intraurban location of service activities; social spaces within urban areas; residential location and urban revitalization; and, women and urban form. Revisions and new sections include discussions of: "edge" cities; polycentric price and population density models; consequences of manufacturing restructuring on the internal structure of urban areas; recent trends in the intraurban location of consumer, producer, and governmental and public service activities; the nature and extent of residential proximity between racial groups; the underclass debate; homelessness; waterfront redevelopment; and women and the urban environment.

The third section, relating to improving livability in urban North America, has been expanded from one to three chapters. The new chapters concern urban transportation, and sustainable urban development. A thoroughly revised chapter from the previous edition concerns local government and the management of urban areas. The local government section has always been a strength of the text, informing the extensive public debate that has occurred in many metropolitan areas about local reorganization and reform. The new urban transport chapter includes extensive discussions of modal choice in monocentric and polycentric urban areas, and the public/private transport debate. The concluding chapter, concerning paths to more sustainable urban development in North America, outlines some of the hard choices that have to be made, and places these choices in the context of future macro-regional urban development.

ACKNOWLEDGMENTS

In acknowledging the help of the many people who contributed to the success of previous editions, and, I hope, the fifth edition, I would like to thank particularly colleagues at Queen's University, Ryerson Polytechnic University and the Ontario Council on Graduate Studies. They, and many friends in the profession, have continued to offer excellent advice and assistance. At Queen's University, Peter Goheen, Eric Moore, Mark Rosenberg, John Holmes, Evelyn Peters, and David Gordon have provided suggestions and comments. At Ryerson Polytechnic University, Dennis Mock has been most supportive, and Ken Jones and Jim Simmons and the staff at the Centre for the Study of Commercial Activity have provided the stimulating type of research environment that has maintained the determination of this Senior Fellow to complete the project.

A special thanks should be accorded the five prepublication reviewers of the fifth edition manuscript who provided great encouragement and many useful suggestions that have been incorporated in the text. They are: Ray Forgianni, Carthage College; Mary Ellen Mazey, Wright State University; Larry D. McCann, University of Victoria; James L. Mulvihill, California State University; Diana Richardson, San Diego State University. Also, and most important, comments and suggestions received over the years from instructors and students who have used the book have been incorporated throughout this (and previous) revisions. Students at Queen's, and more recently Ryerson, have provided highly constructive feedback over the years (including my daughter)!

A special word of thanks to Ross Hough, of the Cartographic Laboratory in the Department of Geography at Queen's University, who prepared the artwork for the fifth edition as he has done a number of the previous editions. George Innes provided the necessary photographic services. Finally, I would like to thank my family and friends who continue to be interested in, and provide support for, my activities in the rather esoteric and focused business of book writing.

Maurice Yeates

Urban North America: A Spatial View

American and Canadian urban areas have much in common. They have been formed fairly recently in world history as a result of massive immigration primarily from Europe and, more recently, from Central America and South and Southeast Asia. The longest continual European settlement is usually regarded as being at the site of what is now known as Quebec City, dating from 1608. Urban development spread primarily from east to west, with urban places invariably spearheading the course of European settlement. Thus, eastern ports, particularly in the United States, have been leading foci of urban development for much of the last two hundred years. New York and Philadelphia, for example, were the largest urban areas in North America by 1790, with 33,000 and 28,000 inhabitants respectively.

In both the United States and Canada, urban growth has been primarily associated with the rise of factory production in the second half of the nineteenth century, and the myriad business and commercial activities that are associated with the implementation of capitalist and market forms of production and consumption. Places such as Chicago, which grew from virtually nothing in 1830 to more than 1 million people by 1890, Los Angeles, which grew from 50,000 in 1890 to nearly 3 million by 1940, and Miami, increasing from 30,000 in 1920 to nearly 2 million in 1990, exemplify the explosive nature of this growth. On the other hand, although rapidity of growth appears to have been a hallmark of North American urban development, the counterpoint during the past thirty years has been metropolitan stagnation in certain parts of North America, particularly some urban areas located in the historic manufacturing heartland of the Middle West and Northeast. Also, nearly all urban areas in North America have declining inner cities, a self-inflicted phenomenon associated with suburbanization and the decentralization of economic activities. The extent of inner-city decay, lower central city compared to suburban income, and racial polarization, varies enormously across urban North America. These symptoms are more prevalent in United States cities than Canadian urban areas.

Apart from common economic bases of development and relatively recent formation based largely on waves of immigration, United States and Canadian urban areas have a similar appearance. They have similar styles of construction and layout, with grid-iron street plans, skyscraper building clusters symbolic of central business areas, limited-access radial and circumferential highways reinforcing and reflecting local government fragmentation, and low-density suburbs providing mainly single-family housing. There is, on the whole, little attempt to conserve what is worth conserving of the old, and every attempt to be new. Automobiles and trucks dominate inter- and intraurban transportation virtually more than in any other part of the world, generating energy demands, environmental impacts, and urban forms that are antithetical to sustainable development. Private transportation has become the prime determinant of the North American urban spatial form.

The viewpoint of this textbook is, therefore, a *spatial* one, with the objective of linking social, economic, and political processes to an understanding of urban patterns (Leitner, 1989). This linkage is vital because an informed citizen must want to know what has been, is, and will be affecting their access to jobs, services, and social activities in urban areas. With such understanding comes an awareness of what needs to be done to improve access to opportunities—but knowledge not only facilitates awareness, it also fosters effective participation in a wide variety of activities directed toward improving the quality of urban life in North America. In making a link between theories and real places, this analysis of urban North America focuses on knowledge, public policy, and action.

POPULATION CLUSTERS AND ASPATIAL PROCESSES

Figure 1.1 provides a rather dramatic view of the distribution of population in the United States and Canada in 1990–1991, as if a picture were taken from outer space at night. This is not, however, a photograph from space. It is a map based on census data, where one small dot represents 1,000 persons. Although there is a wide scattering of dots over the eastern half of the continent, the most noticeable feature is the clustering of people into agglomerations of urban areas. Figure 1.2 shows the way in which the Bureau of the Census aggregates these clusters of people into metropolitan areas in the United States on the basis of counties. These metropolitan areas include 77.5 percent of the population of the United States and 20 percent of the land area. In Canada, where the definition of metropolitan areas is somewhat more restricted, the proportion of the population residing in metropolitan areas is 61 percent (Bourne, 1995).

Defining Metropolitan Areas

The definition of metropolitan areas in the United States may appear rather complicated, but it has an interesting geographical base. A metropolitan statistical area (MSA) consists of a core county containing either a defined city with at least 50,000 people or a number of smaller places totaling at least 100,000 persons (75,000 in New England), along with adjacent communities in counties having a *high degree of economic and social integration with the core* (Bureau of the Census, 1991). For example, the Wichita (Kansas) MSA with a 1990 total population of 485,000 consists of the city of Wichita (population 304,000) in Sedgwick County, and the adjacent counties of Butler and Harvey. In cases where a number of metropolitan areas overlap, as is emphasized by the running together of masses of dots in Figure 1.1 along the northeastern seaboard, around the Great Lakes, coastal Florida, and on the West Coast in southern California, individual component parts of large MSAs (that is, with 1 million or more persons) may be designated primary metropolitan statistical areas (PMSAs). For example, Dade County PMSA (1.9 million), which includes Miami-Hialeah (639,000), combines with Broward County PMSA (1.3 million), which includes Fort Lauderdale-Hollywood (343,000), to form the consolidated metropolitan area (CMSA) of Miami-Fort Lauderdale (3.2 million).

In Canada the definition of census metropolitan areas (CMAs) follows the same geographic base in that they involve an urbanized core and a surrounding area that has a high level of integration with the core. The urbanized core is defined more strictly than in the United States, as a minimum population of 100,000 is required. The level of integration is determined in the context of the labor shed, or range of commuters with respect to the core, whereas the commuting region is defined in terms of aggregations of whole counties and parts of counties. The United States definition is, therefore, much more flexible with respect to the nature of metropolitan areas and polycentricity. This is in part related to the much more important role that the definition of metropolitan areas plays in the definition of target populations with respect to a wide range of federal programs in the United States, where the population of an urban area may, or may not, be entitled to participate in a particular program according to its metropolitan status (Dahmann and Fitzsimmons, 1995).

Thus, the patterns that appear on the "night-time" maps (Figure 1.1) demarcating the clustering of people into urban areas, the merging of clusters in certain areas (such as Miami-Fort Lauderdale CMSA), and the ways in which individual clusters (such as the one delimiting Wichita MSA) spread out with diminishing intensity, represent the outcome of activities of people and businesses operating in geographic space. Businesses and people (residing in households) locate where they do for a variety of reasons. Once located, these businesses, whether they be factories, offices, shops, hospitals, schools, or multiplex

Figure 1.1 The concentration of population in Canada and the United States into clusters, 1990–1991.

(*Source:* Washington, D.C.: Bureau of the Census, Population Division; Ottawa: Statistics Canada, Geography Division.)

Figure 1.1 (continued)

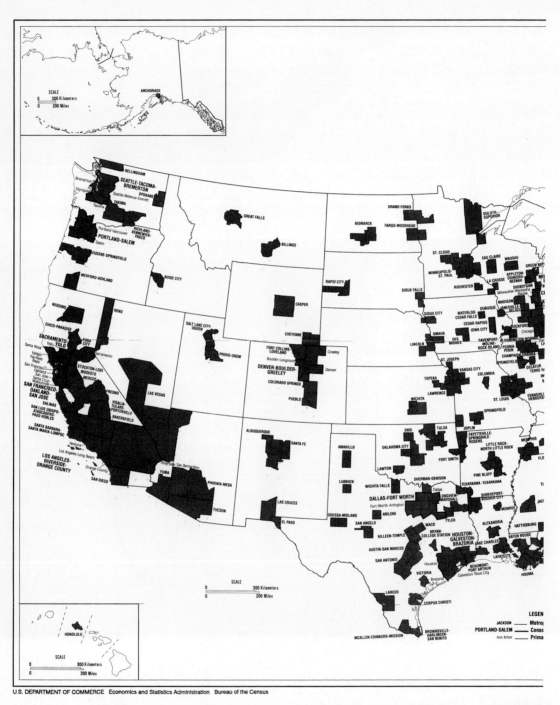

Figure 1.2 Metropolitan areas (MSAs, CMSAs, PMSAs) in the United States as defined by the Bureau of the Census, 1990.

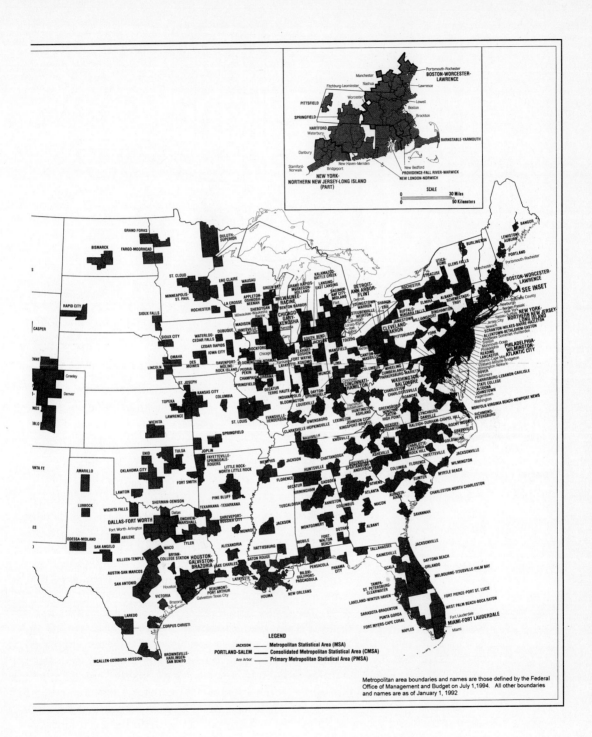

LEGEND

JACKSON —— Metropolitan Statistical Area (MSA)

PORTLAND-SALEM ——— Consolidated Metropolitan Statistical Area (CMSA)

Ann Arbor —— Primary Metropolitan Statistical Area (PMSA)

Metropolitan area boundaries and names are those defined by the Federal Office of Management and Budget on July 1, 1994. All other boundaries and names are as of January 1, 1992.

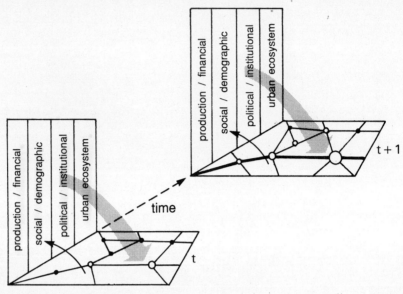

Figure 1.3 Spatial organization is a reflection of the aspatial organization of society.

movie theaters, have to be connected with people. Spatial patterns of people and human activities, which locate where they do as a result (as it turns out) of mostly aspatial processes, generate flows of connecting interlinkages. Spatial patterns cannot exist without these interlinkages. Urbanism is about location *and* connection, for connection gives life to location.

Aspatial Processes Influencing Spatial Patterns

A basic issue concerns, therefore, the processes that may give rise to these patterns. As there are very few, if any, spatial processes *per se,* attention is usually directed toward *aspatial* processes that operate in geographic space. Bourne (1991) defines four broad categories from which these aspatial processes may be derived: production/financial; social/demographic; political/institutional; and the urban ecosystem (Figure 1.3). There is also one spatial category that warrants attention—the existing built environment.

Processes Arising from the Production/Financial System   Urban areas are essentially related to productive efficiency: The agricultural revolution increased farm production and reduced the need for extensive farm labor; the industrial revolution harnessed steam power, and then electricity, to the production of manufactured goods, and with the innovation of the factory system created jobs for displaced rural dwellers in urban places; and the development of sophisticated service and financial industries in metropolitan areas in the twentieth century responded to the needs of society for the production of more services and efficiency in the investment of capital. Thus, any significant changes in the ways in

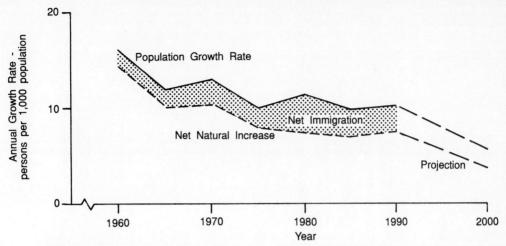

Figure 1.4 Components of population growth rates in the United States 1960 –2000.
(*Source:* Bureau of the Census, 1993, 14–15.)

which production is organized and financed have direct effects on the availability of jobs and where these jobs are generated, and hence differential urban growth.

Recently, there has been a great deal written about the globalization of production and financial industries (Warf and Erickson, 1996), the *restructuring* of manufacturing consequent to competition, and the introduction of new high-technology production and organizational processes (such as flexible manufacturing, just-in-time delivery systems, quality circles, and so forth). These and other related issues will be discussed in greater detail in later chapters. The important point to emphasize at this juncture, however, is that these changes are having a profound effect on the economies of urban areas in North America, and at the same time have reinforced the importance of *local* actions to strengthen the productive efficiencies of individual communities (Hansen, 1991).

Social/Demographic Processes   Changes in the demographic and socioethnic structure of society also have a powerful effect on urban development in North America (Clark, 1987). The components of demographic structure are net natural increase (births-deaths), internal migration, and foreign immigration. During the past fifty years, North America has moved from the high rates of natural increase ("baby-boom"), which fueled post–World War II suburbanization, to low rates of natural increase ("baby-bust") in the 1970s, to the weak "baby boom echo" of the 1980s (Figure 1.4). Declining rates of natural increase since 1960 have led to increasing comparative importance of internal migration and external immigration (legal and illegal) on urban growth (Foot and Stoffman, 1996).

Figure 1.4 emphasizes the increasing importance of *external immigration* on population growth in an era of declining rates of natural increase. Canada and the United States have similar levels of annual immigration, which are the highest in the world. This means net immigration accounts for about 30 percent of

NNI   *NIM*
NFI

population increase = NNI + NIM + NFI

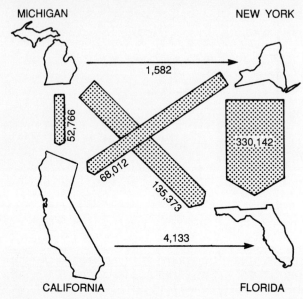

Figure 1.5 Net migration between four states in the United States, 1980–1981 to 1987–1988.
(*Source: The Economist,* June 22, 1991, p.26.)

the population increase in both countries. Immigration as a percentage of population in 1991 is 0.75 in Canada, 0.72 in the United States, and, as the next highest in the world, 0.70 in Australia. Immigration is having a profound effect on the socioethnic structure of North America, changing the continent from one that is predominantly of white European origin to one that is multiethnic (Brimelow, 1995). In 1960, European countries provided 50 percent of the immigrants to North America, but in 1990 Asia and Mexico were the sources of two-thirds of the immigrants. An important feature of immigration is that the bulk of new immigrants locate in the major metropolitan areas. In the United States, 40 percent of the new immigrants locate in Los Angeles, New York, and Chicago, whereas in Canada 60 percent of new immigrants locate in Toronto, Montreal, and Vancouver. Thus the major metropolises of North America are at the leading edge of this *shift to multiethnicity*.

These declining levels of natural increase and changes in the ethno-cultural composition of the population are accompanied by decreasing household size, increasing numbers of single-parent families, and greater recognition of the existence of a variety of lifestyles. As a consequence of decreasing household size (from 3.3 persons per household in 1960 to 2.6 persons per household in 1990), the number of households in North America has increased at a faster rate than the population as a whole. This, coupled with the greater acceptance of a wide array of lifestyles (other than the married male/female with or without children), has led to a demand in urban areas for different types of accommodation and a wider variety of community support arrangements.

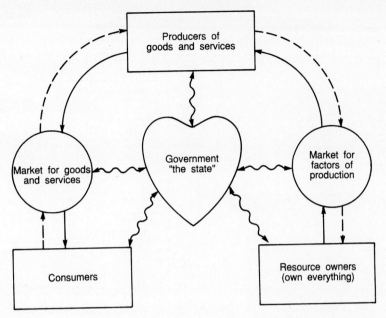

Figure 1.6   A model of the role of government in a market economy.

Declining levels of natural population increase mean that differential urban growth is greatly affected by internal patterns of migration. The effect of *internal migration* is illustrated in the pattern of net migration (net migration = inflows – outflows) between four states in Figure 1.5. This diagram is compiled from Internal Revenue Service data by tracing changes in addresses from year to year on social security number-identified tax return forms. The diagram emphasizes the continuation of the Snow Belt–Sun Belt migration pattern in the United States during the 1980s and, in particular, the movement of people to Florida, a trend which has continued through to the early 1990s (Manson and Groop, 1996).

Political/Institutional Processes   Since the Great Depression of the 1930s and World War II, the "state" has had an increasing effect on urban development in North America. The "state" refers to the myriad legislative and regulative actions of federal governments (United States and Canadian), the states and provinces, and local governments (cities, municipalities, counties, special districts, and so forth). These three tiers of government have different roles. In general, the degree of *direct* intervention decreases with increasing levels of government: Local government has the greatest direct impact; then follows the state/provincial government; the federal governments, finally, have the least direct impact. Indeed, the Canadian constitution specifically allocates responsibility for municipalities to the provincial governments. The situation in the United States is, however, much more flexible, though there are perpetual tus-

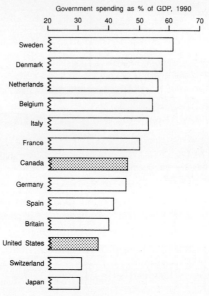

Figure 1.7   Government spending as percentage of GDP in the United States, Canada, and eleven other countries.
(*Source: The Economist,* May 2, 1992, p. 68.)

sles between the states and the federal government concerning jurisdiction and responsibilities (Garber and Imbrosio, 1996).

A diagram of the role of the state in the North American market economy (Figure 1.6) emphasizes that government is expected to be an intervener and regulator in a market economy—to provide for things that the market ignores or imperfectly manages, and protect the public interest. An allocation of goods and services between producers and consumers is achieved through a price-setting mechanism within a market framework. Unfortunately, this market process does not take into account such issues as need, equity of access, and often services that are required for the collective or common good. For example, the "market" for health care in the United States fails to cater to the 30 percent of the population who cannot obtain adequate private health insurance through employment or privately. Likewise, the "market" generally is not good at providing low-cost or free access to collectively required services such as urban parks and public transportation.

Thus, government intervenes in the North American economy as both a provider and regulator. As a provider, the role of government can be gauged from the expenditures of the three levels of government expressed as a percentage of the gross domestic product (GDP), which is presented in comparative terms for thirteen countries in Figure 1.7. Public spending in the United States amounts to about 38 percent of the GDP in 1990, whereas Canada's amounts to 47 percent (reflecting greater public expenditures in Canada on health care and the "social safety net"). This percentage has been increasing, but revenues have not kept apace, resulting in an enormous increase in the national debt of both countries during the 1980s and early 1990s. The burden of this debt now makes it difficult to address some of the terrible problems of social decay and deterioration occurring in the central parts of many urban areas.

As a regulator, government enacts policies that have direct and indirect effects on urban areas. In later chapters, there will be discussion of the ways in which the Interstate and Defense Highways Act (1956) and successive revisions to the Federal Housing Act have had profound direct effects on the provision of certain types of urban infrastructure (built things in urban areas), decentralization, and suburbanization in the United States. Also, there will be a discussion of the ways in which the United States/Canada Automotive Trades Agreement (1965) affected urban growth in central Canada through the creation of jobs in automobile assembly and parts plants.

The Interaction Between Urban Areas and the Environment   Urban areas have an impact on the environment; the environment, in turn, reacts to influence urban development. This inherent feedback cycle reflects the essential theme of the report of the World Commission on Environment and Development established by the United Nations, commonly known as the Bruntland Commission (1987) after its chair Gro Bruntland of Norway: Environmental concerns and economic concerns can and must be mutually supportive. Under this premise, the commission drew attention to many interdependent processes: the interlocked nature of the world's ecology and economy; the need for *intergenerational equity* with respect to resource use and environmental impacts; the absence of proper consideration of interlocking and intergenerational concerns in many public and private decision-making processes; and the need for fundamental change in the ways in which resources are distributed among the people and countries of the world. The thrust of this group of ideas is referred to collectively as "sustainable development."

These ideas are focused in an urban context in the notion of "sustainable cities"—supporting urban growth and development as long as it does not place a burden on future generations (Stren et al., 1991). On examining urban areas, a number of problems become evident: The emission of carbon dioxide (particularly in North America) and methane in cities (particularly in Europe) is placing the atmosphere at risk; the urban consumption of good quality agricultural land places future food production at risk; a lack of sufficient sewage capacity and appropriate secondary level treatment facilities puts groundwater, rivers, lakes, shorelines, and even seas at risk; developed-world garbage is rapidly filling up available landfill sites as people are realizing that disposal facilities of this type put land, streams, and ground water at risk; and hazardous-waste storage and disposal sites have created toxic time bombs, placing the health of current and future generations at risk. Future urban growth and development, therefore, depend on the implementation of adequate processes for sustainability (see Chapter 16).

Aspatial Processes and Urban Growth   Recent developments in the growth experiences of a sample of metropolitan areas in the United States and Canada

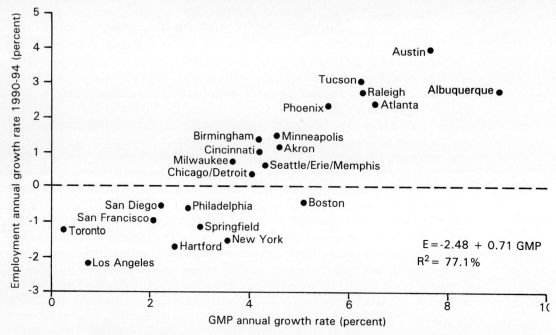

Figure 1.8 The relationship between employment change and changes in gross metropolitan product (GMP) for selected North American metropolitan areas, 1990 – 1994

(*Source:* Modified and re-drawn from GTA, 1990, Fig 2.1, p. 41)

serve to demonstrate the relationship between aspatial processes and spatial patterns and also to introduce some of the themes, such as Snow Belt–Sun Belt shifts and industrial restructuring, which we will discuss in much greater detail in later chapters. The recent growth experiences of twenty-five metropolitan areas are summarized by two economic measures—the 1990–1994 annualized change in employment (E) and gross metropolitan product (GMP)—which may be regarded as indicators of future population growth (Figure 1.8). The employment figures relate to total employment, no distinction being made between full- and part-time employment, or occupation mix. Gross metropolitan product (GMP) is analogous to gross domestic product (GDP). The GMP represents the aggregate value of all goods and services received by or produced within a metropolitan area including that associated with continental and international trade.

The relationship between these two indicators is quite strong (R^2 = 77.1 percent, a perfect association being 100 percent, and no association 0 percent), implying that in the early 1990s employment growth in the sample of metropolitan areas did not generally occur until the annualized GMP was above 2.5 percent. Some metropolitan areas, such as Boston, New York, Philadelphia, and Springfield (MA), incurred negative employment growth rates with GMPs in excess of this general 2.5 percent threshold. Figure 1.8 may be interpreted as providing an urban manifestation of the "jobless economic recovery" during this period following employment downsizing in the private and public sectors. It is quite evident from Table 1.1 that the population growth patterns in the 1990s, as suggested by the economic indicators, will be quite different in some respects from that experienced in the 1980s as a result of the changing nature and regional impact of aspatial influences.

When the twenty-five metropolitan areas in Figure 1.8 are organized regionally in Table 1.1, important spatial variations in the 1980–1990 population growth rates, and the economic indicators of possible current trends, emerge. Metropolises in the Northeast and Midwest generally experienced little population growth in the 1980s. Consequently, the employment trends in the early 1990s are negative for those in the Northeast, or close to zero for most of those in the Midwest. Since the mid-1970s, this basic pattern of limited growth is related primarily to the shift of manufacturing from the metropolises of the Snow Belt to those of the Sun Belt. The negative signs in the migration and manufacturing restructuring columns suggest that in the 1990s most of the urban areas in these regions are experiencing low or negative employment growth as a consequence of industrial restructuring, particularly in the automobile industry. This is one of the main reasons for also hypothesizing continuing population outmigration from these regions. Industrial restructuring and outmigration may well be offset in some places, such as New York, by increasing strength in employment in producer services and immigration.

There are a few anomalies to this hypothesized carryover of many of the growth experiences of the 1980s to the 1990s for most of the metropolises of the Northeast and Midwest. The Toronto CMA, which had a large population increase in the 1980s as a result of tremendous growth in producer services, especially in the managerial, professional, and clerical occupations related to office and provincial government activities, was experiencing a decline in employment in these occupations in the early 1990s. Minneapolis, along with Boston, is one of the few metropolitan areas in the Northeast and Midwest that has had a significant presence in the new information technology industries, but unlike Boston has not incurred a strong negative effect from cutbacks in federal defense expenditures.

In the 1980s, most of the sample of metropolitan areas in the South, Southwest, and west experienced large population increases. These increases reflected the Snow Belt–Sun Belt shift in manufacturing employment, rapid growth of retirement and recreation-related services in many of these urban areas and the regions served by them, acceleration of defense expenditures, and relatively large volumes of population migration (from the Northeast and Midwest) and immigration (particularly from the Latin American countries south of the Rio Grande). The exceptions, Birmingham, as a result of the restructuring of the iron and steel industry in North America, Memphis, becoming overshadowed by Nashville, and San Francisco, where the only great growth outside the MSA in the Bay Area around San José has been associated with Silicon Valley, demonstrate the variety of this growth experience (Bourne, 1995).

The economic indicators for the 1990s suggest that although most of the growth patterns in the metropolitan areas of the South and Southwest may

Table 1.1

Population Change in the 1980s, Economic Change in the Early 1990s, and the Hypothetical Impact of Selected Aspatial Influences on Differential Urban Growth in Selected Metropolitan Areas in the United States and Canada.

Metropolitan Region	Population 1990 ('000s)	Population % 1980/1990	Economic Indicators 1990/1994 annual — Employment	Economic Indicators 1990/1994 annual — GMP	Demography Mig.	Demography Immig.	Economic — Manufacturing Restructuring	Economic — Producer Services	Economic — Defense	Government (provincial/state capitals)
Northeast										
Hartford (MSA)	1,158	7.1	-1.8	2.5	−		−			
New York (PMSA)	16,938	3.0	-1.6	3.6		+	−	+		
Springfield (MSA)	588	3.2	-1.1	3.0			−			
Boston (PMSA)	5,051	6.0	-0.5	5.1	−		−	+	−	
Philadelphia (PMSA)	4,922	2.9	-0.5	2.7			−			−
Midwest										
Detroit (PMSA)	4,267	-2.8	0.3	4.0	−		−			
Chicago (PMSA)	7,411	2.3	0.3	4.0	−		−	+		
Toronto (CMA)	3,893	24.4	-1.2	0.3		+		−		−
Erie (MSA)	276	-1.5	0.5	4.3	−		−			
Milwaukee (PMSA)	1,432	2.5	0.6	3.7			−			
Akron (PMSA)	658	-0.4	1.0	4.7	−		−			
Cincinnati (MSA)	1,526	4.0	0.9	4.2	−					
Minneapolis (MSA)	2,539	15.5	1.3	4.6	+		+			−
South & Southwest										
Austin (MSA)	846	44.6	3.9	7.6	+	+	+	+		+
Albuquerque (MSA)	589	21.4	2.7	9.0	+	+	+			
Tucson (MSA)	667	25.5	3.0	6.2	+	+				
Raleigh (MSA)	856	28.7	2.6	6.3	+		+			+ +
Atlanta (MSA)	2,960	32.5	2.3	6.6	+	+	+	+		+ + +
Phoenix (MSA)	2,238	39.9	2.2	5.6	+	+	−			
Birmingham (MSA)	840	3.0	1.2	4.1			−			
Memphis (MSA)	1,007	7.3	0.5	4.3						
West										
Seattle (PMSA)	2,033	23.1	0.5	4.3	+		+			
San Diego (MSA)	2,498	34.2	-0.5	2.3		+				
San Francisco (PMSA)	1,604	7.7	-1.0	2.1					−	
Los Angeles (PMSA)	8,863	18.5	-2.2	0.7		+		+	−	

(*Source:* Population data from the Bureau of the Census, 1993; Statistics Canada 1992; and economic indicators from Figure 1.8)

continue, growth trends for urban areas in the West may well be quite diverse. The Seattle metropolitan area, with its well-entrenched role in the new information technology industries offsetting the cyclical swings in aircraft manufacturing, appears quite well positioned. San Diego and Los Angeles metro zones, on the other hand, appear the more precarious. Although it is hypothesized in Table 1.1 that these two metropolitan areas, in particular, are continuing to experience significant immigration, the reduction in United States defense expenditures following the disintegration of the former Soviet Union in the early 1990s is hitting the economy of southern California quite hard. An interesting question, therefore, is whether the 1990–1994 economic indicators will prove to have been good predictors of the 1990–2000 metropolitan population growth.

It should be recognized that the possible array of aspatial influences is extremely large. Many of the different aspatial factors work in complicated ways that may reinforce or offset a trend. Furthermore, forces that appear to be extremely powerful in a region during one period may have less effect in another—for example, high energy prices during much of the 1970s stimulated a 45 percent increase in population in the Houston PMSA (1990 population 3.3 million) in 1970–1980, but, with lower relative prices during the 1980s, the decadal growth rate has been 21 percent. Spatial patterns are, therefore, in a constant state of flux as the relative strength and influence of these different aspatial processes change through time (Pred, 1985).

The Effect of Accumulated Urban Infrastructure

Urban areas are landscapes of accumulations of fixed capital—roads, sewers, office buildings, houses, apartment buildings, commercial buildings, retail malls, parks, bridges, power lines, and so forth. These fixed accumulations of capital investment provide the physical framework within which urban life occurs. The important features of urban infrastructure are that it may be owned publicly or privately; that it has accumulated over many decades; and that various elements may not be serviceable or appropriate for current needs.

This stock of fixed capital is vital to the efficient operation of an urban area. One of the major difficulties with urban development in North America is the *fixed* nature of this capital investment. A whole host of legal restrictions, personal attachments, and taxation policies are in place that discourage renovation or replacement. Thus, there has been a strong preference for building anew on "new" urban land rather than renovating, rebuilding, and intensifying the use in existing built areas. For example, often the only parts of inner cities in many urban areas that have been redeveloped are remnants of historic cores, where public–private partnerships have provided for renovation and new construction—usually for office, commercial, and entertainment purposes. On the other hand, large sections of inner cities appear to have been forgotten, isolated, and cut off from the economically active parts of urban areas.

Likewise, there is a part of the urban area—a hidden part—that is rarely noticed and, as a consequence, frequently barely maintained to adequate standards. This hidden part is the infrastructure that provides power, water, and removes sewage and trash. Power systems (electricity and gas) require enormous maintenance. Therefore, it is frequently tempting to fail to reinvest and construct new facilities, leaving the old ones as relics in the urban subscape. Clean water is vital, but people across North America are clearly questioning (through the purchase of bottled water) the acceptability of tap water. Furthermore, the availability of water as well as the efficiency of its use are becoming a major constraint to urban development in some parts of the United States, such as south Florida and the southwestern states.

Practically every drop of water that enters an urban area leaves it via sewage lines. However, in many urban areas sewage lines are ill maintained. Disposal has also become a major constraining factor, for not only does the outflow have to be treated, but the sludge also has to be treated, dried, and removed (frequently to a landfill site). Finally, it appears that in many urban areas the garbage is winning—we know how to collect it, but cannot provide environmentally acceptable means to dispose of it given the current financial situation. Control of this problem appears to require (a) production of less garbage, (b) recycling the recyclable, and (c) incinerating the rest with "clean" incinerators.

Urban areas have to be efficient places for a whole variety of human activities—economic, social, and cultural—if they are to be successful. The accumulated urban infrastructure, visible and hidden, contributes vitally to this efficiency, for it is the stage on which urban life is set. This accumulated fixed investment influences the spatial arrangement of activities within urban areas because some places become, over time, more attractive and feasible for living and economic activities (hence jobs) than others. Thus, urban areas cannot be examined without taking the historical context of accumulation into account—hence the emphasis on the historical aspect of urbanism in this textbook.

THE COMPONENTS OF URBAN ANALYSIS

What is emphasized in a particular field of study relates not only to the general objective, but also to the way in which the inquiry is undertaken. In this book, *the objective is to understand processes that give rise to spatial arrangements of urban phenomena.* This understanding is shaped by the ways in which the objective is pursued, for the methods that are selected influence not only the type of study that is undertaken, but also the phenomena that are regarded as of interest or importance. This is a vital point because issues and phenomena that engage persons working in one mode of inquiry are not necessarily valued in the same way as those working in other modes. The analysis of the spatial arrangement of urban phenomena involves five elements: patterns, philosophical approach, theory, technique, and scale (Table 1.2). The choices made in each of these determine to a large extent not only the type of work that is done, but the area of interest as well.

Table 1.2

Components of Urban Analysis

COMPONENT	APPROACH (EXAMPLES)
1. Patterns	Spatial distributions
2. Philosophical Perspectives	Positivist
	Logical positivism
	Phenomenological
	Humanism
	Structuralism
	Modernism
	Postmodernism
3. Theory	Social human ecology
	Neoclassical economics
	Behavioral approaches
	Classical political economy
	Marxist perspectives
4. Scale	National/Regional/Local
	Urban fields
	Communities
	Neighborhoods
5. Techniques	Maps and Analyses
	Geographic information systems
	Information gathering
	Field work
	Surveys
	Census data
	Techniques for analysis
	Statistical models
	Mathematical models

Urban Patterns

Descriptions, discussions, and analyses in this textbook focus on spatial distributions—the patterns, structure, and organization of urban space. The phenomena that are selected for study are those that have a spatial location. This is the basic reason why much of the second part of the text emanates from the theories, models, and concepts related to the arrangement of different types of land uses within urban areas. As land use definitely involves space, interest focuses on the geographic location of use relative to others, and on interactions among the users. In a sense, a land-use approach provides a way of fixing patterns as the central concern, although the land-use approach is not the only way to do this.

This focus on spatial distributions as they are arrayed into clusters or groups of phenomena—such as retail districts, income-stratified neighborhoods, and technology districts—emphasizes the centrality of regions and places in geographical inquiry. This analysis is not, however, directed toward a mere inventorying of places or some holistic search for "character" that differentiates the place or region from other areas, but is directed toward understanding the *interaction* between local events and processes. It also includes an attempt to link local reactions to global events. The establishment of such linkages is a difficult task not only because of the difference in scales, but also because global events are mediated by local ideologies and cultures (Gertler, 1995). In his study of Pittsburgh, Clark (1989) demonstrates the ways in which changes in the global economy and methods of production have impacted on the economy of the urban area, and have been reformed and facilitated by the existing structure of labor relations, education, and research institutions, in that environment.

Philosophical Approaches

The second element of analysis, the philosophical approach, determines the way in which the analysis is undertaken and the type of evidence that is thought to be meaningful. Throughout the 1960s and 1970s, a *positivist* approach pervaded most studies. A strong element of this approach still permeates the work undertaken in urban geography today. The positivist approach regards "sensed" (that is, measured) experiences as the most acceptable type of information about urban patterns. Thus, many empirical studies use census information (such as census tract data, or household sample surveys) to test ideas about the organization of patterns and interrelationships that may exist in geographic space (Hodge, 1996). It is important, however, to note that empirical information should not be regarded as value free, for it often subsumes the values of the group collecting the information.

Nevertheless, the positivist approach, as reformulated in *logical positivism,* in other words, the scientific method, is used a great deal in urban studies. It provides a way of examining problems that is replicable. This permits evaluation and comparison of information from a variety of studies. Because of the focus on measurable information, the scientific method also places front-and-center in the minds of researchers and readers issues concerning the scope and quality of the information used, and the methods used to analyze the data. Furthermore, the scientific method of theory building and hypothesis testing provides an analytical mode that is used widely in the scientific community and is surrounded by methods, procedures, forms of presentation, and evaluative designs that are commonly understood across a large number of disciplines.

Most studies using this method commence with an outline of a problem to be examined (such as—what are the forces giving rise to gentrification? and, why is gentrification concentrated in certain parts of urban areas?); review the existing theory that provides an understanding of why the problem occurs in the way that it does; restate one or more theoretical propositions into specific hypotheses relating to the issue; test the hypotheses using measured information; and, with these results, re-examine the theory relating to the issue—perhaps postulating new theoretical approaches that might be used to understand the problem. A clearer understanding of the problem can then be used to propose, and support, public policy alternatives related to the issue. As this is the approach to urban analysis preferred by this author, many examples cited in the text are from studies that are logical-positivist in nature.

The restriction of the positivist approach to measured information, however, has led some analysts to apply other philosophical approaches in their work. A number of studies of people's perception of their environment are cast in a *phenomenological* framework, in which information derived from experiences is considered a valid source of knowledge. Ward (1990), in his classic study of children in urban areas, portrays the urban scene through the eyes of children and demonstrates that elements of the landscape hardly noticed by adults are of great interest to children. Experiential information may, in fact, have a lot to do with the way in which individuals at different stages in the life cycle react with their environment. Golant, for example, defines the urban environment of the elderly as experiential—that is, dependent on an individual's "perceptions, feelings, and evaluation" (1984, p. 61).

The application of phenomenology to the study of the ways in which people interact and are influenced by their immediate environment illustrates why this perspective is often considered one of the roots of the *humanistic* approach to spatial inquiry (Tuan, 1976). Humanistic approaches in urban studies are committed to understanding how people respond to and create a way of life (that is, culture) in various urban places. In their study of homeless women in Los Angeles, Rowe and Wolch (1990) make use of experiential information provided by homeless women to illustrate how they construct a daily schedule to live their lives—and hopefully survive under conditions of extreme deprivation. Thus, the ways of life that people are able to develop are shaped by internal (family, ethnicity, education) and external (class, history, location) forces that shape individual responses (Ley, 1990).

Many researchers are pursuing *structuralist* approaches to urban inquiry. In the urban context, the structuralist approach can be regarded as involving an assumption that spatial patterns can only be explained by relating them to the social structure of society, and to the processes that give rise to this social structure. Thus, spatial patterns are regarded as the consequence of social structure and the forces that give rise to this structure. Changes in spatial patterns, therefore, occur as a consequence of changes in underlying societal forces that influence social structure. In this context, Johnston (1983) argues that the evolution of the American urban system is best understood in the context of the changing social structure of society as it metamorphosed from mercantile capitalism, to industrial capitalism, and late capitalism. Similarly, the gendered structure of North American urban areas is examined in Chapter 13 in the context of patriarchal structural processes (Halfacree, 1995).

A recent trend involves attempts at interpreting the late-twentieth-century North American metropolis in the context of *postmodernism* (Soja, 1996; Curry, 1991). Given the profusion of contradictory writing on the subject of postmodernism, it is difficult to say whether this does provide a useful *modus operandi* for urban scholarship (Marden, 1992; Harvey, 1990). In general, *modernity*, which came into being gradually in a variety of social, political, cultural, and economic milieus from the eighteenth century (the Enlightenment), is associated with the search to establish coherent theoretical bases for information and knowledge (Giddens, 1992). It is expressed, for example, in a vision of history as having some direction, in geography as involving processes of a cyclical or evolutionary nature, and in a belief that there are *rational* and *recurring* forces that underlie the many manifestations of the human condition.

Postmodernity, which has evolved in the twentieth century and is gaining pace in its closing decades, is associated particularly with trends in art and architecture (Deutsche, 1990). Postmodernism does not reject modernity. It is, rather, modernity recognizing the irrationalities in itself. The contention is that although there may be various kinds of underlying order to human activities, these activities and "orders" are so diverse, fragmented, and contradictory, that to examine them in the context of assumptions of general rational behavior involves simplifications to the point of absurdity. Furthermore, assumptions of rationality ignore the tendency for rational actions to have *unintended consequences* that undermine the original "rational" behavior (Barnes and Sheppard, 1992). Postmodernism in essence recognizes the need for modernity to come to terms with pluralism in the context of global integration. At the least, in its application to urban analysis, postmodernism emphasizes the complexities of local responses to the fragmentation of space caused by economic restructuring and the globalization of the world's economy.

One of the difficulties that this author has with writing that utilizes a postmodernist mode is that the work often appears to lack a definite theme and coherent structure. This is because the writers purposely reject a structural form in which one discussion logically proceeds to another. Thus, traditional criteria of good writing and research are thrown out of the window because the resulting linearity of the structure is regarded by many postmodernists as excluding important truths. Furthermore, there is a tendency in much postmodernist writing for arguments to be asserted and supported, if at all, by anecdotal and personally experienced evidence, rather than information collected through designed survey and sampling methods. The argument of postmodernists is that information derived from survey and sampling methods excludes important other ways of knowing about a problem or issue. Thus, writers and researchers in modernist, particularly positivist, and postmodernist modes frequently pass like ships in the night in their failure to appreciate each others' contributions (Fainstein, 1992).

Theoretical Bases

The third component of analysis involves the theoretical base of urban studies. There are many different theories involving economic, social, and political processes. This textbook examines and evaluates a variety of them. Theories are usually developed according to the subject matter of concern and the opinion of the original researcher as to which provides the most fruitful path to understanding. Just as there has been considerable debate concerning the most useful philosophical basis for research, so there has been much controversy over the most appropriate theoretical approaches.

For example, one of the earliest theoretical structures used to analyze the social structure of cities is that developed by the Chicago School of Sociology in the 1920s (Park, et al., 1925; Lal, 1990). This group regarded the pattern of social areas within the city as a natural product of change in communities—hence the term *social human ecology*, for the word *ecology* in this context is derived directly from plant biology and indicates that the information relates to groups (that is, communities) of people and households rather than individuals and single-household units. The focus is on the processes leading to change in the racial and ethnic composition of communities. The studies involving this approach are essentially optimistic in nature, emphasizing the advantages of multiculturalism (or pluralism) and the importance of communication for the reduction of racial antagonisms. The main criticisms of this approach are that it does not take into sufficient account class and the processes leading to class formation as they impact on race and ethnic relations (Smith, 1988), because by aggregating data, researchers have tended to ascribe to individuals general characteristics of the aggregations.

Perhaps some of the more vibrant debates have been concerned with economic processes and spatial patterns. Much of the work of the 1960s and 1970s is couched in a positivist framework making use of *neoclassical economic theory*. With this theoretical approach the location of phenomena in geographic space is assumed to be the outcome of a market process (Figure 1.6). Land, housing, and resources in general are allocated between users as a result of prices set in some market framework, with government playing a regulative and providing role when necessary. The wonders of the marketplace outlined in neoclassical theory—with its assumptions of perfect information, perfect competition, rational economic persons, rational firms, and so forth—have been challenged on the grounds that markets of that type do not really exist, and that corporations and the state or government play a far more proactive role than is assumed.

As a consequence, studies using the neoclassical approach either modify the theory by replacing the fully rational economic person or firm with some kind of learning (cognitive) process, as in behavioral studies (Golledge and Timmermans, 1990); or construct different types of markets for different socioeconomic groups, as with the housing submarket approach; or continue to use the theoretical approach in appropriate situations recognizing that model building always involves constraining assumptions, as with land use, land value, and transportation models.

Another outcome has been to reject the theoretical base of neoclassical economics on the grounds that its assumptions (whatever the modifications) are untenable, and the theory is not relevant to the examination of a number of urban issues relating to the apparent tension among the social groups over the allocation of resources and access to opportunities. The result has been a return by a number of writers to the basic tenets of nineteenth-century *classical political economy* (Figure 1.9). In this approach the focus is on the major factors of production (land, labor, and capital) and the interactions between them with respect to the allocation of the proceeds from production. In many urban analyses in the 1970s and 1980s using this approach, land tends to play a minor role, but an important factor of production is added—the state. As has been indicated previously, governments (at the federal, provincial/state, or local level) not only provide the framework within which these groups of players interact, but also act as considerable players themselves (Figure 1.7). Of particular interest

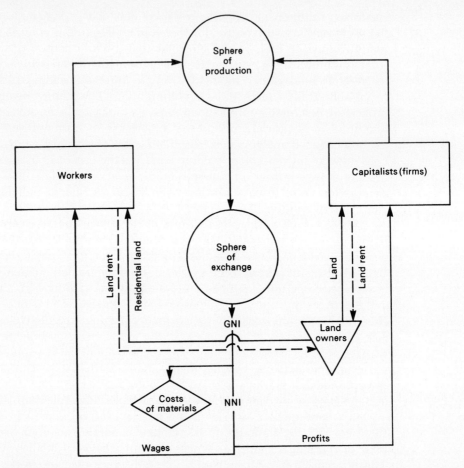

Figure 1.9 A classical political economy view of the way the economy works.

in parts of this textbook (for example, Chapter 5), therefore, are the spatial consequences of interactions among labor, capital, and the state (Barff and Knight, 1988).

Some approaches to analysis within the classical political economy framework have been rooted in the writings of Karl Marx (Harvey, 1989; 1985), which focus particularly on the struggle between capital and labor for the wealth created by production. The writings of Marx, and his followers, remain a powerful force in the world, not because some of the theories are particularly tenable (or implementable), such as the labor theory of value and the materialist conception of history, but because of its strength in analysis and critique of the capitalist mode of production, and its placement, front and center, of *values with universal appeal*—the dignity of labor, the sharing of wealth, and the sharing of decision making. Various aspects of Marx's views—for example, concerning the formation and polarization of classes in capitalist societies; the deleterious consequences to the bulk of the population of the power exerted by the owners of capital and property over those that own comparatively little or none; the consequent impoverishment and dehumanization of workers; and the materialist conception of history which emphasizes that history is the story of production by labor as influenced by economic laws—have been examined in the urban context during the 1970s and 1980s (Peet and Thrift, 1989). The variety of important analyses ranges from those involved with the intermetropolitan location of manufacturing (Walker and Storper, 1981) to inter-class tensions generated during the course of redevelopment in urban areas (Fainstein and Fainstein, 1989).

The Marxist focus on capital–labor relations emphasizes the role of capitalism in class formation through the division of labor. The approach, therefore, has a tendency to interpret urban phenomena through the prism of class formation, thus providing an important dimension that is lacking, for example, in studies using social human ecology approaches. It is also an extremely important vehicle for analyzing the effect of changes in the organization of capital and production on labor. Thus, during a period in which the globalization of capital, new sources of capital, and changes in the organization of production have been quite dramatic, this particular perspective has provided an important way of linking these changes to their impact on people (Storper and Walker, 1989).

Scales of Analysis

One question facing any individual concerned with spatially distributed phenomena concerns the selection of the scale of inquiry. This is obviously true with respect to the mapping of information, but it is also evident in analyses of urban phenomena. The larger the scale of study, the more the personal details become obscured—or, to put this in the context of humanism—the larger the scale of the study, the more likely it is that the responses, actions, and aspirations of individuals may be overlooked. However, it would be difficult and time consuming to examine the major processes influencing North American urban development from the point of view of many small-scale local studies. As a consequence, a balance has to be achieved in the geographical scale at which the material is presented.

The issue of scale is illustrated in part by many different concepts that may be used to define urban areas. Some of these concepts and political definitions are illustrated with respect to Atlanta in Table 1.3. The play area of children is quite restricted. So for them Atlanta undoubtedly consists of, at most, the local block. The high school student is aware of his or her own neighborhood (and the local mall) as Atlanta. The political definition of Atlanta is, however, the City of Atlanta, which in 1990 contains a population of 394,000, more than two-thirds of whom are African American. Somewhat larger than the City of Atlanta (the inner city), and containing some of the suburbs most dependent on employment opportunities in the business areas within the city, is the area served by the Metropolitan Atlanta Regional Transit Authority (MARTA), consisting of DeKalb and Fulton counties, which contain a population of 1.3 million, including the city.

Table **1.3**

Various Political, Legal, and Conceptual Definitions of Atlanta at Different Geographic Scales

CRITERION	DEFINITION	ESTIMATED POPULATION
Finance + business	Atlanta service area	32.6 million
Political	Georgia	6.5 million
Daily urban travel	Urban fields (DUS)	3.5 million
Bureau of Census	MSA	2.96 million
Public transportation	MARTA	1.3 million
Political	City of Atlanta	394,000
Walking distance	Neighborhood	100–2,000
Play area	Block (or street)	20

MARTA is within the metropolitan statistical area (MSA) of Atlanta, the population in 1990 being almost 3 million persons spread over eighteen counties. The *urban field* of Atlanta is more extensive than the MSA (Friedmann, 1973a) as it extends farther than the commuting range to include an area settled by exurbanites and is also used for weekly recreational activities by the urban population. The modern metropolitan area is, in essence, becoming indistinguishable from the urban field, because "instead of the 1941 compact city surrounded by many dispersed farm households and slow-growing rural service centers, we now have a regional city with a lacerated edge surrounded by dispersed exurbanite households and a few farm households" (Russwurm and Bryant, 1984, p. 133). Exurbanized households are those that have left the metropolitan area to "colonize" the immediate rural environment while maintaining employment in urban-based occupations (Davies and Yeates, 1991). The concept of the urban field is therefore now very close to that of the *daily urban system*, which is defined on the basis of the ebb and flow of daily commuting *and* by activity patterns (Berry, 1981). The urban field may, therefore, extend more than 160 kilometers (100 miles) from traditional central city.

In a more political sense, Atlanta is often used coterminously with Georgia itself, because its MSA includes more than 40 percent of the state's population. Thus, many of the actions and policies of the state government are pursued with an eye on their impact on large, urban constituencies. This is true of many state/provincial governments in North America, for in many of these jurisdictions at least 40 percent of the population is located in a large metropolitan region. Finally, since Atlanta is the financial and wholesale center for a large part of the South (Wheeler and Brown, 1985), its sphere of influence extends over the greater part of seven states and involves an estimated 32.6 million people.

Thus, much consideration is needed in selecting a scale for a single empirical study or comparative study. There is no point in comparing apples with oranges—the central city of one MSA with the MSA of another. Even comparing

Figure 1.10 The actual location of United States MSAs, CMSAs, and PMSAs, and Canadian CMAs with a population of 250,000 or more in 1990–1991.

cities is difficult, as some have restricted political boundaries whereas others expand in extent as their urban area grows (Rusk, 1993). As MSAs involve the commuting region of the whole interlinked urban area, these are probably the best units for analyzing the economy of a metropolis. Where MSAs are large (more than 1 million population) and converge to form one massive interlinked region, such as the Chicago-Gary-Lake County CMSA (population 8.2 million) consisting of the Chicago PMSA, Gary-Hammond PMSA, Joliet PMSA, Kenosha PMSA, Lake County PMSA, and Aurora-Elgin PMSA, the total economy is probably best represented by the CMSA.

Techniques for Analysis

The range of techniques used for the analysis and presentation of spatially distributed information is quite diverse. One way in which information may be presented, and interrelationships examined, is with the use of maps as, for example, in *Geographic Information Systems* (Maguire et al., 1991). Maps themselves are perceptions, for they represent choices that have been made by individuals for the presentation of information (Monmonnier, 1991; Tomlin, 1990). Figure 1.10, for example, shows the distribution of metropolitan areas (PMSAs or MSAs in the United States, and CMAs in Canada) that had a population of at least 250,000 in 1990–1991. The map of the continent is equal area, that is, the

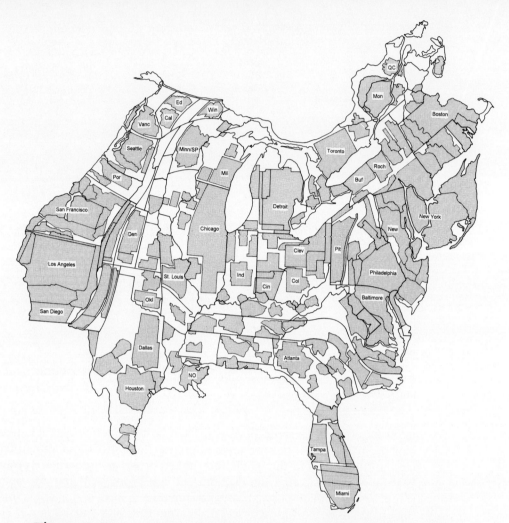

Figure 1.11   An isodemographic map of metropolitan North America.

size of a territory (for example, a state or province) is drawn directly proportional to its actual size on the surface of the continent. However, the urban areas, since they are minute compared with the continent as a whole, are represented simply as points. The map simply shows the actual geographic position of one large urban place relative to other large places.

Figure 1.11 is not based on equal areas, but equal population, for each PMSA, MSA, or CMA, and state/province, is represented according to the size of the population as a percentage of the population of North America. Thus, New York City's PMSA, with 3.1 percent of the population of North America, is represented with 3.1 percent of the map's total area. The map has to be distorted from the "reality" of the shape of the continent in Figure 1.10 in order to fit in the differently sized urban areas, and the distances between places are obviously misrepresented. Nevertheless, Figure 1.11 does emphasize why the spatial aspect of urban phenomena is considered an important area of inquiry—about two-thirds of the population of North America resides in large urban places (that is, in metropolitan areas with populations greater than 250,000), and these are located mainly along the East Coast south to Florida, around the lower Great Lakes, and along the West Coast, particularly in California.

Spatially distributed information also has to be related to the various processes that are thought to give rise to these patterns (Kemp et al., 1992). These relationships are often established through the use of graphs and quantitative techniques of different kinds, as in Figure 1.8. These quantitative techniques range from those of a statistical nature, such as regression, to more general mathematical modeling. Statistical techniques, which are used to investigate the relationships among sets of sample information, are widely used by researchers (Clark and Hosking, 1986). Scott (1983b), for example, uses logit regression analysis and probability models to examine the relationship between the size and location (spatial aspects) of manufacturing plants and the technical structure of production (an outcome of aspatial processes). It should be noted that recent research has been directed toward establishing procedures to address the idiosyncrasies of handling spatially distributed data (Anselin, 1988; Griffith, 1988). Although statistical representations are kept to a minimum in this volume, it should always be remembered that many of the results that are discussed have been produced through such analyses.

Mathematical models, ranging from those of the gravity and interaction type to regional input-output analysis (Haynes and Fotheringham, 1984; Hewings, 1986), have been employed by a large number of researchers. These types of models incorporate basic processes. For example, the gravity model postulates that interaction decreases with distance. Therefore, this type of simple formulation allows the interaction process to be represented in numerical form. The gravity model forms the basis of trip distribution models in transport system forecasting (see Chapter 14). The processes modeled by input-output models relate to intersectoral interactions, wherein urban areas that retain the greatest level of these types of interactions generate the greatest increase in growth from a given investment. Research based on these types of models is used in the next chapter in the discussion of the urban multiplier.

Although many of these statistical and mathematical models are associated with research that is cast in a logical positivist-neoclassical economic theory mold, it should be clear that various techniques for analyzing information must be used regardless of the philosophical perspective or theoretical approach that is incorporated. Information of various types, whether it be sensed or experiential, has to be evaluated with respect to the analysis being undertaken. That is not to imply that the techniques used in analysis must be complex—indeed, the techniques that are used should be the simplest possible that cast light on the relationships being investigated.

CONCLUSION

This chapter has demonstrated that there are a variety of approaches that may be used to analyze patterns of urban phenomena. Choosing from among the major elements of the various approaches is largely a matter of opinion as to which provides the most persuasive line of argument, or the clearest understanding of the issue being considered. Consequently, it is necessary to choose which avenue to take. This choice in effect determines the range of literature that is regarded pertinent. An attempt is made to appreciate contributions to the understanding of urban areas emanating from a variety of modes of study. This textbook is based on the premise that North American urban areas are primarily the product of economic forces, particularly those that pertain to industrial development in the nineteenth and twentieth centuries. These forces have operated in a capitalist and market (or quasi-market) environment. Much of the discussion, particularly with respect to urban systems, recognizes the accumulated impact of long-term fluctuations in the economy on the development of urban systems and the internal growth of urban areas.

By generating an understanding about the way urban areas have grown and work, the text also hopes to contribute to overcoming the ambivalence that many North Americans feel about the urban communities in which most people reside (Garber, 1993). This ambivalence emanates from the mixed views that people have about urban areas—often those in which they themselves do not reside. On the one hand, many urban areas, particularly central cities, are viewed by many people as disaster areas to be avoided at all costs. They are places in which people can live if they have plenty of money, but are difficult environments for the less wealthy. Tom Wolfe's *The Bonfire of the Vanities* (1987) is based on this polarization resting on the assumption that central cities contain two distinct societies—the privileged few who can afford to lead a comfortable life, and the unfortunate majority who cannot. The principal character, Sherman McCoy, seeks "insulation" from the urban reality of crime, poverty, and the intensity of multicultural street contact.

On the other hand, there remains a strong image in many people's minds of the positive ideals of urban existence, epitomized many years ago in some lines of a post–World War I song: "How're you gonna keep 'em down on the farm after they've seen Paree?" These dreams have in common a view that urban areas are places where people's hopes and aspirations for a happy, responsible, and productive life for themselves and their families, can be fulfilled (Wilson, 1991, p. 135 ff). This has been the dream that has attracted immigrants to large multicultural North American urban areas for the past two centuries, and is still attracting people. It is also the powerful dream that is drawing people to rapidly growing metropolises throughout the world. Dreams, of course, are rarely congruent with reality. Thus, one purpose of this textbook is to show what is happening with this reality and to suggest what needs to be done to draw reality closer to the dream.

Part **1**

North American Urban Systems

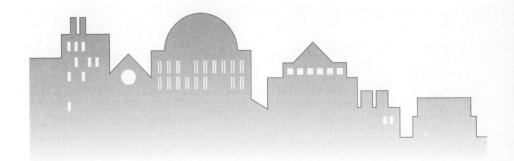

Chapter
2

The Processes of Urbanization

Despite considerable differences in their physical environments and cultural histories, the technically advanced nations of the world have one thing in common—they are all highly urbanized. That is, most of their population resides in urban places. This is a relatively new phenomenon in world history, for prior to 1850 no society could be described as predominantly urbanized, and by 1900 only one—Great Britain—could be so regarded. Yet today, all the industrialized nations are highly urbanized, and urbanization is accelerating rapidly throughout the world, especially among less-developed countries (Figure 2.1)

URBANIZATION

Urbanization is the process whereby an essentially rural society is transformed into a predominantly urban one. Hence, urbanization is usually defined as the proportion of the population of an area (often country) that resides in urban places above a certain population size. Urbanization curves are generally S-shaped (inset, Figure 2.1). So the trends depicted for the three continental areas in Figure 2.1 have to be interpreted in the context of their probable position in the general pattern. The curve representing North America since 1950 would clearly be the top end of an S-shaped curve, whereas Latin America represents the middle, or steepest part, of the S-curve, and Africa is the lowest segment of the S, which precedes the period of greatest increase in urbanization.

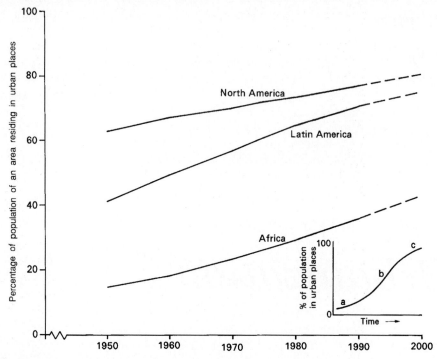

Figure 2.1 Recent trends in urbanization in North America, Latin America, and Africa.

The forces immediately underlying the form of the **S**-shaped curve are demographic, though the demographic trends are in themselves influenced by a host of economic, social, cultural, and political factors.

a) The bottom part of the curve, which has a low positive gradient, is usually related to a combination of high birth rates and high death rates in the area involved. The population is basically rural, though there may be some ports involved with the export of raw materials. Apart from local craft industries, there is little manufacturing and the service industries that exist are limited and focus on retail activities required for daily subsistence.

b) The middle part of the curve, which has a high positive gradient, is associated with high birth rates and declining death rates. The death rates are declining because the economy of the region is growing, per capita incomes are increasing, and significant improvements in health are associated with relatively simple improvements in diet, the availability of clean water, and the provision of basic health care for mothers and young children. The usual reason for economic growth is industrialization. Consequently, the rapidly growing population has associated with it extensive rural-to-urban migration.

c) The top part of the curve, having a shallow positive gradient which becomes flatter as the level of urbanization approaches 100 percent, is associated with low birth rates and low death rates. The death rates have continued to decline along with improvements in nutrition and general health care, but birthrates have also declined as economic and cultural imperatives for large families have become less significant. Furthermore, as the economy has developed and education and skill requirements have increased, per capita education and training has become more expensive. Hence, families make a conscious decision to focus scarce resources on fewer children. The level of urbanization continues to increase even though rural-to-urban migration decreases (because of rural depopulation) but also because larger metropolises become the foci of increasingly complex structures of economic activity.

The S-shaped urbanization curve reflects, therefore, a complex variety of factors that lead inexorably not only to urban growth, but also to metropolitanization.

North American Urbanization

The determination of what proportion of the total population lives in urban places clearly depends on the way places are defined as urban in official statistics. Unfortunately, there is no universally accepted definition of urban places, and considerable variation exists from country to country. For example, in Denmark places with 200 inhabitants are called urban, whereas in Korea places with less than 40,000 are not. In Canada, places are designated for census purposes as urban if they have at least 1,000 inhabitants, whereas 2,500 is the critical figure in the United States. In general, most countries use a base figure between 2,000 and 7,500.

The increase in levels of urbanization in North America since the dates of the first census of the United States and Canada is shown in Figure 2.2. For purposes of comparison, and in accordance with the range used among most countries, places are defined as urban in this textbook if they have at least 5,000 inhabitants. In practically every decade, each census in both countries has reported an increase in the proportion of the total population living in urban places. In the United States, the process of urbanization was quite gradual until about 1820, accelerated through the rest of the nineteenth century and to 1930, slowed during the Depression years of the 1930s and World War II in the 1940s, and then increased steadily during the 1970s and 1980s. It is interesting to note that it was not until 1930 that more than one-half of the population of the United States resided in urban areas of 5,000 or more.

Although the general trend in the urbanization curve for Canada is similar to that for the United States, there are two main differences. First, until recently, urbanization in Canada lagged about twenty years behind the United States. For example, the population of Canada did not become more than 50 percent urban until 1950. As a consequence, although the development of the Canadian urban system can be examined in the context of North American urban development as a whole, there are clearly distinguishing features that require separate attention. Second, the rate of urbanization in Canada since 1950 has been greater than that in the United States, so that by 1990 the proportion of the population in each country residing in places with more than 5,000 people

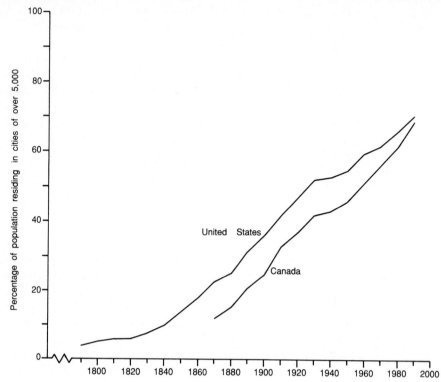

Figure 2.2 Levels of urbanization in the United States since 1790, and Canada since 1871.

was the same—71 percent. Thus, by 1990 the historic twenty-year lag in level of urbanization between the two countries was eliminated. It should be noted that these percentage levels are influenced to a degree by the changing definitions of MSAs and CMAs. When a major census definition revision occurs, as it did in the United States in the 1980s, the urbanization figures increase accordingly.

Regional Variations   The growth and spread of population in any continent is rarely uniform in time and space. Consequently, the level of urbanization often varies considerably in a given period from one region to another. In the United States, for example, the northeastern region—which today is the most highly urbanized area—was more than 50 percent urbanized by 1880; the midwestern and western regions by 1920; while the southern region did not exceed the 50 percent mark before the mid–1950s. Moreover, within each major census region, marked differences occurred in the level of urbanization of individual states at each census period. For example, whereas in Rhode Island and Massachusetts more than half of the population resided in urban areas by 1870, the population of a number of states (such as Idaho, Mississippi, and West Virginia) was still basically nonurban by 1990.

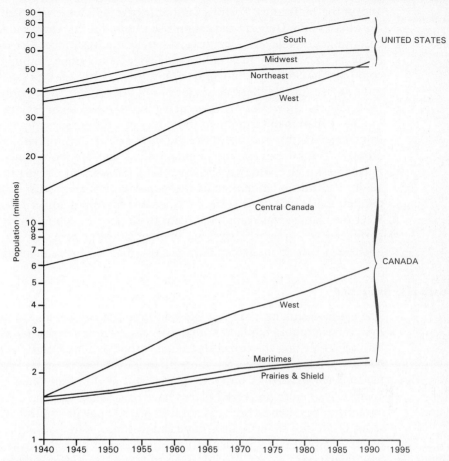

Figure 2.3 Regional variations in the growth of population in North America, 1940–1990.
(*Sources:* Bureau of the Census, 1975; *Historical Statistics of the U.S.: Colonial Times to 1970;* Bureau of the Census, 1993; Statistics Canada, 1991.)

Recent trends in population growth for the four major regions of the United States since 1940, and four major regions in Canada are shown in Figure 2.3. The y-axis of the graph (population in millions) is drawn in logarithmic scale, so that the trend or rate of change in population for the various regions can be identified. The greatest regional growth rates are thus represented by the steepest slopes, and the slower growth rates by the flatter slopes. Declining rates of growth are, of course, demonstrated in curves which slope downwards away from the y-axis.

It is apparent from Figure 2.3 that the greatest growth rates for the past twenty-five years have been experienced in the west of both the United States (including mountain states) and Canada (including Alberta). The slowest growth rates have occurred in the Northeast and Midwest of the United States and in the Maritimes, Prairies, and Shield areas of Canada. The Midwest and Northeast of the United States comprise the traditional manufacturing and agricultural heartlands of the country, whereas the Maritimes, Prairies, and Shield areas of Canada are the traditional fish/wheat/lumber/mining areas of the country. The South of the United States and central Canada (southern Ontario and Quebec) have experienced steady rates of growth fairly continuously since 1950 with the growth of manufacturing in these two large regions.

The urban manifestation of these regional differences in population growth is indicated by the contrast in rates of *metropolitan* development in the different areas. In central Canada, the population has become more concentrated in the four largest metropolitan areas since 1951 (54 percent in 1951 to 59 percent in 1991); and in the United States the population of the South changed from 68 percent metropolitan in 1980 to 71 percent metropolitan in 1990, while in the West the change was from 83 percent to 85 percent. Thus, the regions that experienced the greatest increases in population growth also experienced the greatest growth in the large urban places—the metropolitan areas.

Causes of Urbanization

The cause of urbanization is a key question that can be answered in two ways: first, by examining the factors that give rise to urban areas per se; and second, through analyzing the urban dynamic. Although concentrations of people and dwellings have occurred for more than 14,000 years, more organized urban places have existed in different places and at different times for only 5,000 years. These more organized places have existed because, as a result of diverse circumstances such as imperial control, priestly power, control of irrigation systems, trade, and so forth, they managed to appropriate part of the surplus product of society at large (particularly the agricultural sector) for the support of urban inhabitants.

Surplus product may be defined as the amount of material product in excess of that which is needed to maintain society in its existing state. Private wealth accumulation can occur if a household retains and invests a portion of its self-generated surplus. Through the ages, parts (and occasionally all) of these surpluses have been collected by powerful rulers, or elected governments, in a number of ways—extraction of tithes, taxation, appropriation, and so forth. The extracted surpluses have then been used for a variety of public and private purposes, ranging from improvements in human welfare (such as health, education, etc.), the construction of monuments, palaces, and religious edifices, to the manufacture of armaments for defense and the horrors of war.

The generation of a surplus (through such means as increases in agricultural productivity) and its partial appropriation may result in the formation of urban areas, but does not in itself lead to urbanization. Something more is required. Urban areas have to match agricultural areas as productive units; that is, *urban places have to have a productive dynamic of their own* that generates a surplus, thus increasing wealth, which may, in part, be used to construct urban infrastructure. For much of the last 5,000 years, the productive dynamic of most urban areas was derived from their various roles in: trade, finance, religion, and government; the provision of services for their local hinterlands; and defense. These functions, for the most part, did not generate large metropolises. The great growth of urban areas in the last 200 years, which coincides with urban development in North America, is associated with the industrial revolution and the establishment of urban areas as centers of large-scale manufacturing.

The growth of urban areas as foci of production (of goods *and* services), and hence as centers of employment opportunities in the more developed parts of the world, came just at the right time for unemployed agricultural workers. The agricultural revolution displaced many individuals from rural areas. At the same time, the population began to increase as a result of declining death rates. There were, therefore, both push (from rural areas) and pull (employment opportunities in urban areas) factors that led to large-scale rural-to-urban migration. These, coupled with high rates of net natural population increase and foreign immigration, resulted in rapid urbanization in North America during the past 100 years. During the last fifty years, urban areas in North America have become less important as producers of manufactured goods, and more important as producers of consumer, producer, and public services, ranging all the way from medical services to entertainment. This *postindustrial economy*, with employment-generating requirements that tend to be more knowledge and information based than industrial economies, is creating, as is discussed in later chapters, special tensions for industrial-based North American urban areas.

The Location of Urban Areas

A traditional question asked by urban geographers is why urban areas grew where they did. In discussions of the location of urban places, an important distinction is made between *site* and *situation*. The site is the precise feature of the terrain on which settlement began; situation is its *relative location* to other places and activities in the surrounding region. The built-up areas of today's urban areas usually bear very little relationship to their original sites, which were for the most part selected without any idea of, or consideration for, the scale of urban growth that was to take place there. Site selection was often influenced by factors in the physical environment. Prime considerations were the need for shelter and a good supply of drinking water, freedom from floods, and good defensive positions. Often great care was taken in the selection of the initial site for settlement. This was the case, for example, in the founding of Quebec City, Charleston and Philadelphia, and, later, in locating Washington, D.C., and Ottawa.

More often than not, however, settlements were situated more by chance than anything else. So their subsequent growth had little to do with the resources of the initial site. Madison, the state capital of Wisconsin, is a good case in point. At first sight, the location of this city seems a good example of the role of physical factors; however, the history of the choice of site reveals that the human rather than the physical factor was the most important. In 1836, the Wisconsin Territorial Legislature considered sixteen sites for a future capital, each of them promoted and owned by speculators. Eventually, the decision on the current location was made by a 15-to-11 vote after extensive lobbying that was driven much more by speculator financial interests than any comparative evaluation of site and situation advantages among the sixteen competitors.

Although the sites of initial settlements were significant, their situation was usually the more important locational consideration, since situations had a more direct bearing on the functions performed and on subsequent growth. Physical factors were also important in the situational context, particularly as these related to movement. Thus, many of today's urban giants originated as small settlements on navigable rivers, on good harbors, or at the entrance to passes through difficult terrain. Junction points in the transportation networks of the time were especially important. Therefore, locations were frequently selected at river confluences, bridging points, and crossroads. Particularly important were junctions where different types of transportation came together—the so-called break-of-bulk points at which goods and materials had to be transferred after processing from one means of transportation to another. It is not difficult to appreciate how the initial site and situation could affect subsequent development of urban functions, particularly of these transport-oriented settlements.

As technology advanced, so new locational forces became important; from the start, location became more directly related with the functions settlements were to perform. Thus, with the early growth of manufacturing, new settlements grew up at waterpower sites. The planned mill towns of the Boston Associates in the 1840s at Lowell, Lawrence, Chicopee, and Holyoke in Massachusetts, and at Manchester and Nashua in New Hampshire, are good examples. Later, with the spread of the railroads, junctions of different lines became important growth points. In the West, the operating needs of the railroads dictated the locations where many new settlements grew, for example, at watering stops or in conjunction with car and locomotive repair shops. As coal became increasingly important, new mining settlements grew on the coalfields; and older settlements assumed new significance. Similarly, settlements were founded in conjunction with the opening of new mineral ore deposits. Most recently, with large postretirement populations, a different natural resource—climate and recreational opportunities—has influenced location and subsequent development of settlements to result, for example, in the growth of resort cities in Florida, California, Arizona, and British Columbia.

Interaction Between Urban Areas   Urban areas do not and cannot exist in isolation. Whether we view the developments in transportation and communication as prerequisites or agents for the growth of settlements, both the concentration of surplus product from rural areas into urban places and the development of them as centers of production (for goods and services) necessitates a high level of interaction between urban areas and their hinterlands. Linkages are, therefore, vital. Urban places are *nodes* in networks comprising the movement of goods, services, materials, people, money, credit, and investment, as well as information. These various forms of flows, movements, transactions, and linkages are collectively referred to as *spatial interaction*. The interaction between urban nodes and regions lets us think of them as being organized into a system.

The volume of spatial interaction occurring between nodes in an interconnected system is related to the size (and wealth) of the nodes and the distance and efficiency of the link between them. This may be represented by the gravity model, a model that has been used as the basis for a wide variety of studies of different types of spatial interaction (Christerson, 1994):

$$I_{ij} = \frac{S_i \, S_j}{D^b_{ij}}$$

where, I_{ij} = a number representing the magnitude of interaction between a
place (i) and another place (j);
S_i = mass (could be measured as wealth, or population) of place i;
S_j = mass (could be measured as wealth, or population) of place j;
D_{ij} = the distance between i and j; and,
b = some exponent modifying the distance effect.

This simple model forms the foundation of many complex transportation models for the effect on interaction (I_{ij}) of growth or decline in population, or changes in the friction of distance (as represented by changes in the value of b) may be estimated from changes in the value of S_i, S_j, and/or b (see Chapter 14). For example, if a two-lane road between two urban areas (i and j) is replaced by a four-lane road, trucks and cars can be driven easier and faster between them. Thus, the volume of interaction, as measured by the traffic flow, will increase reflecting a decrease in the friction of distance (which can be represented by a decrease in the value of the exponent b).

Thus, some changes in interaction are brought about as a result of alterations to the structure of transport networks, which alter the relative location of urban areas and their importance as nodes in the network. Some places become more accessible, others perhaps less so. Travel times and costs of movement are reduced as the heavier effects of distance are decreased. In fact, forms of electronic communication, which can now connect information around the world to a personal computer in any office or home has, for some types of flows (such as ideas, information, stock prices, graphics, and so forth), led to the *annihilation of space*. Conversely and ironically, increased traffic and a general deterioration of roads and bridges in many metropolises around North America has increased congestion (or the friction of distance) for many of those same workers who are now instantaneously connected to the world through their computers.

Interaction, then, plays a number of crucial roles in shaping the form and structure of urban systems, as it does in the internal structure of urban areas. Four of these roles are particularly important. First, in the same way that the market economy is integrated through price-mediating mechanisms, so interaction performs a spatial *integrating* role. Because people and activities are separated from each other in space, interaction is essential to link them. Thus, it is through the movements, flows, and transactions between urban areas and regions that their functions and activities are coordinated and joined into a unified whole—the space economy. Second, interaction permits *differentiation* within urban systems because it allows certain urban areas to concentrate on particular economic activities. Thus, Detroit concentrates on the production of automobiles, whereas Los Angeles has a focus in the production of films and television shows. Interaction is thus necessary for a spatial division of labor, and makes possible the functional specialization of urban areas and regions.

Third, and most important, interaction is the medium of spatial *organization*. The various kinds of movements, flows, and transactions between urban areas are the expressions of linkages, relationships, and, above all, interdependencies within urban systems. Thus, what goes on in one place or in one part of the system often affects what goes on in other places and other parts of the system. Because of relationships and interdependencies, urban areas are articulated into the complex functional whole we refer to as the *urban system,* in which today's largest metropolitan centers have traditionally played a dominant organizing role. Lastly, interaction is extremely important in bringing about *change* and the reorganization of spatial relationships within urban systems. The diffusion of ideas and information, and migration, are key ways by which changes are transmitted through the urban system.

THE SPREAD AND GROWTH OF URBAN AREAS

The spread of settlements and their growth in North America has been dependent on a complex set of processes operating in space. These may be defined as innovation diffusion, migration, the flow of capital, and the spread and concentration of decision making and control. Corresponding to each of those processes are spatial patterns: the spatial distribution of power; systems of activity location; networks of capital sources and flows; and settlement patterns. Urbanization, as will be demonstrated with respect to the United States in Chapter 3 and Canada in Chapter 4, is a complex of processes. Therefore, the geographic patterns that are generated in the course of urban development are multi-faceted.

The Spread of Innovation

Much of North American urban development is related to technological *invention* and *innovation*, not only with respect to the growth of urban areas and the establishment of certain regions as "motors" of development, but also to the decline of certain areas (von Tunzelmann, 1995, pp. 185–218). Invention refers to scientific discovery and to any combination of existing knowledge that has practical use, especially in production. Not all inventions and new knowledge lead to production—in fact, only a small proportion do. However, many ideas and discoveries may be combined and used with others in production at some later date. For example, knowledge concerning lasers has been around for a number of decades, but it was not until the 1980s that use of laser technology appeared in surgery, ophthalmology, and the preparation of micro-circuits. In other words, scientific discoveries concerning lasers in the 1950s led to *innovation*—the application of an invention for commercial purposes—in the 1980s. Similarly, knowledge of internal combustion engines was quite well known before the end of the nineteenth century, but it was Henry Ford who developed techniques for marrying this knowledge to the production of low-cost automobiles in the 1920s.

Large urban areas are the most likely places for invention and innovation to occur and spread or diffuse to other places, usually progressively smaller ones (Howells, 1984). For example, in Canada the greatest focus of inventive activity, measured on a per capita basis, is in Toronto and Montreal (Ceh and Hecht, 1990), which are also the centers of finance and manufacturing in the country. The ongoing *restructuring* of productive activity in the world reflects the diffusion process as new ideas and methods of organization spread to other places from the more populous and wealthy areas to smaller places. Along with this, the foci of invention and innovation may also change over time as new foci of invention and innovation are established. For example, the large metropolises of the northeastern and midwestern United States have been joined by the San Francisco Bay Area (for example, Silicon Valley), Los Angeles, and the Puget Sound region (Seattle-Vancouver) as centers of invention and innovation.

Migration

Migration has always played an important role in the growth of urban areas. During the nineteenth century, the growth in urban population depended largely on the migration of people from rural areas to growing urban centers, and on immigration from abroad. Today, rural-to-urban migration is no longer significant to North American urban growth, but the movement of people between urban areas with a general drift toward the larger places, and immigration from abroad, are the most significant features (Moore and Rosenberg, 1995). The North American population is the most mobile in the world: People often move long distances for new job opportunities, better lifestyles, or even simply to reinvent themselves.

Although the economic opportunities of urban areas have exerted the greatest pull on migrants, the social advantages of urban life have also played a contributing role. The prospective excitement of urban life has exerted a persuasive influence on migration patterns and strongly influenced rapid urban growth. The wide range of cultural facilities found in the larger metropolitan areas today is important because it encourages a highly educated cadre to live in them. It is from the ranks of these people that key personnel are recruited; thus, the larger cities have become increasingly attractive as employment locations in the rapidly growing administrative and service sectors of the economy.

Migration has also played a significant role throughout history as a mechanism in the diffusion of innovations, culture, and social customs. As such, it has been an important part of the process of spatial change. For example, migration makes urban areas more cosmopolitan; it also contributes to many of their challenges, especially social ones. Migration is also important in bringing about differential growth in the urban system. With North American birth rates close to the replacement level, migration and immigration become the chief determinants of urban growth. The rapid urbanization of California since the 1950s and the more recent growth of large urban areas in the Southwest and Florida are two examples that have been cited.

Capital Flows

The word *capital* is used in many ways. For our purposes, three possible meanings are useful: (1) capital in the form of tangible assets, such as machinery, buildings, and inventory; (2) intangible forms of capital, including legal instruments (such as stock certificates and warehouse receipts) that represent claims to ownership of tangible goods, and also including money intended for investment in such assets; and (3) human capital, or people who are educated, trained, or in some way equipped for any type of productive or creative activity that has present (or future) value. These three types of capital are required in various combinations for all types of production and the provision of services. The use of each of these types of capital entails a cost, payment of which can take the form of rentals and maintenance fees; interest and dividends; or fees (such as tuition fees) and taxes devoted to the support of education.

An important point with respect to these forms of capital is that some are more mobile (between urban areas and countries) than others. Capital in the form of tangible assets tends to be relatively fixed—for example, urban infrastructure. However, all fixed assets depreciate over time, some (like computers) a great deal faster than others (like bridges). Human capital is also relatively fixed in location with respect to country-of-residence, and usually even within countries. In some respects, however, fixed capital is becoming more mobile: The formation of trade areas, such as the common labor market in the European Union and the increased movement of professional workers between the United States and Canada under the North America Free Trade Agreement, is leading to the breakdown of some barriers to mobility.

Intangible assets, by contrast, are highly mobile. The global integration of financial markets has made this type of capital even more mobile. Figure 2.4 demonstrates that the increase in *rate* of outflow (not absolute magnitude by value) of intangible capital related to transnational corporations was twice that of tangible exports during the latter half of the 1980s. This great change occurred with the extensive integration by computers of world financial markets by the mid–1980s.

This increased rate of capital mobility has coupled with it significant changes in the debt/investment situation in both Canada and the United States, which are causing negative effects on urban development in North America.

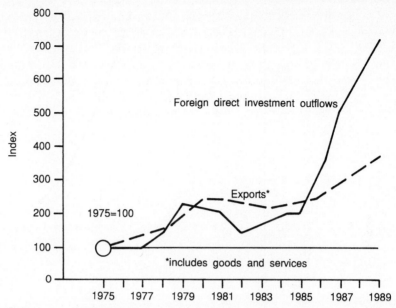

Figure 2.4 A comparison of the increase of foreign capital investments by transnational corporations compared with total world exports since 1975. (*Source:* United Nations Center on Transnational Corporations.)

The change with respect to accumulated debt is illustrated by the United States' federal debt in Figure 2.5. The national debt situation in Canada is somewhat worse. Briefly, Figure 2.5 indicates that since World War II the accumulated national debt in the United States, expressed as a percentage of gross national product, declined until about 1980. Since that time it has increased and is now approaching 60 percent. This means that more and more of the United States federal budget is involved with interest payments related to loans needed to carry the debt. Thus, whereas in 1980 interest payments accounted for 9 percent of federal expenditures, in 1990 the proportion was 15 percent—the third-largest expenditure category after defense (25 percent) and social security (21 percent). As a consequence, much political attention in both the United States and Canada in the 1990s has been devoted to curbing the rate of increase in debt.

This increasing absorption of capital to service national debt in Canada and the United States has occurred, not entirely coincidentally, at a time when both countries are failing to generate sufficient intangible capital (through savings) to service their own requirements. In 1984, the United States changed from being a country that generated sufficient capital for its own needs to one that imported capital. As a consequence, there has been a reluctance to invest in urban infrastructure, which is illustrated by a decline in investment in highways and education facilities (in constant dollars) in the United States since 1970 (Figure 2.6). The slight increase, for maintenance

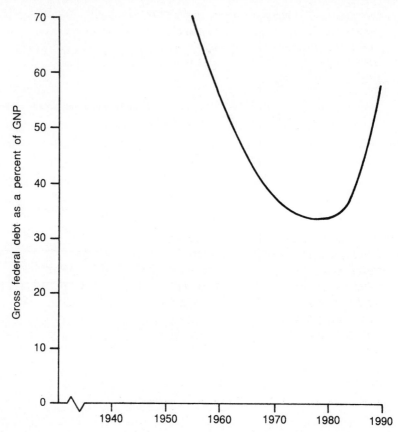

Figure 2.5 Change in accumulated United States federal debt as a percentage of GNP.
(*Source:* Bureau of the Census, 1993, p. 315.)

purposes, during the mid 1980s, was offset by recessionary related declines in the early 1990s.

The Concentration of Capital An important result of differential urban growth is the development of a *hierarchy* of urban areas based on the size of their economies and the degree of control that they exert over their surrounding regions and other (usually smaller) places. Since urban areas are the focus for their surrounding regions, each can be considered the center of a *nodal region* (trade or tributary area), the extent of which is related to its position in the hierarchy. Metropolises at the top of the hierarchy dominate not only their immediate nodal regions, but also other urban areas and thus indirectly their immediate nodal regions as well. Borchert (1983; 1978) has demonstrated the *integration* of the United States urban hierarchy through an examination of correspondent bank linkages. This is a revealing way of measuring the strength of relations among urban areas and, therefore, a good measure of dominance

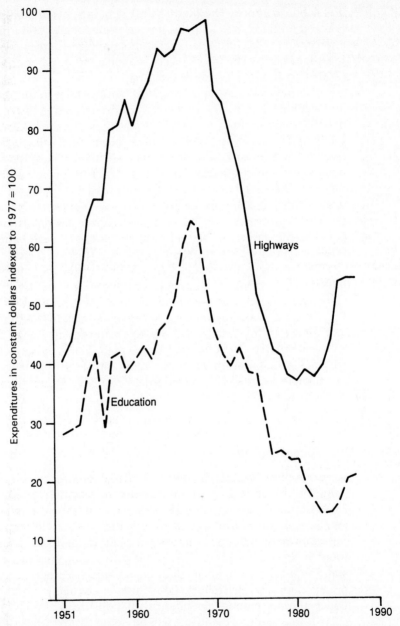

Figure 2.6 Per capita public construction expenditures in the United States from 1951.

(*Source:* Modified from Chatterjee and Abu Hasnath, 1991, Figure 2, p. 48.)

and subdominance. The hierarchical structure postulated in Figure 2.7, which is based on Borchert's analysis for the United States and information in Semple and Green (1983) for Canada, presents the three highest levels and combines a fourth and fifth. The lower levels, which could number another five (see Chapter 6) are not included in the diagram.

The highest-order center is New York, which is the focus of the financial and stock exchange activity on the continent, and is the headquarters location of the largest number of major corporations (Green, 1995). The data in Table 2.1 include only those MSAs/PMSAs that ranked among the twenty metropolises with the most headquarter offices in 1989. The information indicates that headquarter offices (as a surrogate for "corporate control") appear to be becoming more dispersed in the United States, with a large loss occurring with respect to New York. But, it must be noted that New York remains dominant, for the total *assets* controlled by the corporate offices headquartered in the PMSA is equal to the asset total for the next ten highest metropolitan areas combined (Holloway and Wheeler, 1991, p. 59).

The second-order centers, linked directly to the corporate financial community of New York City, are major regional foci in North America. San Francisco and Los Angeles dominate in the West; Chicago and Detroit in the Midwest; Boston, Washington, D.C., and Philadelphia in the Northeast; and Toronto and Montreal in Canada. The change columns in Table 2.1 suggest that the greatest relative declines are being experienced by Detroit and Los Angeles whereas Atlanta, Minneapolis-St. Paul, and Boston are experiencing the greatest relative increases in position. In Canada, Montreal is losing relative to Toronto and Vancouver.

A Model of Urban System Growth

An attempt to synthesize the four spatial processes comprising urbanization (innovation diffusion, migration, capital flows, and control) is shown in Figure 2.8. Two kinds of geographical space are identified: *core* and *periphery*. Core regions are defined as areas that have "a high capacity for generating and absorbing innovative change," whereas peripheral regions are those whose "development path is determined chiefly by core region institutions with which they stand in a relation of substantial dependency" (Friedmann, 1973, p. 67). Combined, core and periphery constitute a complete spatial system, between which the autonomy–dependence relationship is the key to the dynamics of growth. In Chapters 3 and 4 it will be noted that North America has developed, by the mid–1990s, more than a single core and a rather complicated pattern of peripheries.

Innovations (N) diffusing outward from the core are considered the *motor* in the growth process (Figure 2.8). They force the spatial system through successive structural transformations by the processes of innovation diffusion (D), control (C), migration (M), and investment (I). These are indicated by arrows of

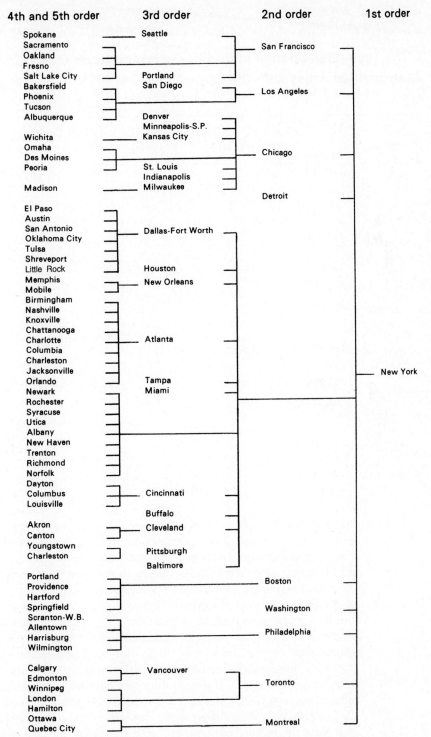

Figure 2.7 The hierarchical structure of the larger metropolises in the North American urban system.

Table 2.1

The 1969–1989 Change in Number of Headquarter Offices in the Twenty Metropolitan Areas with the Largest Number of Headquarter Offices in the United States.

MSA/PMSA	1990 POPULATION (THOUSANDS)	HEADQUARTER OFFICES 1989	1979	1969	GAIN/LOSS 1969–89
New York*	8,547	86	132	187	−101
Chicago	7,411	58	66	71	− 13
Dallas	2,676	26	19	14	+ 12
Los Angeles	8,863	25	32	34	− 9
Boston	5,051	24	16	17	+ 7
Minneapolis-St. Paul	2,539	23	24	21	+ 2
Houston	3,322	22	20	11	+ 11
Bridgeport*	444	21	29	10	+ 11
Cleveland	2,202	19	20	23	− 4
Atlanta	2,960	18	9	6	+ 12
San Francisco	1,604	17	19	23	− 6
Philadelphia	4,922	15	20	27	− 12
St. Louis	2,493	15	17	17	− 2
Pittsburgh	2,395	15	20	17	− 2
Newark*	1,824	14	9	10	+ 4
Detroit	4,267	13	21	21	− 8
Hartford	1,158	12	13	10	+ 2
San Jose	1,498	11	8	2	+ 9
Richmond	866	11	6	5	+ 6
Cincinnati	1,526	11	9	7	+ 4

(*Sources:* Table compiled from Ward (1994), Table 4, p. 476; Yeates (1990), Table 1.5, p. 23; and Bureau of the Census (1993), p. 37.)
*Part of New York-Northern New Jersey-Long Island CMSA. The New York PMSA includes New York City (7.3 m) and White Plains plus suburban Westchester County (1.2 m).

different width representing asymmetry in the autonomy–dependency relationship. For example, the volume of controlling decisions emanating from the core may cause a net outflow of capital from the periphery, which in turn intensifies migration from the periphery to the core in response to the changing location of economic and social opportunities.

At the same time, however, the continued diffusion of innovations from core to periphery could lead to conditions demanding a partial restructuring of its dependency relationships. Continued growth of the system requires, of course, that the tensions built up in this way be resolved. Each of the four processes brings about changes in spatial patterns in both core, or heartland (H), and periphery (P). For example, innovation diffusion (D) alters the sociocultural pattern (SC); control (C) processes the pattern of power relationships (OP); migration (M) influences the settlement pattern (S); and investment (I) structures the pattern of economic activities (EA).

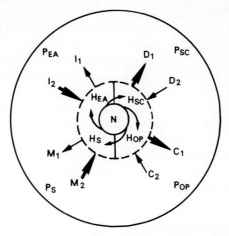

Figure 2.8 A core-periphery model of urban system development.
(*Source:* This figure is modified from the original in Friedmann, 1973, p. 68, by permission of Sage Publications Inc.)

THE URBAN HIERARCHY

In the previous section, the notion of an urban hierarchy has been introduced with respect to financial and corporate control. The concept of the urban hierarchy postulates a ranking of cities into groups on the basis of their economic, social, and administrative importance. The concept can, therefore, be thought of as a vertical dimension that complements the spatial dimension, for urban places have not only a location in space, but also a position at a particular level in the hierarchy. In general, places that have the greatest economic, social, or administrative importance usually tend to be the largest in population size as well. Thus, a simple indicator of rank in the hierarchy, particularly in the case of service centers (see Chapter 6), is population size.

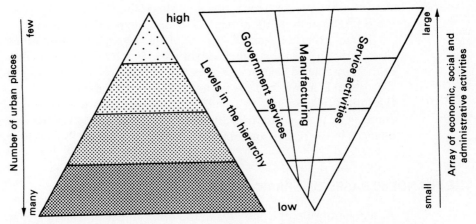

Figure 2.9 Hierarchies of size classes and functions.

In its structure, the hierarchy of urban areas can be likened to a pyramid (Figure 2.9). The pyramidal shape results from the fact that there is a greater number of smaller places than larger ones in the urban system. Thus, the number of places at each level (or order) in the hierarchy decreases as one moves upwards through it. Conversely, the economic, social, and political importance of urban areas is akin to an upside-down pyramid (Figure 2.9), with the few places at the top exhibiting the largest number of different roles, some of which are quite specialized (Beauregard, 1989). As a consequence, it is often useful to analyze the structure of the hierarchy in terms of some group of specialized activities, or some measure (such as money) that summarizes a number of activities, as has been done in Figure 2.7. The discussion in Chapter 6 demonstrates that one of the bases of a hierarchy is the distribution of consumer services among urban places.

The Importance of the Urban Hierarchy

The urban hierarchy, which is an outgrowth of the nature of the economic system and political structure in which it has developed, is important for three reasons. In the first place, it appears to be a relatively efficient structure for transmitting new technology, managerial expertise, and other economic, social, and cultural innovations from the centers of activity to the more peripheral areas. The hierarchy is, therefore, a useful agent for diffusion (Hansen, 1992, p. 301). Goods, materials, services, people, funds, decisions, and new ideas diffuse outward and downward from heartland metropolises to those in the hinterland.

Secondly, as noted previously, the hierarchy permits specialization, division of labor, and differentiation of economic activities. New York City specializes in financial and economic activities that pervade the nation. Washington, D.C., is the national political center, Detroit the focus of the national automobile industry, Toronto the financial center for the Canadian subsystem, and so forth. This hierarchical structure has perhaps developed as a result of the hierarchical nature of business enterprises in North America, for many industries have developed a size distribution of firms similar in form to the distribution sizes of urban places. This is not at all surprising, for it simply reemphasizes the interlocked nature of the urban and economic-political systems.

Finally, the hierarchical nature of the urban system demonstrates that a major influence on the size of urban areas is the extent and economic strength of the hinterland. Each metropolis has to function in a manner appropriate to the size and character of the hinterland. It is, therefore, possible to speculate on the future relative growth of different metropolises partly on the basis of past changes in their position in the hierarchy, and partly on the basis of present changes in the nature and extent of their hinterlands.

THE ECONOMIC BASES OF URBAN AREAS

As a general rule, the basis for urban areas in North America is economic, and their population size is largely a reflection of the economic activities they perform. These economic activities range from services to administration and manufacturing. Change in the population size of urban areas is thus, in some way or another, related to, and perhaps even conditioned by, corresponding changes in their economic structure. A fundamental concern, therefore, is the way in which population change and economic structure are related. This relationship is influenced by the size of the *urban* multiplier and the *productivity* of the urban area.

The Urban Multiplier

The idea of the multiplier is used a great deal in the social sciences. It is crucial to economics (the Keynesian multiplier) and to planning (the regional multiplier). In general, the multiplier refers to the *amount of increase in income, jobs, or spending generated beyond that which was originally created by a given investment.* A given increase in investment in something, let's say a new factory, results not only in the jobs created in the factory, but in jobs elsewhere in the local community as well. The people employed in the new factory purchase local services (such as education and health care); they also have to buy food and other goods. All this creates more jobs. In turn, income generated by people in these jobs is also spent locally. So through interdependence the jobs in the new factory are multiplied into many other jobs. There are two main ways the multiplier concept has been used in urban studies. A relatively simple way is the economic base multiplier, and a more sophisticated approach is the input-output multiplier.

The Economic Base Approach   The economic base approach divides the economy of an urban area into two parts. The basic ("city-supporting" or traded) sector refers to the goods and services produced within an urban area but sold beyond its borders; the nonbasic ("city-serving" or derivative) sector refers to the goods and services produced and sold within the urban area itself. The rationale for the approach is that the existence and growth of an urban area depends on the goods and services it produces and sells beyond its borders—the basic component (Farness, 1989). Furthermore, the size of the nonbasic component is viewed as being related in some way to the size of the basic component:

$$\frac{\text{Total activity in}}{\text{the city (TA)}} = \frac{\text{Total basic}}{\text{activities (BA)}} + \frac{\text{Total nonbasic}}{\text{activities (NBA)}}$$

For analytical purposes, activity should be defined as income (or expenditures), but is frequently defined in terms of employment. Thus, if total income in an urban area (TA) is $400 million, and the income from BA is $100 million, then the income from NBA is $300 million. This example also suggests that every $1.00 in BA generates $3.00 in NBA. Therefore, if new BA activity is generated, providing an income of $50 million, and the basic-to-nonbasic ratio of 1:3 stays constant, NBA will increase by $150 million, and TA activity will consequently increase by $200 million (200 = 50 + 150). The multiplier, which indicates the growth in TA as a result of an increase in BA, is 4 (the sum of the basic-nonbasic ratio, or TA/BA).

There are, of course, conceptual and computational difficulties with the economic base concept and its empirical applications (Bloomquist, 1988). The conceptual difficulties relate to the assumption of the split between basic and nonbasic activities, the primacy given by the approach to BA in the growth (or decline) of a city, and the problems related to defining the spatial unit involved. The computational difficulties concern how the basic and nonbasic components may be defined, and the fact that employment is often used rather than income because money flows are difficult to obtain.

The basic and nonbasic components are usually estimated at either the micro level (small scale) for individual places, or at the macro level (large scale) for a number of urban areas. Micro-level studies require detailed analyses of each activity in an urban area to determine which portions of an enterprise or activity are basic and nonbasic (Gibson and Wordon, 1981). Macro-level studies use aggregates of urban places and estimate for each economic sector the proportion that is basic (traded) or nonbasic (derivative). For example, in the Greater Toronto Area (see Figure 10.9), which includes a number of communities, 87 percent of the manufacturing sector is estimated to be basic, 36 percent of distribution services, 52 percent of financial and insurance services, 51 percent of business services, and 100 percent of tourist activities (Table 2.2). Manufacturing dominated the basic component with 60 percent of the traded economy, while tourism, which is 100 percent basic, provides only 2 percent of the traded economy.

The Multiplier and Urban Size   A common method used at the macro level to estimate the basic and nonbasic components is the minimum requirements approach (Moore and Jacobson, 1984). The minimum requirements approach estimates for each sector of economic activity (such as manufacturing, or finance and insurance) the proportion of activity in urban areas of the same general population size that is usually *nonbasic*. These nonbasic proportions, which are regarded as minimum requirements for serving the urban areas themselves, are then added together for each sector to yield the proportion of total employment in an urban area that can be regarded as nonbasic. These totals can then be plotted on a graph against urban population size to demonstrate that the nonbasic share of an urban area's employment increases with the population size (Figure 2.10). This relationship also demonstrates that the population size of an urban area also increases with the size of the multiplier (which is scaled at the top of Figure 2.10).

The multiplier increases with urban population size because, whereas the inhabitants of small urban areas may have to go to larger places for certain special goods or services, the inhabitants of large places invariably have access to nearly all the goods and services they require within their own metropolis. There is, therefore, a greater leakage of money from small places than from large metropolises. The shift in the plotted lines from Figure 1.6 suggests that this leakage is increasing over time as large metropolitan areas become more dominant. For example, the graphs indicate that the multiplier for cities of 1 million population size increased from 2.2 (calculated as 100/100 – 55) in 1950 to 2.7 in 1980. In the case of the Greater Toronto Area, which has a population

Table 2.2

The Relative Importance of Different Sectors to the Basic (or Traded) Component of the Economy of the Greater Toronto Area.

ECONOMIC SECTOR	% BASIC	BASIC VALUE ADDED	BASIC EMPLOYMENT	VALUE ADDED/ EMPLOYEE	% OF BASIC ECONOMY
Manufacturing	87%	$ 21.0 *billion*	327,000	$ 64,000	60%
Distribution Services	36	7.1	239,000	30,000	17
Finance and Insurance	52	4.9	47,000	105,000	11
Other Business Services	51	3.2	14,000	230,000	8
Tourism	100	0.7	26,000	27,000	2
Totals		36.9 *billion*	653,000		98%

(*Source:* GTA (1995), p. 51.)

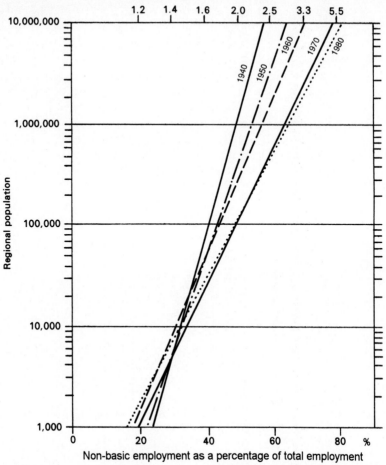

Figure 2.10 The changing relationship between the proportion of a United States urban region's employment that is nonbasic, estimated using the minimum requirements approach, and population size.
(*Source:* Moore and Jacobson, 1984, p. 223.)

of 4.5 million, the basic-nonbasic ratio is estimated to be 1:3 and the multiplier 4 (GTA, 1995, p. 49).

The Input-Output Multiplier The input-output (I/O) approach does not assume a distinction between basic and nonbasic components of urban economic activities (Hinojosa and Pigozzi, 1988). Instead, the I/O approach recognizes that all aspects of the economy of a nation, region, or urban area are interrelated and interdependent (Miller and Blair, 1985). Thus, for example, an increase in the amount of housing construction in an urban area generates increases in activity in services of all kinds, not only within the metropolis in which the growth is occurring, but elsewhere in the nation (and world) where activities are linked in some way to those of the housing industry in the growth area. The I/O multiplier for a particular urban area will, therefore, relate the increase in total output (or income) to an increase in output (or investment) in a particular growth sector (such as housing). In a study of the Los Angeles, California, economy, Wolch and Geiger (1986) use I/O multipliers to demonstrate that the not-for-profit sector (such as philanthropic organizations and universities) have an enormous impact on the regional economy.

For example, using interregional, intersectoral I/O tables for the Canadian economy, what would be the impact of a $10 million increase in output in the transportation industry of Montreal? According to the linkages modeled in the interregional, intersectoral I/O table (Yeates, 1975), the *total* impact on all sectors in Montreal, Quebec, and other regions would be something like this (numbers rounded):

Direct	Indirect

Direct		Indirect										
10	+	10	+	10	+	2	+	10	+	0.7	+	15
		(Montreal)		(Quebec)		(West)		(Ontario)		(Maritimes)		(Foreign)

$$= \$57.7 \text{ million}$$

The direct increase in output is $10 million in the transportation industry, which indirectly generates an additional $10 million worth of output over all sectors (housing, services, other manufacturing, etc.) in Montreal. Thus, the multiplier for Montreal is 2.0 (Direct + Indirect/Direct). The indirect impact elsewhere in Quebec was $10 million; thus, the multiplier for Quebec as a whole is 3.0. The indirect impact on West Canada was $2 million, on Ontario $10 million, on the Maritimes only $.7 million, and in foreign areas (mainly the United States), $15 million. The Canadian multiplier for the increase in output in Montreal is therefore 4.27, and the total multiplier over all sectors in all regions (including foreign parts) is 5.77.

The model described above assesses the multiplier effect on *output* of a change in output in a particular sector of activity on all sectors. But, what about the multiplier effect of an increase (or decrease) in *different activities* on employment and income in a region as well? Conway (1990) provides a good example of the use of an input-output model (the Washington I/O model) to answer this question with respect to the economy of the State of Washington. Table 2.3 indicates that a unit change in the wood products industry results in the greatest output, employment, and income multiplier effects in the state. The greatest impact on jobs is, however, experienced by changes in the wood product industry and aerospace. In this latter regard, "aerospace" refers primarily to Boeing and the myriad of companies that interconnect with it in the Seattle-Tacoma CMSA (Erikson, 1974).

The Productivity of Urban Areas

One of the interesting questions facing urban analysts is why some metropolises tend to get larger and larger and some large ones even decline in population. There is a view that when urban areas grow above a certain size (a popula-

Table 2.3

Selected Output, Employment, and Income Multipliers from the Washington Input-Output Model, 1982[a]

	OUTPUT MULTIPLIER	EMPLOYMENT MULTIPLIER	INCOME MULTIPLIER
Agriculture	1.97	1.84	2.07
Wood products	2.65	3.92	2.91
Aerospace	1.47	2.42	1.60
Trade	1.81	1.71	1.78

[a]Total change in economy per unit change in industry.

tion figure of 250,000 has often been cited) the chances of continued growth improve and contraction becomes unlikely. Above this size, a growth mechanism comes into being, like a ratchet, which locks in past growth and inhibits contraction. Several reasons have been suggested for this comforting notion of a size-based urban ratchet with respect to urban growth.

First, as urban places become larger, their economic structure usually becomes more diversified. This increased level of diversification tends to ensure that local growth rates coverage around the national average. Second, as places become larger, they also gain in political power and are thus able to exert greater political pressure as they bargain for government contracts and support. A third factor is the large amount of fixed capital that has been invested in larger urban areas, which means that they have better developed roads, sewage and water systems, communications networks, public transportation, and other infrastructure that attracts more firms. A fourth factor is the relative ease of doing business in a large urban community with many services, education and training facilities, and suppliers. Finally, large urban areas tend to attract creative, entrepreneurial, and innovative persons who collectively provide continuous sources of stimulation to the local economy.

Much evidence, however, indicates that growth is related to the tension of opposing forces of the economies and diseconomies of urban size. If the factors cited above with respect to the urban ratchet can generate production economies, then size and age can bring with them certain diseconomies. Large urban areas can become congested, older infrastructure may be inappropriate for newer types of economic activities, and a large population may contain within it deep economic, social, and cultural fissures that are inimical to capital investment. Even more probable, economic factors that were advantageous to growth at one time period may no longer be relevant in another time period. In short, there may be a time (as is indicated in Figure 2.11) when diseconomies of size may begin to offset the economies, and output per capita in an

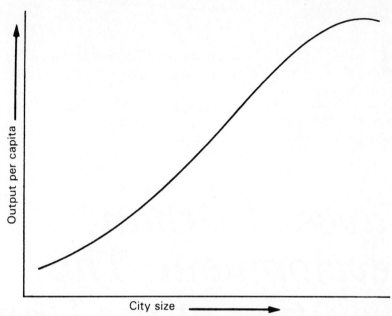

Figure 2.11 The suggested relationship between the productivity of cities and city size.

urban area decreases. When this occurs, economic activities may not be replaced when they close down, and some businesses may leave and move elsewhere. This is followed by a decline in population. During economic recessions, plant closings are greatest in those areas where diseconomies have become the greatest.

CONCLUSION

The processes of urbanization are, therefore, exceedingly complex. They are related to the developmental state of a society, the type of society that has evolved, and the ways in which these processes are translated to the location of settlements in geographic space and differential urban growth. Although it is not really possible to predict urban size, there is abundant evidence to indicate that there are a variety of possible paths of growth and even decline. The population size of an urban area is related to the strength of its economic base. The most important feature of urban development in North America is that urban areas have had to have a productive role in their nation's economy to flourish. This productive role is reflected in its position in the urban hierarchy and function within a complex core-periphery framework. In the following two chapters these concepts are examined in the context of the influence of long-term fluctuations in the North American economy and technological change.

Waves of Urban Development: The Evolution of the United States Urban System

Urban development in North America is directly related to economic development and the generation of surpluses from both the urban and rural spheres of production. Although it is not possible to calculate this "surplus," it is possible to show how wealth within the nation has changed during the past 125 years. Figure 3.1 graphs the changes in the gross national product (GNP, the sum of all goods and services produced) in the United States in 1958 dollars per capita since 1870. Similar information is available for Canada, where the trend line is about the same as that in Figure 3.1, except that depressions/recessions seem to be deeper in Canada and last a little longer (Yeates, 1985a; McInnis, 1990). In looking at the general change in GNP per capita, it becomes evident that even though the theoretical amount necessary to maintain society undoubtedly increases through time, primarily as a result of accumulation and increasing basic living standards, the possibility of surplus generation also increases enormously. After all, society as a whole is seven times wealthier in the 1990s than it was in the 1890s.

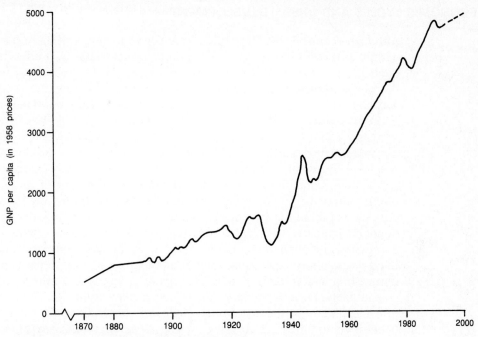

Figure 3.1 One hundred and twenty-five years of change in GNP per capita (1958 constant dollars) in the United States.
(*Source:* Compiled from data in Bureau of the Census, 1993, p. 455; and 1975.)

The generation of a surplus during a particular time period, which may be appropriated in some way to finance urban infrastructure, relates quite directly to immediate changes in wealth. *Urban infrastructure* refers to all those built things that comprise urban physical space, such as houses, roads, sewer and water systems, office buildings, manufacturing plants, and so forth. If the GNP per capita increases considerably over a few years, then the amount of surplus generated may be comparatively large. So all kinds of public and private investments may occur or be planned. On the other hand, if the GNP per capita increases slowly, or decreases, it is quite likely that the surplus will be small or non-existent. Likewise, the money (or capital) available for urban development will be small or nonexistent. It is for this reason that the fluctuations seen in Figure 3.1 are extremely important: When the rate of growth in GNP per capita is quite high, urban development booms, but when the growth levels off or decreases, then urban development slows down considerably or reaches a standstill. These fluctuations have generated considerable interest in the possible existence of a number of types of economic cycles (Tylecote, 1992), and in the relation of these cycles to investment in urban (Harvey, 1985, pp. 17–21) and regional (Marshall, 1987) development.

ECONOMIC CYCLES AND URBAN DEVELOPMENT

Although economic cycles vary enormously in length, and there may be cycles within long-term fluctuations, economic cycles have a similar internal sequence. The bottom of the cycle is the *trough*, where unemployment is high, the rate of return on investments is low, and there is much unused industrial capacity. This is followed by a period of *expansion*, where the rate of unemployment decreases and income, consumer spending, and business investment all increase. The *peak* of the cycle occurs when there is a high use of industrial capacity and labor, and the rate of return on investments in productive capacity and labor begins to level off. There may well be a shift in investments to more speculative, consumption-oriented ventures. As the rate of increase in economic output begins to decelerate and the rate of return on investment decreases, a contraction may occur, leading to a *recession*. When demand and production decrease, the recession may intensify into a deep trough (or depression) if a crisis in confidence occurs as a result of overspeculation. During a recession or depression, unemployment rates become high, opportunities diminish, and standards of living for many people and families may decrease considerably. Periods of expansion and peaks have tended to be the longest part of cycles. Many of the economic policies of government since the 1930s have been directed toward controlling these cycles, lengthening expansions, and attempting to protect people against the worst ravages of downswings when they do occur.

There are two types of cycles—those induced by *endogenous* forces, and those induced by *exogenous* forces.

1. An endogenous cycle, such as an inventory cycle or a business cycle, is one that is self-generating. That is, the forces that create an upswing also contain within them characteristics that induce a downswing, which in turn generates an upswing.

   a) Inventory cycles, which have been documented for some time (Kitchen, 1923), tend to be of three to four years duration, and are related to the immediate adjustments in inventories that businesses make in response to short-term fluctuations in sales. An overstock in inventory in one time period is compensated by reduced orders for the next.

   b) Business, or investment cycles, tend to be about seven to eleven years duration, and appear to be related to changes in fixed capital investments by businesses in response to perceived and expected changes in rates of return (profit) on capital investments. The difference between the inventory and business cycles is that the former applies to nonfixed assets, whereas the latter assets of a more fixed nature, such as plant and machinery (van Duijn, 1983). These fluctuations have a considerable impact on short-term urban development as fixed capital investments invariably give rise to short-term changes in the demand for business and office space, as well as housing.

2. Exogenous cycles are induced by forces "outside" the immediate framework of the economy.

   a) Inventory and business cycles may well have been replaced in the last forty years by political cycles reflecting governmental periods of office in the United States of four years. Incumbent governments may well attempt to lower interest rates (and/or increase the money supply) prior to an election to boost the economy, and increase interest rates (and/or reduce the money supply) afterwards—thus creating upturns and downturns in the economy.

   b) Building cycles, of about fifteen to twenty-five years duration (Kuznets, 1961), involve longer-term fluctuations in periods of construction activity, and appear to be related to major periods of immigration and high rates of population increase (Berry, 1991b).

   c) Although there is some debate concerning whether long waves of forty or sixty years duration, identified by Kondratieff (1935), actually exist (Mager, 1987), the general consensus is that there are long-term fluctuations in national economies of some fifty years duration within which the Kuznets building cycles are embedded. For example, Berry in his analysis of a variety of United States economic data for a 200-year period, concludes that "within the inherently high noise levels of history, prices and economic growth move in synchronized rhythms . . . in approximately 55 year waves within which 25–30 year cycles are embedded" (Berry, 1991a, p.10). These long waves are related to a mix of fundamental economic transitions, giving rise to a "new economic paradigm" during each epoch (von Tunzelmann, 1995, pp. 95-100). This new economic paradigm is stimulated by a combination of the following exogenous characteristics:

      i) Emphasis on different manufactured products—for example, information-processing equipment rather than steel products

      ii) A concentration of new enabling technologies—for example, the microchip and recombinant DNA technology

      iii) Different forms of industrial organization—for example, just-in-time systems, and "quality circles"

      iv) New geographical patterns for the location of capital investment—for example, low rates of capital investment and disinvestment in parts of the North and East, and new investment in the South and West

      v) Major international conflicts—for example, World War I and World War II

      vi) Changes in population growth, household formation rates, and the size of the workforce—the dominant recent demographic force in North America being the 1945–1963 baby boom, followed by the baby bust and a baby boom echo.

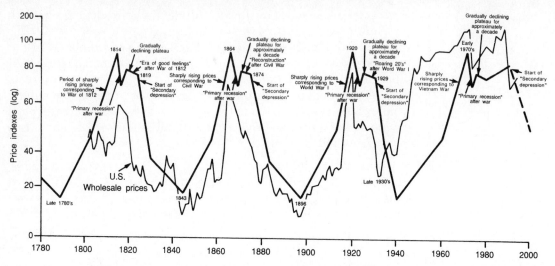

Figure 3.2 Kondratieff long waves and index of United States wholesale prices, 1780–date.
(*Source:* Mager, 1987, Figure 2.1.)

A confluence of exogenous forces such as these at different times results in major changes in the structure of national (and global) economies, and thus quite different periods of urban development.

Long Waves and Major Eras of Urban Development

The method used by Kondratieff (1935) to determine the periodicity and length of the long waves was to transform information related to prices of raw materials and some manufactured products, volumes of trade, and interest rates into deviations from a general trend, and then to use nine-year moving averages to smooth out the inventory and business cycles to estimate the long-term cyclical patterns. The information that he used pertained to France, the United Kingdom, the United States, and Germany. A diagram of the time line of the cycles identified by Kondratieff is presented in Figure 3.2, with the years given on the horizontal axis. The heavy black line in the diagram represents the general model of the three long waves described by Kondratieff using information up to the mid-1930s; the lighter line indicates the close relationship between one important economic index (United States wholesale prices) and the model. Recent long waves have been added, based on the work of Mager (1987) and Berry (1991a), for the twentieth century. These long waves are superimposed on the general trend in GNP per capita in Figure 3.3 to emphasize that the fluctuations occur in the general context of increasing North American wealth.

Major Eras of Urban Development   On the basis of these long waves, which embrace the period from mercantilism to capitalism, it is possible to define four eras, and perhaps foresee a fifth. These are:

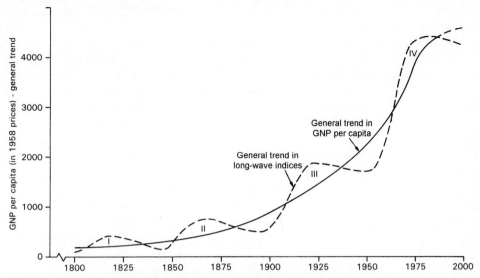

Figure 3.3 Conceptual graph of Kondratieff long waves superimposed on the general trend in GNP per capita, 1800–date.

1. *Frontier mercantile (to 1845),* in which wealth is created through trade and the accumulation of tangible wealth such as land and gold. In this era, profits produced by merchants through trade are used to purchase consumer goods and services and are also reinvested in other trade ventures.

2. *Early industrial capitalism (1845–1895),* which witnessed the establishment of manufacturing and processing plants by entrepreneurs using capital initially derived from trade (or imported from Europe), but more and more from the profits from the sale of manufactured goods. Most of these plants were relatively small scale and locally owned, and their growth was nurtured behind protectionist trade barriers. Perhaps the most important factors stimulating growth in this period are technology in the form of the steam engine and railroad, the availability of cheap immigrant labor in the wake of the famines in Ireland and elsewhere in Europe during the 1840s, the opening up of the rich agricultural lands of the Midwest of the United States and southwest Ontario, and the stimulus to industrial production in the Northeast associated with the United States Civil War (1861–1865).

3. *National industrial capitalism (1895–1945)* is characterized by the emergence of powerful national corporations (Chandler, 1977), large-scale assembly line manufacturing, and the production of a vast array of consumer items. As national corporations became stronger, the United States moved from a protectionist industrial policy to a free-trade position. Technological change was one of the chief stimulants of the era, with the harnessing of electricity and the invention of the electric motor, the development of fine steels, the telephone, and the invention of the internal combustion engine, which led to the growth of the assembly line automobile industry after 1920. Growth

was also stimulated by the enormous influx of immigrants between 1885 and 1911 (more than 25 million), which both stimulated aggregate demand and held down wages during the early part of the period. There were also great increases in agricultural productivity. Although inequality of income between classes was widening in the United States throughout much of the nineteenth century, particularly from 1880 on, the gap began to narrow after 1914 as the growth rate of the labor force began to decrease and skilled and semiskilled labor was more in demand. Hence the emergence of the *consumer* society. The concentration of economic and financial power to the United States, and the growth of manufacturing industry in Canada, was greatly enhanced by World War I.

4. *Mature industrial capitalism (1945–2000?),* an era which commenced with North America as the leading financial and industrial center in the world, and concluded with North America sharing this role with the European Union and Japan. During this era, many of the United States national corporations merged or expanded into large multinational corporations and achieved dominant positions with respect to the North American market and, in some cases, the world market. Two of the prime stimulants for this level of domination were the preeminent position assumed by the United States following World War II, and the number of technological innovations stimulated by the war effort that were transferred to domestic production, particularly those connected with the aircraft industry and electronics. Also, raw materials (particularly oil from the Middle East) were relatively low in price. But growth was also stimulated to an enormous extent by a high rate of population growth (the "baby boom") that lasted for fifteen years (from 1947 to 1962) and by the large-scale migration of African Americans (and some whites) to the inner cities of the large metropolitan areas. All these factors combined—greater wealth, a further narrowing of income disparities, technological change, demographic pressures—helped to generate a tremendous demand for services, and consequently for service employment. This period witnessed the emergence of the *service society.* Toward the end of the era the dominance of the United States multinational corporations is being challenged or replaced in many sectors by Japanese and European (particularly German) conglomerates, many of whom pioneered higher-quality products, new production processes, and more efficient forms of industrial organization.

Each of the eras just described generated particular forms of urban system organization. One approach to a description of the various forms of urban organization has been provided by Borchert (1967; 1972) in his classic analyses of the growth of the United States urban system. In his studies, he defined epochs which turn out to be from the middle (or close to the peak) of each of the long waves defined in Figure 3.2, and focused primarily on the impact of transport innovation on the changing urban system (Berry, 1991b).

Hall and Preston (1988) have also focused on the impact of *new information technologies* (NITs) as having a driving impact on long waves, and hence a significant influence on present and future urban development. Like Berry

(1991b) and Mager (1987), they predict that the next long wave, or fifth Kondratieff, will commence around the year 2000 and will be driven by new information technologies. One suspects that the evidence required to accept the hypothesis of the onset of new long wave is akin to the data requirements for accepting (or rejecting) the proposition of global warming—decades of evidence are required and society will probably not be convinced it is in a new era until it is half-way through. Something different *is* arising again. Therefore, we are uneasy because we are unsure as to what the impacts will be.

URBAN SYSTEM DEVELOPMENT IN THE UNITED STATES

The evolution of the United States urban system is described and discussed in terms of the eras defined by the long waves (Table 3.1), with the mapped representation of the system occurring at or just after the peak for each era. Thus, the frontier mercantile era is represented by a map of the size distribution of urban places for 1830–1831, early industrial capitalism by a map for 1870–1871, and so forth. The basic theme to be emphasized is the way changes in the structure and organization of the urban system reflect the major socioeconomic forces of each era. The most recent era is discussed in the greatest detail because it is during this period that the population of North America has become concentrated in large metropolitan regions.

The Frontier Mercantile Era, 1790–1845

The economy of North America throughout the eighteenth and early nineteenth centuries was based almost entirely on agriculture and the export of staple products. Even by 1830, only 9 percent of the population of the United States resided in places of 5,000 or more, and about 3 percent in the area that became Canada in 1867. The major economic function of most cities was *mercantile,* involving the import of manufactured products from Europe and the export to Europe of fish, furs, timber, and agricultural products. Manufacturing, which until 1775 had essentially been forbidden in the English colonies (and remained discouraged in Canada up to 1876) was still very much in its infancy. It was characterized by handicrafts rather than machines, and organized by households and small workshops rather than factories. The major industry was cotton textiles. By 1815 a total of 170 mills were operating in port towns and villages (such as Lowell and New Bedford, Massachusetts) based on hydropower sites in New England (Vance, 1977). Ports were also sites of considerable shipbuilding and repair.

The largest urban areas during this era, consequently, were found along the Atlantic coast (Figure 3.4), with New York (202,600), Philadelphia, Pennsylvania (161,400), Baltimore, Maryland (80,600), and Boston, Massachusetts (61,400) dominating the trade of the original thirteen colonies and forming the heartland of the developing continent. Each of these cities was competing for the new hinterland opening up west of the Appalachian Mountains. The three next largest urban areas were located along the route that traditionally led into

Table 3.1

Compilation of Terminology Related to the Evolution of the North American Urban System

LONG WAVES		BORCHERT (1967; 1972)		HALL/PRESTON (1988)		TYPE OF CITY
Era	Dates (From/to)	Era	Dates (peak to peak)	Era	Dates (From/to)	(Chapter 7)
Frontier-Mercantile	1790–1845	Sail/wagon	–1820			Mercantile
Early Industrial Capitalism	1845–1895	Steam/iron rail	1820–1870	Mechanical	1846–1895	Classic industrial
National Industrial Capitalism	1895–1945	Steam/steel	1870–1920	Electrical	1896–1947	Metropolitan
Mature Industrial Capitalism	1945–1995/2000?	Auto/truck/air	1920–1970	Electronic	1948–2003	Suburban

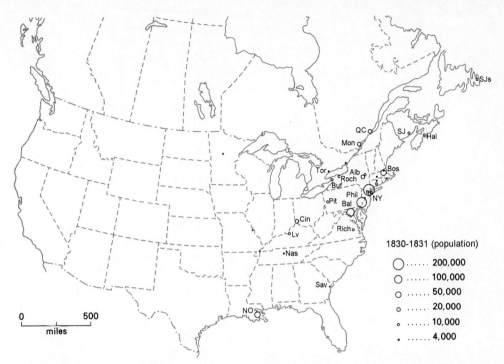

Figure 3.4 The relative size and location of all urban areas in North America with a population of 4,000 or more in 1830–1831.

the heart of North America. At the Mississippi Delta, New Orleans (46,300) received the produce from the hinterland, but the difficulty of passage upriver impeded the flow of two-way traffic. Quebec City (34,700) and Montreal (41,000), located along the St. Lawrence River, controlled the northern route into the region, but the relative growth of these two cities was already suffering from the loss of hinterland that resulted from the independence of the United States from British colonial rule.

Most of the urban areas located in the hinterland, which stretched as far west as the Mississippi River, were small in size and served as centers of commerce for the surrounding agricultural settlements. In 1830, Cincinnati, Ohio, was the largest of these interior towns (24,800). It acted as a service center for the rich Ohio Valley, with its merchants importing manufactured products and agricultural supplies directly across the Appalachians from the East Coast ports and exporting products down the Ohio and Mississippi rivers via New Orleans to Europe. Chicago was just beginning its rapid growth during the last fifteen years of the era.

Urban System Organization During the Frontier Mercantile Era A general model of the spatial organization of the urban system in the frontier mercantile era is presented in Figure 3.5. A number of ports along the East Coast become *gateways* that serve as intermediary trade centers between the staple

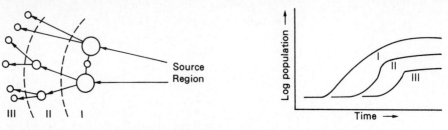

Note: I, II, III refer to level in a size hierarchy

Figure **3.5** The frontier mercantile era: model of urban system organization and paths of population growth for different level centers.
(*Source:* Simmons, 1978, p. 64.)

and agricultural-goods producing hinterland and the foreign market to which the area is tied (Johnston, 1982, p. 70). The main flow of information and industrial products is from the foreign market, through the gateway, to the hinterland; staples and agricultural commodities flow in the reverse direction.

Therefore, particular directions of interaction and urban size arrangements develop. The largest and highest-order urban places are along the coast, with second-order centers farther inland, connecting the ports to the resource towns (or lower-order centers) of the frontier. Growth is achieved by extending the hinterland into unexploited territory or by capturing part of adjacent areas. In this sense, the Treaty of Paris (1783) was incredibly profitable to United States urban areas, as the British ceded much of the Great Lakes hinterland of Montreal. The entrepreneurs of New York City then solidified their control of the hinterland with the opening in 1825 of the Erie Canal, which linked the Hudson River to Lake Erie at Buffalo and hence the whole of the Midwest around the lower Great Lakes.

Early Industrial Capitalism, 1845–1895

Whereas during the frontier mercantile era urban areas were usually directly dependent on the hinterland for their wealth, during the industrialist-capitalist era urban places themselves became the focus of wealth-creating production, though they were still dependent on the agricultural and resource-producing areas for food and materials. Industrial capitalism involves investments by the owners of capital in labor and equipment to produce goods. Market forces also dictate, for the most part, that the goods be produced at competitive prices. There is, therefore, a continuous drive for greater productivity, that is, output per worker hour. This was achieved through the development of the factory system (the breaking down of manufacturing into specialized routine tasks) and the realization of economies of scale through increases in factory size.

The era of early industrial capitalism embraces the transformation of the United States economy from one that was based primarily on trade in resources to one that processed resources into manufactured products. Chicago became the chief center for processing agricultural products: "As grassland gave way to pasture, and pasture to feedlot, the general tendency was for people to replace natural systems with systems regulated principally by the human economy" (Cronon,

1991, p. 223). The steam engine was undoubtedly the key technological innovation enhancing productivity in farms, factories, and transportation. Whereas in 1850 there were about 125,000 factories and workshops employing 1 million production workers, by 1900 there were more than 500,000 factories and workshops employing almost 5 million workers (Bureau of the Census, 1975, p. 666).

Most important, however, was the use of the steam engine in barges (up the Mississippi and Missouri rivers), ships, and trains. For example, in 1850 there were 9,000 miles of railroad track in the United States, but by 1870 the rails covered 53,000 miles. The network linked all the cities of the East Coast, spread inland most extensively through the Midwest around the lower Great Lakes, and by 1869 reached Oakland, California, on San Francisco Bay. The building of canals and canalized rivers was also vital to the transport of heavy goods because the iron rail would not support weighty loads. As a consequence of these improvements, the basic framework of the United States transportation system was in place by the end of the era, the gross "friction of distance" between places had been substantially reduced, and a more efficient integrated space was emerging within which resources could be mobilized and products distributed.

The Emerging Urban System   Between 1850 and 1890, the population of the United States increased from 23 million to 63 million. Simultaneously, the proportion of the population residing in places of 5,000 persons or more increased from 14 percent to almost 32 percent (Table 3.2). The urban population increased, therefore, by about 16.6 million. By 1870, at the peak of the long wave, New York City (Manhattan) had become the dominant city in North America (Figure 3.6a) and, when combined with Brooklyn, had a population in excess of 1.2 million. The level of dominance is indicated by financial flows as measured by correspondent bank linkages (Figure 3.6b). By 1981, New York had become the financial center of the United States, with the rest of the developing country linked to it either directly, or indirectly through Chicago (Conzen, 1977). Analysis of these bank linkages suggests that a simple interconnected hierarchy of urban areas had emerged, involving the gateway cities of the East Coast, New Orleans in the South, San Francisco in the West, Chicago and St. Louis in the Midwest, and Pittsburgh and the cities of the Ohio Valley. Most of these cities also developed a considerable manufacturing base that was nurtured by protective tariffs, stimulated through the demands of the Civil War, and made competitive in part by the labor provided by thousands of poor immigrants.

Urban areas inland in locations convenient to both rail and water transportation grew phenomenally. St. Louis on the Mississippi River, connected by rail with the East and by steam paddleboat and barge with New Orleans and the Gulf of Mexico, grew from a population of 6,000 in 1830 to 311,000 in 1870. Chicago became the hub of the midwestern rail system and also had an all-water transport route via the lakes and the Erie Canal to the Atlantic, giving rise to its self-proclaimed status as the "greatest primary grain port in the world." The population of Chicago grew to almost 300,000 by 1870. Cincinnati, the other big inland city of the Midwest, had grown less dramatically, reaching a population of 216,000 by 1870. Its economy, based on manufacturing (particularly pork meat packing), commerce, and transportation, was more diversified than Chicago's.

Table 3.2

The Population of the United States since 1790 and Canada since 1871, and Level of Urbanization

| Date | UNITED STATES | | | CANADA | |
	Number of Places over 5,000	Total Population (millions)	Percent Urban	Number of Places over 5,000	Total Population (Millions)	Percent Urban
1790	12	3.9	3.4			
1800	21	5.3	5.2			
1810	28	7.2	6.3			
1820	35	9.6	6.2			
1830	56	12.9	7.8			
1840	85	17.0	9.8			
1850	147	23.2	13.9			
1860	229	31.4	17.9			
1870	354	38.6	22.9	21	3.7	12.2
1880	472	50.2	24.9	38	4.3	15.3
1890	694	62.9	31.5	47	4.8	21.0
1900	905	75.9	35.9	63	5.3	24.6
1910	1,202	91.9	41.6	91	7.2	32.6
1920	1,467	105.7	47.1	110	8.8	36.6
1930	1,803	122.7	52.3	137	10.4	41.7
1940	2,042	131.6	52.7	152	11.5	43.0
1950	2,449	150.7	54.5	207	14.0	45.4
1960	3,293	179.3	59.8	259	18.2	50.9
1970	4,140	203.2	61.7	355	21.6	56.2
1980	5,084	226.6	66.2	409	24.3	61.8
1990	5,831	248.7	71.2	437	27.3	70.5

(*Sources*: Bureau of the Census (1975, 1993); Leach (1983); Statistics Canada (1992).)
Note: Number for places of 5,000+ for 1990 is estimated from published aggregated information to provide consistency with previous years.

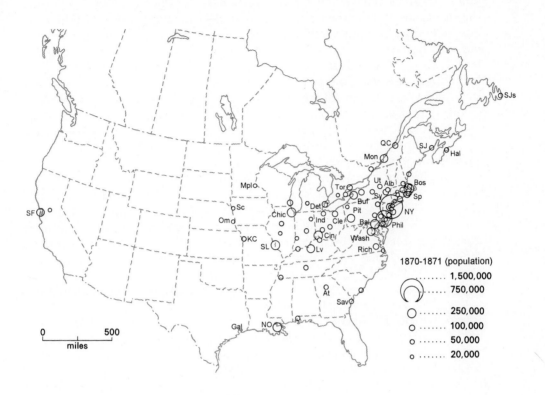

(a)

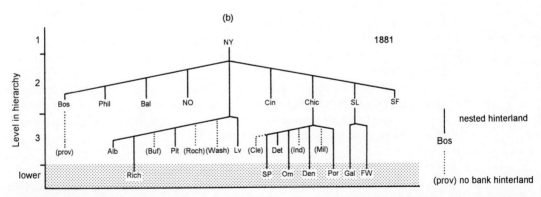

(b)

Figure 3.6 Urban size and interaction by the middle of the early industrial era: (a) the relative size and location of all urban areas in North America with a population of 20,000 or more in 1870–1871; and, (b) the hierarchy of urban areas in the United States as defined by correspondent bank linkages.

(*Source:* Part (b) redrawn from Conzen, 1977, Figure 9, p. 106.)

Urban System Organization During the Era of Early Industrial Capitalism When an indigenous manufacturing base developed, it began to change the spatial structure of the urban system quite significantly. Although the economy of many urban areas was still largely devoted to trade and services, a number were now devoted primarily to manufacturing. A feature of the early industrial era was that urban places grew rapidly, but tended toward a more hierarchical size distribution because the economy became organized around the needs of production taking place in the growing domestic economy. Manufacturing brought the concentration of labor and materials in a few locations, and financial requirements the need for an organized system of access to capital. These financial and industrial activities generated further growth through agglomeration and concentration.

Furthermore, the larger places related directly to a hinterland which included a number of smaller places that formed part of the region's industrial complex. Thus, a rank-size distribution arises from the self-organizing hierarchical structure of cities consequent to the effects of agglomeration and concentration, distance between urban areas at different levels of service provision, and the locational choice behavior of individuals, households, and firms (Hoag and Max, 1995). The rank-order distribution changes as an urban system evolves. These complex ideas will be addressed in greater detail in subsequent chapters.

Rank-Size Distributions Rank-size models can be used to highlight a number of trends related to the populations of urban areas in a region (Alperovich, 1993). The *rank-size rule* says that if urban areas (defined in some way, such as MSAs or CMAs) are plotted on double-log paper in descending order according to their population sizes on the vertical axis, and by their corresponding rank on the horizontal axis, the resulting pattern often forms a smooth progression approximating a straight line (Jones and Lewis, 1990). The largest places in 1830, 1870, 1920, 1970, and 1990 are plotted in Figure 3.7, and the simple equation for the rank-size rule is given. The rank-size rule is useful in that it provides a summary of the outcome of a multitude of forces that influence the size of places in urban systems, and illustrates the aggregate effect of these changing through time.

During the mercantile frontier era, it was quite possible to have a number of urban gateways of roughly similar size—hence an urban rank-size distribution for a country could be *stepped*, with one or two large urban areas, a few regional centers, and many frontier outposts. Or, if there was only one gateway, the rank-size distribution might be primate, reflecting the dominance of the highest-order place. In the industrial era, the rank-size distribution is more likely to converge on a straight line as a result of the hierarchical organization of capital flows and the agglomerative effect of the concentration of financial, manufacturing, and business activities in a few major national and regional centers. A number of intermediate-sized places, and many smaller centers, have specific roles in a particular region's production and consumption systems. This is beginning to become evident in the plot for 1870 in Figure 3.7.

The spatial patterns of urban places and interactions that develop are also usually quite different from those of the previous era (Figure 3.8). The largest

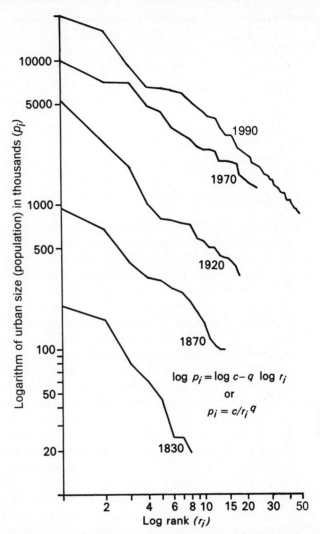

Figure 3.7   Rank-size distribution of major urban areas (by population) in the United States, 1830, 1870, 1920, 1970, 1990. As urban systems develop, the rank-size distributions become quite linear and less steep. (Definitions of urban areas change frequently. The 1970 definition involves SMSAs, whereas the 1990 definitions involve MSAs and CMSAs, which are larger units than SMSAs.)

and highest-order places contain a combination of manufacturing and business functions and are influenced by the business cycle, but the effect is dampened because of the large number of possible offsetting growth–decline activities that exist. The smaller places are, however, influenced a great deal by the business cycle, for they have a limited range of activities. So when the demand for the output of the major employment sector decreases, massive local unemployment and great hardship can occur.

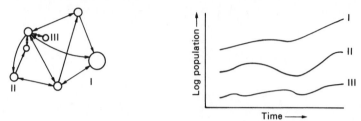

Figure 3.8 The early industrial capitalism era: model of urban system organization and paths of population growth for different level centers.
(*Source:* Simmons, 1978, p. 64.)

National Industrial Capitalism, 1895–1945

The period from 1895 to 1945 was one of full-scale industrialization in which the United States economy completed its transition from an agricultural and mercantile base to an industrial-capitalist one, and in which the urban pattern jelled. Financiers perceived that the greatest return on investments was to be gained from the manufacturing sector. These investments in turn helped to generate great increases in wealth in the country. Although the various stimulants that came together at the beginning of the period have been outlined previously, it should be recognized that World War I, coming in the middle of the cycle, accelerated the growth of manufacturing, and at the same time weakened the financial strength of Europe while enhancing that of the United States. World War II, coming at the end of the era, not only helped drag the economy out of the Great Depression, but effectively set the stage for the next long wave.

One of the chief characteristics of this era were larger and larger units of production. These were made possible by a significant change in the organization of industrial and commercial enterprises that occurred after 1870, when limited liability joint-stock companies were permitted to form. Backed by the finance houses (such as J. P. Morgan), these companies aided the development of market-oriented manufacturing on a scale never before experienced. By 1910, 540 companies employed in excess of 1,000 workers each, and the trend was clearly toward larger units. A brake to this growth came in the form of the Sherman Anti-Trust Act of 1890, but in spite of this law, the increasingly complex system of economic organization allowed functional consolidation to occur. Therefore, by the turn of the century, a number of gigantic corporations had already come into being. These included Standard Oil (from the J. D. Rockefeller Corporation) and US Steel (from the Carnegie Corporation).

Urban growth based on manufacturing would have been impossible, however, without the extension and consolidation of the railroad network. By 1919, there were some 240,000 miles of track in operation in the United States after gauges had been standardized. Steel rails replaced iron, permitting the construction of larger locomotives and heavier equipment, and hence longer and cheaper hauls of raw materials, particularly coal. The average freight charge per ton was reduced from 3.31 cents in 1865 to 0.70 cents in 1892, whereas the average length of haul increased from 110 miles per ton in 1882 to 250 miles per ton in 1910 (Pred, 1965). The integration of the United States economy really occurred, therefore, during the third Kondratieff long wave.

As a consequence, the number of people living in population centers of 5,000 or more increased by 50 million during the era, mostly in urban areas that contained the new or expanding manufacturing plants. The number of persons engaged in manufacturing increased from about 5 million in 1900 to nearly 10 million by 1940 (at the threshold of the World War II manufacturing buildup), and the average number of employees per manufacturing establishment increased from about twenty-four to fifty-five during the same period. There was, therefore, a continuous drive by entrepreneurs to achieve ever greater economies of scale. The big change that became evident during this period was that whereas in the mercantile era profits were generated in the process of trade, through price markups and so forth, in the manufacturing eras the level of profit could be determined at the point of production. The owners of capital, therefore, tried to select locations that minimized their costs, particularly for labor, raw materials, and transportation; and also to develop methods of production (such as assembly line manufacturing) that achieved lower costs per unit of output.

Costs with respect to labor involve a number of components, among them wages and the infrastructure of the urban environment within which the factory is located. The drive to keep the cost of labor as low as possible led to enormous poverty and exploitation in industrial places, which became endemic during the recessionary periods of building cycles. The United States trade union movement was active and expanded during this era as a reaction to this situation: The unions' goals were higher wage levels, job security, improved health and pension benefits, and better working conditions. Environmental and living conditions in working-class areas of industrial cities were often quite squalid, for in the drive to keep production costs down and to maximize profits, there was little incentive for the owners of capital to invest in improving housing and neighborhood conditions in the parts of cities in which they did not live. There has been a continuous series of political pressures ever since to make government at the local, state, and national levels more responsive to the need for social services, improvements in the infrastructure of urban areas, maintenance of public health, and adequate housing standards.

The Urban System During the Era of National Industrial Capitalism

The outcome of this enormous industrial expansion was continued growth in some of the leading urban areas of the early industrial era, and dramatic growth in cities that specialized in large-scale manufacturing and finance. The urban areas on the East Coast of the United States continued to grow rapidly in population, with the New York metropolitan area quadrupling in size to 5.3 million by 1920, Philadelphia nearly tripling in size to 1.8 million, and Boston and Baltimore maintaining similar rates of growth. In all, twenty-eight cities, extending in a 500-mile arc from Lynn, Massachusetts, to Norfolk, Virginia, had populations exceeding 100,000 in 1920 (Figure 3.9a).

Even more fascinating is the growth in urban areas in the Midwest and lower Great Lakes area, with Chicago moving from a position of fifth-largest

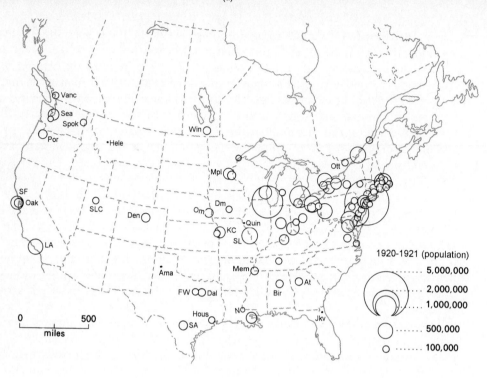

(a)

1920-1921 (population)

········ 5,000,000

········ 2,000,000

········ 1,000,000

········ 500,000

········ 100,000

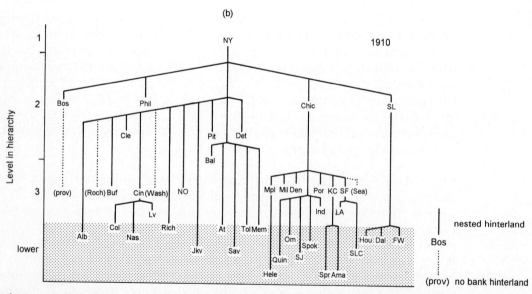

(b)

Figure 3.9 Urban size and interaction by the middle of the national industrial era: (a) the relative size and location of all urban areas in North America with a population of 100,000 or more in 1920–1921; and, (b) the hierarchy of urban areas in the United States as defined by correspondent bank linkages.

(*Source:* Part (b) redrawn from Conzen, 1977, Figure 9, p. 106.)

city in North America in 1870 to second-largest in 1920. Its ninefold increase in urban population (to 2.7 million) was a result of its leading position in rail transportation, and commercial, warehouse, financial, and manufacturing activities. Also notable is the increase in the number of cities along the West Coast with populations in excess of 100,000, led particularly by Los Angeles, California. Finally, there is some development of larger urban areas in the South and in Texas, although this was only in an early stage by 1920.

The growth of industrial urban areas per se was most evident around the lower Great Lakes. Detroit, Cleveland, and Pittsburgh moved to positions of fourth-, fifth-, and ninth-largest city in the United States by 1920. Detroit, chosen by Ford and Durant (General Motors) as the home of the automobile industry, grew along with the adoption of the automobile as a prime means of transport, particularly for social and recreational purposes for the middle and upper classes. There were 8,000 automobiles registered in the United States in 1900; by 1920 there were 8 million. Cleveland, well served by rail and Great Lakes transportation, developed an iron and steel industry as part of its iron ore transport function, and later became the center of the burgeoning petroleum and petrochemical industry (J. D. Rockefeller). Pittsburgh, located among coalfields, became the center of iron and steel production (Andrew Carnegie).

The Spatial Organization of the Urban System During the Era of National Industrial Capitalism   Two main elements of the spatial organization of the urban system under national industrial capitalism became evident by the middle of the era. First, the size distribution of the largest urban places became more hierarchical and, second, a highly interconnected manufacturing core (or heartland) controlled the economy of the nation (Borchert, 1991). The pattern of correspondent bank linkages modeled in Figure 3.9b emphasizes the dominance of New York as the financial hub of the nation, and the importance of Chicago as the major subnational center serving the Middle West and West. The rank-size information in Figure 3.7 further demonstrates that the largest places became more dominant, reflecting their status as leaders in the organization and financing of the industrial economy, as well as their significant presence in manufacturing. The nature of social, economic, demographic, and political relationships that develop between a core and periphery have been outlined in Chapter 2.

This type of core-periphery situation, as it had developed in the United States at the peak of the Kondratieff third long wave, is presented in diagram form in Figure 3.10 in which the three levels of the size hierarchy of the previous era are replaced by four. The core is demarcated by a dashed line, with the periphery (or hinterland) relating to it in a dependent manner. The interurban flows within the core are numerous, as the industrial and multifunctional nature of the urban economies generates complex sets of relationships. The interurban flows between the core and the regional and small, resource-producing centers of the periphery are uncomplicated: manufactured goods from the core, raw materials from the periphery. The growth paths of the variously sized urban areas are represented as being greatest for places in the core, where capital can be organized for investment in productive capacity.

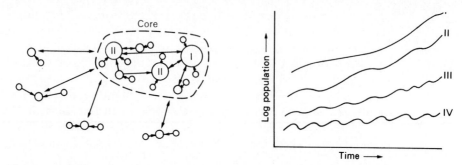

Figure 3.10 The era of national industrial capitalism: model of urban system organization and paths of population growth for different level centers.

Mature Industrial Capitalism, 1945–Date

Though there was a significant recession following World War I, the Great Depression of the 1930s has been the deepest experienced so far during the capitalist era. Because it was such a terrible time for many people, a major effort has been made by governments in the second half of the twentieth century to prevent a similar situation from recurring. Although an economic recovery was underway in the United States by 1935, it was the exigencies of World War II that provided much of the technological and productive stimulus for the prolonged period of growth that occurred during the first half of the fourth long wave. This unusual period of continuous and rapid growth from 1949 to 1973 (Figure 3.1), during which per capita wealth (in constant dollars) increased by over two-thirds, followed by a second period from 1973 to date during which growth has been interrupted by three recessions, has led Hobsbawm (1994) to divide the period into two parts—the "Golden Age" and the "Landslide."

Given the distinctive impact that these two periods have had on urban development in North America, the changing nature of the United States urban system will be examined in the context of these two parts of the long swing (as will the Canadian urban system in the next chapter). In essence, the first half reflects a period during which the United States dominated economically (and increasingly in terms of pop culture) the noncommunist Western World, whereas the second half involves a period during which this "dollar hegemony" is challenged, particularly by Japan and Germany. A large part of the entire era is also dominated by geopolitical ramifications of the Cold War between a United States–led Western Bloc, and a Soviet Union–led Communist Bloc, which reached a symbolic conclusion with the fall of the Berlin Wall in 1989. The importance of this Cold War era as far as urban development is concerned is that defense requirements not only diverted funds away from necessary social and infrastructure developments, the deficit financing involved also necessitated higher interest rates which also raised the cost of urban development.

Mature Industrial Capitalism: United States–Led, 1945–1973 The United States emerged from World War II as the most powerful country in the world.

At that time its industrial corporations were in an extraordinarily dominant position with respect not only to products and markets, but to financial and potential governmental support as well. Furthermore, the United States dollar had become the preferred medium of international finance. The net result was (1) a shift in the concentration of activity to larger and larger corporations, (2) a tendency for the largest corporations to try to set prices through institutional and organizational arrangements rather than purely through market forces, and (3) dominance of United States-owned multiple-plant corporations in the world economy. This tendency for concentration in large corporations is illustrated by the fact that in 1973 the 500 largest industrial corporations accounted for 80 percent of all corporate profits in the United States, and together employed more than 15.5 million individuals.

Within the framework of United States corporate dominance, a number of factors combined together to stimulate the large increase in wealth during the era. The population began to increase at a phenomenal rate (by more than 60 million in two decades) as a result primarily of high rates of national increase (the "baby boom") which reached a peak in 1959 and declined thereafter. Given that the housing stock (and urban infrastructure) had grown little during the depression years of the 1930s and the war years of the 1940s, the baby boom added an unusual urgency to the need for housing and services within urban areas, and also generated a tremendous increase in consumer spending.

The war years had also stimulated an enormous amount of research and development (primarily at governmental expense) into new technologies, or the enhancement of existing technologies, which resulted in the release of numerous technical innovations into the economy at the war's conclusion. These technologies related to aviation, synthetic materials of all kinds, communications, and the harnessing of nuclear energy. The United States government continued, after the war, its increased practice of working closely with private industry and universities in targeted areas of applied and theoretical research and development.

Given the steady increase in wealth during this period, and the experience of the central government in "managing" the economic (and social) environment of the nation during the war years, it is not surprising that the period also witnessed an increase in the role of the government in society. This increase was desired by the population, which wanted an improvement in social security, the social safety net, and improved access to education and housing. Thus, whereas the previous eras are characterized by limited government involvement and influence in the national economy, the first half of the fourth long wave was a time of increasing governmental involvement which the booming economy afforded. United States federal expenditures comprised 10 percent of the GNP in 1940, but 16 percent by 1950 and 20 percent by 1970—one-fifth of the total economy (Bureau of the Census, 1993, p. 331). This federal figure excludes state and local government expenditures, which comprised 11 percent of the GNP in 1970.

As a consequence, the geographical direction of government expenditures has an increasingly significant impact on growth, and hence differential urban

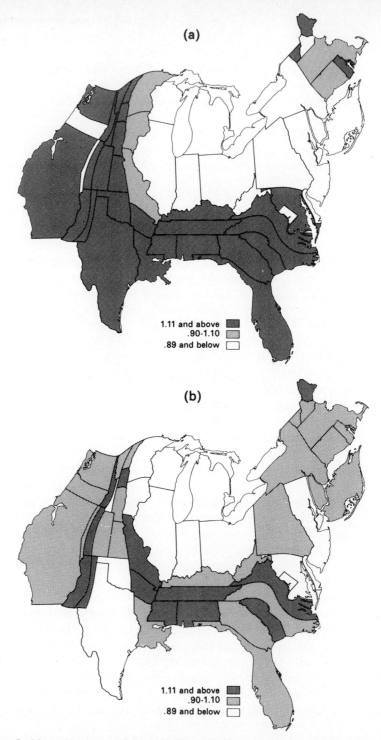

Figure 3.11 The flows of United States federal funds (a) 1965–1967, and (b) 1982–1984, expressed in federal expenditures per dollar of federal taxes, by state. (Note: This is an isodemographic map—the states are drawn, therefore, proportionate to their population size.)

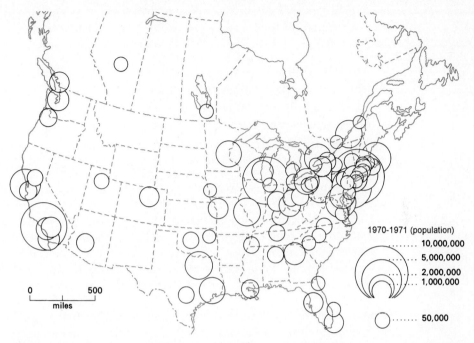

Figure 3.12 The relative size and location of all metropolitan areas (SMSAs and CMAs) in North America with a population of 500,000 or more in 1970–1971.

growth, among the various regions. Figure 3.11 illustrates the spatial distribution (by state) of federal expenditures per dollar of federal taxes for 1965–1967. It is quite noticeable that the core, or heartland area, receives far less in federal expenditures than it paid in taxes, whereas the hinterland states received considerably more. For example, Michigan received $.58 in federal expenditures for every $1.00 its taxpayers paid in federal taxes in 1965–1967, whereas Georgia received $1.52. There was, therefore, a federally induced net drain of monies away from the traditional heartland to the hinterland, that is, from the North to the South and West.

The transition to a service society, apparent in the relative increase in federal government expenditures, is also reflected in increasing expenditures in the tertiary sector (wholesale and retail trade; finance, insurance, and real estate; and services, including education and hospitals). This sector has always provided a large component of the GNP, but in these boom years the contribution increased from 35 percent in 1950 to 42 percent in 1970. This was accompanied by a large increase in service sector employment, particularly among women. The great increase in participation of women in the paid labor force began to occur in the early 1970s.

The Development of the Urban System to 1973 Figure 3.12 illustrates the relative size and distribution of all SMSAs and CMAs in North America that had a population of 500,000 or more in 1970–1971. The changes in the distribution of population that occurred between 1920 and 1970 can be characterized as involving:

1. An increase in scale of urbanization, leading to the emergence of super-metropolitan and megalopolitan urban formations.
2. The deconcentration of continental urban growth, resulting in metropolitan urban developments across much of the continent.
3. The spectacular growth of urban areas beyond the old central cities that contained most of the urban population in 1920.

The general deconcentration and metropolitanization of the population between 1920 and 1970 is directly related to the way in which automobile, truck, and air transportation shaped urban development. In particular, the widespread adoption of the automobile facilitated the decentralization of population in metropolitan areas. The construction of federally financed highways in the 1930s, the Pennsylvania, Ohio, and Indiana turnpikes in the 1950s, and the high-capacity, limited-access Interstate Highway system in the 1960s and 1970s resulted in a new network of surface transportation that allowed the truck to replace the railroad and created a continental market that could be well served from many different locations. At the same time, the rapid growth of air transportation in the 1960s and early 1970s virtually eliminated rail passenger transport and made all metropolitan areas highly accessible to each other.

The most noticeable feature of Figure 3.12 is the extensive concentration of urban population in a 500-mile arc along the East Coast of the United States. This region, which in 1970 contained 43 million people, or one-fifth the population of the United States, has been defined as a *megalopolis* (Gottman, 1961). The urban areas in the region have merged into each other, either directly through physical expansion, or indirectly as a result of overlapping commuting fields and rapidly expanding exurban dormitory towns. Another transnational supermetropolitan region, the Great Lakes megalopolis, is in the process of forming. This megalopolis extends from Milwaukee and Chicago on the west, through Detroit and Cleveland, to Pittsburgh, Buffalo, and Toronto on the east. By 1970 it contained an urban population of more than 40 million people.

Probably the most publicized growth in the North American urban system during this era occurred in the South, in Texas, and along the West Coast. Much of this area has been referred to as the Sun Belt (in contrast to the Snow Belt or Rust Belt of the Midwest and Northeast), and there is no doubt that warmth has a certain attraction. Nowhere is this more apparent than in Florida, where four major metropolitan areas grew rapidly in response to favorable climatic conditions that stimulated the development of retirement communities and internationally advertised recreational attractions. There was also growth of manufacturing in the South associated with the shift from the Northeast of the textile and furniture industries, and marked by the emergence of Atlanta as a major financial and business center for the region. Urban areas along the Gulf Coast, Texas, and in Oklahoma grew in population due to the development of oil fields and petrochemicals, and later because of the electronics and aerospace industries. Dallas, in particular, emerged as a center of finance for the region.

The rapid growth of Houston, Texas, exemplifies quite clearly the importance of entrepreneurial manipulation of federal and local governments, local institutions, and public and private capital to stimulate growth in Sun Belt cities. For example, much of the initial economic dynamism in the city during this era is related to the completion of the Houston ship canal by the United States Bureau of Public Works and to a decision by the National Aeronautics and Space Administration to locate the Manned Spacecraft Center in the metropolis.

Even more dramatic was the growth of Los Angeles from a city of nearly 580,000 people in 1920 to become the second-largest metropolis on the continent, with a population of more than 7 million in 1970. By this date, Los Angeles has become part of a supermetropolitan area that contains 12 million people, extending some 250 miles from Santa Barbara in the north to San Diego in the south. The economic base of this extensive region is derived from oil, food processing, recreation, film and television, petrochemicals, electronics, and aerospace activities. A smaller metropolitan area of 4.5 million developed around San Francisco Bay based on a variety of manufacturing, food processing, electronics, finance, commerce, and port-related activities. Farther north along the West Coast, the cities around the Puget Sound developed yet another integrated metropolitan region of 3.5 million that focused on Seattle, Washington, and its port and aircraft industries.

One of the significant features of the fourth long wave is *suburbanization*, which will be discussed in greater detail in Chapter 7. The suburbanization process has been a part of the North American urban scene since the middle of the nineteenth century. Suburban developments can be interpreted as the spatial expression of the class divisions that emerged with the more complex divisions of labor that occurred as the capitalist era evolved and urban areas grew outward (Walker, 1981; Edel et al., 1984). The period between 1945 and 1973 was one in which the automobile shaped the suburbs, for the automobile provided people with a personal means of transportation, freed the population from reliance on public transportation, and permitted expansion in residential areas with lower population densities and more arcadian environments.

As a consequence, between 1950 and 1970, suburban areas throughout the United States increased in population by 60 percent, whereas the central cities increased by only 32 percent. The increase in population of the central cities was largely attributable to an increase in the African American population which, by 1970, comprised 22.5 percent of the population of the central cities. The African American population had migrated in large numbers from the South to the industrial cities of the Northeast and Middle West, and the port cities of the West, in response to the jobs created during the war and in the boom period following the peace. Thus, by 1973 the United States was well on its way to a situation in which the great majority of the nonwhite population lived in the central cities and the vast majority of the white population lived in non-central-city locations. During the 1960s and early 1970s, a number of these central cities became the arena for the expression of much legitimate African

American rage over a lack of employment opportunities, poor housing, and a wide array of socioeconomic inequalities.

The Spatial Organization of the Urban System by 1973   As the urban system became ever more highly interconnected between 1945 and 1973, the initial advantages of urban locations in the Northeast and lower Great Lakes area became less important. In fact, throughout the entire era the population growth of these large regions was less than that in the rest of the United States (see Chapter 16). Furthermore, the growth rates of the larger places in the Snow Belt began to level off, and in a few cases to decline, whereas regional centers in the Sun Belt gained considerable population. This is reflected in Figure 2.7, which diagrams the pattern of linkages between major metropolitan areas in North America (Borchert, 1991; Yeates, 1994). A number of subnational centers, such as San Francisco, Los Angeles, Dallas-Fort Worth, and Atlanta, have emerged to complement the role of Chicago in the Middle West.

As a result, the rank-size distribution for 1970 begins to become less steep as the regional centers in general, and those in the Sun Belt in particular, begin to assume greater importance (Figure 3.7). Given that urban development in North America is related in large part to the interurban organization of financial flows and corporate decision making, it is not surprising that there are a number of studies attributing the growth of these regional centers to the corporate headquarters that have been relocated to them (Semple and Phipps, 1982; Stephens and Holly, 1981; Wheeler and Brown, 1985).

The general model of urban system organization that emerges during the 1945–1973 period is thus more complex than the one that preceded it (Figure 3.13). A core-periphery arrangement remains, with the megalopolitan transformations giving rise to corridors of high-volume, interurban transportation. The regional centers develop their own strong hinterlands, and in one or two areas "incipient cores" begin to develop due to the concentration of particular local growth stimuli. The net result is that the population growth rate of the second-level centers (Figure 3.13)—the regional centers—is far greater than that of the highest-level centers, and the growth rates of the third- and fourth-level centers are also quite large. Davies (1988) documents these types of variations in growth rates at different levels in the urban hierarchy in his study of urban system development in Texas.

Mature Industrial Capitalism: Global, 1973–Date   The second half of the fourth long wave is associated with: a marked decrease in the international dominance of United States corporations; a rise in areas of economic strength in Japan, Southeast Asia and the European Union; the establishment of a highly integrated world financial market facilitating rapid international capital flows; considerable transnational economic growth–recession instability; and the emergence of new products and process technologies. These changes have led to the development of a more highly integrated world economy in which the United States economic leadership role is not as strong as in the immediate post–World War II decades. A number of them have also led to a relative decline in the im-

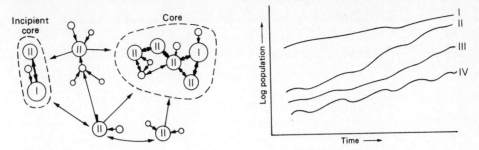

Figure 3.13 Model of urban system organization and paths of urban population growth, circa 1970.

portance of manufacturing, and an increase in the importance of a rapidly expanding service or *postindustrial economy* (Bell, 1976; and see Chapter 6).

Within the United States, employment in governmental and tertiary activities continued to increase, but employment in manufacturing began to decrease as industries began to restructure in response to price and product quality competition from foreign producers (see Chapter 5). Furthermore, the period from 1973 onward is one in which the population growth rate declined significantly and the population began to age quite significantly. This is important not only because an aging population influences consumption patterns and the flow of funds for social services, but also because senior citizen expenditures directly affect interregional capital transfers. For example, in the ten-year period between 1980 and 1990 the share of payments from *private* pension plans in the GNP increased from 1.4 percent to 2.5 percent.

The decrease in the population growth rate is associated with a great increase in the number of women in the paid labor force, particularly in the service sector—the participation rate of women in the paid labor force increased from about 40 percent in 1970 to 60 percent in 1990. Associated with this shift are social demands related to day care for young children, after-school care, single-parent housing and support services, maternal-paternal leave, equal employment opportunities, and comparable pay for work of comparable value. Also, now that the baby boomers are in the job market, household (particularly one- and two-person households) formation rates have increased.

The Development of the United States Urban System, 1973–Date The socioeconomic trends outlined in the previous section have had (and are having) a significant impact on the development of the United States urban system in the last two decades of the twentieth century. These spatial impacts can be summarized as (1) macro-level counterurbanization and Snow Belt–Sun Belt shifts; (2) regional concentration, and (3) local deconcentration.

The term *counterurbanization* has been used by Berry (1976) to imply a reversal of a trend that had existed up to the early 1970s. Whereas urbanization is a process of population concentration into large (i.e., metropolitan) urban areas, counterurbanization is a process of population deconcentration, in which

Table 3.3

Percent Population Change in United States Urban Areas, by City Size, for Three Time Intervals, 1960–1990

SIZE GROUPING	1960–1970	1970–1980	1980–1990
1 million or more	7.4	−6.9	14.3
500,000–1 million	17.1	−16.2	−7.3
250,000–500,000	−2.7	12.3	20.3
100,000–250,000	21.9	19.4	15.7
50,000–100,000	29.6	8.6	20.4
10,000–50,000	−19.7	−14.7	7.4

(*Source:* Bureau of the Census (1993), Table 45.)

the population spreads out from large (i.e., metropolitan) to small nonmetropolitan places or rural areas. Thus, urbanization is a process involving increasing size, increasing density, and greater heterogeneity (or mixing) of people of different racial groups and social classes. Conversely, counterurbanization describes a state in which the growth of the largest places begins to decrease while that of smaller places and rural areas begins to increase, in the process decreasing urban densities and population heterogeneity, which leads to greater separation between different races and social classes.

The counterurbanization proposition appears to be substantiated by some, but not all of the evidence. The slightly conflicting nature of the evidence is perhaps crystallized in Table 3.3. The urbanization concept implies that the larger places should, in general, experience population growth, while the smaller places might incur slow growth or decrease in population. In general, this appears to be the case for the decade up to 1970. Between 1970 and 1980, the largest places experienced considerable decreases in population, while the mid-size towns and cities exhibited quite large increases—hence counterurbanization! Conversely, in the 1980–1990 period, the largest metropolitan areas increased in population, whereas those between 0.5 million and 1 million continued to decrease. There is no doubt that urban areas in the mid-sized population range have experienced the most significant continued growth.

Thus, it appears that while the 1960s were years in which urbanization continued apace in nearly all large metropolitan areas, in the 1970s growth rates were much diminished, and in the 1980s the larger metropolises generally experienced population increases (Champion, 1989). Frey (1990) suggests that during the 1970–1990 period smaller places in nonmetropolitan counties adjacent to metropolitan areas grew faster than the larger places in similar locations and that this is just part of the new urban revival. This general view is endorsed by Taaffe and colleagues (1992) who demonstrate that much of the nonmetropolitan growth in southern Ohio is related to transport corridors leading to metropolitan centers, such as Columbus, Cincinnati, and Huntingdon (West Vir-

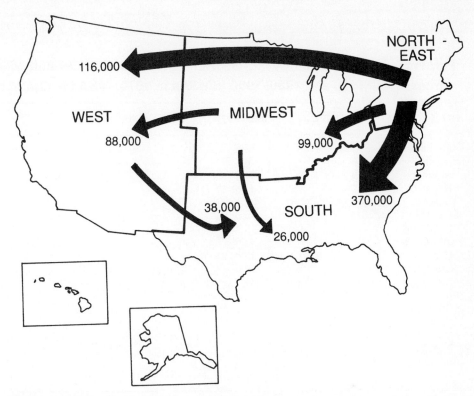

Figure 3.14 Net migration flows between major regions of the United States in 1990–1991.
(*Source:* After Gober, 1993, Figure 6.)

ginia)-Ashland (Kentucky). Thus, much nonmetropolitan growth is in reality the outer edge of metropolitan expansion in an age in which distance has been annihilated. This notion is reconfirmed clearly by Savitch and colleagues (1993) who emphasize that there remain strong economic ties between central cities, suburbs, and more distant areas.

The fact that metropolitan areas in the Sun Belt have been experiencing population growth rates greater than those in the Snow Belt was a feature that began to emerge during the 1960s, but the Sun Belt–Snow Belt population shift has been a feature of the second half of the fourth long wave. The pattern of net migration illustrated for the major regions of the United States for 1990–1992 in Figure 3.14 has been a general feature now for almost a quarter of a century. This is reflected in the quite different metropolitan growth rates exhibited in Table 3.4: Many of the Snow Belt metropolitan areas have either decreased in size or grown at a fairly low rate for three decades (even when fairly extensive definitions of their area are used). It is interesting to note that some large metropolitan areas in the Sun Belt are now exhibiting a decrease in the rate of population growth due to the spreading out of population beyond their metropolitan boundaries.

This deconcentration, or spreading out, of population at the local level has

Table 3.4

Percent Population Change in Selected Sun Belt and Snow Belt Metropolitan Areas, 1960–1970 and 1980–1990 (SMSAs in 1970; MSAs or CMSAs in 1990)

AREA	1960–1970	1980–1990
United States	13.3	32.5
Sun Belt		
Atlanta	36.7	32.5
Dallas-Ft. Worth	39.9	32.6
Denver	32.1	13.6
Houston	40.0	11.1
Los Angeles	16.4	18.5
Miami	35.6	19.2
Phoenix	45.8	40.6
San Antonio	20.6	21.5
San Diego	31.4	34.2
Snow Belt		
Chicago	12.2	0.2
Cleveland	8.1	–3.6
Detroit	11.6	–2.4
New York	8.2	3.3
Philadelphia	10.9	3.0
St. Louis	12.3	2.8

(*Source:* Bureau of the Census (1993; 1973).)

been occurring for many decades. So a continuation of the trend is not unexpected. The population of the largest metropolitan areas, that is, those with more than 250,000 people, had changed from predominantly central-city dwellers to more than 50 percent suburban by 1970. Between 1970 and 1990, this suburbanization continued apace and, by 1990, *70 percent of the population of the twenty largest metropolises was suburban,* that is, non-central city located. Furthermore, a number of people have been relocating from metropolitan areas to the countryside (part of the counterurbanization trend), involving a migration process known as *exurbanization* (Davis, 1993). Exurbanization, involving people who used to live in urban areas who have relocated to rural areas but who still have at least one wage/salary member of the household employed in an urban area, can comprise one-tenth of rural nonfarm populations (Davies and Yeates, 1991).

One important characteristic of the central city–suburban distinction is that many central cities have now become the location of certain racial and demographic minorities (O'Hare, 1992). Table 3.5 indicates the proportion of central-city populations that are in the African, Asian, and Hispanic American minority groups. Central cities also provide the residential location for a significant share of the elderly population who generally remain rooted to the homes, neighborhoods, and communities that they have inhabited for most of their lives (Gober,

Table 3.5

Metropolitan and Central City Populations, and Race and Ethnicity, for the Twenty Largest United States Metropolitan Areas, 1990

Urban Region	POPULATION (THOUSANDS)		PERCENT OF CENTRAL CITY POPULATION				
	Metropolitan Region (PMSA or MSA)	Central City	White	African American	Asian American	Hispanic American	
New York[1]	16,938	7,323	40	29	7	24	
Los Angeles[2]	8,863	3,485	36	14	10	40	
Chicago	7,411	2,784	37	39	4	20	
Boston	5,051	574	58	26	5	11	
Philadelphia	4,922	1,586	51	40	3	6	
Detroit	4,285	1,028	20	76	1	3	
Washington	4,223	607	27	66	2	5	
Houston	3,322	1,631	40	28	4	28	
Dallas	2,676	1,007	47	30	2	21	
Minneapolis/ St. Paul	2,583	540	78	10	6	3	
San Diego	2,498	1,111	57	9	12	21	
St. Louis	2,493	397	50	48	1	1	
Pittsburgh	2,395	370	70	26	2	1	
Baltimore	2,382	736	39	59	1	1	
Phoenix	2,238	983	70	5	2	20	
Cleveland	2,213	506	47	47	1	5	
Oakland	2,083	372	26	44	15	14	
Tampa	2,068	280	57	25	2	15	
Seattle	2,033	516	72	10	12	4	
Miami	1,977	359	9	27	1	63	

(*Source:* Calculated by author from Bureau of the Census (1993), Tables 42 and 46.)
Note: Percentages may not equal 100 because of other minorities (American Indian, etc.).
[1]Including Bronx, Manhattan, Brooklyn, Staten Island, Newark.
[2]Including Los Angeles and Long Beach.

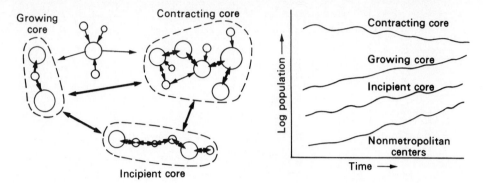

Figure 3.15 Model of United States urban system organization and paths of urban population change, circa 1995.

1993). This is, of course, another manifestation of the "decreasing heterogeneity" aspect of the counterurbanization concept.

The Spatial Organization of the Urban System There are, therefore, a number of different trends occurring in the spatial organization of the United States urban system during the last quarter of the twentieth century. These trends, presented diagrammatically in Figure 3.15, and in isodemographic form in Figure 1.8, can be thus summarized:

1. Certain major metropolitan areas have become transaction centers linking parts of the nation to the network of international capital markets. In the United States, these world cities appear to be New York, Chicago, San Francisco, Los Angeles, Miami, and Dallas-Fort Worth (Friedmann and Wolff, 1982; Gottman and Harper, 1990). New York has retained its dominant position in the world in institutional stock ownership (Green, 1995).

2. The old manufacturing heartland appears to be contracting toward the more diversified economies of New York, Boston, Chicago, and Philadelphia.

3. The California urban region, extending from Los Angeles-San Diego through to San Francisco-Oakland-San José, has established itself firmly as a core of technological (Silicon Valley) and cultural innovation.

4. An incipient core appears to be developing along the Gulf Coast based on the petrochemical industry and the new information-processing and space technologies.

5. With the decline in the birth rate, immigration (Martin and Midgley, 1994) and intercity migration have become much more important factors in differential urban growth (Geyer and Kontuly, 1993). Whereas in the 1960s internal migration accounted for about 10 percent of the relative change in population in the United States, in the 1980s it accounted for almost 30 percent. As a consequence, the growth rate of metropolitan areas can fluctuate quite markedly from year to year according to underlying economic fortunes. For example, with the decrease in oil prices (and consequent investment in re-

lated enterprises) in 1985–1986 there was a considerable downturn in the economies of the Gulf cities and a sharp decrease in their rates of growth. Likewise, the reduction in expenditures on defense in the 1990s is having a significant effect on the economy of southern California.

CONCLUSION

When studying the spatial structure of the United States urban system during the latter half of the twentieth century, we have to consider a number of important emergent features. The urban system at the end of the fourth long wave is quite different from that at the end of the third (1945). Rather than one major core, there are now several of megalopolitan appearance, each containing one or more major metropolises serving as conduits for international capital. Many of these major metropolises have merged to form continuous corridors of urban development, the Boston-Washington megalopolis being the most famous, but with others developed between Chicago-Detroit-Pittsburgh and between San Diego-Los Angeles-San Francisco. With almost universal automobile ownership and the massive construction of intraurban and interurban controlled highways and limited-access expressways, suburbanization has been so extensive that urban areas are decentralized in form, and perhaps counterurbanization is taking place.

The social consequences of this massive decentralization have, in some instances, been horrendous, particularly with respect to the traditional central city which, in many ways, has become an anachronism. However, central cities, in many cases, are still closely entwined with the economic health and social stability of the urban region of which they are a part. The United States urban system in the 1990s is, therefore, quite different from that existing in the 1940s. In general, it has become a highly interlinked metropolitan society of considerable complexity, containing 83 percent of the population, in which the advantages of urban agglomeration with respect to production, innovation, and consumption can be realized in all regions and in parts of most urban areas.

Chapter

4

The Evolution of the Canadian Urban System

Urban development in Canada occurred more slowly than in the United States because until Confederation in 1867 the economy of the country was structured to serve the needs of a European (particularly British) industrial-capitalist economy by providing raw materials. Therefore, it took some time for Canada to establish a more industrially based economy. Thus, urbanization in Canada reflects the shift in the nature of the economy to a more industrial base that *began* to gain momentum particularly following Confederation. Figure 2.2, in fact, suggests that the colonial nature of the Canadian economy throughout much of the nineteenth century resulted in about a twenty year lag, as compared with the United States, in the forging of an urban-industrial nation. Nevertheless, urban development in Canada may be examined in the context of the long waves of economic growth. In particular, the transformation of the nation to urban-industrial, when it did occur, was swift as a consequence of exogenous growth forces that coalesced toward the end of the nineteenth century and the beginning of the twentieth century, which then compounded during the third and fourth long waves.

URBAN DEVELOPMENT IN MERCANTILE AND STAPLES ECONOMIES

Throughout the frontier-mercantile (1790–1845) and most of the early industrial-capitalist (1845–1895) eras, British North America, which became Canada in 1867, basically depended on the export of staples (fish, fur, lumber, cereals) and imported manufactured products. This is not to imply that some basic domestic industrial activities did not develop. There were workshops and, particularly toward the latter half of the nineteenth century, factories for the production of textiles, furniture, and transport equipment. Industrial development did not, however, occur anywhere near as rapidly as in the United States because British trade laws provided no protection or encouragement for fledgling manufacturing industries in Canada. This may be contrasted with the United States where one of the factors leading to the United States Civil War (1861–1865) related to the higher protective tariffs against European manufactured imports being levied by the United States government to protect domestic industries. Hence the urban growth that occurred in British North America developed around the production and organization of a mercantile staples economy that was controlled and financed from the United Kingdom (Simmons, 1991, p. 105).

The Urban System in the Mercantile Era

At the turn of the nineteenth century (1800), the population of British North America, which totaled no more than 340,000 people, was located in rural areas particularly adjacent to the banks of the St. Lawrence River and in pockets along the east coast, the so-called Maritimes. Quebec City (8,000 population), Halifax (8,000), Montreal (6,000), St. John's (3,000), Saint John (2,000), and Fredericton (1,000), were the main urban nodes (McCann and Smith, 1991, p. 73). These urban nodes existed pretty much in isolation from each other, though the mainly French settlements along the St. Lawrence were interconnected by the river and its tributaries, which served as the major highway.

By 1831, however, some limited urban development was beginning to occur (Table 4.1). This had been stimulated by the expansion of trans-Atlantic trade

Table 4.1

Changes in the Rank-Order of Top-Ten Urban Places in Canada, 1831, 1871, 1921, 1971, and 1991 (Population figures in parentheses in thousands)

1831	1871	1921	1971	1991
Montreal (40)	Montreal (115)	Montreal (724)	Montreal (2,743)	Toronto (3,893)
Quebec City (34)	Quebec City (60)	Toronto (611)	Toronto (2,628)	Montreal (3,127)
Halifax (14)	Toronto (59)	Winnipeg (228)	Vancouver (1,082)	Vancouver (1,603)
St. John's (12)	Saint John (41)	Vancouver (223)	Ottawa-Hull (603)	Ottawa-Hull (920)
Saint John (10)	Halifax (30)	Hamilton (114)	Winnipeg (540)	Edmonton (840)
Kingston (4)	Hamilton (27)	Ottawa (108)	Hamilton (498)	Calgary (754)
Toronto (4)	Ottawa (24)	Quebec City (95)	Edmonton (496)	Winnipeg (652)
	St. John's (23)	Calgary (63)	Quebec City (481)	Quebec (646)
	London (18)	London (61)	Calgary (403)	Hamilton (600)
	Kingston (12)	Edmonton (59)	Windsor (259)	London (382)

(*Sources*: Canada Year Book (1924, Table 37; 1984, Table 2.10); Gentilcore and Matthews (1993), Plate 10; and Statistics Canada (1993) *Population and Dwelling Counts* Cat. No. 93–303.)

and immigration following the end of the Napoleonic Wars in Europe in 1815. At the end of this series of wars, a period of inflation occurred in western Europe, resulting in a depression that left many people jobless and destitute. As a consequence, a considerable number of people emigrated from the British Isles to British North America—100,000 between 1816 and 1823, and 225,000 between 1829 and 1835 (Gentilcore and Matthews, 1993). These were quite high volumes given the limited carrying capacity and hazardous nature of the sea transport of the day.

The rank order of major urban places in 1831 in Table 4.1 reflects the urban situation modeled for the mercantile period for the United States in Figure 3.5. Each of the five largest urban places served as links, or gateways, between a resource-producing hinterland, with the resources frequently supplied by indigenous peoples. Furs, fish, and lumber were transshipped to Europe in return for manufactured goods. The financing and control of the trade was primarily from London and Liverpool (United Kingdom). Gradually, however, as the economy of British North America developed with increasing immigration and settlement, and mercantilism transformed into capitalism, the urban nodes began to be defined more in terms of the hinterland they served than their linkages with the controlling British economy and political administration. This took time, and is mainly the story of the era that followed.

The Urban System in the Era of Early Industrial Capitalism, 1845–1895

The latter half of the nineteenth century was a period in which British North America *began* to transform itself into an urban-industrial economy and a nation. Although this transformation was in many ways quite dramatic, it did not involve the spectacular urban expansion and growth that occurred during the same era in the United States. The population in 1851 of the area that was to become Canada in 1867 was about 2.4 million, and in 1891, at the end of the era, it had doubled to 4.8 million. The population of the United States during the same period increased nearly threefold. Basically, the era of early industrial capitalism in Canada was one in which the elements required for an urban-industrial economy were established, and the main areas of population concentration were delineated (Gentilcore and Matthews, 1993, Plate 29).

What were these elements? The first and most important element was people. Canada did not experience the same consistency and volume of immigration as the United States. The era commenced with a wave of immigration mainly from the British Isles between 1845 and 1854 of about 450,000 people, which greatly stimulated urban growth and expansion into Upper Canada (Ontario today). However, there was also considerable emigration, so much so that from 1870 to 1901 outmigration from Canada (mainly to the United States) exceeded immigration. The population maintained its growth due to an extremely high birth rate (and declining death rate).

The second most important element was Confederation (1867), which allowed the new country to *begin* to break its colonial ties with the United Kingdom. This, in economic terms, is manifested by the implementation of the National Policy of 1879, one aspect of which was the introduction of protective tariffs to nurture industrial activities. Unfortunately, rapid growth following independence and the National Policy did not occur because the period between 1873 and 1895 (the latter half of the second long wave) was characterized by a decrease in prices for raw materials—the staples of the Canadian economy. These decreasing prices favored those countries, such as the United States, the United Kingdom, and Germany, that had already established a significant manufacturing base which could profit considerably from cheap material inputs, and low-cost food (particularly wheat) for their urban-based industrial labor forces.

An equally important feature of the era was that economic life in Canada, as elsewhere, was transformed by the introduction of the steam engine. The steam engine made possible western expansion and integration of the new nation. Prior to the railroad, water transport via the St. Lawrence and Great Lakes was the chief means of transport for goods and passengers. The first phase of the necessary canals bypassing rapids and waterfalls (such as Niagara Falls) in the lower Great Lakes and the St. Lawrence had been completed by 1852, providing water transport for commercial shipping from Quebec City to Windsor (Jackson and Wilson, 1992). The water system was not, however, operable in the winter months. One feature of the National Policy was federal stimulation for the establishment of a transcontinental railroad on Canadian territory, a project that was fulfilled by 1885. By that time, Quebec, Ontario, and the Maritime Provinces were served reasonably well by an integrated railway system that focused primarily on Montreal and secondarily on Toronto. Thus, in 1890, nearly 18,000 miles of rail lines were in operation within Canada (Urquhart, 1965, p. 532).

A fourth element in this era that has had a profound impact on urban development in Canada in the twentieth century, and that has provided a continuing fundamental matter for internal political debate, is the increasing linkage of the Canadian economy with that of the United States. In 1868, 47 percent of all imports (by dollar value) into Canada came from the United Kingdom, and 35 percent from the United States. By 1890, the percentages had almost reversed, with 45 percent of all imports by value coming from the United States, and 36% coming from the United Kingdom. The direction of exports—raw materials to the United Kingdom—remained similar throughout the period. This interlinking of the United States and Canadian economies was, however, only beginning during the era, for even by 1900 the vast majority (85 percent) of foreign capital being invested in Canada still came from the United Kingdom.

The Spatial Organization of the Urban System By the middle of the era most of the basic nodes of the Canadian urban system were in place. Furthermore, with the integration of water transport between the Maritimes, Quebec, and Ontario, and the burgeoning railway system (and, toward the end of the era, the electric telegraph), Canadian urban development *began* to be organized around the larger cities. Montreal and Toronto were emerging as the hubs of the national railroad system (Canadian Pacific being headquartered in Montreal); Montreal was the major conduit for foreign capital into Canada, though Toronto was beginning to take a financial role; Ottawa, the nation's capital, was

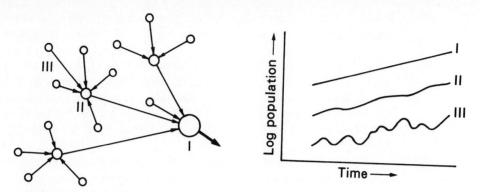

Figure 4.1 A model of urban system organization in Canada during the era of early industrial capitalism.
(*Source:* After Simmons, 1978, p. 64.)

beginning to grow; and Hamilton was becoming the focus of domestic iron, machinery, and machine tool manufacturing industries. The Maritime ports of Saint John, Halifax, and St. John's were beginning their relative decline compared with urban growth in central and, later, western Canada.

The organization of the urban system became much as represented in Figure 4.1. Rather than many *independent* gateways into and from the Canadian-space economy (captured diagrammatically in the mercantile model in Figure 3.5), a few are emerging as major regional centers around which are a number of other settlements related to these places in a hierarchical fashion. Small towns (level III), usually located along railroad branch lines, serve as collecting points for raw materials, which are shipped to subregional centers (level II), and then to major regional centers (level I) for partial processing and export. Information, news, and goods are transmitted or transported to the major centers and from these down the hierarchy of settlements within the regions they are serving (Goheen, 1990). The growth paths of the largest places are fairly smooth, but the growth patterns for single-industry resource towns (level III) can be quite uneven, going from boom to bust and, perhaps, with commodity price fluctuations, even boom again within a few years. By 1890, Montreal (population 200,000) and Toronto (175,000) had become level I commercial centers serving the entire country.

URBAN DEVELOPMENT IN THE ERA OF INDUSTRIAL CAPITALISM, 1895–1945

The first half of the twentieth century is the critical period in the development of urbanism in Canada. It was the period during which the spatial pattern of the system of cities jelled, the transformation of the Canadian economy from a staples orientation to a more balanced basis of dependence on both raw material production *and* manufacturing occurred (Figure 4.2), a domestic core-periphery pattern of interurban relations formed, transnational urban connections became focused much more on the United States than the United Kingdom and

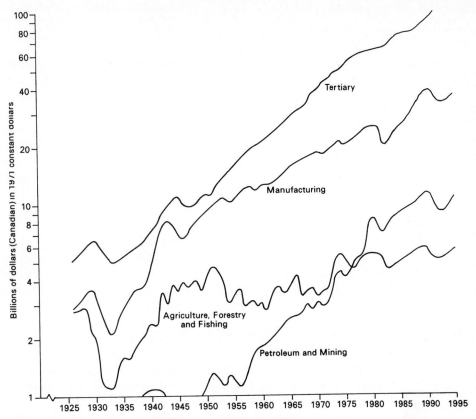

Figure 4.2 The contribution of major sectors of economic activity to the Canadian GDP, 1926–1994 (in constant 1971 dollars).
(*Source:* Leacy, 1983; Statistics Canada, 1994.)

Europe, and a massive shift of population from rural to urban areas took place. During this era, the population of Canada nearly tripled, whereas that of the United States doubled, and the urban population increased by 6 million (Table 3.2).

The major driving forces behind this transformation were those detailed previously (see Chapter 3) with respect to this long wave: the application of electricity to production and a wide variety of consumer goods (radio, telephone, etc.); the shift to steel rails, which reduced long-distance rail transport costs for heavy and bulky raw materials; and the development of the internal combustion engine for trucks and personal transport. These major groups of innovations were of particular vital importance for the development and integration of the Canadian economy for they *reduced the friction of distance:* They allowed a country extending some 4,500 miles from east to west, with most of the population spread out linearly along the United States-Canada border, to become interconnected.

Just as the second long wave commenced with a period of significant immigration, so did the third. The period from 1896 to 1913 was the first since Confederation to experience positive net immigration. Three and one-half million people came to Canada, and about 2.5 million left, implying that about one-third of Canada's population growth during the period was due directly to immigration. Whereas the previous waves of immigration had consisted overwhelmingly of persons from the British Isles, this wave involved individuals and families not only from the British Isles (65 percent), but also central and eastern Europe (20 percent), and northern and western Europe (10 percent). The British went primarily to the urban areas, particularly Toronto, Montreal, London, Hamilton, Windsor, Winnipeg, and Vancouver, whereas those from continental Europe located chiefly in rural areas in Alberta, Saskatchewan, and Manitoba. Railroads had opened up the Prairies, and the growing demand for cheap wheat to help feed workers in European industrial cities stimulated western homesteading.

With the developing transport network focusing on Montreal and Toronto, and greater integration of regional economies to major regional centers, it is not surprising that during the first three decades of the twentieth century many small local manufacturing firms were amalgamated or absorbed into large national concerns, headquartered in or close to Montreal and Toronto. This was facilitated by the tendency for United States-headquartered firms to locate branch plants, particularly those owned by the rapidly expanding United States automobile companies, in southwestern Ontario in such places as Oshawa, Toronto, and Windsor. Whereas in 1900 14 percent of direct foreign capital invested in Canada came from United States-headquartered companies, by 1930 60 percent of foreign direct investment came from the United States and most of this was invested in manufacturing in the Montreal, Toronto, and other urban areas in southwestern Ontario. Thus, within thirty years a British-dominated economy was becoming an American-dominated economy, and the primary direction for the interaction was through Toronto.

There is also no doubt that World War I had a significant impact on the development of Canadian manufacturing industries—particularly in textiles and transport equipment. The major impact of the 1914–1918 war years were, however, on Canadian raw material production, particularly mining, cereals, wood, and fish. There was, however, a shift toward domestic processing (such as canning and smelting), which generated local businesses in single-industry towns. Thus, the war helped to stimulate the Canadian economy and had a considerable impact on urban growth on the Canadian Shield, with Sudbury becoming the leading producer in the world for nickel used for armaments, and service towns in the Prairies (Ring et. al., 1993).

The period of great growth in the Canadian economy, and consequent urban development, came to an abrupt halt during the Great Depression of the 1930s. The depression was deeper and more lasting in Canada than the United States because of the raw material export basis of the economy. When markets for primary products collapsed, vast parts of the country, particularly single-industry and service towns, were severely affected. Even larger urban areas with more diversified economies were badly affected, a situation which was exacerbated by the trek of some unemployed male workers to the cities looking for work and food. Urban areas were not prepared for the kinds of support that were in general required. So, as in the United States, a number of municipalities went bankrupt. The tragedy really only came to an end with the onset of World War II and national deficit-inducing expenditures associated with the war effort. The most significant impact of the Great Depression on the nation, and urban areas, was to make society aware of the need for: (1) strong national financial policies to dampen the extremes of economic and business cycles; and (2) broadening the purview of national, provincial, and municipal governments to include a variety of social insurance and support services (see Chapter 15).

A Latent Core-Periphery Framework

By 1921, the structure of the Canadian urban system was in place. Although much differential urban growth (and decline) has occurred subsequently throughout the twentieth century, the spatial pattern has remained much as it was by that date. Montreal and Toronto had emerged as major metropolises serving the nation as a whole, with Montreal the headquarters of many financial activities and headquarters of a number of national enterprises, such as the Canadian Pacific Railway. Toronto was also becoming an important financial center, particularly with respect to investments in manufacturing and resource development, and in 1919 had become the location of the headquarters of the Canadian National Railway. By 1921, it had developed a more diversified economy than Montreal, with a national outreach.

Given the agglomerative propensities for national financial and business headquarters to concentrate in a limited number of national metropolises, one interesting aspect of urban development in Canada during the twentieth century has been the Montreal–Toronto rivalry (Dagenais, 1969; Yeates, 1994). In effect, Montreal and Toronto had become twin hubs of a core region within central Canada. This region, extending 1,000 kilometers (625 miles) from Quebec City to Windsor (Ontario), and bordered to the north by the Canadian Shield and the south by the lower Great Lakes and the Canada-United States border, comprises a core around which an eastern or Maritime and western or Prairie periphery is arranged (Figure 4.3). This "growing core," though it remained in many ways linked to the British economy, was becoming integrated into the rapidly growing manufacturing heartland of the United States centered on Chicago, Detroit, and New York.

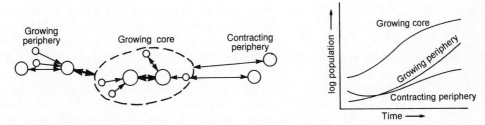

Figure 4.3 A model of Canadian urban system development and paths of population growth by the mid-1920s.

The Eastern Periphery   Table 4.1 highlights the relative decline of urban areas in the eastern periphery, or Maritime Provinces. Saint John, Halifax, and St. John's disappear from the ranking of top-ten cities in 1921. The coming of iron ships and marine steam engines had undermined the advantage that the region had for the building of sailing ships (Marr and Paterson, 1980, p.426). The region was excessively dependent on lumber and fisheries industries, but as prime timber became scarce there was a protracted decline of local economies. Furthermore, with improved water transport direct to Toronto as well as Montreal, the growth of manufacturing in central Canada, and the expansion of agriculture in the Prairies, the export roles of their ports were greatly diminished. Thus industrial activities that had developed based on local raw materials became less competitive with producers elsewhere, and were some distance from the manufacturing heartland of North America (Figure 5.6). This slow, but inevitable, comparative (with the rest of Canada) deindustrialization of the Atlantic Provinces has been pervasive throughout the twentieth century despite numerous federal and provincially funded development programs.

The Western Periphery   Conversely, the western periphery, involving the Prairies and lower Fraser Valley to the Pacific Coast, had been growing rapidly, with Winnipeg, Vancouver, Calgary, and Edmonton joining the group of largest cities. Winnipeg had become the transportation center of the Prairies and the focus of the wheat trade, whereas Vancouver had become an important port city, particularly with respect to the export of lumber, forest products, and cereals. The nature of the Prairies and the west as a raw-material producing periphery was identified at the beginning of the era in 1897 by the federal government with the establishment of the Crow's Nest Pass Agreement, which later, with some modifications, became known as the Western Grain Transportation Subsidy (Marr and Paterson, 1980, pp. 332–334).

This program, which lasted (with modifications) until 1995, provided subsidies for the rail companies, and hence cheap freight rates to Prairie farmers who generally paid between 30 percent and 50 percent of the nominal rate for shipping their grain by rail to export points at the lakehead (Lake Superior) and on the west coast. In its original form, it also subsidized the transport of settlers to the west. Although the grain subsidies were established to encourage farming on the Prairies, foster western rail service, and promote western and lakehead (Lake Superior) port activity, it also had the effect of limiting the establishment of grain-associated industries, such as milling and food processing, in urban places on the Prairies because it was artificially cheap to export the raw material. (See "Least Cost Theory" in Chapter 5.)

Winnipeg had emerged by 1921 as the third-largest city in Canada, serving as the rail and financial gateway linking an extensive grain-producing hinterland with Toronto and Montreal, and from there to European markets (Nader, 1975). The growth of Winnipeg is directly related to the spread of agriculture, the extension of branch railway lines connecting it to this hinterland, its role as

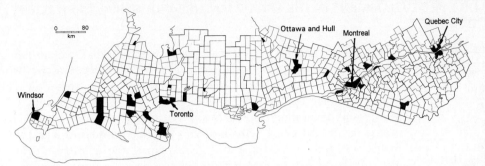

Figure 4.4 Census subdivisions defined as urban in the Windsor-Quebec City axis (population densities greater than sixty persons per square kilometer) in the core in 1921.

a finance and marketing center for Prairie agriculture, and its crucial location linking the west to central Canada. In 1891, there were 600,000 hectares of land in principal grain crops on the Prairies—by 1908 there were 3.8 million hectares. From 1918 to 1928, wheat was Canada's most important export (by dollar value). The growth of Winnipeg was thus tied in many ways to grain, and as the importance of the grain industry to the total Canadian economy gradually decreased after 1928, so did the relative growth of Winnipeg.

The Growing Core The "growing core" was thus becoming the manufacturing, financial, and wholesale-service center for the nation, and the periphery the producer of raw materials for partial processing in the core or direct export. An interesting indicator of this growing importance of manufacturing is the domestic production of steel ingots and steel castings, which had increased from 13,000 tons in 1890 to 1.5 million tons in 1930. This output was the basis of the development of the transport, machine, and consumer durables industries in central Canada. An urban manifestation of this manufacturing development is the iron, steel, and transport equipment manufacturing town of Hamilton, which had become the fifth-largest city in the country by 1921.

The core region was thus the first part of the country to be urbanized: By 1921, just slightly more than 50 percent of the population of central Canada resided in areas that can be defined fairly strictly as urban (Figure 4.4). In fact, 21 percent of the population of the country resided within 65 kilometers (40 miles) of the centers of Montreal and Toronto. This was, therefore, a *period of concentration* of population into urban areas within the core—indicative of the massive rural-to-urban drift of the population that was well underway. The urban population had not, however, spread out much beyond the confines of the politically defined central-city limits. Thus, urban areas were clearly defined. There was little urban sprawl as such, and any decentralization that did occur was related closely to the location of commuter rail lines.

URBAN DEVELOPMENT IN THE ERA OF MATURE INDUSTRIAL CAPITALISM, 1945–DATE

There is no doubt that the springboard for economic development and urban growth in Canada during the first two decades of the era was World War II and the economic and social demands following the cessation of hostilities. The exigencies of the war stimulated an enormous amount of applied research and development, which was of great importance to Canada. Improvements in air and truck transportation provided a quantum leap in efficiency in the movement of people and goods across the country. Furthermore, the application of more sophisticated electronics to communications greatly improved interaction between businesses and people in the different regions. By the middle of the era, telephones, radios, and television sets were in almost every home, providing a level of immediate access to local, national, and international communities that was unknown fifty years before. In a country as large as Canada, with a thinly spread population outside the major metropolises, such improvements in communications and transport were extremely important.

Bases of Urban Growth

Following the turmoil in postwar Europe, immigrants streamed to Canada in great numbers (Figure 4.5a,b). Between 1951 and 1971, annual net immigration was the highest it had ever been, with the total for the period exceeding 1.6 million persons. One-quarter of the immigrants came from the British Isles, 20 percent from Italy, 15 percent from Germany, and about 30 percent from countries and territories in Eastern Europe occupied by the former Soviet Union. This large number of people from southern, central, and eastern Europe accelerated the change in the ethnic structure of the nation. A country that had been dominated by descendants of French settlers in Quebec, and primarily English, Scottish, and Irish settlers in the rest of the country, began to be pressed by the requirements of a more multicultural populace. Furthermore, the vast majority of these settlers located in urban areas, with 50 percent going to Toronto, Montreal, and Vancouver. This was, therefore, a metropolitan-focused immigration.

Coinciding with these high volumes of immigration was a high rate of *natural increase* (Figure 4.5a). The boom in the rate of natural increase lasted from 1946 to 1966, reaching a peak in 1959. It is this phenomenon that has done much to drive consumption and service requirements in Canada during the fourth long wave. The great increase in population fostered an immediate demand for housing, which in turn generated high demand for household goods and medical and social services. The large increases in numbers of children required many more schools and teachers, and, by the 1960s, many more university student places. By the 1970s, there began to be pressure on jobs, by the late 1980s to early 1990s many mid-life crises—and so on! There will, of course, be tremendous pressure on pensions and social services for the elderly, which will begin to reach a peak by about the year 2010 (Foot and Stoffman, 1996).

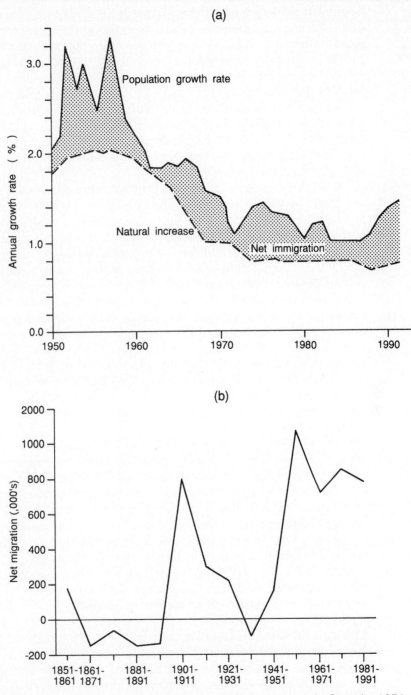

Figure 4.5 (a) The components of population growth in Canada, 1951–1991, and (b) the importance of net migration in the 1951–1991 period compared with previous periods.

(*Source:* Up-dated from Simmons, 1991, p. 118; Statistics Canada, Catalogue 11–402E.)

Economic Linkages and Urban Growth   Canada came out of the war as one of the leading *manufacturing* countries in the world. This position was consolidated in the immediate postwar years as indicated in Figure 4.2, which graphs the change in value of output (in constant 1971 Canadian dollars) in tertiary activities, manufacturing, and primary industries (divided into agriculture, forestry, and fishing; and petroleum and mining). Note that the graphs are presented on a semi-logarithmic scale, which means that the slopes of the curves indicate the rates of change at different time periods. The growth in manufacturing from 1947 to 1977 was, apart from a few hiccups, virtually continuous. After that time, there have been three significant business cycle downturns, followed by short-lasting recoveries.

Manufacturing was stimulated primarily through branch plant production in Canada by United States-owned corporations. By 1961, 60 percent of manufacturing industries in Canada were controlled by companies headquartered outside Canada, and 44 percent by United States-based organizations (Kerr et al., 1990, Plate 51). Much of this manufacturing growth occurred in southwestern Ontario, as one-third of all American branch plants were located in metropolitan Toronto. Even greater United States investment followed the 1965 signing of the Automotive Trades Agreement ("Auto Pact") between Canada and the United States. This agreement resulted in a rationalization of the North American automobile industry and the establishment of a large number of parts and assembly plants in the region within 100 kilometers of Toronto, extending southwest in Ontario to St. Thomas and Windsor (Holmes, 1992). In 1989, the Auto Pact was essentially replaced by a Free Trade Agreement between Canada and the United States, which provided a framework for the liberalization of trade, transport, and service activities, and a revamped mechanism for resolving trade disparities.

Also of significance during this era was the relatively stable contribution of agriculture, forestry, and fishing to the economy as compared with the rapid growth in the value of output in petroleum and natural gas after 1950 (Figure 4.2). Increases in productivity in agriculture and the development of agribusiness further reduced the demand for labor on farms and led to outmigration from agricultural areas (Kerr et al., 1990, Plate 59), particularly in the provinces of Saskatchewan and Manitoba, and to a concomitant decrease in the population and range of provision in local service centers. On the other hand, the rapid expansion of the *petroleum and natural gas* industry, which is strongly linked to markets in central Canada and the United States, generated considerable growth in the larger urban centers in Alberta.

Service Activities and Urban Growth   One of the major features of Canada during the era involves the growth of tertiary activities (Figure 4.2) and the transition to a postindustrial or *service society*. Tertiary activities include employment in wholesale, retail, insurance, finance, health, education, government, defense, and personal services. Of particular significance in the era has been the expansion of health, social, and educational services, that is, those funded from the public purse. Most of these tertiary activities are urban ori-

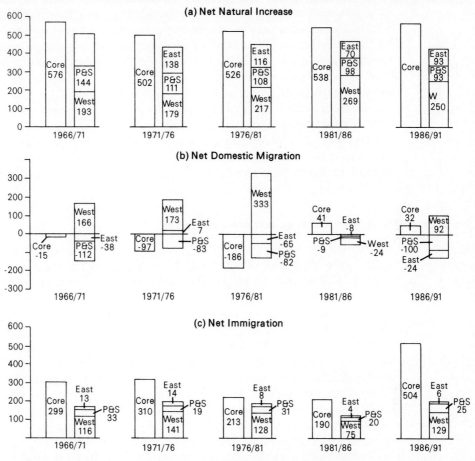

Figure 4.6 Components of core-periphery demographic change, 1966–1991;
(a) net natural increase; (b) net domestic migration; and (c) net immigration.

ented, particularly those that are publicly provided. Thus, the growth of larger
urban areas has been stimulated a great deal by the expansion of the public sector, and is, of course, being threatened during the 1990s by the attempt of governments at all levels to balance budgets and control runaway costs, particularly in the health, education, and social service sectors.

Canadians have, during the last half of the era, become extremely supportive of the establishment of *national standards* for, and universal access to, health
care and social benefits (including unemployment insurance). In order that all
parts of the country, rich and poor, may have relatively equal access to reason-
able-quality medical, educational, and social services, the federal government
redistributes a portion of the national wealth (collected in taxes) from the relatively richer provinces (usually Ontario, Alberta, and British Columbia) to the
relatively poorer provinces (the Maritime Provinces, Manitoba, Saskatchewan,
and sometimes Quebec) in the form of transfer payments and benefits.

The question that arises, in terms of urban growth, is whether this redistribution of wealth discourages people from moving to areas of opportunity. Figure 4.6 indicates that there has been a great volume of internal immigration (net domestic migration) from the Prairie Provinces (Manitoba and Saskatchewan) and Shield (P&S), and the Maritime Provinces (east) to the west (British Columbia and Alberta) and the core (southern Ontario and Quebec). The general weight of accumulated evidence (Day and Winer, 1994) suggests that people do *not* tend to move if local unemployment insurance benefits are comparatively good, provincial and local government taxes are comparatively low, and transfer payments (per capita) are relatively high. Benevolent federal social policies can, therefore, discourage people from moving from declining areas to growing areas.

The Changing Spatial Organization of the Urban System

During the last half of the twentieth century, a strong core region developed, characterized by a shift of population to the largest urban places (see Tables 4.1 and 4.3). The era was also marked by the continuance of a strong staples economy, particularly in wood and wood production, oil and gas, and cereals, in the periphery. Whereas the core became highly interlinked and relatively compact, urban development in the periphery is relatively dispersed with a few urban centers serving extensive regions. The patterns of urban growth experienced in the urban centers within the core and periphery reflect the differences in the complexities of their economies. The larger urban places, particularly those in the core, are less affected by economic cycles than those in the periphery. The smaller places, with less diverse economic bases are more affected by economic fluctuations, but single-industry towns in the periphery are particularly subject to "boom and bust" phenomena (Clemenson, 1992; Everitt and Gill, 1993).

Two periods of urban development can be recognized in this era in the context of changing core-periphery relations:

1. A period of consolidation (1945–1975) of a domestic core-periphery structure during a time of almost continuous economic growth and accumulation when *real* per capita expenditures increased at an average rate of 3.1 percent per annum and annual corporate profits averaged 11.5 percent of the GDP.

2. A period of refocusing of the core and periphery, from 1975 to date, during which *real* per capita expenditures, and hence possibilities for capital accumulation, increased on average at a rate half (1.5 percent per annum) that during the period of consolidation, and annual corporate profits averaged 9.5 percent of the GDP.

The Period of Consolidation of the Core-Periphery Urban Structure

During the period from 1945 to 1971, the population of the core increased at a rate 50 percent greater than that in the periphery, and the population in areas defined as urban almost doubled. By 1971, central Canada had become highly interconnected intraregionally by air, truck, rail, and water, and interregionally to the periphery. For example, the St. Lawrence Seaway, a joint Canada–United States megaproject, was completed in 1959, allowing vessels of 26 feet draft to penetrate 1,600 kilometers (1,000 miles) from the St. Lawrence River into the heart of the continent. In 1971, the core contained 55 percent of the population of the country, received 61 percent of the total national income, and generated more than 80 percent of Canadian patents and trademarks (Table 4.2). The Windsor–Quebec City corridor, therefore, had become the location of innovation in Canada, and had associated with it higher levels of per capita income (core 25 percent greater than the periphery) and greater population growth than the periphery.

Symptomatic of this core role is the fact that the area had more than 70 percent of the manufacturing jobs in the country, and that the manufacturing plants generated more than 75 percent of national value-added from manufacturing. Finally, although the area had only 13 percent of the land in farmland, this land provided more than one-third of the nation's farm revenue. The productivity of the core is, therefore, related to high rates of innovation, growth in manufacturing, and productivity in agriculture. Although the core's share of service employment reflects its share of total population, it does have about 67,000 more jobs in the tertiary sector than would be expected on the basis of population alone. This undoubtedly reflects the concentration of office and financial "producer" services in Toronto and Montreal.

The "twin-hub" nature of the organization of the core is illustrated by traffic flow data in Figure 4.7. This pattern is replicated by a variety of other indicators—telephone calls, rail service, within-core air passenger trips, and the movement of manufactured goods and partly processed materials (Yeates, 1985). The information in Figure 4.7 not only clearly identifies the arrangement of interaction around the two links, and the linkages between these two metropolises, but also the major highways that link the core economy to that of the United States. The pattern of traffic is mainly linear in a generally east-west reciprocal direction.

The way in which the periphery is interconnected with the core is illustrated in Figure 4.8a, which plots the primary and secondary airline traffic outflows from metropolitan areas, in percentages of the total outflow from each CMA in 1971. The basic pattern emphasized in the first-ranked outflows is the link between the periphery and the core, primarily Toronto, with Montreal and Toronto forming a reciprocal pair with the highest volume of traffic in the country. Regional centers, particularly Vancouver, Winnipeg, and Halifax, serve as intermediaries linking the periphery to the core. The importance of Montreal as a hub for the eastern periphery is emphasized in the pattern of second-rank outflows. But the most interesting feature of the pattern of second-ranked outflows is that New York is the focus of flows from Toronto and Montreal. The primary and secondary outflow patterns emphasize the centrality of the domestic core in the core-periphery arrangement, the role of regional centers in the periphery as intermediaries, the strong link between the two metropolises of the domestic core, and the linkage of these two metropolises with the primary United States metropolis of the major extraterritorial core.

Table 4.2

Some Indicators of the Changing Role of the Windsor-Quebec City Core Area in Canada, 1951–1991

INDICATORS	CORE	PERIPHERY	PERCENT IN CORE
Area (000s km²)			
Total area	176	9,799	1.8
Occupied area[1]	176	1,099	14.0
Population			
1951 (000s)	7,272	6,737	51.9
1961	9,745	8,583	53.2
1971	11,920	9,648	55.3
1981	13,194	11,154	54.2
1991	14,684	12,320	54.4
Total National Income[2]			
1961 ($ millions)	12,813	8,666	59.7
1970	30,946	19,873	60.9
1981	125,967	108,027	53.8
1991	262,686	203,008	56.4
Manufacturing Employment[3]			
1961 (000s)	986	366	72.9
1971	1,173	464	71.7
1980	1,355	495	73.2
1990	1,374	495	73.6
Value Added[3]			
1961 ($ millions)	8,129	2,803	73.4
1971	16,132	5,286	75.6
1980	50,486	19,409	72.2
1990	96,524	34,409	73.7
Farm Cash Receipts[4]			
1966 ($ millions)	1,533	2,741	35.9
1971	1,747	2,766	38.7
1981	6,356	12,325	34.0
Hectares in Farmland[5]			
1966 (000s)	10,023	60,444	14.2
1971	8,950	59,714	13.0
1981	8,075	57,815	12.0
Tertiary Employment[6]			
1971 (000s)	2,821	2,159	56.6
1981	4,393	3,407	56.3
Canadian Patents/Trademarks[7]			
1951 (percent)			77.7
1961			82.5
1971			82.6
1981			71.2
1986			77.1

[1]The estimate of "occupied" area of Canada is taken from L. D. McCann (1982), p. 8.

[2]Estimates based on total income from taxable and nontaxable returns in *Taxation Statistics* (1963, 1972, 1983, and 1993), for census divisions. Estimates for partial census divisions are prorated according to population share.

[3]Statistics Canada, *Manufacturing Industries, Series G* (1964 and 1973); Statistics Canada, *Manufacturing Industries of Canada: Sub-Provincial Areas* (1980 and 1990) prorated.

[4]Statistics Canada, *Farm Cash Receipts*, 27, 4; 32, 4; and 43, 12. Estimates for partial census divisions based on proration related to farm area.

[5]Statistics Canada, *Number and Area of Census Farms* (1971 and 1981).

[6]Statistics Canada, *Labour Force, 15 years and over, by Industry Type, by Census Subdivision, by Residence, 1981,* Microfiche CTE 81B35, and a special tabulation for 1971 (6591-PO1826-213-1971). The tertiary employment category includes the labor force in; transportation and communications; trade, wholesale and retail; finance, insurance, and real estate; community, business and personal services; and public administration and defense.

[7]Ceh and Hecht (1990). Data compiled from Canadian government patent (1951–1971) and trademark (1981 and 1986) records.

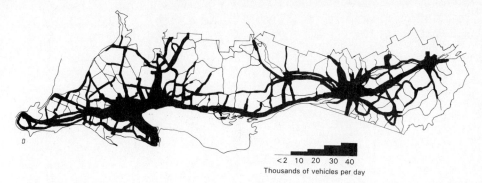

<2 10 20 30 40
Thousands of vehicles per day

Figure 4.7 Traffic volumes on the main highways in the Windsor-Quebec City axis.

Differential Urban Growth in the Period of Consolidation Five main features of urban growth in Canada may be highlighted in this period of core-periphery consolidation:

1. Although the population of the country concentrates in the core, with the areas within 65 kilometers of Montreal and Toronto having 30 percent of the nation's population in 1971, there has been considerable urban growth in the west, with concomitant comparable decline in the Prairies and in the east. In particular, Vancouver, Edmonton, and Calgary exhibit rapid growth, whereas Winnipeg, Halifax, Saint John, and St. John's have grown at a much slower pace.

2. There is the emergence of an almost contiguous macro-urban strip extending from Windsor to Quebec City (Figure 4.9). Not only have the urban areas around Montreal and Toronto coalesced, but the whole area between has become one highly integrated corridor in terms of both flows (of goods, people, and information) and physical sprawl. This merging of metropolitan and smaller urban communities has given rise to the corridor of urban development which is connected to the sprawling lower Great Lakes megalopolis (Leman and Leman, 1976).

3. The greatest surge in metropolitan growth occurred around Toronto and into southwestern Ontario. This is related primarily to the growth of the *automobile assembly and parts* industries following the Auto Pact, and the gradual shift of financial and office activities from Montreal to Toronto.

4. At the same time as population concentrated into metropolitan areas, there has also been a spreading out, or *suburbanization*, of the population and economic activities, within all urban areas. The population of Canada increased by 54 percent (or 8 million persons) in two decades between 1951 and 1971, when much of the increase was housed in suburban locations. The outward spread of suburban areas in Canada did not, however, proceed as rapidly as the United States, primarily because national mortgage insurance schemes to encourage new homeownership in Canada were not as generous as those

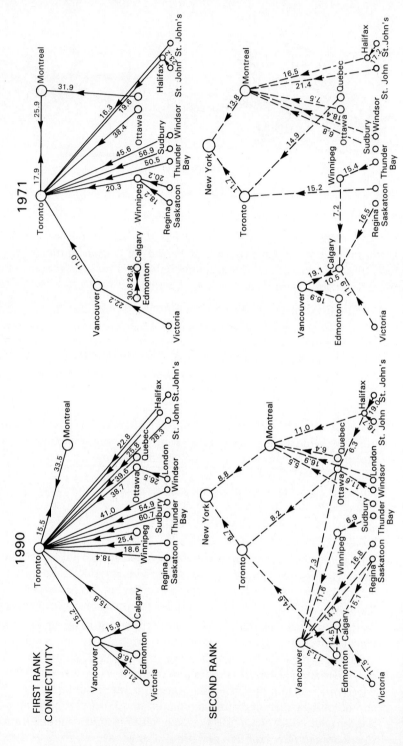

Figure 4.8 Changing metropolitan connectivity in Canada, (a) 1971 and (b) 1991 (percentage of air passenger outflows from Canadian CMAs, United States and domestic flights).

Figure 4.9 The extent of urban development in the Windsor-Quebec City axis, 1981.

in the United States, and the deduction of mortgage interest from personal income tax is not allowed (see Chapters 7 and 12 for more discussion of this point). However, the years between 1951 and 1976 witnessed the greatest period of suburban development in Canada (Miron, 1988; 1993).

5. As a corollary to this suburbanization, *inner cities* either grew little or declined. Simultaneously, a considerable resorting of the population occurred as the outward-migrating middle class left behind the older generation, the less wealthy, and the recent immigrant populations. Nevertheless, central cities in Canada did not experience the same level of decline as those in the United States because there was not the same level of population resorting by race, and neither was there the same comparative volume of suburbanization of people and economic activities (Mercer and Goldberg, 1986).

The Period of Refocusing of the Core-Periphery Urban Structure As in the United States, since 1976 many national and transnational events affecting the Canadian economy have had an impact on urban development. These changes are related to: (1) the slowdown in the natural rate of population increase, which has meant that immigration and interprovincial and interurban migration have become more important demographic determinants of economic growth; (2) social changes, particularly with respect to the increased participation of women in the paid labor force, and the changing nature of household formation; and (3) national, continental, and global economic changes that have led to more focused economic development, and, consequently, more focused urban growth.

Demographic Changes In the post-1976 period of low rates of natural population increase (Figure 4.5), differential urban growth is influenced primarily by factors influencing internal migration and the destination of immigrants. The chief factor influencing internal migration is job availability and wage rates with the flows dampened somewhat as a consequence of national and provincial social policies (Day and Winer, 1994). The ethnic character of immigrants has also changed considerably. During the 1980s and early 1990s, about 50 percent of immigrants to Canada came from Asia, and only 30 percent from Europe (Ley and

Bourne, 1993). Also, whereas in the 1945–1976 period rural-to-urban migration remained a significant factor in urban growth, rural-urban drift has not been a significant factor since 1975. Indeed, in some parts of the country, particularly around Toronto, Montreal, and Vancouver, exurbanization (or urban-to-rural migration) has become a noticeable trend (Davies and Yeates, 1991).

The effects of net natural population increase, net domestic migration, and net immigration on core-periphery population growth are indicated in Figure 4.6 for five-year time periods between 1966 and 1991. The amount of net natural population increase in the core and periphery is generally in accordance with the share of the national population in the two parts. If anything, the rate of net natural increase in the core is slightly greater than that in the periphery.

There have, however, been considerable fluctuations in the geographic patterns of net domestic migration, and *net immigration*. These flows are, of course, influenced directly by employment opportunities, wage rates, and perceptions regarding the hospitable nature of different urban environments. An indication of the strengthening nature of the core is that the traditional westward drift of population in the country to British Columbia and Alberta decreased during the 1980s, and the core actually gained in net domestic migrants. This gain was entirely in the Ontario part of the core, for whereas both Ontario and Quebec lost population through domestic migration up to 1981; during the 1980s, Ontario has gained and Quebec lost. Although the major metropolitan areas in the core and the west have always been the favored first destinations of new immigrants, during the 1980s the core was by far the most favored destination with one-third of the new core immigrants locating in Quebec and two-thirds in Ontario. Data on the first destinations of new immigrants suggest that two out of every three new immigrants locate in Toronto, Montreal, or Vancouver, with Toronto receiving by far the largest number (Moore and Rosenberg, 1995)

Women in the Paid Labor Force   Since 1975, there has been a rapid increase in the *participation rate* of women in the paid labor force (Ghalam, 1993). Figure 4.10 shows that whereas the participation rate for men (percent of all men ages 15 to 64) has remained fairly steady though declining somewhat during the recession of the early 1990s, the participation rate (percent of all women ages 15 to 64) has increased from 45 percent to nearly 60 percent while exhibiting the same recessionary downturns. Although many of the jobs occupied by women are part time, or involve lower wages than men's jobs (an issue discussed in Chapter 13), this much greater involvement of women outside the home nevertheless is having a significant social impact in such areas as child care, after-school care, and so forth. The implications are magnified by the great increase in the participation rate for women (two- and one-parent families) whose youngest child is less than 16 years of age. By 1993, the participation rate for mothers with at least one dependent child had reached 70 percent.

The implications of these changes are enormous, particularly as the "historic model" for urban growth and development is basically male oriented. Women not only may want to work in the paid labor force, many have to work to provide for their families. Urban society is only just beginning to come to realize the ways in which this fundamental social and cultural change is influ-

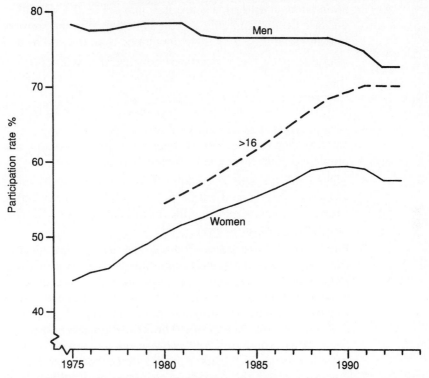

Figure 4.10   Labor force participation rates for men and women, and women with youngest child less than 16 years of age, 1975–1993.
(*Source:* Statistics Canada, 1994.)

encing the structure of the city. In the context of differential urban growth in a system of cities, it is quite evident that families need to locate where the possibilities of jobs for both males and females are high. This invariably means that job seekers increasingly turn to the larger and more diversified metropolitan areas where there is a greater array of employment opportunities for both women and men.

Economic Changes   As was noted in Chapter 3, a major change occurred during the 1970s with respect to the nature of the world economy. To summarize, what appears to have happened is that:

1. *New highly productive cores* have become consolidated in western Europe and Japan, as well as in California and Southeast Asia.

2. *Capital and information have become extremely mobile* as a highly interconnected world capital market has developed around the major international financial centers of New York, Tokyo, and London, with regional metropolises, such as Los Angeles, San Francisco, Chicago, Miami, Paris, Frankfurt, Rio de Janeiro, Toronto, Sydney, Hong Kong, Bangkok, and Singapore, serving as capital conduits to these major centers.

3. *Multi-national* enterprises are key players in the volatility of world capital, because these companies owe less allegiance to the countries and communities in which they may have developed than they do to major shareholders.

4. Manufacturing has *restructured* considerably with the introduction of new high-tech production and organizational processes (such as flexible manufacturing, just-in-time delivery systems, and so forth), which have embedded within them quality-maximizing and cost-minimizing goals.

5. *Service activities* involving health care and maintenance, finance, producer services, and the delivery of social services now occupy a larger share of the employed labor force than previously (Coffey, 1994).

6. The *relative prices of raw materials*, as compared with other products and services, have fluctuated wildly due to political factors and economic restructuring. This is having a significant impact on the viability of resource towns (Barnes and Hayter, 1994).

7. Finally, the *United States—Canada Free Trade Agreement* (1989), which was followed by NAFTA (1994) including Mexico, is drawing the economies of the United States and Canada even closer. Whereas in the United States the effect of the agreement is limited to a few large border states, in Canada the impact is virtually nationwide, though having a particular impact on the regions around Vancouver and Toronto (Hayward and Erickson, 1995). Thus, changes in production and process technologies, such as those involving manufacturing (see Chapter 5) that affect the United States economy instantaneously affect the Canadian economy primarily through the urban conduits of Toronto and Vancouver.

Differential Urban Growth in the Period of Refocusing of the Core-Periphery Structure   Table 4.3, in which Canadian metropolitan areas are subdivided into those located in the core and those in the periphery, indicates the postulated connections (again with + and – signs as in Table 1.1) between differential urban growth and certain demographic, social, and economic aspatial factors during the 1986–1991 period (Bourne and Olvet, 1995). Cities located in the periphery are subdivided into those on the Maritime east coast, those located on the Prairies or on the Canadian Shield, and those in the west. The three general aspatial components are defined conceptually as: manufacturing restructuring; the effect of changes in the prices and volume of raw materials (such as wheat, oil, natural gas, and wood); producer services (including finance, insurance, computing and education services); and the effect of the emergence of production-financial centers in Southeast Asia and around the Pacific Rim. The political-institutional influence is defined in terms of the seats of government in the provinces that experienced the greatest economic growth during the 1980s, and the effect of the federal policy of transfer payments.

There are four distinct geographic patterns exhibited by the 1986–1991 urban growth rates: metropolitan-nonmetropolitan (and rural); core-periphery; within core; and within periphery. For the country as a whole, there has been a continued shift of population from nonmetropolitan and rural areas to the twenty-five metropolises (Bourne and Olvet, 1995; Coffey, 1994). In 1986, these metropolises accounted in aggregate for 57.5 percent of the nation's population. By 1991, the share had reached 58.9 percent. This is accounted for primarily by the fact that the largest metropolitan areas are the destinations of new immigrants. With respect to core-periphery contrasts, the signals, as represented by individual metropolitan growth rates, are mixed. In general, as Canada's postindustrial economy developed, the strength of the core in producer service industries as well as manufacturing yielded higher population growth rates for core metropolises than those in the periphery.

There are distinct differences in urban growth patterns within both the core and the periphery. In the core, the major metropolises in Quebec grew at a rate just below the national average, whereas most of those in Ontario grew at a rate above the national average. Most of the gains in employment in producer services and high-tech manufacturing were experienced by metropolises in Ontario (Bathelt and Hecht, 1990). Thus, the core of the country, which traditionally involved both southern Ontario and southern Quebec organized around the twin metropolises of Toronto and Montreal, is now contracting to southern Ontario. The focus is Toronto, which has become the conduit for capital into Canada, connecting the nation with the major financial centers in the world. It has become the chief location for the Canadian headquarters of businesses operating in Canada (Rice and Semple, 1993).

Contrasts in urban growth rates within the traditional periphery are equally as interesting. The metropolises of the Prairies, Shield, and Maritimes have, apart from Halifax, growth rates that are below the national average. Population decline has undoubtedly been averted through large transfers of support funds as part of federal equalization and other social policies (Simmons, 1984). On the one hand, because the prices of staple products relative to those of manufacturing and services declined during the 1980s, the cities and metropolitan areas in the raw material-producing part of the country experienced considerable outmigration. On the other hand, the metropolises in the west incurred considerable population increase as exports of natural gas and wood to the United States remained strong. Also, exports of raw materials from the west to the rapidly expanding economies of the Pacific Rim countries, particularly Japan, increased apace.

One of the interesting features about Canadian urban growth is its volatility, particularly in metropolises located in the periphery (Simmons, 1991). For example, for the three five-year periods between 1976 and 1991, the population growth rates for Saskatoon have been –5.8 percent, 14.6 percent, and 4.7 percent. These rates are directly related to the relative prices of natural resources, particularly wheat, in the province of Saskatchewan. The larger metropolises experience less fluctuation in growth rates because they have more diverse economies, whereas those with smaller populations have less diverse economies and are, therefore, likely to experience greater variations. For example, during the 1990s, as world prices for cereals increase and international support subsidies and trade barriers decrease, Saskatoon, Regina, and Winnipeg may experience increased population growth related to increasing wealth in their hinterlands.

Table 4.3

Population Change in CMAs in Canada, 1986–1991 and the Impact of Selected Aspatial Factors on Differential Urban Growth

METROPOLITAN AREAS	POPULATION (IN THOUSANDS)		PERCENT CHANGE	DEMOGRAPHIC	
	1986	1991	1986–1991	Net Migration	Net Immigration
CORE					
Toronto (Ont.)	3,432	3,893	13.4	+	+
Ottawa-Hull (Ont/Que)	819	920	12.4	+	+
Hamilton (Ont.)	557	600	7.7		+
London (Ont.)	342	382	11.5	+	
St. Catharines/Niagara (Ont.)	343	365	6.2		
Kitchener/Waterloo (Ont.)	311	356	14.5	+	+
Windsor (Ont.)	254	262	3.2		
Oshawa/Whitby (Ont.)	203	240	18.0	+	
Kingston (Ont.)	122	136	11.5	+	
Sherbrooke (Que.)	130	139	7.1		
Montreal (Que.)	2,921	3,127	7.0	–	+
Quebec City (Que.)	03	646	7.0		
Trois Rivières (Que.)	129	136	5.8	–	
PERIPHERY					
Maritimes					
Halifax (N.S.)	296	321	8.3		
St. John's (Nflnd)	162	172	6.2		
St. John (N.B.)	121	125	3.1	–	
Prairies and Shield					
Winnipeg (Man.)	625	652	4.3	–	+
Saskatoon (Sask.)	201	210	4.7	–	
Regina (Sask.)	187	191	2.8	–	
Chicoutimi-Jonquière (Que.)	158	161	1.6	–	
Sudbury (Ont.)	149	158	5.9		
Thunder Bay (Ont.)	122	124	1.8	–	
West					
Vancouver (B.C.)	1,381	1,603	16.1	+	+
Edmonton (Alb.)	774	840	8.5		
Calgary (Alb.)	671	754	12.3		
Victoria (B.C.)	255	289	12.8	+	
CANADA	25,309	27,297	7.9		

	ECONOMIC			POLITICAL/INSTITUTIONAL	
Manufacturing Restructuring	Pacific Rim	Staples	Producer Services	Government	Equalization
+			+	+	−
			+	+	−
−			+		−
−					−
+			+		−
−					−
+					−
			+		
−				+	
				+	
		−			
				+	+
		−			+
		−			+
		−			+
		−			+
		−			+
		−			
		−			−
		−			−
	+	+			−
		+		+	−
	+	+			−
	+			+	−

The contraction of the core to the Toronto and southwestern Ontario area has been complemented by the development of incipient cores around Vancouver-Victoria (Seattle), and Edmonton-Calgary. These incipient cores have developed in part because of their oil, gas, and lumber resource base, but also in part because they are becoming more and more interlinked with the growing economies around the Pacific Rim, particularly California, Japan, Southeast Asia, and China. The growth of Vancouver is being stimulated particularly by its connection with the burgeoning Pacific Rim. Thus, the Canadian economy, and therefore urban development, is being influenced not only by the traditional core areas of the midwestern and northeastern United States and western Europe, but also by the new cores in California and Japan, and the rapidly expanding consumer markets of China and Southeast Asia.

This refocusing of core-periphery relations in Canada is reflected in the changing pattern of airline passenger outflows (United States and domestic flights) from Canadian CMAs in Figure 4.8b for 1990. The first-rank connectivities indicate that between 1971 and 1990 Toronto has strengthened its position as the leading focus of travel for all the metropolitan areas in the country. In general, links in the ascending hierarchy have strengthened (Montreal to Toronto, Vancouver to Toronto, Toronto to New York), whereas links down the hierarchy have weakened. Furthermore, both the first- and second-rank connectivities indicate the strengthening of Vancouver as the hub of an incipient core on the west coast. Montreal remains the focus of second-ranked outflows in the east.

Canada has, therefore, moved rapidly from a simple core-periphery structure where southern Ontario and southern Quebec, comprising central Canada, formed the core and the rest of the country the periphery, to a more complex situation in the 1990s. The core has contracted to south and southwestern Ontario, with Toronto serving as the focus of financial, service, and industrial development. An incipient core is developing in the western part of the old periphery, but this core is in effect part of the Pacific Northwest urban agglomeration (see Chapter 16) and has strong links with the United States West Coast as well as central Canada (Davies and Donoghue, 1993). Montreal is becoming a regional center, serving the distinct society of Quebec. The CMAs located in the Maritime Provinces are linked primarily to Toronto, and only secondarily to Montreal.

CONCLUSION: URBANISM IN CANADA AND THE UNITED STATES

Although United States and Canadian urban areas have much in common in terms of bases of development, relatively recent formation, similar styles of urban construction, grid-iron street layouts, large-scale dependence on the automobile, high levels of suburbanization and sprawl, immigrant populations, ethnic diversity, internal migration, and the social structure of society, Canada is different from the United States (Lipset, 1990). These differences are generating an interesting debate on the ways in which they are reflected in the nature of urban development in the two countries (Mercer and Goldberg, 1986; Ewing,

1992; Garber and Imbrosio, 1996). The questions are: How great are the differences, particularly as they are manifested in urban structure and livability? And why do they exist? Although these questions are addressed in various chapters, for example, Chapter 8 with respect to urban population densities, and Chapter 11 with respect to the social structure of urban areas, some observations concerning these differences may be useful at this juncture.

Although both the United States and Canada are federal states, in Canada the responsibility for local and municipal government is specifically delegated by the British North American Act to the provinces. The United States constitution does not really address local government except to regard it as a matter of local self-determination. Thus, in the absence of a clear delegation of authority, the federal government in the United States can be more interventionist, if it so wishes, with respect to local governments than the federal government in Canada. For example, there are many entitlement programs in the United States that refer to defined metropolitan areas.

Although it is not really clear which system has been most beneficial to urban areas, undoubtedly the Canadian system has made it easier for provincial governments to encourage the formation of regional governments and undertake regional transportation planning, thus making the provision of services between central cities and suburbs more equal (see Chapter 15). On the other hand, politicians at the provincial level have not wanted these larger regional governments to have strong leadership, particularly powerful mayors, because such persons may challenge political power at the provincial level. For example, a strong mayor (if there were one) of metropolitan Winnipeg would speak for 600,000 people in a province of 1 million persons. In the United States, on the other hand, it is possible to have a "strong mayor" situation if the local jurisdiction prefers that model (for example, Chicago, New York, and Seattle).

In general, the provinces of Canada are much more powerful than the states in the United States, and one province, Quebec, for the last twenty years has had a government either in power, or leading the opposition, that is seeking secession from Canada in one form or another. The French-speaking population of Canada, which is concentrated in Quebec and is the dominant sociocultural group in the province, is one of the two (along with English-Canada) "founding peoples" of the nation. Recently, the federal government and a number of provincial governments have recognized, in various ways, that the indigenous peoples of Canada (Indians, Innuit, and so forth) should also be to a large degree self-governing within the nation.

A series of constitutional crises concerning Quebec, the most recent vote in Quebec concerning separation of the province from Canada having been held in October 1995 (the outcome was "no," by a small margin), has reinforced shifts to Ontario in the location of people and economic activities. The underlying trend, however, for the concentration of financial and business activities to Toronto has been underway for half a century. Rapid urban growth in the Canadian west is less influenced by these constitutional crises as it is related more to general economic growth in places around the Pacific Rim and Southeast Asia, and fluctuations in the prices of raw materials, particularly oil, natural gas, lumber, and wood products.

Although poverty, violence, and, at times, social discord, exist in both American and Canadian urban life, they are not as extreme in Canada as the United States. Much of the reason for this is the existence in Canada of universal and more generous social programs funded by federal, provincial, and local governments. In those areas where national standards of service provision have been agreed, such as health care and unemployment insurance, access in times of individual or family need has come to be regarded as a social right. Provincial public housing policies in Canada have not resulted in as many social disasters as in the United States because they have been better conceived and managed, and do not appear to be as racially directed. There is no conception of a constitutional right to bear arms in Canada. Therefore, accumulated gun ownership is nowhere near as ubiquitous as in the United States. Thus, though substance abuse (e.g., drugs, alcohol, etc.) exists in both countries, the lethal mixture of guns *and* drugs is not as prevalent in Canadian urban areas as those in the United States. As a consequence, there are fewer "no go" zones in Canadian urban areas than those in the United States.

In Chapter 11, it will be noted that Canadian urban areas tend to be more heterogeneous than those in the United States, particularly with respect to racial mix. Although there are notable racial and ethnic concentrations in Canada's largest metropolises, segregation, particularly blacks from non-Hispanic whites (and increasingly Hispanic whites) is much more pervasive in United States than Canadian urban areas. Racial segregation has been exacerbated in the United States where urban areas are more suburbanized and decentralized, and a large proportion of African Americans are in effect left behind in parts of inner cities that are crumbling. It would seem that every urban social problem in the United States has a racial dimension (Beauregard, 1993). Although rights of equal access to employment and educational opportunities, certain services, and due process are assured in both Canada and the United States, there would appear to be more equality of access to services of similar quality, particularly education and health care, in Canada than the United States. Thus, levels of service disparity between suburbs and central cities are not quite as apparent in Canadian as United States urban areas.

Finally, the closer merging of the Unites States and Canadian economies as implemented by the North American Free Trade Agreement, which reinforces the economic integration of the two countries that has been taking place over the last one hundred years, serves also as a powerful influence for the harmonization of social policies. Although the black-white and inner city-suburban contrasts are fundamental to the differences between Canadian and United States cities, urban areas in both countries respond similarly to the same sets of economic and social forces. Urban areas in Canada and the United States share a similar experience of recent growth, exhibit similar patterns of regional growth and stagnation, and are being reshaped in a similar fashion by common global forces (Frisken, 1994). In the meantime, each country is following its own pattern of social policy development in response to social and economic change, and the differences in social policies between the two countries go a long way toward explaining some of general variations in livability between urban areas in the two countries.

Chapter

5

Urban Areas as Centers of Manufacturing

The relationship between urban growth and manufacturing in North America is extremely important. The rapid growth of urban areas in the United States and Canada, which has occurred within an evolving core-periphery framework, is related primarily to the geographical concentration of industrialization. Industrialization, in turn, is characterized by a series of periodic shifts in the centers of manufacturing growth as a consequence of the development of new products and new production processes (Storper and Walker, 1989). For example, the Snow Belt–Sun Belt shifts of the last few decades of the present century were in large part related to the shift of capital away from the industrial cores of the Northeast and Midwest to the South and West (Smith and Dennis, 1987). This shift in capital may have occurred as a consequence of investment decisions made in the private sector (Pudup, 1992) or the public sector—that is, by the "state" (Markusen, 1991; Crump, 1993). Manufacturing is a vital element in the economic base of most urban areas, either directly as a basic industry, or indirectly through the multiplier effect and through interdependencies generated by forward and backward linkages (Daly et al., 1993).

THE IMPORTANCE AND NATURE OF MANUFACTURING

Urbanization in North America has been both the product of, and inextricably intertwined with, economic growth. Some time ago, Kaldor (1966) presented a number of propositions concerning economic growth in a country. These have

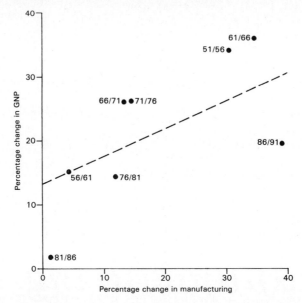

Figure 5.1 The relationship between percentage change in GNP and percentage change in manufacturing, Canada, 1956–1991. (Note: An R^2 of 100 percent would be a perfect correlation, and an R^2 of 0 percent would be no correlation.)

subsequently been the subject of much debate (Thirwall, 1983; McCombie, 1983). The most basic of Kaldor's propositions is that the faster the rate of growth of manufacturing in a country, the faster the rate of national economic growth (as measured, for example, by GNP). This is not just because manufacturing forms a part of GNP, but because the wealth generated by high productivity in manufacturing induces economic activity elsewhere in the economy. For example, even in an economy such as Canada's, which is still strongly dependent on the production of raw materials (see Figure 4.2), the relationship between change in GNP (in constant dollars) between 1951 and 1991 (in five-year intervals) is strongly related to the growth of manufacturing (Figure 5.1). Manufacturing in North America is, therefore, a most important engine of growth (Page and Walker, 1991), and consequently a most important stimulant to urban growth (and, on the down side, to urban decline).

Examining Manufacturing

Manufacturing, as an activity, may be analyzed in three main ways: with respect to the *components* of the industry, to the *value added,* and to the *stage* of the production process involved. There are two main components to manufacturing—durable and nondurable goods. The durable goods sector (Table 5.1) consists mainly of firms and plants that manufacture products that are expected to exist in use for at least three years. In North America, the urban areas involved in this type of production are located primarily in the traditional Man-

Table 5.1

Durable and Nondurable Goods Manufacturing Sectors: Percent of Total Output by Sector, 1990

SECTOR	PERCENT (%)
Durable goods	56.0
Machinery, except electrical	10.0
Electric and electronic equipment	10.0
Fabricated metal products	7.0
Motor vehicles and equipment	5.2
Primary metals industries	4.6
Lumber and wood products	3.3
Instruments and related products	3.2
Stone, clay, and glass	2.7
Others	10.0
Nondurable goods	44.0
Chemicals and allied products	10.2
Food and kindred products	8.5
Printing and publishing	7.0
Paper and allied products	4.9
Petroleum and coal	3.5
Apparel and other textile	2.5
Textile mills	2.1
Tobacco products	1.7
Others	3.6

(*Source:* Bureau of the Census (1993), p. 741.)

ufacturing Belt that extends from eastern Iowa to southern New England and southern Ontario (Figure 5.2). In those places in which iron and steel, farm equipment, automobiles, and electrical machinery are produced, there has been considerable instability and some decline as a result of excess capacity, an aging capital stock, and foreign competition (Negrey and Zickel, 1994).

The nondurable goods sector consists of those industries that produce manufactured products that are consumed rather quickly when used. Urban areas specializing in the manufacture of nondurable goods are located in the New England area, the Interstate 95 corridor from New York City to Washington, D.C., and in a broad belt along the southeastern Piedmont Plateau, from Virginia to North Carolina and South Carolina (Figure 5.3). This last area has been growing the fastest in recent years, as the textile and processed foods industries move from the Manufacturing Belt to this lower-wage area (Johnson, 1985; Angel and Mitchell, 1991).

A second way of looking at manufacturing is with respect to the value added in the production process. *Value-added* may be defined as the value of the goods produced less the cost of the materials, supplies, fuel, and electricity

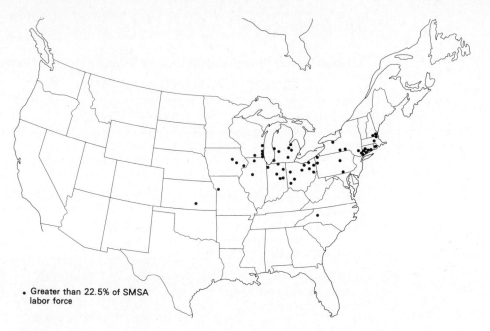

• Greater than 22.5% of SMSA
  labor force

Figure 5.2   The location of durable goods manufacturing in the United
States.
(*Source:* Archer and White, 1985, p. 134.)

used in the manufacturing process. Low value-added industries tend to have
low capital labor ratios and relatively low wage rates, whereas high value-added
industries tend to have high capital–labor ratios and higher wage rates. In
Canada, more than 70 percent of the country's manufacturing is located in
southern Ontario and Quebec, and within that area there are distinct differences in the location of high value-added and low value-added industries (Figure 5.4). High value-added industries are located primarily in southern Ontario
and Montreal; low value-added industries in southern Quebec. Norcliffe and
Stevens (1979) explain the origins of this difference in terms of the initial advantage of a wealthier agricultural base in Ontario, and the much higher level
of foreign investment in Ontario than in Quebec.

A third way of looking at manufacturing—one that will be used extensively
later in this chapter—is with respect to the stage of the production process that
is involved. Production can be divided into three main stages. The first is *processing,* which involves the conversion of various types of raw materials into
more useful products, such as copper ore into copper ingots or wire. The second stage is that of *fabrication,* which describes the conversion of processed
products into a form that can either be used directly, or can be used as part of
another product. For example, the copper wire may be fabricated into bunches
of wires used for control and information mechanisms in automobiles. The final stage is that of *integration,* which, as the label implies, involves the bringing
together of a number of fabricated materials into some finally assembled prod-

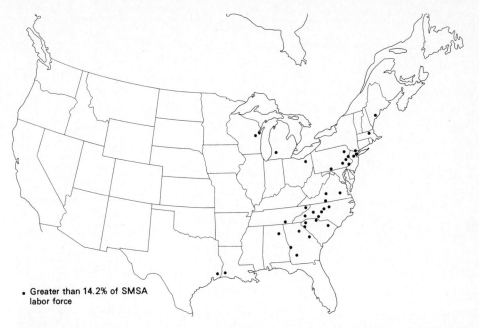

• Greater than 14.2% of SMSA
  labor force

Figure 5.3 The location of nondurable goods manufacturing in the United States.
(*Source:* Archer and White, 1985, p. 135.)

uct. For example, the bunches of electrical wiring may be brought together with the rest of the subassembled pieces for final assembly of an automobile. Some idea of the complexity of this three-stage process in the multinational operations of a major automobile company is illustrated in Figure 5.5, in which the arrows indicate the major directions of flows of the products of the fabrication stage (3 and 4) and the outcome of the integration stage (1 and 2). Note that integration occurs only at one or two locations within each major market, such as New Stanton, Pennsylvania, for Volkswagen in the United States.

Thus manufacturing is extremely important not only for the economic health of urban areas in which such activity is dominant, but also for the economy of North America as a whole. The changes in manufacturing that are taking place are not, however, resulting in increases in manufacturing employment. Even though worker productivity and the number of plant establishments has increased (Table 5.2), the contribution of manufacturing to the GNP has been decreasing since the mid–1960s as a consequence of the emergence of the service society and competition from other countries. Though it must be recognized that 1982 and 1991 were low points in recessions, the number of production workers has decreased by more than 2 million during the past two decades whereas the number of other employees in manufacturing (sales, office, design, etc.) has actually increased. In fact, part of the increase in productivity that has occurred has undoubtedly been due to production worker lay-offs (Rees, 1992).

(a)

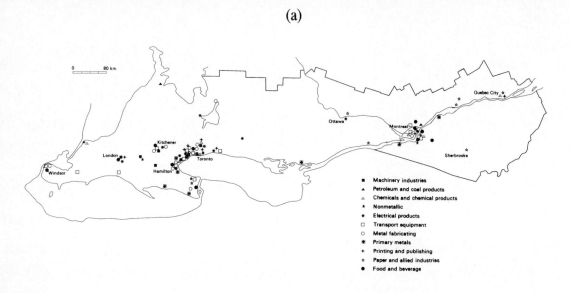

(b)

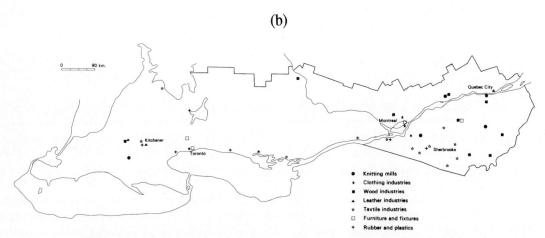

***Figure* 5.4** The location of (a) high value-added, and (b) low value-added manufacturing industries in central Canada.

REGIONAL CHANGES IN THE LOCATION OF MANUFACTURING

One of the major changes that has occurred in North America over the past twenty-five years has been in the regional concentration of manufacturing. This type of change has occurred in the United States much more than in Canada. In Canada, more than 70 percent of manufacturing employment and value-added created in the process of manufacturing is located in the metropol-

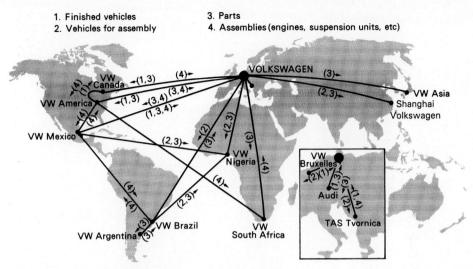

1. Finished vehicles
2. Vehicles for assembly
3. Parts
4. Assemblies (engines, suspension units, etc)

Figure 5.5 The worldwide supply and assembly network for Volkswagen.
(*Source:* The Economist, Aug. 24, 1985, p. 74.)

itan areas and cities of southern Ontario and Quebec. This is the manufacturing heartland of the country, and there has been little decentralization of manufacturing beyond that region (see Table 4.2), though there has been some growth of manufacturing in such western cities as Winnipeg, Vancouver, Edmonton, and Calgary.

In the United States, however, there has been considerable deconcentration of industry away from the traditional industrial heartland of North America—the *Manufacturing Belt* (outlined in Figure 5.6). This has raised the question of

Table 5.2

Manufacturing in the United States: Some Summary Statistics, 1963–1991

ITEM	1963	1972	1982	1991
Percent of GNP from Manufacturing[a]	29	26	22	18
Number of Establishments[b]	102	114	123	NA
All Employees[c]	17.0	19.0	19.1	18.1
Production Workers[c]	12.2	13.5	12.4	11.5
Value-Added per Production-Worker Hour[d]	26.58	29.97	41.72	48.2

[a]GNP = Gross National Product (includes domestic production and trade).

[b]Establishments in thousands (with 20 or more employees).

[c]Employees and production workers expressed as annual average in millions.

[d]In 1987 constant dollars (calculated by author).

(*Source:* Bureau of the Census (1993), *Statistical Abstract of the United States,* 1993, Washington, D.C., U.S. Dept. of Commerce, p. 742.)

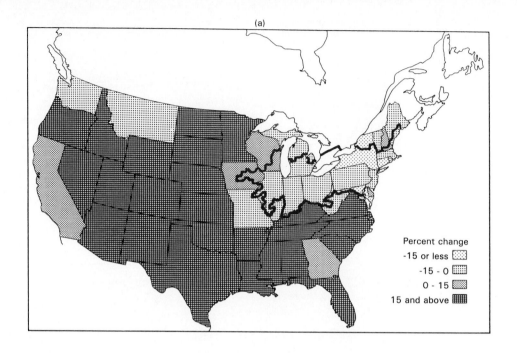

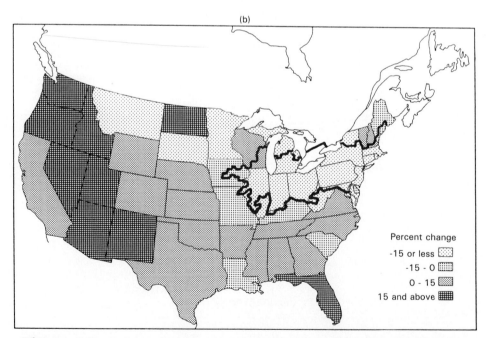

Figure 5.6 Percent change in manufacturing employment by state (a) 1966–1977, and (b) 1977–1991. (Heavy black line demarcates the traditional Manufacturing Belt.)

deindustrialization (Bluestone and Harrison, 1982) in the Manufacturing Belt as a result of the loss of jobs. In the late 1950s, about two-thirds of all employment in manufacturing in the United States was located in the states that formed part of the Manufacturing Belt. The types of manufacturing that predominated at that time involved the production of transport equipment (automobiles and trucks), chemicals, machinery, primary metals (iron and steel), and textiles and clothing (in New England and New York). This two-thirds' share remained fairly steady up to 1968 when it began to decrease rather rapidly, so that by 1991 the Manufacturing Belt contained only 48 percent of all employees in manufacturing in the country.

Snow Belt–Sun Belt Shifts

Of greater concern than the change in share is the change in actual number of jobs. Information in Table 5.2 emphasizes the fact that *total* employment in manufacturing in the United States has changed little over the past fifteen years, apart from fluctuations up and down with the business cycle. There has not been an overall trend for growth, or decline. There has, in effect, been a *comparative shift* in manufacturing employment, for while employment has decreased in the Manufacturing Belt it has been increasing elsewhere. In the ten years between 1967 and 1977 the Manufacturing Belt lost 1.6 million jobs in manufacturing, whereas the urban areas outside this area gained 1.8 million jobs. Hence the great interest since the mid–1970s in Snow Belt–Sun Belt shifts (Sawers and Tabb, 1984; Markusen, 1991; and others).

In the ten years preceding 1977, the only states that experienced a decrease in manufacturing employment (with the exception of Washington) were in the Manufacturing Belt, and all of the states which had a high growth rate (greater than 15 percent) were located beyond this area (Figure 5.6a). The state with the largest manufacturing in 1977, California, experienced a growth rate of 10 percent for the same ten-year period. Thus, the concentration of decline was quite remarkable. Ultimately, it led Keinath (1985, p. 222) to conclude that there was increasing divergence between the core and the periphery in a postindustrial age: As the core is fragmenting, various regions within the periphery are developing more diversified economies.

The period between 1977 and the early 1990s witnessed two significant recessions which have accentuated the Sun Belt–Snow Belt contrasts. It must be recalled that whereas the year 1977 marks the recovery from the rather mild recession of the mid-1970s, 1991 marks the bottom of the recession of the early 1990s. Thus, national loss of 1.5 million manufacturing jobs between 1977 and 1991 has a strong cyclical component, for the total number of jobs recovered to roughly pre-recession levels by 1995. However, the recession hit the old Manufacturing Belt particularly hard: Between 1977 and 1991, this area lost 3 million manufacturing jobs, while the Sun Belt area gained 1.5 million jobs. The Sun Belt–Snow Belt pattern of growth and decline (particularly the areas where the decline is greater than the national decrease of 8 percent) is particularly well demonstrated in Figure 5.6b.

Reasons for Snow Belt–Sun Belt Shifts   This shift away from the old Manufacturing Belt has to do with (1) the changing nature of the relationship between the old core and the new cores in what was the periphery, and (2) the establishment of new "technology districts" (Storper, 1992) or "technopoles" (Scott, 1993). The traditional core-periphery model, as has been described previously, has the core as the location of industrial innovation and the focus of all kinds of fabricating activities, which, in the past, have grown much faster than resource-processing industries (e.g., leather, tobacco products). The periphery, on the other hand, is supposed to be the focus of resource-based activities, and generally imports fabricated products. The high productivity levels of the core result in a generally higher standard of living than in the periphery.

Given the generally higher wage levels, the core becomes vulnerable in time to the flight of labor-intensive (i.e., mostly nondurable) industries (Casetti, 1984). Thus the shoe, textile, and clothing industries have shifted from New England and New York to parts of the South. Deconcentration of some elements of manufacturing has occurred as a result of differential wage rates and levels of unionization, which are generally less in the Sun Belt. Angel and Mitchell (1991) note that although the rate of growth of real wages in the South during the 1980s was greater than that for the North, regional wage disparity remains an important factor influencing greater job growth in the South.

The second basis for deconcentration is more complex as it stems from technological change. Norton and Rees (1979) argue that the strength of the traditional Manufacturing Belt stemmed from its leadership in the machine tool industry, which was instrumental in the development of assembly line manufacturing. Page and Walker (1991), for example, note the great importance of the implementation of the Ford assembly line production process in the 1920s to the development of manufacturing in Detroit, Chicago, and other cities of the Midwest. This technology is now universal, but the new technologies that are generating new industries are based on integrated circuits (Henderson, 1989) and biochemical engineering.

Furthermore, industries emanating from these newer enabling technologies, in order to be successful, have to be involved with continuous product innovation, which Storper (1992) describes as "product-based technology learning" (PBTL). Successful industries of this type, while tuned closely into global developments, tend to be regionally clustered where they can, in a very real sense, learn from each other as a consequence of labor mobility and industrial spin-offs (see also Chapter 9). This kind of dynamism exists in clusters in all regions across North America, exemplified, for example, in such industries as microelectronics (Dobbage and Rees, 1993) and personal computer assembly (Angel and Engstrom, 1995).

LEAST-COST THEORY AND THE LOCATION OF MANUFACTURING

Given these general patterns in the location of manufacturing, and some of the broad interregional changes that have been occurring, it is appropriate to examine some general theories relating to industrial location. The earliest body of theory relating to the geography of manufacturing is *location theory,* which attempts to interrelate the changes that are taking place in the economy to changes in the location of manufacturing. As Webber (1984) emphasizes, there are two main parts to the theory. The first part involves an understanding of the way the economy works, and the second an attempt to link that understanding to the historical evolution of manufacturing through the mechanisms of least-cost theory.

The Assumed Nature of the Economy

An appreciation of the importance of location theory requires an understanding (albeit brief) of the way that the economy works in the geographical area in which the theory is to be developed. The most fundamental point to reiterate is that the North American economy is basically capitalist, which in the context of manufacturing means that:

1. Production is organized by the owners of firms. Although expenditures from the public purse constitute a substantial share of the GNP, both Canada and the United States have a much higher level of private ownership of the means of production than most countries in the world. For example, in many European countries, telecommunications, electricity production and distribution, railways, airplane manufacture, and a large share of steel production, are in public ownership.

2. The objective of the owners of the means of production is to sell the output at a profit. As a consequence of this imperative, in a competitive environment there is a continuous need to increase productivity. The corollary to this is that the owners of firms may seek to decrease the level of competitiveness by various means in order to decrease the relentless drive for higher productivity.

3. The owners of the firms use capital (i.e., money) to buy raw materials, labor, machinery, and so forth to produce the output. This capital is created primarily through household savings and undistributed profits. A fundamental crisis facing capitalism since 1975 has been the low rate of savings in all countries except Japan and Germany, and the accumulation of massive deficits in the public sector which have led to huge needs for public borrowing.

4. The allocation of all these resources (capital, labor, raw materials, and the output from production) is achieved through a series of markets that set money prices (or values) for the goods, skills, and production which are being purchased or sold. These markets are usually imperfect; that is, buyers and sellers usually do not have perfect information. There may well be a dominant influence in the market, and there are invariably segmented or discriminatory markets of one kind or another. Nevertheless, there is some kind of market.

In this type of capitalist environment, decisions are not, on the whole, made by some central organization on the basis of certain national or regional

Table 5.3

Summary of the Factors Influencing the Comparative Advantage of a Firm

TYPE	VARIABLE
Technological Change	Level of technical expertise
	Product innovation
	Product quality
Human Resources	Labor costs
	Labor skills
	Management skills
Financial Capital	Access to working capital
	Investment capital
Product Price	Price
	Customer access to credit
Spatial Linkages	Proximity to suppliers
	Proximity to markets
	Distribution networks

(*Source:* After Dobbage and Rees (1993).)

objectives which are put into effect by planners. Decisions, and in this context, *location decisions to purchase or build new manufacturing establishments*, are made at particular times by the owners of firms on the basis of the information that they have at that time. A summary of the factors influencing the comparative advantage of a particular firm is presented in Table 5.3. The location decisions may, therefore, be reasonably sensible (i.e., close to profit maximizing) at the time they were made, but as conditions change they may become less sensible (or, on the other hand, even more appropriate). The location of a plant must, therefore, be examined not only in the context of its appropriateness in todays environment, but also in the context of its historical rationale.

LEAST-COST THEORY

The first person to attempt a general theory of manufacturing location was Alfred Weber (1909; 1929). Weber defined locational factors as those "forces which operate as economic causes of location." These forces are classified in his analysis in a number of ways.

1. *General and special factors:* General factors are applicable to all industry, such as capital costs, whereas special factors are those pertaining to a given industry, such as perishability. General and special factors are excluded from the theory.

2. *Regional factors:* The regional factors, which establish the fundamental locational framework of an industry over a broad area, are transport costs and

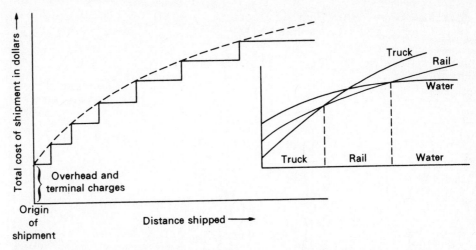

Figure 5.7 The relationship between transportation costs and distance for selected modes of transportation.

labor costs. Weber regarded transport costs as the prime determinant of interregional location, and labor costs as the "first distortion."

Weber's location theory, therefore, focuses on factors which impact at the regional level. In this context, he gave prime importance to transport costs, and labor costs. His method of analysis was to isolate what he regarded as the basic causative factors, examine the impact of these, and then relax the assumptions progressively to introduce other locational factors. Following this method, it is assumed that the owners of firms are seeking locations for production that will incur the least cost. There are three components of costs to the firm: (1) the costs of inputs, whether these be in the form of raw materials or partially fabricated goods; (2) costs of processing, such as labor and energy, which are called production costs; and (3) costs of delivery of goods. As a consequence, least-cost theory (following Weber, 1929) involves analyzing the influence of transportation costs (1 and 3) and production costs (2) on location.

Minimizing Transport Costs Transport costs for freight increase with distance, but at a decreasing rate, because rate structures are usually set by zones. A certain amount is paid for the first few miles, an additional amount for the next few, and so on, with the width of each zone increasing with distance from the point of shipment (Figure 5.7). There are also considerable fixed charges; that is, charges that do not vary with distance and have to be paid regardless of length of trip. These include the costs of loading and unloading and are usually referred to as *terminal charges*. A third characteristic of transport costs is that the magnitude of the terminal charges and the steepness of the cost curves are related to the means of transport that is used. The differences between truck, rail, and water transport for bulky goods are generally as is shown in the inset

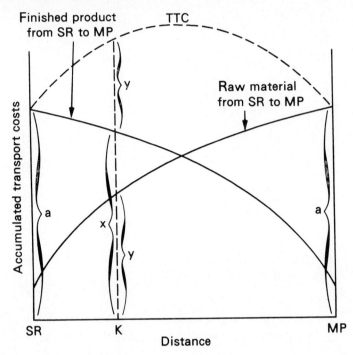

Figure 5.8 The effect of equal transportation costs for raw materials and finished products, assuming a curvilinear relationship between transportation costs and distance.

in Figure 5.7—water transport (if available) is generally more cost efficient for long hauls, rail for medium-length hauls, and trucks for shorter distances.

The variable and fixed nature of transport costs can be applied to a simple situation involving a firm that wishes to build a plant to: (1) manufacture a product that requires one material input from one source (SR); and (2) sell its output in one area MP (e.g., a given urban place). In this example it is also assumed that: (3) the material source (SR) and the market (MP) are separated but connected by one means of transport; and (4), initially, there is no *weight loss* incurred in the raw material in the process of manufacture, that is, one hundred units of a raw material will yield one hundred units of a finished product. In situations of this type there are two possible least-cost locations—SR and MP. All other sites between these two locations are more expensive because they incur an additional set of terminal charges as well as the disadvantages of two shorter hauls, rather than one that is longer. Thus, in Figure 5.8 the total transport cost (TTC) of manufacturing at SR and sending the finished product to MP is cost *a* and sending the raw material from SR to MP for manufacture is also cost *a*. However, if the firm were to manufacture its product at place K, it would also have to pay to ship the raw materials and to unload it (cost *y*), and would incur more costs to ship the finished product from K to MP (cost *x*).

However, transport costs are usually higher per unit weight for finished products than for raw materials. One important reason for this difference is

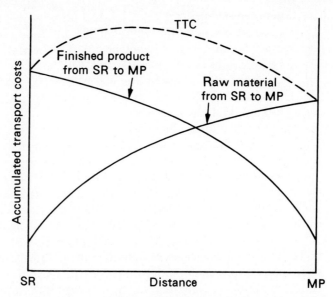

Figure 5.9 The effect of higher transportation costs for finished products than for raw materials, assuming a curvilinear relationship between transport costs and distance. (Compare this diagram with Figure 5.8.)

that as manufactured goods increase in value through the process of manufacture, transportation costs constitute a lower proportion of the total value of the shipment than they did for the raw materials. Manufactured goods thus are said to be capable of bearing higher transportation charges. Also, it is often the policy of national governments to subsidize transportation charges for raw materials produced some distance from ports, in order to promote access to foreign markets (e.g., the Crow Rate in Canada, which was used for many years to provide low rates for the transportation of wheat and other cereals to lakehead and west coast ports). As a result, the transportation cost curve for finished products is steeper than that for raw materials, and the least-cost location for manufacture will be at the marketplace (Figure 5.9). Thus, if a given quantity of raw material yields the same quantity of finished product, the industry will locate at the marketplace, where the total cost of transportation will be lowest.

The Effect of Variations in Weight Loss A given quantity of raw material does not, however, usually yield the same quantity (by weight) of finished product. There is, invariably, a great deal of weight loss in manufacturing. This has to be taken into account in the simple locational model being discussed. As the location of manufacturing will be at either SR or MP, assume that the freight rates for manufactured product (P) average $20 per kilogram per kilometer for the total distance between SR and MP, and that the freight rate for the raw material (R) averages $15 per kilogram per kilometer for the same distance. Also assume that 1,000 kilograms of raw material R are to be processed into finished product P.

If the manufacturing plant is located at MP, it will always have to transport 1,000 kilograms of materials to MP. Thus, the total transportation cost will be $15,000 per kilometer. If the weight loss is 20 percent and the firm locates at SR, 1,000 kilograms of raw material will yield 800 kilograms of finished product, and the total cost of shipping this will be $16,000 per kilometer. Thus, with a weight loss of 20 percent, the firm will gain an advantage of $1,000 per kilometer by locating at the marketplace.

However, with the weight loss of 40 percent, 1,000 kilograms of R yield 600 kilograms of P. Because the cost of transporting this amount at the manufactured product rate is $12,000 per kilometer, the firm will gain an advantage of $3,000 per kilometer if it locates at the raw material site. Consequently, given the transportation cost structure for finished products and raw materials: In general, the higher the weight loss, the more likely a firm will locate at the source of inputs, if only transportation costs and weight loss are considered.

Hence, the importance of the distinction between the processing, fabrication, and integration stages of manufacturing. Processing plants are quite likely to be found at raw material locations—particularly those involved with the concentration of ores or the manufacture of ingots of various kinds, whereas fabrication and integration is more likely to occur where the major markets are located. For example, INCO (International Nickel Company), a transnational firm headquartered in Toronto with $2.5 billion annual sales and 15,700 employees in twenty countries worldwide (INCO, 1994), has its manufacturing enterprises in North America distributed very much according to this model. Nickel and copper are smelted (i.e., processed) in peripheral areas where the raw material is mined, such as Sudbury (Ontario) and Thompson (Manitoba). Some of the metals are then sent to plants at Huntington (West Virginia), Burrough (Kentucky), at the edge of the major market—where high-performance alloys are fabricated, and blades, rings, discs, and so forth, are produced for integration into aero and gas turbine engines elsewhere in the Manufacturing Belt.

Transshipment Points It is, however, common for more than one transport mode to be employed. When this occurs, the goods or materials have to be transferred from one type of transportation to another, as at a port. Transshipment points are also break-of-bulk points, because the goods being shipped usually have to be broken down into smaller units for subsequent transportation, for the contents of an ocean-going vessel would fill many railroad cars or trucks. Transshipment and break-of-bulk points also incur additional costs, such as extra terminal charges for loading and unloading. Therefore, transshipment points (which in North America were often "gateway" ports) attract manufacturers because some additional costs can be avoided even though the freight rate in shipping the manufactured product to the market may be higher. The attraction of these locations is particularly strong for those types of manufacturing using raw materials with a high weight loss. Moreover, firms often choose break-of-bulk sites, and especially ports, because raw materials can be collected there easily from many locations. In fact, before the development of the railroad, inland cites could not grow very large because transportation costs were too high.

The widespread use of containers for the transport of fabricated products has, however, resulted in significant shifts in the location of plants involved with the final integration of goods. Containers greatly facilitate loading and unloading. Thus, they reduce enormously the costs of transshipment. As a consequence, break-of-bulk considerations have become less relevant for material at a late stage in the production sequence, and plants involved with the integration of products are more likely to be located within the major markets. An example of this is the location of a Chrysler/Mitsubishi assembly plant in Lexington (Kentucky), using engines and electrical components from Japan, and chassis, bodies and other parts from North American plants (Kenney and Florida, 1992).

The Decreasing Influence of Transport Costs   One of the major features of transport costs is that they have been declining steadily over the past four decades in importance as a component of total manufacturing costs. This decline in importance may be attributed to:

1. *The change in the composition of the manufacturing sector:* Over the past few decades, the "heavy" industries (iron and steel, metal fabrication, etc.) have declined relative to the rapid growth of the "light" industries (electronics, clothing, etc.). The "heavy" industries tend to be more sensitive to transport cost variations than the "light" industries.

2. *The reduction in the importance of raw material locations:* Fewer raw materials are now needed per unit of output. Consequently, raw material sites have become less relevant to the locational decision. For example, "at the turn of the nineteenth century, six tons of coal were needed to smelt one ton of iron, whereas today about one-and-a-quarter tons of coal are needed or, in the case of electric furnaces, no coal at all at the place of manufacture" (Norcliffe, 1975, p. 22). Hence emerged the highly resource-efficient "maxi-mills," integrated operations that process iron ore into a wide variety of carbon-steel products; and non-coal burning "mini-mills" for the production of steel (particularly steel rods) in the South and Southwest (OhUallachàin, 1993).

3. *The development of transport technology:* Many developments in the new techniques of transportation, such as containers, piggy-back railroad cars and trucks, telecommunications, long-distance transmission lines, and welded pipelines, have made it easier to transport goods, energy, and information a great deal cheaper.

Thus, even though energy costs have risen quite considerably over the past ten years, the effect on the transport bills of firms has not been as great as might have been expected.

Production Costs   Given the declining relative importance of transport costs in the location decision of the firm, the importance in least-cost theory of spatial variations in *production costs,* and the way in which these production costs may be influenced by *agglomeration,* has risen accordingly. There are two types of production costs: labor and nonlabor. Nonlabor costs, which involve such things as geographic variations in the price of land, taxes, and heating and air-conditioning costs, and so forth, do not vary that much over North America, and they are not a large enough proportion of the costs of the firm to have much influence on location decisions. Webber (1984) estimates that nonlabor costs comprise only about 2 percent of total value-added in manufacturing. The availability of national and local government incentives may serve to lower nonlabor production costs at certain locations, but in the medium term (four to eight years) a location has to be viable because governmental subsidies cannot be provided indefinitely.

Variations in Labor Costs   Although labor costs account for a decreasing share of total value-added in manufacturing (down from 52 percent in 1963 to 40 percent in 1991), it is variations in the costs of management, wage rates (and the structure of union agreements), and the availability of different types of labor that have become the single most important factor influencing location decisions. Wage rate variations are quite high as they vary not only geographically, but according to type of industry (high-wage versus low-wage industry), and within an industry. For example, average hourly earnings by production workers in manufacturing in Arkansas in 1992 were 20 percent lower than the national average ($11.45), whereas those in Ohio were 18 percent above; workers in textile mill production earned 35 percent less than the national average, whereas those in the aircraft industry received 32 percent more; and workers in the apparel and textile industries in New York earned at least $1.00 an hour more than those in the same industry in Alabama.

These variations in hourly earnings are, of course, related to many factors. The most important influence on average hourly earnings is productivity (operationally defined in Table 5.2 as value-added per production worker hour), which is determined by level of skills, education, and the efficiency of the productive capital (machinery, etc.) and production processes available to the worker. Some industries, such as aircraft manufacturing, demand a high level of skills and education, whereas others, such as textile industries, can use less-skilled labor. It is often argued that amenities, such as a nice climate, affect wage levels, in that people are prepared to work for lower wage rates in more attractive or warm locations. Wage rates in Florida are, for example, 16 percent lower than the national average. It is also claimed that wage rate differences are related to levels of unionization (high in states and industries with strong union representation), though it would seem that a limited argument for wage rates, even in unionized industries, have to be related to productivity and industry competitiveness.

The basic point, however, is that given these types of regional, industry, and within-industry variations in wage rates (which change over time), and the profit imperative that exists in North American economy, manufacturing firms will tend to seek locations for new plants in lower-labor-cost areas, provided that productivity is satisfactory. They will also tend to expand production in existing plants in low-labor-cost areas, provided that the differences in productivity are not too great. Hence, we will find the greater-growth manufacturing industries, particularly those that are labor intensive, in the generally lower-wage-rate Sun Belt states (Angel and Mitchell, 1991).

The Influence of Economies of Agglomeration   Economies of agglomeration are savings in per unit production costs that can be made by a firm when it operates along with a large group of economic activities in a relatively small compact geographic area, such as a city (Petrakos, 1992). Agglomeration economies consist of two types: localization and urbanization.

Localization economies arise when firms in the same industry, or closely related industries, concentrate at the same place because they can take advantage of forward and backward linkages and pools of labor that have similar and easily transferable skills. An example of such a situation is the semiconductor industry in Silicon Valley, Santa Clara County, California, where the leaders of the industry (e.g., Hewlett-Packard, National Semiconductor Corporation) are surrounded by a "dense concentration of smaller electronics related suppliers and producers" (Saxenian, 1984, p. 164). All these industries take advantage of the highly trained research-oriented labor force focusing around the major universities in the area, and the abundant supply of labor (chiefly recent immigrants) for low-wage-rate assembly tasks.

Localization economies may, therefore, give rise to geographic clusters of similar firms, which have been described as "industrial districts" (Scott, 1988a; Harrison, 1992). The basic premise of such districts is that the firms operating within them achieve cost savings as a result of proximity to suppliers and various types of interfirm collaborative action (Saxenian, 1992). The responsiveness of different industries to these types of situations varies greatly. Gertler (1995a) demonstrates that, in Ontario, proximity of users to producers of advanced process technologies is necessary for successful implementation of the technologies to occur. Similarly, Dobbage and Rees (1993) conclude that companies involved with microelectronics and computer-aided manufacturing in the United States perceive a significant advantage in locating close to markets and suppliers. On the other hand, Appold (1995) suggests that in the metalworking sector in the United States the proximity of producers to customers does not enhance the performance of firms.

Urbanization economies consist of a number of features in urban areas, which can act together to create a situation that may facilitate lower business costs.

1. *Social fixed overhead capital*: The availability of existing infrastructure in larger urban areas is a particularly important influence on the location of small and medium-sized plants. Infrastructure involves all those aspects of the physical environment that facilitate the operation of a plant, such as the availability of roads (particularly, limited-access highways), waste disposal facilities, water, power, freight depots, airports, and housing. It must be recognized that society as a whole must pay for the establishment and maintenance of social fixed overhead capital. So no one element (e.g., manufacturing firms fleeing the inner city) should be allowed to opt out of paying its share (such as for education).

2. *External economies* realized from better information and face-to-face contact: Large urban areas provide an enhanced information environment, which is particularly attractive for the headquarter office, planning, financial, and R&D (research and development) activities. This is particularly the case when there are many small- and intermediate-size companies in an urban area rather than a few large ones who might tend to provide most corporate services in-house (Norton, 1992).

3. *Economies of supply and distribution*: Firms which locate in large urban areas are able to realize economies through the availability of such services as freight forwarders and large transport agencies; wholesale suppliers, warehousing and storage facilities; accounting, advertising, marketing, and financial agencies; and the sheer size of local markets.

Agglomeration economies may turn into diseconomies and lead to declining productivity in the largest urban areas as costs rise and efficiency is reduced (Mooman, 1986). Transportation, congestion, pollution, lack of reinvestment in infrastructure, crime, and so forth may begin to raise costs to a level sufficient to encourage firms to move from the urban region altogether and establish manufacturing plants at other places. Beeson (1990) suggests that productive efficiency in large metropolitan areas in the United States in the 1970s and 1980s has decreased as a result of urbanization diseconomies and high wage rates. Aggregate productive efficiency in urban areas, however, changes quite quickly. Employment and wages grew much more rapidly in the southern and western regions of the United States in the 1970s and 1980s, but in the 1990s the western regions are faring poorly relative to the rest of the country (Mills and Lubuele, 1995).

INDUSTRIAL RESTRUCTURING

Although the basic principles of least-cost theory are important, they must be viewed within the context of the tremendous changes that have been occurring in manufacturing in North America, Japan, Southeast Asia, and western Europe since the 1970s (Dicken, 1992). Modern manufacturing has been affected greatly by two fundamental events. First, an increasing concentration of ownership among highly competitive national and transnational corporations means that the location of a particular manufacturing plant must be viewed in the context of the structure and decision making processes of large organizations. Second, rapid product and process innovation, stimulated primarily by a new generation of enabling technologies, such as the microchip in information processing and recombinant DNA technology in biochemical engineering, are having as fundamental an impact on manufacturing and manufacturing location as the steam engine and the evolution of industrial capitalism in the late eighteenth century, or the harnessing of electricity and the maturing of capital markets at the end of the nineteenth century. Manufacturing is undergoing a "restructuring," involving a drive for quantum leaps in the quality and productivity of existing production, and the development and introduction of new products and production processes, such as flexible production and just-in-time delivery systems (Best, 1990). The growth of manufacturing in California and Texas, along with the resurgence of industrial activity in parts of New England, suggests that the newer industries have greater locational flexibility (Gertler, 1992).

Manufacturing's New Era

Just as in the discussion of location theory it was essential to sketch out the nature of the national economy in which location decisions are made, so in the case of fundamental restructuring it is necessary to *outline* the political, social, and economic context of some elements of the transition. The discussion will be couched in the context of political economy; that is, it will involve the changing relationships between capital (the "business" community), labor, and (instead of land) the state (or national, state/provincial, and local governments). The transition relates to the changes that began to permeate through manufacturing in North America during the last half of the era of mature industrial capitalism (Trachte and Ross, 1985). The discussion is presented, therefore, in terms of changes between the United States–led period of the 1945–1973 era and the post–1973 period when United States industrial leadership gradually weakened as manufacturing strengthened globally.

Changing Capital to Capital Relations   In the context of manufacturing firms, and the relationships between them, the 1945–1973 period may be defined as one in which manufacturing in the world was dominated by a number of leading large firms, predominantly (though not by any means entirely) headquartered in the United States. These companies had, over a number of decades, reached a *modus operandi* that reduced price competition to a minimal amount by focusing instead on product differentiation and marketing. The consequence of this was relative stability in the short term (three to five years) in market shares among the major producers of goods. Productivity throughout the era increased considerably as a result of a movement to larger and larger scale plants producing larger and larger quantities of standardized consumer products. A considerable proportion of these products were absorbed by the internal domestic market, but during the 1960s markets were increasingly being sought beyond the national borders for the escalating volumes of output. This situation provides the context for Holmes's (1983) discussion of the integration of the North American auto industries in the 1960s.

The search for new markets occurred at the time that major European Economic Community and Japanese producers began to move from domestic markets and also, in the case of Japan, from the nurturing influence of Pentagon purchases during the Korean War (and, later, the War in Vietnam), to the rich North American market (Schoenberger, 1985; Chang, 1989). Hence, during the mid–1960s and 1970s, when United States manufacturing corporations were looking for new markets, Japanese and European producers moved more strongly than before into the North American market. They did this through

Table 5.4

The World's Top Ten Producers of Motor Vehicles, 1990

COMPANY	COUNTRY	WORLDWIDE PRODUCTION (MILLIONS OF VEHICLES)
General Motors	United States	7.0
Ford	United States	5.4
Toyota	Japan	4.7
Volkswagen	Germany	3.1
Nissan	Japan	3.1
Fiat	Italy	2.5
Peugeot-Citroën	France	2.1
Honda	Japan	2.0
Mitsubishi Motors	Japan	1.9
Renault	France	1.8
Mazda	Japan	1.6
Chrysler	United States	1.5

(*Source: The Economist* (1992), October 17, "The Endless Road.")

price competition, new products (such as VCRs), and a differentiation of product in style and quality (Schoenberger, 1988a). One element of this differentiation is a greater variety of products of particular types, which is achieved through lower-quantity production runs based on decreasing plant sizes and greater output flexibility within plants. The net result in the post–1973 period is highly unstable market shares among the major manufacturing corporations in the world that operate either directly, or through wholly or partially owned subsidiaries, in many different countries.

The Automobile Industry An example of the changing form of business organization that results from greater competition is in the evolving structure of the automobile industry, an evolution which is inextricably entwined with cities in North America. A car is the second-largest purchase (after a home) for most people, and provides freedom and access to employment (directly or indirectly), but also threatens to choke many cities. In 1970, 27 percent of the total world output of motor vehicles (29.3 million) was produced in the United States, but by 1990, when the world output at best had increased to 48.1 million units, the United States share had decreased to 20 percent. Thus, it was not that American production had declined, but other centers of manufacturing had become major players in the industry (Table 5.4).

The emergence of these non-United States corporations and their ability not only to dominate their domestic markets but also to compete successfully in the North American market (which, after the European Union, is the second-largest automobile market in the world), is mainly related to their greater competitiveness. Production and process technologies have changed radically in the industry from the time that Henry Ford transformed a high-price-per-vehicle craft industry into a low-price-per-vehicle factory production system with his *marriage of interchangeable parts to the moving assembly line* (a production system known as "Fordism") in Detroit in 1913. By the mid–1950s, this system of mass production had become highly refined and organized, making intensive use of machinery, producing a limited range of models in each factory (Table 5.5), and was being copied elsewhere, particularly in the postwar European auto industry.

During the 1960s, the Japanese began to implement a different production system than that pioneered by Ford, following quality control principles advocated by W. Edwards Deming that has come to be known as the Toyota production system (Halberstam, 1986, pp. 312–320). This production innovation, involving concepts like flexible production, and just-in-time delivery (Linge, 1991) from subcontractors and to markets (Holmes, 1986), and an impressive attention to quality in the flexible production process, has led to a complete restructuring of the North American automobile industry. The flexible production system replaces the assembly line with a team, or quality circle, in which each worker in the team performs a number of tasks rather than just one as in the assembly line process. The worker, therefore, operates in a multi-tasked rather than single-tasked environment. Each production worker is the "customer" of the worker in the "circle" or line before, taking only from the previous worker what is needed to complete that particular job. The role of the previous worker is then to replace what is taken. This means that products are "pulled" through the production system in response to demand (one aspect of which is a demand for quality) rather than "pushing" products through a system which accepts a certain proportion of quality defects—which is what the scientific mass production system was doing. The process also allows for a degree of process innovation, product development, and product differentiation during the course of production—hence the term *flexible production* to describe the process.

The result was that the higher-quality and competitively priced Japanese products provided a severe challenge to the United States domestic industry: In 1965 imported motor vehicles accounted for 6 percent of domestic United States sales, but five years later this share increased to 24 percent, and since 1970 this import proportion has remained roughly the same. Even with the restructuring of the North American auto industry that has taken place—restructuring in terms of adapting some of the new methods of Japanese "lean" production modified from the Toyota production system—there remain significant differences in productivity (Table 5.6). In particular, although the number of hours it takes United States automakers to manufacture a car, and the quality defects, are down, Japanese automakers, on the average, still lead the field. This incessant drive for productivity, quality, and variety in production (leading to greater flexibility in production by the end of the century) is characteristic of other major industries as well, ranging from clothing and textiles, to steel (Florida and Kenney, 1992).

Table 5.5

Summary of the Evolution of Automobile Making

PRODUCTION CHARACTERISTICS	CRAFT INDUSTRY (EARLY 1900s)	FORD MASS PRODUCTION (1920s)	SCIENTIFIC MASS PRODUCTION (1950s)	JAPANESE LEAN PRODUCTION (1980s)	COMPUTER-CONTROLLED MACHINE TOOLS (1990s)	FLEXIBLE PRODUCTION (2000–)
Number of Machines in a Typical Factory	3	50	150	150	50	30
Number of Different Models Made	infinite	3	10	15	100	infinite

(*Source: The Economist* (1992), Oct. 17, "The Endless Road.")

Table 5.6

Productivity Characteristics of Automobile Production in the United States, Japan, and Europe

AVERAGE FOR AUTO PLANTS IN:	JAPAN	UNITED STATES	EUROPE
Performance			
Productivity (hours per car)	16.8	25.1	32.6
Quality (defects per 100 cars)	60	82	97
Factory Layout			
Factory space (per sq. ft, per car, per year)	5.7	7.8	7.8
Size of repair area (as % of assembly space)	4.1	12.9	14.4
Stocks (for eight sample parts)	0.2	2.9	2.0
Employees			
Workforce in teams (%)	69.3	17.3	0.6
Suggestions (per employee per year)	61.6	0.4	0.4
Number of job classifications	12	67	15
Training of new workers (hours)	380	46	173
Automation (% of process automated)			
Welding	86	76	77
Painting	55	34	38
Assembly	2	1	3

(*Source: The Economist* (1992), Oct. 17, "The Endless Road.")

Changing Capital-to-Labor Relations   Although it should be clear that one way in which this greater competition has come about is through imaginative innovations in production processes that have increased productivity (output per worker hour), prices have also been contained (or reduced) in part through major efforts to control the cost of labor and of material inputs. During the first half (1945–1973) of the era of mature industrial capitalism, the manufacturing labor force became split into two basic components. One set of manufacturing workers became highly unionized, and the others remained nonunionized. Although the proportion of total manufacturing employment in union membership was never as high as in western Europe (by 1964 almost 50 percent of manufacturing employment was unionized in the United States), nevertheless the labor conditions that were negotiated set standards for the much of the rest of the workforce.

From the time of the negotiation of the landmark Ford Motor Company/United Auto Workers (UAW) collective agreement in 1947, many of the larger corporations and unions created what amounted to highly favorable (when compared with the nonunion sector) working conditions for unionized workers. In return for increases in productivity through scale economies, the workforce in these privileged industries (such as, motor vehicles, iron and steel, petrochemicals, and aircraft production) achieved large increases in wage rates, reductions in the length of the work week, good fringe benefits (health care, pensions), and a level of job security through complex job demarcation and seniority preference ("last hired, first fired"). The Ford/UAW 1947 agreement, which set the pattern, became the cornerstone of postwar American prosperity as it guaranteed union labor high wages which, in effect, drove the post–World War II mass consumer market. The metropolitan areas in which these unionized industries were located—Detroit, New York, Chicago, Baltimore, Philadelphia, Pittsburgh, Cleveland, Seattle—prospered during this period. It must, however, be remembered that half of all manufacturing workers in the United States were *not* members of unions, and in most cases their wage rates, fringe benefits, job security, and so forth were a great deal less than those in most of the unionized industries.

Unfortunately, this model of worker security and wage compensation could not be sustained in the face of growing price and quality competition from other world producers—particularly Japan and Germany. The United States model was based on increasing volume of output and escalating prices in a relatively fixed market share environment. The new methods of production and new products pioneered elsewhere, which still involved higher levels of worker job security and, in the German case, compensation (fringe benefits) than the United States, required a drastic response. As we have seen, this response has been primarily for corporations to shift manufacturing to low-wage areas within or outside the United States or, in a special case, to a completely uncontrolled free-trade corridor on the Mexican side to the United States–Mexico border to gain access to low-wage Mexican workers slaving away in the so-called "maquiladoras" (South, 1990).

Changes in Unionization and Collective Agreements Another response has been to encourage unions to reduce previously negotiated benefits and levels of job demarcation (see "number of job classifications" in Table 5.6), partially in response to the multi-tasking requirements of flexible production systems. This "encouragement" has been undertaken by employers in the private and public sectors, one tactic being to play one region off against another (Holmes and Rusonik, 1991) which has led to fragmentation, splits, and inter-union friction. If the 1947 Ford/UAW agreement was the benchmark agreement typifying the first part of the postwar era, then President Reagan's breaking of the PATCO strike (air traffic controllers' union) in 1982 highlighted the realities of the present period. By 1990, union membership in manufacturing in the United States had dropped to 4.6 million (down from 8.3 million in 1964), the membership comprising only 23 percent of the labor force.

It should, however, be emphasized that the shift of certain elements of manufacturing to less regulated low-wage areas has led to a new international division of labor (Schoenberger, 1988b). Some countries or some regions within countries are effectively allocated low-wage aspects of manufacturing, while the home bases retain high-wage professional, technical, finance, management, and research and development aspects of the multinational enterprise. This is, in part, the consequence of a production culture that is produced rather than inherited (Gertler, 1995b). The propensity of a workforce to operate in a team environment, to care for machinery, to learn to operate and maintain software and different types of production systems, to undertake job retraining (see

Table 5.7

Change in Productivity and Compensation in Manufacturing, 1970–1991 (average annual percent): United States, Canada, Germany, Japan, and the United Kingdom

INDEX AND YEAR	UNITED STATES	CANADA	GERMANY*	JAPAN	UNITED KINGDOM
Output per hour					
1970–1975	3.4	3.4	5.1	6.8	4.0
1975–1980	1.7	2.7	3.9	8.6	1.4
1980–1985	3.7	2.8	3.5	5.9	5.7
1985–1991	2.9	0.3	1.8	4.6	4.0
Compensation per hour					
1970–1975	8.2	10.9	12.3	20.3	19.5
1975–1980	9.2	10.8	8.1	7.2	16.3
1980–1985	5.9	8.3	5.3	4.7	9.2
1985–1991	4.0	4.9	5.3	4.9	8.0

(*Source:* Bureau of the Census (1987), *Statistical Abstract of the United States.* Washington, D.C.: United States Department of Commerce, p. 813; and, (1993), p. 863.)
*Former West Germany.

Table 5.6) and so forth, are characteristics produced by the state and local institutions and the corporations for which employees work (Saxenian, 1994). The continuous large increases in productivity (output per hour) as compared with compensation per hour (Table 5.7) for Japan and, to a lesser degree, Germany (the 1985–1991 figures for Germany are affected by the absorption of the former East Germany into West Germany) in comparison with the figures for the United States and Canada are indicative of this point. The restructuring of manufacturing, if it is to be successful for all concerned, requires close examination of national and corporate behaviors that are favorable to the required new industrial production processes and practices.

Changing Capital-to-State Relations   In the period between 1945 and 1975, the role of the state in North American society increased enormously. In terms of capital (and labor) this role can be interpreted as involving three elements. First, with the support of the business community, the state has provided (or organized) ways of delivering increasing levels of benefits and security to labor through such means as Social Security, Medicare, and the whole group of programs known as the "social safety net." We have seen that, in a number of ways, particularly health care and unemployment insurance, the Canadian "social safety net" is somewhat more wide ranging and generous than that in the United States.

Second, through regulation and directed government expenditures, the state has effectively ensured minimum market shares for the largest companies. We have seen in Chapter 3 how federal expenditures in the United States have influenced the growth of different regions in the United States, and in Chapter 4 that equalization payments in Canada have provided financial support for the less wealthy parts of the country. This kind of support has been even more directed in some cases, such as with respect to the Chrysler bail-out, federal support for indebted cities such as New York City, and continuous financial support for the aircraft industry which, quite literally, would be grounded without development support through defense expenditures and subsidies for sales through guaranteed government loans.

Finally, there has been considerable direct involvement by the state in the research and development departments of high-technology industries, through increased expenditures for "defense," medical research, and so forth. Markusen and colleagues (1991), for example, argue that defense expenditures in the United States have created an extensive zone across the United States (in the pattern of a "Gun Belt") involved with military aerospace and electronics production. This belt extends from New England to Long Island south to Florida and Alabama, across Texas and Arizona to California and north to Seattle, with Colorado serving as an outlier. It involves companies such as Raytheon (Patriot missiles, Boston); United Technologies (engines, Hartford); Grumman (jets, Bethpage, New York); IBM and Martin Marietta (systems procurement and software, Washington, D.C.); MacDonnell-Douglas, Lockheed, Hughes, and Rockwell (airplanes, missiles, defense electronics, naval, in Los Angeles and San Diego); Boeing (aircraft, missiles, Seattle); Digital Equipment, Ford Aerospace, Kaman Corporation (computers, nuclear physics, Colorado Springs). This government-stimulated aspect of manufacturing was located (in large part) away from the traditional manufacturing areas as a result of lobbying and political decisions, one aspect of which was to be some distance from states where strong unions prevailed. In the 1990s, as military expenditure cutbacks occur, many of the defense-related industries in areas such as Los Angeles and San Diego are closing plants, and the labor force is being laid off, retired early, and shifting to newer high-technology growth industries or the service sector.

Deregulation and Investment Capital   With the transition to greater competition and the restructuring that this has generated, some of these capital-to-state relations have changed considerably. First, greater international competition between large multinational corporations has led to a drive for *deregulation* as the underwriting of market shares by the state has crumbled. Airline deregulation is an example of this, a latest step being a new "Open Skies" policy between the United States and Canada, permitting airlines based in one country to operate a domestic service in the other. Associated with this has been the globalization of financial markets, and the consequent gradual deregulation of various aspects of the financial industry. The largest and most powerful banks in the world are no longer located in the United States, which has a decentralized state-restricted banking structure which seems out of phase with the highly centralized megabanks located in other countries.

Second, because of the particular concentrated nature of product and process innovation that has occurred in North America since 1973 and the rapid depreciation of old existing plants and infrastructure, there is an excessive need for new capital that has given larger cash-rich companies the whip-hand in negotiations concerning the location of plants. For example, OhUallachàin (1993) indicates that raw steel output in the United States peaked in 1973 at 151 million tons, and by 1989 had declined to 98 million tons. This was due to increased competition from foreign mills, the substitution of plastics and other materials such as ceramics, and the increased imports of automobiles. As Japanese auto manufacturers began to establish auto plants in the Midwest (Kenney and Florida, 1992), companies headquartered in Japan began to invest in their own steel plants (maxi mills), or provide financial capital and expertise for the modernization of plants that are owned by companies headquartered in the United States, to supply the new factories.

The Location of Multiplant Manufacturing Firms

As manufacturing firms restructure, they evolve more complex forms of organization. In small, single-plant firms, the management function may take place within the manufacturing plant or in the same town, but as the firm gets larger and establishes other plants, various functional specialties emerge, each of these with different locational characteristics. Humphrys (1982) has proposed a fairly simple model that captures the nature of this complexity and emphasizes the different locational forces that are involved in general with each element (Figure 5.10). The model delineates three main components: the organizational structure of firms; the general nature of associate labor markets; and the locational attributes of the various components of the firm. The model emphasizes that the organizational structure of the firm creates a spatial division of labor with managers and white collar workers concentrating in certain areas and blue-collar workers in others.

The Location of Headquarter Offices and Research and Development

The central position of the headquarter office or management function of the firm is indicated by Figure 5.10. Decisions emanate from this office to both the R&D and production functions of the firm, which, in turn provide feedback that is integrated, to a lesser degree into the decision-making apparatus. The headquarters office is involved with the financial, legal, policy-making, and strategic decisions, and the personnel involved are the salaried and managerial staff. This managerial force is supported by a wide array of specialists (lawyers, accountants, and so forth) and general support staff (such as secretaries). The basic locational principle for this aspect of the firm is the necessity for good access to all types of information and finance, which is facilitated by situations to promote face-to-face contact among executives of different companies. The headquarters function is, therefore, invariably located in a leading metropolis, traditionally in a downtown site close to or within the financial community.

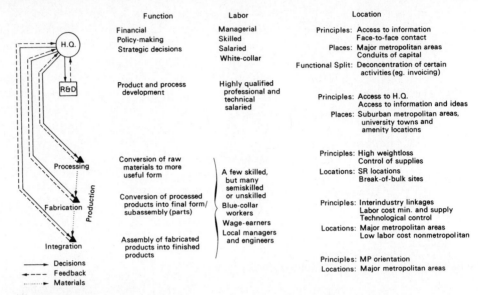

Figure 5.10 The interrelationships among multiple-plant functions, type of labor market, and location.
(*Source:* Humphrys, 1982, Figure 4.9, p. 322.)

Some management functions may be decentralized to suburban locations within the field of the major metropolis, particularly those that are of a routine nature (such as invoicing) and are not too sensitive within the power structure of the firm. One function that is of crucial importance to the modern firm but has its own particular locational imperatives, is research and development (Howells, 1984). Research and development activities involve highly qualified professional and technical staff and, although the headquarters office requires considerable control over their activities and enormous feedback from them, the most efficient locations for R&D are in an environment that fosters creativity and are attractive to well-trained, high-energy individuals. Suburban research parks and places within the vicinity of one or more research-oriented universities appear to be the preferred locations. Also, there must be a high level of accessibility to transportation (airports) for the high-priced staff! It is important to be aware that the R&D and headquarter functions are closely entwined. As a consequence, the R&D activity invariably is concentrated within the same country as the headquarters office.

The Location of Production Facilities The Humphrys (1982) model clearly identifies the range of types of production with which a multiple-plant firm may be involved: processing, fabrication, and integration. It is envisaged that, on the whole, materials flow from processing to fabrication and integration, although plants in all stages require machinery as well as material inputs. Each of the three types of production involves both technically skilled and unskilled blue-collar workers, nearly all of whom are paid on an hourly basis. There are also usually local managers and plant engineers who are salaried workers.

Processing Plants   Plants involved in processing are invariably dealing with materials that incur considerable weight loss in manufacturing. The locations that are selected are, therefore, invariably raw material locations or break-of-bulk sites. There is also a tendency for large multinational corporations to place processing facilities within the vicinity of raw material locations in order to secure control of supplies. Dome Petroleum, Nova Corporation, and Dow Chemical established polyethylene, ethylene oxide, and vinyl acetate plants (respectively) in Alberta to this end (Chapman, 1985).

Fabricating Plants   Plants involved in fabrication are invariably dealing with inputs that incur minimal weight loss in manufacturing but that do involve bulk. Consequently, a number of other locational forces come into consideration: interindustry linkages, costs of labor and supplies, and technological control.

1. *Interindustry linkages*: The traditional argument is that as the output of plants involved in fabrication is directed toward plants involved in final assembly (integration), they will tend to locate within the proximity of the final markets. However, transport costs have become less of a constraint. So such locations are no longer determinate. Nevertheless, the complex set of interindustry linkages which have developed between plants in towns and cities in Michigan, Ohio, Indiana, and Ontario have served to maintain strength in motor vehicle and general manufacturing within this area (Stafford and Wu, 1992).

2. *Labor costs and supplies*: There are three aspects of labor to be taken into account—cost, quality, and the way it is organized. For those plants that require unskilled or semi-skilled labor, the firm will invariably search for low-wage-rate locations. South (1990) demonstrates that many fabricating plants seeking low wage, nonunionized labor in an area free of environmental regulations, have located in the *maquiladoras* on the Mexican side of the United States–Mexico border. These plants, established mainly by United States–headquartered multinational companies, are involved with export-oriented production (hence the designation "maquiladoras") primarily to the United States. Some fabricating industries that require skilled labor tend to be more constrained in location, such as fabricated metal products in Pittsburgh (Pennsylvania) and Hamilton (Ontario).

3. *Technological control*: Many aspects of fabrication involve the production of materials that are the outcome of much research and development. The components have been patented, but even with this protection companies

may be loath to have the producing plant move into a site that could be beyond the control of the head office. These types of plants may well be located in the vicinity of the headquarters office. This occurs often in the pharmaceutical drug industry, where firms like to maintain secrecy and control over the development of new drugs.

Plants Involved with Integration   The third aspect of production, the integration of fabricated products into finished products for the consumer market, is very much market oriented in location. The output is bulky, often fragile, and therefore reasonably expensive to ship. As a consequence, integrated assembly plants are usually found within the major urban regions that have been developing in North America since 1945 (Chapters 3 and 4). It is significant that all the major automobile companies in the world have located at least one final assembly plant within these regions.

The Effect of Restructuring on the Location of Manufacturing

Manufacturing restructuring, as a response to price and quality competition, has had significant locational repercussions. These locational repercussions involve (a) intermetropolitan growth, stagnation, or decline; and (b) intrametropolitan plant and employment shifts involving primarily decentralization to suburban areas or nonmetropolitan locations (Barkley and Hinschberger, 1992). Intermetropolitan changes will be examined in this section, intrametropolitan shifts are discussed in Chapter 9. Negrey and Zickel (1994) examine changes in manufacturing employment and population for 140 of the largest metropolitan areas in the United States for the 1970–1987 period, during which time the combined population for these areas increased by 24.9 percent (compared with 18.4 percent for the United States as a whole), and manufacturing employment increased 12.6 percent (compared with little change for the United States as a whole). In general, Negrey and Zickel (1994) define five groups of metropolitan areas based on changes in employment and population, with one anomalous case, Boston, which is placed by the author in the first group.

Classic Deindustrializing Metropolises   The classic deindustrializing metropolises have in general decreased in population size by 5.2 percent and incurred a large loss in manufacturing employment for the period under consideration (Table 5.8). There has been widespread disinvestment in manufacturing with rates of growth in the other employment groups well below the average for all metropolises. As there are few alternatives to manufacturing (e.g., tourism), the economies of these cities are not diverse enough to compensate. So the population decreases as manufacturing declines. Some people leave to look for employment elsewhere, and there is little immigration. Boston is an anomaly be-

Table 5.8

Types of Manufacturing Metropolises: Average Rates of Change in Population (1970–1987) and Employment (1972–1987) by Area Type

EMPLOYMENT GROUP	DEINDUSTRIALIZING	STABLE CENTERS	SLOW GROWTH	SERVICE CENTERS	NEW MANUFACTURING	ALL 140
Population	-5.2	7.0	13.2	36.4	49.9	24.9
Manufacturing	-29.3	-17.8	27.2	-13.9	52.0	12.6
Nonagricultural	8.2	24.5	54.1	34.1	78.1	47.1
FIRE*	49.9	57.9	107.8	71.2	111.7	83.7
Trade	27.7	42.1	70.8	47.7	93.2	63.4
Services	65.0	90.2	107.4	110.8	146.5	112.4
Number of metropolises	19	38	11	15	56	140

(*Source:* Extracted from Negrey and Zickel (1994), p. 34.)

Note: Boston excluded from table, but included in total.

FIRE = finance, insurance, real estate.

Figure 5.11 Location of classic deindustrializing metropolises, United States.

cause even though its population declined, manufacturing remained fairly stable. The metropolises grouped in this category are plainly in the historic core, or old Manufacturing Belt of the United States (Figure 5.11). With each successive recession since 1970, the metropolises in this group have experienced closure or downsizing of plants, and little employment growth has occurred during upswings. There has, however, been considerable restructuring of some of the main industries, such as automobile production, as well as growth of some newer high-technology industries, which has resulted in increases in output and product marketability. But this has not been accompanied by much increase in manufacturing employment.

Stable Urban Areas in Transition   Metropolises in this group consist of those in which manufacturing declined (Figure 5.12), but population grew, indicating that their economies are more diverse than the classic deindustrializing metropolises. In one sense they can be regarded as in transition—compensating for losses in employment in manufacturing by some gains in services and other nonagricultural activities. The location of these urban areas is mainly in the Northeast and Midwest, but the pattern is much more spread with outliers in the South and West. In the next chapter it is noted that San Francisco and Chicago are "world cities," important in international finance, and are, therefore, not dependent on manufacturing. Similarly, Omaha, Kansas City, Des Moines, Memphis, Birmingham, Indianapolis, Columbus, Cincinnati, and Philadelphia are important regional centers providing retail, commercial, financial, and wholesale services for extensive hinterlands.

Figure 5.12 Location of stable manufacturing centers in transition, United States.

Slow-Growth Urban Areas Slow-growth urban areas are those in which manufacturing employment grew quite quickly, but the population grew at a much lower rate (Figure 5.13). Most of the metropolises in this small group are located in the Northeast, a few in the Upper Midwest, with Wichita as an outlier. Some of these cities are sites of major research-oriented universities (Ann Arbor, Madison, Minneapolis-St. Paul, Hartford), while most of those in the Northeast are basically older manufacturing centers that have retained a "niche" product. Wichita, as the headquarters of Gates Learjet, has an employment pattern which fluctuates considerably in response to changes in the aerospace industry, particularly the market for small luxury aircraft. These slow-growth centers also had high rates of growth in finance, insurance, and other activities in the services sector.

Growing Service Centers This group of fifteen metropolitan areas has experienced quite rapid population growth despite a decline in manufacturing which averaged about 14 percent for the group as a whole (Figure 5.14). In these urban areas, the growth in other activities, particularly in services and finance/insurance, has offset decreases in manufacturing employment. The majority of the centers are located in the South, with Atlanta, Dallas, New Orleans, Houston, and Corpus Christi as major regional centers, comparable to Denver at the junction of the High Plains and Rocky Mountains. Orange County, a suburban metropolis in the extensive Los Angeles region, has experienced extraordinary growth as a result of decentralization (see Chapter 7). Atlantic City has achieved recent growth consequent to legalized gambling, and York is a smaller

Figure 5.13 Location of slow-growth manufacturing centers, United States.

Figure 5.14 Location of growing service centers with little manufacturing, United States.

Figure 5.15 Location of new manufacturing centers, United States.

service center in Pennsylvania. In the West, Eugene-Springfield is a university town and major service center for Oregon, while Oakland has benefited particularly from great growth in trans-Pacific trade.

New Centers of Manufacturing This large group of metropolitan areas is plainly the spatial antithesis of the first and second groups—the classic deindustrializing and stable centers. Whereas the deindustrializing groups are in the Northeast and Midwest, the new manufacturing centers are located primarily in the South and West (Figure 5.15). Fifteen of the fifty-six metropolises are in California alone, and another nineteen in the South. In general, both the population and manufacturing employment of these metropolises increased 50 percent (Table 5.8) during the 1970–1987 period. Responding to, and reflective of, the great increase in manufacturing employment, employment in the services, trade, and finance/insurance (along with real estate) also increased at a rapid rate as well. These are metropolises that benefited directly from the upswing in the business cycle during the latter half of the 1980s, and also by the increase in defense expenditures, force-fed by the big increases in federal budget deficits during the Reagan years. The economies of these cities thus experienced a jolt during the recession and defence cutbacks of the early 1990s. Nevertheless, they remain centers of manufacturing expansion, particularly in the consumer goods, computer, microelectronics, and general high-technology industries.

CONCLUSION

The distribution of manufacturing among urban areas in North America is, therefore, the result of many interacting forces that have operated in different ways over the past two hundred years. The real growth in manufacturing occurred from about 1830, and for much of the time the locational factors that had the greatest influence were those that are considered to be in the realm of least-cost theory (that is, transport, production, and labor costs). There have, however, been particular periods of restructuring when manufacturing in general has had to respond, through massive new capital investments in new products and production processes, to price, product, and technological competition. One obviously occurred in the 1920s with the marriage of the innovation of interchangeable parts to assembly-type factory production and processes that permitted the realization of greater scale economies.

The most recent era has witnessed another wave of technological and process innovation. This is coupled with a change in the dominant position of North American manufacturing industries in the world. The phrase "global competition" captures a situation in which large multiplant, and often multinational, corporations have responded to domestic and international price and product competition through significant process and product innovation. These companies may well have many aspects of their production processes concentrated in industrial districts, or they may locate fabrication processes in different regions or countries in order to take advantage of low labor costs or special skills, and less regulation with respect to social, economic, and environmental matters.

Thus, corporations headquartered in one region or country may well have a large part of their processing, fabrication, or integrative productive capacity located in a different region (such as the South) or part of the world—particularly transferring aspects of production to low-wage areas, and other aspects (such as assembly) close to major markets. The home territory, therefore, retains the organizational (that is, control) and developmental aspects of the enterprise, whereas the other territories receive jobs for their lower-wage "blue-collar" workers. This type of situation is leading to a "new international division of labor" (Schoenberger, 1988b) with certain parts of the leading industrial nations retaining wealth, education, and control, and other parts (particularly in the poorer regions or countries) undertaking the more routine and menial tasks in conditions of greater social and economic uncertainty. As a consequence, the socioeconomic conditions of, and employment opportunities for, production workers in the older industrial areas (such as in the United States Northeast and Midwest) severely deteriorate, while the opportunities and employment conditions for workers in the newer industrializing regions improve but do not attain the levels enjoyed previously by workers (particularly those benefiting from good collective agreements) in the older industrial areas.

Chapter

6

Urban Areas as Centers of Service Activities

Although manufacturing is an extremely important element of the economic base of urban areas and provides the underlying rationale for many aspects of differential urban growth, service (or tertiary) activities contribute the greatest and increasing share of GNP (see, for example, Figure 4.2 for an illustration of the increasing importance of service activities in Canada). In the United States, service activities accounted for 72 percent of the GNP (and total employment) in 1990—up from a 65 percent share of GNP (and total employment) in 1980. This translated into an increase of 30 million jobs in the service sector in the ten-year period, as compared with a loss of 1 million jobs in manufacturing and construction (Daniels, 1991). This shift to a postindustrial economy is having a profound influence on North American urban development.

In the case of Pittsburgh, an urban area whose economy has been traditionally based on manufacturing, the shift to tertiary activities has meant that manufacturing employment decreased from 226,000 in 1979 to 128,000 in 1986, with a concomitant increase in service employment (Lamonde and Martineau, 1992). Thus, high-wage manufacturing jobs (averaging $25,000 per year in 1986) were lost and replaced in part by lower-wage service sector jobs (averaging $19,000 per year in 1986). Also, the jobs that were lost are generally occupied by males, whereas service sector jobs are more equally occupied by both men and women. The growth of service sector activities, therefore, has to be viewed in the context of industrial restructuring, decreasing earnings per person, and increasing participation of women in the paid labor force.

161

TYPES OF TERTIARY ACTIVITIES

There are, in general, three broad types of tertiary activities:

Consumer services
Producer services
Governmental and public services

This threefold distinction is important because it incorporates a range in types of end uses, which have, in general, different financial bases, and are associated with different theoretical explanations concerning their location and allocation among urban areas in North America.

Consumer services are those involving distributive activities, where the sale of the product achieves, directly or indirectly, its end use of final demand. *Retail trade* obviously involves the sale of manufactured products to final consumers. Also, *wholesale trade* involves the distribution of products from warehousing facilities to retailers. Sometimes wholesalers sell directly to consumers without going through retail outlets—this often occurs with mail-order firms. Other activities included within the consumer category are those involved with *entertainment and recreation* (such as restaurants and theme parks) and various *personal services* to private households or individuals (such as beauty salons and auto repair).

Producer services are those that mainly provide inputs to, or are linked with, other economic activities (Daniels, 1993c). They are, therefore, part of the production process. Included are *transportation, communication, and other public utilities* (such as electricity, water, gas); *finance, insurance, and real estate*; and *business services* such as advertising, information technology functions and systems, personnel supply services, consulting firms involved with management/planning/research and development, and security services. The general view is that as the economy shifts from a "Fordist" mode of industrial organization (assembly line production involving long product runs and economies of scale—see Chapter 5) to more flexible production there is an increasing need for more advanced producer services (Sjoholt, 1994). In an uncertain economic environment with fewer opportunities for achieving great economies of scale, more management and process innovations and routines are being sought outside the firm in the producer services sector. Thus, in this regard, the distinction between "manufacturing" and "producer services" is highly blurred.

Governmental and public services are generally provided, or supported to a considerable extent, from public funds. Thus, public administration—the means by which the government in effect attempts to regulate the capitalist economy, to ameliorate some of its less acceptable outcomes (such as poverty), to provide for those in need through the social safety net (i.e., unemployment insurance and Social Security), and to provide services that enhance the general welfare and productivity of the nation (such as education)—is an extremely large component of this category. Included also is the general health care system, which in Canada is supported from public funds, and in the United States from both public (Medicaid and Medicare) and private sources, usually in the form of individual or group insurance. Employment in public administration and general public services expanded greatly during the 1945–1973 period, but during the 1990s there are increasing attempts to bring this high rate of growth under control. Much of the service sector public policy debate is devoted to increasing efficiency in this component of activity—particularly in health care provision and governmental administration.

Employment Change in the Tertiary Sector

The increasing importance of service activities in the United States is indicated in Table 6.1, with respect to employment changes since 1970. It will be recalled that the decennial dates occur before (1970) or just at the commencement of (1980, 1990) business recessions. As has been discussed previously (Chapter 5), manufacturing employment remained fairly constant throughout the twenty-year period, while employment as a whole increased by 50 percent, being led by the service industries.

In Table 6.1, the service employment sector is divided into the three main components discussed in the previous section. This division into consumer, producer, and governmental plus public employment involves a re-aggregation of published statistical information that is not quite as precise as would be wished. For example, the producer services group includes employment in real estate, much of which is involved in final demand actions of households (i.e., home purchases), but this cannot be disaggregated from the finance/insurance subgrouping. There is a consistency, however, in that the same definitions are used for the three dates. In general, it is believed that the inclusion of certain elements of employment in inappropriate categories may even out "in the wash," given that some aspects of entertainment and recreation, for example, are linked as much to marketing in the producer services sector (e.g., major league sports) as they are to consumer services related to households.

Table 6.1

The Increasing Importance of Service Activities: United States Employment (in Thousands) by Industry, 1970–1990

INDUSTRY	1970 NUMBER (%)	1980 NUMBER (%)	1990 NUMBER (%)	% CHANGE 1970–90
Manufacturing	20,746 (26.4)	21,942 (22.1)	21,184 (18.0)	2.1
Consumer Services	21,251 (27.0)	26,419 (26.6)	32,468 (27.5)	52.8
Producer Services	10,283 (13.1)	15,804 (15.9)	22,737 (19.3)	121.1
Government & Public	16,951 (21.5)	24,419 (24.6)	29,726 (25.2)	75.4
Total Employment	78,678	99,303	117,914	49.9

(*Source:* Compiled by the author from the Bureau of the Census (1993), p. 409, legal services being divided equally between consumer and producer services (government lawyers being included in government and public).)

Note: Total employment includes construction, agriculture, and mining.

The interesting feature of Table 6.1 is the rapid growth of the producer services group, which is led by the information technology industries. Employment in this broad group of activities (which ranges from data processing to software development and network organization) has increased fourfold *since 1980*, and by the early 1990s provided almost as many jobs as auto repairs and servicing! The finance, insurance, real estate grouping also grew quite rapidly, doubling in employment in the twenty-year period. Employment in private security agencies serving the business sector also increased quite rapidly, an interesting commentary on the fortress behavior of North American society as it developed during the last quarter of the twentieth century.

Employment in the governmental and public services also increased at a rate greater than the average (Table 6.1). This is almost entirely related to the doubling of employment in the health services industry (including hospitals), which now comprises the third-largest provider of employment in the nation (following manufacturing and retail trade) and involves a large proportion of female workers (above 75 percent). The educational sector (elementary through to colleges and universities) and public administration grew at a rate comparable to the increase in the nation's population.

The increase in employment that occurred in the consumer goods industry is related primarily to the growth of retail and wholesale trade, automobile services and repairs, entertainment and recreation, and legal services. The retail trade group, in particular, is extremely large with total employment in 1992 almost equal to that in manufacturing as a whole (19.6 million jobs). Many of these jobs are, however, part time and involve annual earnings per person that are, on average, one-half those in manufacturing.

There are interesting contrasts in the gender and ethnic pattern of employment along the three main components of the service industries (Table 6.2). Whereas manufacturing has a large proportion of male workers, the governmental and public sector, with a much larger total employment, has a large proportion of female workers. The rapidly growing health sector, with non-doctor

Table 6.2

Gender and Ethnic Employment by Industry: United States, 1992

INDUSTRY	TOTAL (IN THOUSANDS)	% FEMALE	% AFRICAN AMERICAN	% HISPANIC AMERICAN
Manufacturing	19,972	32.9	10.4	8.3
Consumer Services	32,800	47.9	9.0	9.0
Producer Services	21,693	39.4	11.6	7.6
Governmental & Public	32,050	61.9	12.3	5.3
Total Employment	117,598	45.7	10.1	7.6
Population (1990)	248,710	51.3	12.1	9.0

(*Source:* Compiled by the author from Bureau of the Census (1993), p. 409 and pp. 30–31.)

Table 6.3

Montreal: Change in Experienced Labor Force (in thousands) by Industry Type, 1971–1991

INDUSTRY TYPE	1971 NUMBER (%)	1991 NUMBER (%)	% CHANGE 1971–1991
Manufacturing	302 (28.0)	303 (18.6)	0.3
Consumer Services	164 (15.2)	302 (18.6)	83.9
Producer Services	163 (15.1)	316 (19.4)	93.5
Governmental & Public	206 (19.1)	349 (21.5)	69.4
Montreal*	1080	1626	50.6
Canada*	8627	14220	64.8

*Includes construction, transportation and communications, primary industries (agriculture, forestry, fishing, mining) and "other."

Note: In this compilation, the experienced labor force in producer services includes that in wholesaling.

(*Source:* Compiled by the author from data in Coffey (1994), Table 2.1, p. 35.)

wage levels similar to those of production workers in manufacturing, is heavily female-employment oriented. The rapidly growing producer services sector is 60 percent male, but tending toward a more even gender split. The only industries that have an equitable representation of African Americans are in governmental and public services, and producer services, and Hispanic Americans have significant representation in the consumer services industries, particularly those involving personal services.

This general pattern of stagnation or decline in manufacturing, contrasted with growth in the tertiary industries, is well reflected in the particular case of Montreal (population in 1991, 3.1 million). This metropolis has the second-largest concentration of manufacturing employment in Canada, but this aspect of its economy has been declining both absolutely and in relative terms since 1971 (Table 6.3). Growth in the service industries has compensated for this, but the growth in producer services somewhat less than might be expected on the basis of the growth of producer services in the United States and the leading position of Montreal in Quebec and Canada. On the other hand, growth in consumer services and in governmental and public employment is greater than might be expected. The decline in manufacturing and the less-than-expected growth in producer services are undoubtedly related to the comparative shifts in high value-added manufacturing and financial activities in the metropolitan Toronto area, which have been discussed previously in Chapter 4.

THE INTERURBAN LOCATION OF CONSUMER SERVICES

Consumer services, involving the distribution and exchange of goods and services, are the most ubiquitous of all tertiary activities. Without exception, a considerable portion of the economic activity of settlements is concerned with the collection of goods from, or the distribution of goods and services to, the people living in the settlement and the surrounding area. For some settlements this marketing activity may be overshadowed by other more specialized roles, such as manufacturing, but for innumerable others, particularly smaller settlements in agricultural regions, the provision of consumer services to the local settlement population and the surrounding area is the basis of their existence. Towns and cities that provide this consumer service function for themselves and the surrounding region are called *central places*. In North America, such towns were established, on the whole, *in advance or at the same time* as agricultural settlement occurred, and they served as links to the developed part of the continent. Hence, most urban places formed part of an expanding and complex set of distribution centers.

The Central Place Model

The *central place* model addresses issues related to the *location* of settlements involved with consumer services in geographic space, and the *allocation* of consumer services among these settlements (King, 1984; Semple, 1985; Berry and Parr, 1988). The model was developed by Christaller (1933; 1966) and is based on detailed empirical study of the settlement pattern and service structure of southern Germany between 1929 and 1932. This had been an agricultural region for many centuries and had changed slowly as the West European economy evolved from feudalism through mercantilism to industrial capitalism. It was (and is), therefore, an entirely different situation to the relatively quick agricultural and urban-led settlement of large parts of North America (Page and Walker, 1991).

Thus, the locational aspects of the model, which addresses questions concerning *where central places will locate and how large they will be,* is based on empirical work in a part of the world that had a fairly evenly spread agricultural population for many centuries. These locational aspects, which Christaller characterizes as involving a system of central places located on the lattice points of equilateral triangles with hexagonal hinterlands, are, therefore, of less relevance in the North American situation than the allocational aspects of the model which provide a rationale for the *distribution of consumer services among central places*. The allocational aspects of the model also illuminate the way in which the consumer service aspect of the economies of urban areas will change as they, and the regions they serve, grow (or decline) in population and wealth (Prescott and Vandenbroucke, 1992). The following discussion will, therefore, focus on the allocational rather than the locational aspects of the model.

Concepts Concerning Central Places

Central places are defined as urban areas having *central functions;* that is, retail and commercial establishments which provide consumer services for the inhabitants of urban areas and the surrounding region that they serve (Beguin, 1992). Central places must, therefore, be highly accessible to their tributary areas, and if this accessibility changes in some way (through rail closures or road improvements) there will be a concomitant impact on the number and range of central functions found in a central place.

The central functions provide consumer services for the population of urban areas and the people in the surrounding region that it serves. The surrounding region may include both a rural agricultural population and people collected together in smaller places. Directly analogous to the logic of economic base theory, Christaller (1966, p. 18) emphasizes that the portion of consumer service activity that becomes important is that fraction which is not serving the needs of the urban area itself (city-serving) but the population within its region (city-supporting): The larger the proportion of central-function activity devoted to the population of the surrounding region, the higher the *centrality* of the central place. This link between the central place model and economic base theory is important because it emphasizes that the consumer service sector can, in the case of urban areas with large hinterlands, provide considerable basic employment (Mulligan and Kim, 1991).

A *business type* (or central good) is one particular type of central function (for example, a grocery) which is located at a central place to serve the population of the central place and that of the surrounding region. Some business types are of a higher order; that is, they need a large market base (population ¥ income level) in order to survive, whereas others may be of a lower order because they need a smaller market to exist. The highest-order business type in a central place, therefore, does two things: It determines the order of a central place and the extent of the hinterland. The radius of the hinterland is in part determined by the density of the market, which is usually defined operationally as the population density. For any given market density, higher-order business types have a larger *range;* that is, people will be prepared to travel farther for the high-order business type than the low-order business type. By definition, settlements offering high-order business types will be high-order central places.

The order of a business type can, therefore, be defined in terms of the minimum level of demand required to support one establishment of a particular business type. Following Berry and Garrison (1958), this minimum is named the *threshold*. Thus, the threshold for grocery stores would be the minimum level of demand (operationally defined as population) required to keep one grocery store in business. The demand emanating from the central place and the surrounding region may be much greater than the threshold, in which case there may be many more than one establishment of that particular business type in the central place.

Christaller (1966, pp. 60–64) proceeds to construct a *central place hierarchy* based on the concepts of range and threshold and on the assumption that the thresholds cluster into a number of discrete groups. It is these groups that define the order in the hierarchy of central places. Thus, in Table 6.4 the highest order business types are 1 and 2, the thresholds of which cluster at a size that is significantly larger than the thresholds for 3 and 4, which in turn cluster at a size that is significantly larger than the thresholds for 5, 6, and 7, and so forth. Consequently, business types 1 and 2 define the places in which they locate as the highest-order central places, which have not only business types 1 and 2, but all the other business types as well. Similarly, business types 8, 9, 10, and 11

Table 6.4

The Supply of Twenty Business Types from G Central Places

CENTRAL PLACES	BUSINESS TYPES (CENTRAL GOODS)																			
	(HIGH)																			(LOW)
	1	2*	3	4*	5	6	7*	8	9	10	11*	12	13	14	15*	16	17	18*	19	20
A	X	X	X	X	X	X	X	X	X	X	X	X	X	X	X	X	X	X	X	X
B		X	X	X	X	X	X	X	X	X	X	X	X	X	X	X	X	X	X	X
C					X	X	X	X	X	X	X	X	X	X	X	X	X	X	X	X
D								X	X	X	X	X	X	X	X	X	X	X	X	X
E												X	X	X	X	X	X	X	X	X
F																X	X	X	X	X
G																			X	X

* Hierarchical marginal business type (or central good).

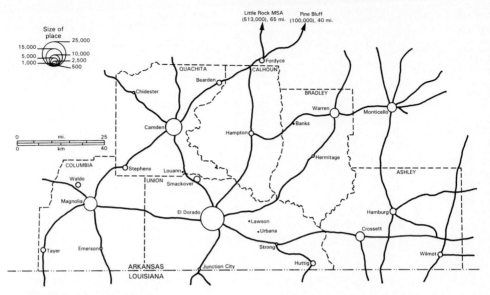

Figure 6.1   Southern Arkansas: settlements and major roads in the six-county El Dorado region.

cluster around a threshold value which is significantly smaller than that for business types 5, 6, and 7, but significantly larger than that for the cluster of business types 12, 13, 14, and 15. Certain business types at the margin between two levels in the hierarchy are referred to as the hierarchical marginal business type because it may be just possible for them to be provided from the next lower level of center in the hierarchy.

The Central Place System in Southern Arkansas   Some of the basic concepts of the central place model can be illustrated with respect to the allocation of consumer services among a number of small urban areas in southern Arkansas. El Dorado (population 24,000) is the largest service center in a six-county area comprising three commuting zones focusing on Magnolia, Crossett, and El Dorado (Figure 6.1). The region has a population of 145,000, with an average population density of thirty persons per square mile. The chief economic activities are associated with oil extracting and refining, agriculture and forestry, and a wide range of consumer and public services. Thirty percent of the population is African American (Arkansas Institute for Economic Advancement, 1993, various tables). In general, the population is quite poor. The state ranks forty-eighth in per capita income in the United States. Compared to that, the El Dorado area has an economic growth rate well below the average for all counties in Arkansas and the United States as a whole. The region borders on Louisiana, where the population of adjacent counties with similar economic and social conditions lies within the commuting zones of Monroe, Ruston, and Shreveport.

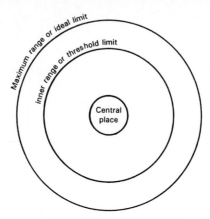

Figure 6.2   The maximum range, inner range, and complementary region of a particular business type in a central place.

The key element in establishing a hierarchy is the order of business types, which can be determined from threshold values (Table 6.4). High-order business types have large thresholds and are found only in the largest urban places, whereas the lowest-order business types are found in all centers of all sizes. The perimeter of an area around a central place which contains enough population (including the central place) to support one establishment of a given business type (such as a bakery) defines the *inner range* of the goods sold by that particular business type, while the *maximum range* (or *ideal limit*) is the theoretical boundary beyond which no person will travel to obtain that good (Figure 6.2). Thus, the complementary region of a central place must lie between the inner range and the maximum range of the good for its associated business type to be in existence at that place. Many studies of change revolve around the reasons for the expansion or contraction of complementary regions.

Measuring thresholds   Thus, the number of business types (B) and the number of establishments (E) in a particular central place are related to the size of the population of the place (P) plus that of its associated complementary region (R). These relationships can be expressed in symbolic from as follows:

$$(1) \quad E_i = f_1 (P_i + R_i) \qquad \text{where } i = 1,2,3, \ldots ,n$$
$$(2) \quad B_i = f_2 (P_i + R_i)$$

for *n* central places. The symbols f_1 and f_2 mean "some mathematical relationship," which is determined empirically later. For any particular business type (j) the number of establishments in a particular central place can be expressed by the relationship:

$$(3) \quad {}_jE_i = f_3 (P_i + R_i) \qquad \text{where } j = 1,2,3, \ldots ,m$$

for *n* central places and *m* business types.

In most research situations it is extremely difficult to measure R_1, and so the equations are usually modified to exclude that element. This modification

means that for business types in small central places the thresholds are usually underestimated, whereas in larger places where the R_i component may be proportionately small, the modification is not so serious. For this reason it is wise to use a method for calculating thresholds of a particular business type that takes into account the number of establishments in large as well as small centers. Such a method, using simple regression analysis, was first introduced by Berry and Garrison (1958) and used by many others, including Yeates and Lloyd (1970) in Ontario in an economic impact study.

The method can be illustrated using data obtained from local telephone directories, field work, and the Arkansas Institute for Economic Advancement (1993), for a number of basic services collated in the format of Table 6.4. The number of establishments of a particular business type (e.g., grocers/supermarkets) are plotted on a graph with respect to the associated population of the central place. Then the "best-fit" line is fitted to the scatter of points using least-squares regression. In the case of grocery establishments, as with all other business types in this analysis, the best-fitting line most conforming to the assumptions of regression analysis (such as linearity, and homoscedasticity) is of the form $Y = a X^b$ which can be expressed in linear form in common logarithms as $\log (Y) = \log a + b \log(X)$. A logarithmic equation of this form is often used for measuring the ratio of the rate of change in Y to the rate of change in X, the rate of change, known as elasticity, being the regression coefficient b.

An application of regression analysis to the distribution of grocery stores among central places is presented in Figure 6.3 where the axes are in logarithmic scale. The least squares best-fit line is of the form:

$$\log (\text{\# grocers})_i = -1.14 + 0.56 (\log \text{pop.})_i$$

and the R^2 value is 82 percent (100 percent would indicate a perfect fit with the entire scatter of points located along the best-fit line). With this equation it is possible to fit not only the line through the points, but also to calculate the *threshold*, for this will be where the number of grocery stores is 1. The threshold value, as can be seen from the graph, is 107 persons.

Clusters of Thresholds   The regression method is used to calculate the thresholds for eighteen other basic service types in central places in southern Arkansas, and these are plotted along a line nomogram in Figure 6.4. The threshold values for these nineteen basic and most common services in the region vary between 107 and 2,340 (for optometrists), with three main clusters:

  Cluster 1: average threshold 171—grocers to restaurants and fast food
  Cluster 2: average threshold 746—auto repairs to attorneys
  Cluster 3: average threshold 1,778—dentists to optometrists

It should be noted that the higher-order cluster indicates the population size at which basic health services begin to be available.

Given the spread of thresholds in each cluster, the order of central places (as defined by this particular set of basic services) is as follows:

  Group A: central places with a population size up to 260 persons

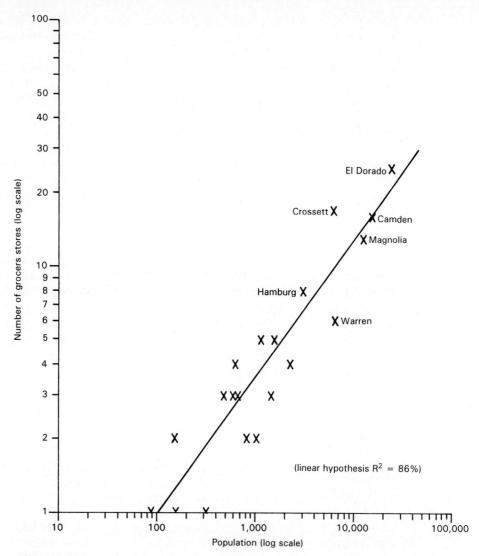

Figure 6.3 The relationship between the number of grocery stores and population in central places in southern Arkansas, 1993.

> Group B: central places with a population size from 450 to 1,000 persons
> Group C: central places with a population size from 1,300 persons

The proportionate number of business types of a given cluster rank found in centers of each size group are listed in Table 6.5. The table can be interpreted by reading either down the columns or across the rows. Reading down column C, it is evident that all the larger central places (population greater than 1,300) have all of the higher-order basic consumer business types listed in Figure 6.2. The smallest places (A), have none of the higher-order basic business

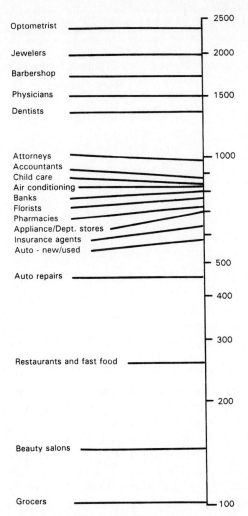

Figure 6.4 The tendency for threshold values to cluster for nineteen services in central places in southern Arkansas, 1993.

types, only 4 percent of the middle group (cluster 2) and none of the higher-order business types. If the model worked perfectly, each of the diagonal cells and the entries to the right would be 100 percent, and the entries to the left of the diagonal (which have 4, 0, and 18) would all be zero.

The Southern Arkansas Central Place Hierarchy in Regional Context   The list of service activities included in this example is restricted to those most commonly required for day-to-day living, and does not distinguish establishments by the quality and range of their service provision. For example, the general

Table 6.5

Proportion of Business Types of a Given Cluster Found in Centers of a Certain Size Group

CLASS OF CENTRAL FUNCTION	THRESHOLD CLUSTER RANK	AVERAGE THRESHOLD	SIZE OF URBAN CENTER		
			A (SMALL)	B	C (LARGE)
			PERCENT OF BUSINESS TYPES		
Lowest-order business types	1	171	60	100	100
	2	746	4	51	100
Higher-order business types	3	1,778	0	18	100

category "restaurants and fast foods" does not distinguish establishments offering Cajun cuisine from those providing fast foods. Likewise, the all-inclusive category "appliance/department stores" includes establishments ranging from independently owned family enterprises to chain retail department stores providing everyday wear consumers wishing to purchase upscale specialty goods and supplies go to a higher order central place outside the El Dorado region for sufficient choice.

This is illustrated in Figure 6.5, which provides graphs indicating the relationship between the total number of retail (Figure 6.5a) and wholesale (Figure 6.5b) establishments in the larger central places (20,000 population and more) and the local population served, which is operationally defined as the population of the county in which the place is located. Although Memphis (Tennessee) provides a considerable array of consumer service facilities for central western Arkansas, the central places within the state have a good level of service provision. Information in Figure 6.5a suggests a threshold value for the first establishment (of any business type) of 165, fortuitously close to the average (171 persons) threshold estimated for the lowest-order business types (Table 6.5). Wholesale activities generally define higher-order central places, and the threshold for the first wholesale establishment is calculated to be 640 persons.

The graphs (Figure 6.5) indicate that Little Rock, with its many diverse governmental, health care, and industrial activities, is, apart from being the state capital, also the premier central place in the state. The other places form groups that are at a lower level in the consumer service provision hierarchy. This information can be coordinated with that in Figure 2.7 and Fik and Mulligan (1990, Table 3) to suggest a central place hierarchy in Arkansas (and immediately adjacent states) in general, and southern Arkansas in particular, as indicated in Table 6.6. The level above "major metropolitan" is "national metropolitan," involving places like New York, Los Angeles, and Chicago, which have national coverage. Arkansas consists of places in a five-level hierarchy, with Little Rock the highest-order central place, and at the lower end many

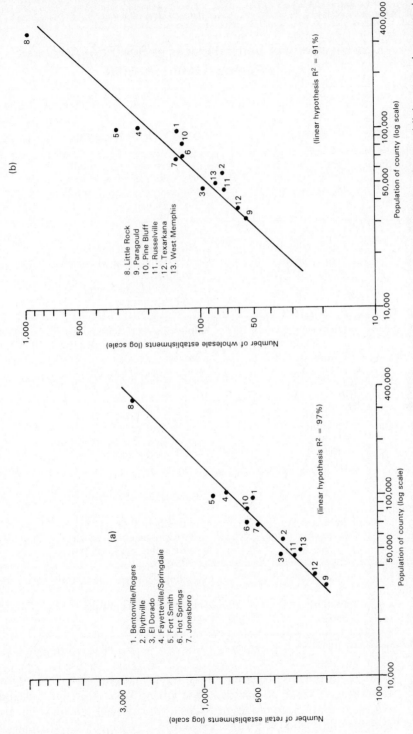

Figure 6.5 The relationship between the number of (a) retail stores, and (b) wholesale establishments in central places, and the local population served, Arkansas, 1990.

Table 6.6

The Hierarchy of Central Places in Southern Arkansas in a Regional Context, 1990

TYPE OF CENTER	EXAMPLES	CONSUMER SERVICES	POPULATION SIZE RANGE OF CENTERS
Major Metropolitan	(Dallas-Fort Worth) (New Orleans) (St. Louis)	Complete wholesale/retail Wide array of producer services Wide aray of governmental and public services	1–4 million
Metropolitan	Little Rock (Memphis)–West Memphis (Oklahoma City) (Nashville)	All wholesale/retail Large department stores Greater than 1,000 wholesale establishments Greater than 2,000 retail establishments	400,000–1,000,000
Regional Arkansas	Fort Smith Fayetteville-Springdale Pine Bluff El Dorado Hot Springs	Wholesale/retail with good representation of specialty stores 50–400 wholesale establishments 200–1400 retail establishments	20,000–200,000
Town El Dorado region	Camden Magnolia Warren Crossett Hamburg	Full range of convenience stores for everyday needs 5–50 wholesale establishments 50–200 retail establishments	3,000–15,000
Village El Dorado region	Smackover Hampton Stephens Junction City Waldo	Limited range of convenience stores 5–15 business types 10–40 retail establishments 1–4 wholesale establishments	500–2,500
Hamlet El Dorado region	Chidester Emerson Louann Banks	Occasional convenience stores 1–3 business types 2–5 retail establishments	less than 500 persons

Note: Places in parentheses not located in Arkansas.

small hamlets and villages serving the rural community. This hierarchy of central places forms a "stepped" population rank-size distribution which suggests that the central place model provides a conceptual basis for the urban rank-size model (White, 1978).

Changes in the Central Place Hierarchy

An interesting example of the ways in which both spatial and aspatial changes in society are reflected in the evolution of central places in a region is provided by Stabler (1987) with respect to Saskatchewan in Canada. This province has a population of 986,000, 37 percent of which is located in rural areas. The two largest urban areas are Regina and Saskatoon (see Table 1.1). There are many small places serving a highly dispersed agricultural community that is traditionally involved with wheat production and the cultivation of other commercial cereals. Apart from agricultural products, there are also considerable exports from the province of potash and nickel.

On the basis of detailed information pertaining to retail and wholesale activities (similar to the information in Figure 6.3), Stabler (1987) defines a sixfold hierarchy of central places ranging from "minimum convenience" (or hamlets) to "primary wholesale–retail" (or regional centers). Then on the basis of the number of activities and services in each town for 1961 and 1981, he places each of the 598 urban centers in one of the six categories for both time periods. He is, therefore, able to trace the change in the hierarchy of trade centers in Saskatchewan over a twenty-year time period. This sixfold hierarchy has been re-aggregated to a fourfold system in Table 6.7 for simplicity reasons and to be consistent with that developed previously for comparative purposes.

In general, it can be observed from Table 6.7 that: (1) the larger places experienced an increase in the amount of service provision, whereas the smaller places (villages and hamlets) remained static; and (2) there has been a downward shift of central places in the hierarchy—that is, some towns have become villages, and many villages have become hamlets. This type of downward shift had been noted in earlier studies (e.g., Hodge, 1965). Thus, while 288 places were classified as villages in 1961, 166 were so classified in 1981; but whereas 271 places were classified as hamlets in 1961, 400 were so classified in 1981.

Table 6.7

Change in the Number of Central Places at Different Levels in the Hierarchy: Saskatchewan, 1961–1981

LEVEL OF CENTER	NUMBER OF CENTRAL PLACES		AVERAGE POPULATION		AVERAGE NUMBER OF ESTABLISHMENTS	
	1961	1981	1961	1981	1961	1981
Regional	2	2	103,834	158,379	1,526	3,438
Town	37	30	4,784	6,680	123	243
Village	288	166	421	677	23	28
Hamlet	271	400	121	125	6	5

(*Source:* Data extracted and re-aggregated by author from Stabler (1987), Table 2, p. 230.)

The largest places, particularly Regina (the provincial capital) and Saskatoon on the transcontinental rail lines, have grown at the expense of the smaller places.

Reasons for Changes in the Hierarchy   The change in the central place hierarchy in Saskatchewan relates to changes in agricultural production, transportation, and shopping behavior during the 1961–1981 time period. These changes are examples of the general type of factors that influence positively or negatively all central place systems:

1. *Changes in market potential*: This is manifested in Saskatchewan by a continuation of the consolidation of farm holdings (thus increasing farm size) and perpetuation of the substitution of capital for labor that had commenced some decades previously (see Chapters 2 and 4). Thus, the agricultural labor force decreased considerably, and the population density in rural areas continued to diminish. As a consequence, the market demand was no longer there in 1981 to support the number of villages that had existed in 1961, and many business establishments closed shop.

2. *Transportation changes*: The hierarchy was also affected enormously by an extensive upgrading of the provincial road network which, along with the spread of multiple car ownership that occurred in the 1950s and 1960s, made it easier for people to travel farther for goods and services. This was aided by the closing down of rail branch lines and many local grain elevator terminals by the railway companies to reduce costs: The improved roads now made it possible for farmers to transport grain to larger terminals at greater distance. Local grain elevators, and the economic and social services associated with it, were the *raison d'être* for many villages. The development of the interstate highway system, and the associated "rationalization" (which means closing down) of extensive parts of the rail system in the United States has had a similar effect in cereal-producing areas.

3. *Changes in shopping behavior*: Road transport developments and the rapid dissemination of information through improved communication technologies (television, newspapers, telephones, and so forth) have facilitated *multipurpose shopping*. Multipurpose shopping involves an element of recreation: Individuals or families no longer go to their local central place for everyday convenience goods, but combine trips for these purchases with weekly trips to higher-order centers for specialized shopping and recreational or social purposes. These activities are now found in climate-controlled planned shopping malls in larger towns and regional centers. Rushton (1971), Eaton and Lipsey (1982), and Mulligan (1987) clearly demonstrate that the insertion of multipurpose shopping in the central place model results in the favoring of one-stop shopping facilities over smaller ones (see Chapter 10). Eaton and Lipsey (1982) in fact argue that multipurpose shopping provides the most persuasive behavioral rationale for the central place model which had been developed almost intuitively on the basis of empirical analysis by Christaller (1933; 1966).

4. *Changes in merchandising technology*: It should be noted that new retail formats located in "big boxes," introduced since the mid–1980s, are now having a huge impact on the hierarchy of service provision (Jones et al., 1994). "Category killers" (such as sports, or shoe marts) are leading to restructuring of the range of retail services provided in shopping centers, and severely impacting department stores which have been the main type of "anchors" for malls. Also, some of the new "big boxes" are pioneering new methods of inventory control and price targeting with low-cost delivery of a wide range of products. These new retail formats are invariably located in the larger urban markets, thus providing additional reasons for people to travel from smaller to larger places for their consumer needs.

THE INTERURBAN LOCATION OF PRODUCER SERVICES

It should be emphasized at this juncture that the threefold consumer-producer-public definition of tertiary activities used in this chapter is but one of several classifications that might be used. Martinelli (1991, pp. 17–18) suggests that the most useful classifications are according to:

1. The function or place of the service within the socioeconomic system—in which case services are grouped according to the "sphere" to which they belong (see Figure 1.7). For example, the "sphere of production" would involve manufacturing and all associated economic activities. The "sphere of exchange" would involve all those activities involved with the final demand of products.

2. The type of demand—which focuses on the market for services. In this case, services are producer services when they are inputs to other economic activities, and consumer services when they are consumed by individuals and households.

3. The form of the supply—who are the providers? Services may be supplied by the government, private businesses, nonprofit organizations, quasi-governmental organizations, self-employed persons, and so forth.

Although most published information provided by governmental statistical agencies tends to follow (3), a hybrid approach provides better coverage for governmental and public services. As a consequence, the approach in this chapter involves (2) and (3) (Daniels, 1991a).

Producer services, which essentially provide inputs to other production and consumption processes, are involved with the circulation (distribution) of capital, information, people, and goods. The distributive infrastructure includes: banking, transportation and related services, communications and information transfer, and real estate. Business services include: research and development, insurance, nonbanking financial services, legal services, accounting and financial advising, technical and professional services, advertising and public relations, marketing services, management and human relations consulting, word processing and data processing, security services, services to buildings, and equipment rental (Martinelli, 1991, Table 2.1). Most studies relating to producers services involve some of these categories, for example, Harrington and colleagues (1991) and Etlinger and Clay (1991), but few involve all.

One of the most all-embracing studies estimates the total value of international transactions of private services (including banking, securities, insurance carriers and agents, holding companies, business services, motion pictures, legal services, and professional services), all transportation and passenger fares, private education, and private health services (BEA, 1991). By 1990, this broad group of producer services contributed 8 percent of total United States exports, the magnitude of producer good exports ($31.8bn) almost equaled that of agricultural exports, and the volume of producer goods exports had increased during the 1980s at an annual rate of 13.3 percent. Producer goods industries are, therefore, not only growing rapidly in employment, they are also becoming an important element in international trade.

In this context of trade in producer services, Drennan (1991) demonstrates, with respect to a more narrowly defined set of producer services (i.e., the "private services" listed above), that the largest "gateway" CMSAs—New York, Los Angeles, Chicago, and San Francisco-Oakland—are the locations of businesses that generate one-quarter of total producer services earnings in the United States. Producer services are, therefore, significant city-supporting (or "export") components of many metropolitan economies (OhUallacháin and Reid, 1991; Nowlan, 1994) and they can play an important role in more general economic development (Hansen, 1990).

Factors Influencing the Concentration of Producer Services

The various studies of producer services in different parts of the world (Daniels and Moulaert, 1991; Daniels, 1993) have emphasized, with respect to urban systems, that producer services have (1) tended to concentrate in the major metropolitan centers, thus reinforcing their dominance and control over the rest of the economy; and (2) interaction between urban places does not appear to be organized in so clearly a hierarchical fashion with respect to producer services as it does with respect to consumer services (Illeris, 1991). The central place model, including its behavioral modifications, is therefore much less useful for analyzing the location of producer services. Coffey and Bailly (1991) suggest that the metropolitan concentration of producer services is the consequence of a need for "massive external economies of scale" (p. 108) with respect to requirements for appropriately skilled human resources and opportunities for backward and forward linkages.

Human Resources   Human resources are probably the principal factor with respect to the location of producer services, not only because they are more labor intensive than manufacturing, but also because of the labor qualifications requirement. On the one hand, the creative and technical aspect of producer services requires highly educated labor. On the other hand, the more routine office aspects require a pool of labor with lower qualifications. Metropolises, particularly those with a range of postsecondary educational facilities, provide the most appropriate variety of skilled pools of labor, particularly in the current era of multiple-breadwinner households. The probability of two persons obtaining satisfactory employment is far higher in large urban areas than smaller ones.

Backward and Forward Linkages Apart from appropriately skilled labor, producer services require access to the "sources or creators of knowledge, information and technical ability" (Coffey and Bailly, 1991, p. 108). These *backward linkages* include those of a face-to-face variety as well as those generated through networks to universities, complementary companies, research institutes, hardware producers, and so forth. Opportunities for an array of linkages of this type are greater in large metropolises, particularly those that have highly developed telecommunications services infrastructure involving widespread penetration of fiber optic cable, digital switching, television cabling, high-speed voice and data transmission lines, and "intelligent buildings"; that is, buildings constructed to cater for and buffer against modern electronics and with sufficient easy-access conduits for future recabling (Schmandt et al, 1990).

Forward linkages are important because a large portion (perhaps one-half) of the output of the producer services sector is purchased by other economic activities. These purchases, particularly in manufacturing, are controlled by headquarters (or regional) offices. Thus, as administrative and other activities tend to locate in metropolitan areas, and corporate headquarters tend to locate in the largest places (Table 2.1), producer service companies must locate accordingly to maximize business opportunities. The restructuring of business activity is reinforcing this trend as many activities that used to be conducted "in-house" are being let out or privatized.

The "World City" Hypothesis

The restructuring of manufacturing, the integration of the world's financial system (Warfe, 1989), and the concomitant growth of producer services industries, has led Friedmann (1986) to speculate that at one extreme there is developing a group of "world cities" that serve as the places for the control of capital and investments. These world cities tend to have a number of characteristics in common:

1. A world city is essentially a product of forces that emanate as much from international as national linkages. The internal structure of a world city, which arises from the divisions of its labor force, the economic activities it contains, and so forth (see Chapter 11), are, therefore, the product of supranational rather than national forces. This can obviously create tensions, particularly when supranational forces are not quite congruent with national forces (e.g., national financial or social policies).

2. These key centers serve as "basing points" not only for the pricing of capital (e.g., interest rates), but also for the ways in which production and distribution are organized. They are locations of the headquarters of major multinational corporations and the whole panoply of producer service enterprises

Table 6.8

World Cities

CORE COUNTRIES		SEMI-PERIPHERY COUNTRIES	
Primary Cities	Secondary Cities	Primary Cities	Secondary Cities
London	Brussels		
Paris	Milan		
Rotterdam	Vienna		
Frankfurt	Madrid		
Zurich			
			Johannesburg
New York	Toronto	São Paulo	Buenos Aires
Chicago	Miami		Rio de Janeiro
Los Angeles	Houston		Caracas
	San Francisco		Mexico City
Tokyo	Sydney	Singapore	Hong Kong
			Taipei
			Manila
			Bangkok
			Seoul

(*Source:* Friedmann (1986), p. 72.)

that set trends not only in the ways production is financed and organized, but also in products and tastes.

3. World cities, therefore, tend to be

- Major sites for the concentration and accumulation of capital

- Points of destination for large numbers of immigrants (both international and domestic) seeking employment in a dynamic economic environment

- Places of intense social and class-spatial polarization because of the large amounts of wealth that may be generated in, or associated with, jobs in international finance and business as compared with the domestic wage pattern in the local sector

- Metropolises where the social costs (e.g., for health, education, assimilation, multicultural harmony, unemployment, homelessness) that are incurred *may* exceed the fiscal capabilities of the local tax base.

The primary and secondary world cities as defined by Friedmann (1986, p. 72) are listed in Table 6.8. Of course, this list itself is conjectural, subsuming many debates revolving around the "core/semi-periphery/periphery distinction" (Simon, 1993), the perception of the role of various metropolises in different parts of the world, and the rapidity of change in South and Southeast Asia and the Chinese mainland. Nevertheless, one quarter of the cities listed are located in

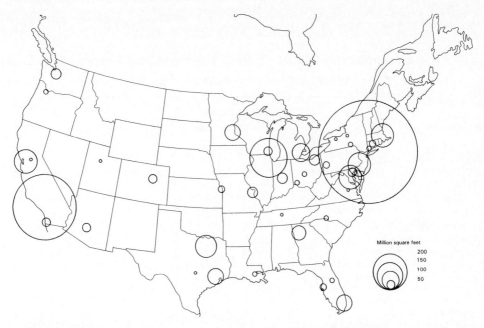

Figure 6.6 The location of primary occupied office space in the United States, 1990.
(*Source:* Sui and Wheeler, 1993, Figure 2, p. 37.)

North America, and it is these that serve as the major links between the North American urban system and the world economy.

The pivotal role of the United States financial and economic system in the world's economy is having a powerful impact on the nature of its urban system (Warf and Erickson, 1996). The preeminent "world city" is the New York CMSA, which, on the basis of a variety of indicators (headquarters offices, international finance, stock exchange value and volume, bank income and assets) ranks as "the single most important source of economic power and decision-making in the United States and, indeed, the world" (Wheeler, 1990, p. 379). The dominance of the New York metropolitan area in office activities is indicated in Figure 6.6, which shows the amount of office space in the leading metropolitan areas in the United States. New York has almost twice the occupied office space of Los Angeles, which is the next largest office center.

The Concentration of Producer Services

Although producer services are concentrated in the largest urban areas, there is a slow spread of these services to smaller places. OhUallacháin and Reid (1991) present data for business and personal services for 1976 and 1986 by metropolitan size class (Table 6.9). The information shows that while there has been a considerable employment increase in these specific categories of producer service activities, and the largest metropolises (those with greater than 1 million

Table 6.9

The Geographical Distribution of Business and Professional Services Employment[a] by Size of MSA Population

Metropolitan Category	EMPLOYMENT (THOUSANDS)		PERCENT SHARE	
	1976	1986	1976	1986
Total Metropolitan	2,665.9	5,641.9	96.4	93.7
Large Metropolitan[b]	1,916.4	3,947.1	69.3	65.5
Intermediate Metropolitan[c]	734.7	1,661.0	26.6	27.6
Small Metropolitan[d]	14.8	33.8	0.5	0.6
Nonmetropolitan	100.6	380.8	3.6	6.3
U.S. Total	2,766.5	6,022.7	100.0	100.0

[a]Standard industrial classification (SIC) codes 73 and 89.

[b]Population of metropolitan area greater than 1 million.

[c]Population of metropolitan areas between 100,000 and 1 million.

[d]Population less than 100,000.

(*Source:* OhUallacháin and Reid (1991), Table 1, p. 261.)

population) retain a share of national employment greater than their 1986 population share (65.5 percent of business and professional employment compared with 53 percent of the national population), nevertheless the intermediate-sized and nonmetropolitan areas have been increasing their share. The growth of employment in intermediate and small towns may well be related to the proliferation of small private firms in the new economy which appear to be distributed ubiquitously throughout the urban hierarchy (Lyons, 1995). The non-metropolitan growth of producer services reflects the outward spread of economic activities in large CMSAs and MSAs to adjacent nonmetropolitan counties (Cook, 1993).

More than 50 percent of the employment in business and professional services in the United States is concentrated in fifteen large metropolitan areas (Figure 6.7). These are located either along the northeastern seaboard from Boston to Washington, D.C., (Gottman's, 1961, "megalopolis"), the lower Great Lakes area, and in California. The others outside these macro-urban regions are in Dallas, Houston, and Atlanta (Hick and Nivin, 1996). The metropolitan areas that have experienced a growth rate in this type of employment greater than the national average are either within the regions dominated by New York (Newark, Bridgeport, Long Island) and Los Angeles (Anaheim), or have significant subnational regions in which they dominate (Dallas, Minneapolis), or have particular industries of national and international importance, such as Detroit (auto industry) and Washington, D.C., (political power center). Coffey and

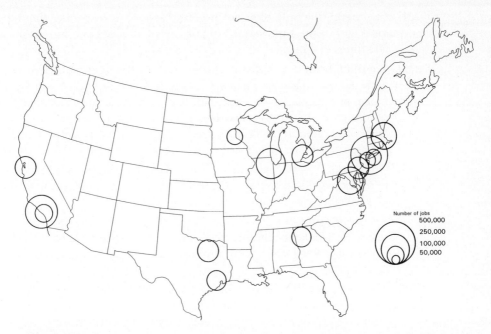

Figure 6.7 The fifteen metropolitan areas in which is located 50 percent of the employment in business and professional services in the United States, 1990.

Polèse (1987) note that producer services are most sensitive to the needs of the firms they serve. So as a consequence they tend to "move up" the urban hierarchy in their location, following the locational pull of headquarters offices.

Trade in Producer Services

Coffey and Polèse (1987), however, postulate that whereas some types of producer services (such as business and professional services) may locate in a hierarchical fashion and distribute their activity (trade) through the urban system following a hierarchical distribution path, others (such as computer engineering) may follow a distinctly nonhierarchical pattern and trade across the hierarchy. Esparza and Krmenec (1994) examined this proposition in a detailed study of the trade patterns for a sample of Chicago-based firms in a range of producer goods industries: accounting (44 firms), advertising (39), computer and data processing (50), engineering and architecture (42), and management and public relations (84).

The findings of their study are important when viewed in the context of the economies of scale theoretical argument outlined earlier.

1. In general, the bulk of sales and purchases with respect to these 259 firms is local within the Midwest. Fifty percent of all sales and purchases are within 500 miles of Chicago.

2. There is, however, a distinct difference in the geographical extent of trade between purchases and sales. Chicago firms tend, on the whole, to buy (i.e., make purchases) from local suppliers. This is quite consistent with economic base theory discussed in Chapter 2 and the central place model—high-order centers, which invariably have large populations, exhibit a high degree of self-sufficiency because they have the services they need.

3. On the other hand, sales are more widely distributed, with some sales going to other "world cities," and some to lower-order centers in the Midwest service hierarchy of which Chicago is the apex.

4. The proportion of sales that are local (i.e., within the Chicago metropolitan region) varies considerably between business types—accounting has 89 percent of its sales local, whereas management and public relations has 61 percent local.

5. The degree of linkage of sales to other world cities is inversely related to local sales—accounting is least world city oriented, whereas management and public relations firms have sales to firms in Los Angeles, New York, San Francisco, London, Mexico City, Paris, São Paulo, and Tokyo.

Trade in producer service industries in the United States, therefore, flows through a two-level organization of cities. Some types of producer services (management and public relations, engineering and architecture, computer and data processing) sell locally and to other firms in world cities, while others (accounting and advertising) appear to have the bulk of their sales local. This latter group, therefore, tends to interact with other firms in a hierarchical manner through the United States urban system, while the other groups trade within the system and across it both nationally and internationally. This is, of course, to a certain extent related to the specific location of Chicago with respect to United States manufacturing. For example, advertising firms have considerable sales to local manufacturing industries. The pattern of flows does, however, reinforce the growing importance of the producer services sector in international trade, and with respect to the economic base of metropolitan areas.

THE INTERURBAN LOCATION OF GOVERNMENTAL AND PUBLIC SERVICES

It has been noted previously (Table 6.1) that governmental and public services in the United States employ one-quarter of the total labor force, and that this share is increasing. This employment category involves a similar proportion (23.3 percent) of the 1991 employed labor force in Canada. The main types of employment involved are:

Governmental services: Federal, state or provincial, and local (including special districts).

Educational services: Primary, junior, secondary, postsecondary, and technical/professional.

Health and social services: Including all types of social agencies, hospitals, health care, and therapies

These three types involve roughly similar shares of the labor force. In Canada, for example, education involves 6.8 percent of the employed labor force, government-related activities 7.8 percent, and health and social services 8.6 percent.

The Production and Consumption of Governmental and Public Services

The interurban location of governmental and public services is directly related to the degree to which production and consumption are locationally and temporarily inseparable (OhUallacháin and Reid, 1993, p. 253). For example, health care services, and often social services, have to be provided where the people are, and have to be available when they are needed. The only exception would be for certain specialized hospital services (e.g., heart by-pass operations) which may require particular expertise and recovery facilities. Likewise, equal access to educational facilities to the secondary level are required (by law) to be available in the immediate home vicinity (or "immediate" by school bus) for children and young adults. Postsecondary educational facilities, on the other hand, although they may be available locally are not required to be, and have often been located by state/provincial governments, or philanthropists in places some distance from major population centers (e.g., State College, Pennsylvania; Ithaca, New York; or Sackville, New Brunswick, Canada). Some governmental services have to be available locally to serve the immediate needs of the population (e.g., post offices, welfare services), while others of a policy or regulative nature may well be located in national or state/provincial capitals.

Thus, while the aggregate supply of governmental and public services tends to be distributed according to the demand (i.e., population), there are some distinct subtleties that occur when production and consumption can be locationally separated—and the bringing together of the two involves the movement of people, or the transmission (by mail, telephone/fax, or electronic) of information. Figure 6.8 shows the relationship between employment in governmental and public services and metropolitan population for twenty-eight metropolitan areas with a population greater than 100,000 plus two urban areas with a population of less than 100,000 that are provincial capitals (data from Statistics Canada, 1993). The size relationship is quite strong, endorsing the strong locational production–consumption link.

Variations in the Production–Consumption Link There are, however, a number of urban areas which provide services where the immediate production–consumption link is not as strong. These are identified by *concentration indices* (sometimes called *location quotients*) in Table 6.10. Concentration indices express the ratio between the actual proportion of national employment in a

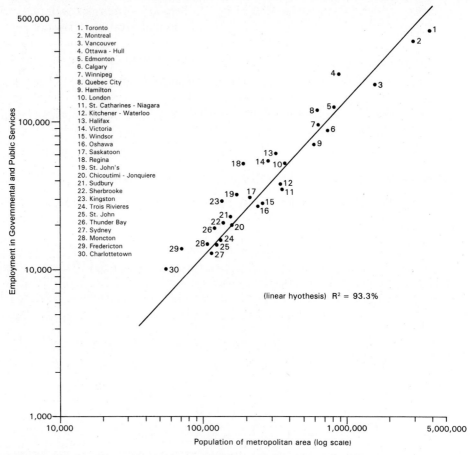

Figure 6.8 The relationship between employment in governmental and public services and metropolitan population, Canada, 1991.

category existing in a metropolitan area compared with the proportion that would be expected to exist on the basis of that urban area's share of the national population. Thus, in the case of Regina (population 191,692) its share of national employment in governmental services is 11,470/1,111,385 = 1.03 percent, whereas its share of the country's population is 191,692/27,296,859 = 0.70 percent. Thus the concentration index for Regina in governmental services is 1.03/0.70 = 1.47. Regina has, in effect, 47 percent more employment in governmental services than would be expected on the basis of its population size alone. If the index is greater than 1.0, there is a concentration of the activity to which the index relates in that metropolitan area. If the index is less than 1.0, the urban area has less employment in that category that one might expect on the basis of population size alone.

The ten metropolises with the largest indices in each of the three categories of governmental and public service employment are listed in Table 6.10. At the outset, it must be noted that the largest metropolises, those with 1 million or

Table 6.10

Concentration Indices for Government, Education, and Health Care and Social Services for Major Metropolitan Areas in Canada, 1991

GOVERNMENT SERVICES		EDUCATION SERVICES		HEALTH CARE AND SOCIAL SERVICES	
Metropolis	Index	Metropolis	Index	Metropolis	Index
Ottawa-Hull**	3.36	Kingston	1.94	Sherbrooke	1.51
Fredericton*	2.46	Fredericton	1.54	Kingston	1.50
Halifax*	2.27	Saskatoon	1.51	London	1.46
Victoria*	2.11	Sherbrooke	1.51	St. John's	1.40
Quebec City*	2.08	St. John's	1.36	Victoria	1.39
Charlottetown*	1.99	Charlottetown	1.30	Thunder Bay	1.37
St. John's*	1.86	Sudbury	1.29	Quebec City	1.34
Kingston	1.86	Kitchener-Waterloo	1.27	Halifax	1.32
Regina*	1.47	London	1.26	Saskatoon	1.31
Thunder Bay	1.35	Ottawa-Hull⎫ Quebec City⎭	1.19	Winnipeg⎫ Regina ⎭	1.29

(*Source:* Indices calculated by author from Statistics Canada (1993).)
**National capital.
*Provincial capitals

more population (Toronto, Montreal, Vancouver), do not appear in the table. The economies of the largest places are usually quite diversified. So the indices for the various employment categories tend to be around unity for these metropolises. Nevertheless, some interesting variations in the generally close production–consumption link can be noted:

1. Governmental services are more highly concentrated in a few places than either education or health care and social services. The indices for the metropolises with the greatest concentrations are far greater in the governmental services category than in the other two. This is because government employment, although spread across the country, tends to be concentrated in the national capital (Ottawa-Hull) and the provincial capitals. Seven of the ten provincial capitals are represented as having high concentrations—the only ones missing are Edmonton (1.23), Winnipeg (1.15), and Toronto (0.76). Although Toronto has the second-largest amount of employment in governmental services (121,175 persons) after Ottawa-Hull (126,025 persons) in the country, this figure is not as large as might be expected on the basis of its population size and role as a provincial capital. Toronto has a highly diversified economy with, for example, 18 percent of the manufacturing employment in the country (a concentration index of 1.30 for manufacturing employment). Thus, as the locational decision with respect to national and provincial/state capitals in North America tends to be a historical

artifact (i.e., some person or group of elites made the locational decision many years ago), the distribution of governmental services is a historical artifact also, particularly for those services for which the production–consumption link is quite elastic.

2. While educational services appear to be much more closely linked to the distribution of population than governmental services, a few places do have much greater employment in education services than might be expected based on their size alone. This appears to be largely related to the placement by governments of postsecondary institutions. Each of the eleven metropolises with high concentrations of employment in educational services has, at least two postsecondary institutions (universities or community colleges). One of the best ways for a community to foster local development is for it to garner a viable postsecondary institution.

3. Similarly, although health care and social service employment have a close consumption–production link, a few places have greater concentrations of employment in this category than might be expected based on population size. Each of the metropolises on the list have postsecondary institutions, which include medicine or professional schools in the health sciences (e.g., nursing, rehabilitation therapy). The only exception is Thunder Bay, which has hospital facilities to serve an extensive area of northwestern Ontario.

A few places appear to be extraordinarily "well blessed" with much greater employment than expected in all three categories (Quebec City, Kingston) and the national and provincial capitals appear in at least two categories. The phrase "well blessed" is used because these types of service employment appear to be remarkably resilient to business recessions, though they may be less resilient during the current era of governmental downsizing (Kearns, 1995). Kingston is an interesting case because it has not only a number of postsecondary institutions and teaching hospitals, but it is also the location of an army base, federal and provincial correctional service facilities (prisons), and the administrative office for the Ontario Hospitalization Insurance Programme (OHIP)—most of these located in the area as a result of political decisions.

CONCLUSION

The distribution of service activities among urban areas is, therefore, the consequence of a number of locational factors. Consumer services are generally located in a hierarchical fashion among urban areas, giving rise to a hierarchy of central places with respect to these types of consumer activities. Low-order business types are located ubiquitously among most central places, whereas the higher-order business types, particularly those involving wholesale activities are located in the higher-order central places, which are usually the largest cities serving the most extensive regions. The highest-order central places not only have the highest-order business types, but also the lower-order business types as well. The array and quality of business types available in central places is related to consumer demand. Thus, in the illustrative example chosen (southern Arkansas), where per capita income is among the lowest in North America, service provision is related strongly to local needs and affordability. Change in the hierarchy and number of central places is related primarily to transport improvements, increasing wealth, and the evolution of consumer behavior.

Producer services that are purchased by other economic activities, are located in the larger metropolises (particularly "gateway" and "world cities") because of the need for external economies of scale. These external economies relate to the need for high-quality well-trained human resources, factors of production that take time to develop, are extremely mobile, and are attracted to well-serviced locations with good amenities and access to the nation and world beyond. Also emphasized as important locational requirements are forward and backward linkages—forward linkages to markets, and backward linkages to the sources and creators of knowledge and information. These locational principles including the need by most two-adult person households for joint employment opportunities, virtually dictates location in the vicinity of large metropolises, even in these days of extraordinarily flexible electronic communication.

The production–consumption link is emphasized particularly with respect to the location of governmental and public services. Even though these services, as is illustrated by the Canadian example, are distributed much in accordance with the immediate needs of consumers, there are clear variations where the production–consumption link is less strong. These variations occur particularly with respect to earlier political decisions concerning the location of national and state/provincial capitals, the seats of county governments and local county court houses, as well as other aspects of education, government administration, and national defense.

Part **2**

Internal Structure of Urban Areas

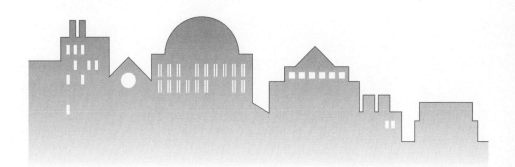

Chapter
7

Internal Structure and Deconcentration

Although the vast size, spaciousness, and grandeur of the various land-scapes of North America, stretching, in a paraphrase of the Woody Guthrie classic, from the Atlantic to the Pacific and the southern Gulf to the Arctic shores, magnetically attract attention, the economic and cultural life of the population is predominantly urban centered. These are the hothouses which foment creativity and industry, ferment change, and fragment daily life. They are an amalgam of the striving of immigrants, the hopes of young people and families, the desolation of the poor, the despair of the homeless, the struggles of mortgaged-to-the-hilt suburban households, and the complacency of the well established. The characteristic features of these urban areas—rectangular grids, superimposed auto-choked highways, suburban sprawl, edge cities, frag-ile downtown cores, declining inner cities, high-rise buildings, new and per-sistent ethnic communities, lifestyle neighborhoods—provide, in the aggre-gate, the elements that distinguish North American urban areas from those in other parts of the world (Rybczynski, 1995). This chapter describes the ways in which the locational attributes of these characteristics have developed and accumulated.

MAJOR PERIODS OF ACCUMULATION OF URBAN INFRASTRUCTURE

The spatial growth of urban areas occurs in surges (Olson, 1979). There are times when the pace of development is rapid, and there are others when it is less rapid or even stagnant. These surges occur in about twenty-year cycles (the shorter Kuznet, or building cycles) that are embedded within the Kondratieff long waves. There appear to be two or three peaks per long wave: One follows the onset of growth in a new era, preceded by a contraction in accordance with a business recession, followed by one or two other surges of growth which can lead to overspeculation and overbuilding toward the end of the era. These building cycles are remarkably persistent (Grenadier, 1995), as is indicated in the data (Figure 7.1) on the number of housing starts for Baltimore since 1775 and Montreal since 1851. Both of these rather similar cities exhibit the same number of peaks and troughs (eight) for the time period since 1851.

The information on housing starts for Baltimore is particularly interesting because of the period of time that is covered by the data. In terms of volume of housing starts, and hence pace of growth, the four eras identified in Figure 3.2 and Table 3.1 are quite discernible in Baltimore (as are the later three eras for Montreal):

1. *Mercantile* towns and cities predominated until about 1845, characterized, in the case of Baltimore, by fewer than 800 housing starts per year.

2. The *classic industrial* urban area is generally typical of the period from 1845 to 1895, when industrialization and immigration generated significant urban expansion. In Baltimore the general level of housing starts increased to between 1,000 and 2,000 per year. Montreal, on the other hand, reflecting the limitations of a mercantile and staples economy, did not incur such rapid growth (even though it was the leading metropolis in Canada at that time, see Chapter 4). Its housing starts generally numbered 700 to 800 per year.

3. The *metropolitan* era extended from 1895 to 1945, when there are two clear peaks: One at the beginning of the twentieth century prior to World War I, and the other during the Roaring Twenties prior to the Great Depression and World War II. During these two housing construction booms, starts in both Baltimore and Montreal ranged from 2,000 to 5,000 per year, reflecting high levels of immigration, rapid population growth, and industrialization.

4. The era of greatest *suburban* growth and deconcentration of both people and economic activities occurred between 1945 and the last decade of the twentieth century. The term *suburban* carries the same implication in this era as it had in the more limited periods of deconcentration in the previous eras: the population is nonfarm residential; the households are primarily middle or upper income; there is a considerable distance between residences and workplaces; and the housing densities are relatively low compared with the older parts of the central cities (Jackson, 1985, p.11). The rapid suburbanization is indicated by the quantum leap in housing starts in the extended

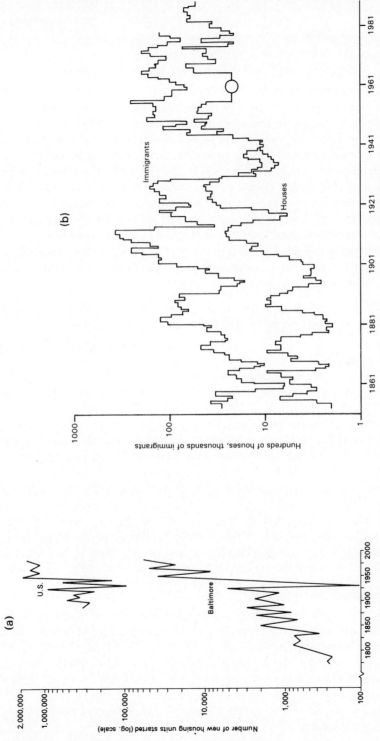

Figure 7.1 Number of new housing units starts in (a) the United States, 1885–1985, and Baltimore, 1775–1980; and (b) Montreal, 1851–1986.

(*Sources:* Bureau of the Census, 1975, 1993; Olson, 1979, pp. 558–559; Olson and Kobayashi, 1993, p. 139.)

Baltimore CMA to between 10,000 and 50,000 per year. The data for Montreal do not reflect much of the suburbanization that took place because the building permit data relate to a more limited definition of its region (Ile de Montreal). Suburbanization has been as extensive around Montreal as elsewhere in North America (Reid and Yeates, 1991).

The important feature of the information in Figure 4.2 is the difference in the number of housing starts between the eras, reflecting the quantum variation between the eras in the magnitudes of urban growth.

Furthermore, as each era is influenced by a different combination of social, economic, transportation, governmental, and cultural (style, design, etc.) forces, urban areas become accretions reflecting the various strengths of these factors at different times. For example, Art Deco style and decoration, which came to the fore in the 1920s and 1930s with the flourishing of modernism in architecture and the increased pace of life exemplified by the automobile and the transcontinental express train, can be seen in vestige form in certain buildings and districts (such as the Chrysler Building in Manhattan and South Miami Beach) today. Also, as new construction invariably (but not always) takes place on new land, these reflections occur most clearly at the fringe of urban areas during each era.

Urban Areas During the Mercantile Era

During the frontier mercantile era, urban areas were, on the whole, quite small and compact. Only toward the end of the era, when industrialization became more prevalent, did some places become quite large. New York City reached a population of 100,000 by 1820, Philadelphia by 1830, and Baltimore and New Orleans by 1840. The rapidity of growth in New York City was such that in 1811 the New York Commissioners Plan recommended a grid layout for newly developing areas beyond the original Dutch and English colonial settlements in lower Manhattan. This expedient street pattern became the model for the physical organization of most urban areas in North America (Washington, D.C., being a notable variant). The internal structure of most places reflected the dominant mercantile basis: In "gateway" cities the port, shipbuilding and repair yards, and places of business of the financial institutions, merchants, and shopkeepers, were all clustered close together. Urban areas inland, which existed to link with the agricultural and resource-based hinterland, had a similar nucleation of service activities around which residences clustered in a compact fashion.

The reason for this compactness was that the costs of transportation for people and goods were very high. The only sources of power for transport were either people or horses. Thus, prior to the 1840s, the geographic extent of North American urban areas was controlled primarily by the distance people could walk to work, shops, and social-recreational activities. As a consequence, most people tended to live either close to or within the same building as their place of work and they did not move around urban areas nearly as much as people do today.

Warner (1962, p. 19) notes that in Boston in 1850 (when it reached a population of 100,000) "streets of the well-to-do lay hard by workers' barracks and tenements of the poor; many artisans kept shop and home in the same building or street; and factories, wharves, and offices were but a few blocks from middle-class homes." Some spatial separation of classes did, of course, exist, for the more wealthy were able to afford carriages and reside in more salubrious surroundings farther away from the sounds and smells of commerce and the workshops. For nearly everybody, however, a lack of cheap intraurban transport bound urban areas together in a compact unit, and as few people traveled frequently between districts within larger urban areas each district formed its own identity.

Urban Areas During the Early Industrial Era

During the era of early industrial capitalism industrial cities became characteristic in North America. Many urban areas during this period were based primarily (though by no means completely) on manufacturing and related activities such as mining and refining. The rhythm of urban life, therefore, became determined by the location of the factories, the organization of the workday, and the social relations that developed as society became influenced by the owner-manager-professional-worker distinctions that became more prominent with increased specialization. The internal structure of urban areas, therefore, reflects the imperatives and outcomes of these rhythms and social relations through: (1) the need to provide more space for the population and economic activities of the rapidly growing urban areas; (2) the spatial separation of economic activities; and (3) the spatial separation of social classes.

Urban Space and Transport Innovations   The imperative to provide more urban space was not met well for much of the nineteenth century. The only way in which urban areas could be made less compact was with improvements to intraurban transport: Movement had to be easier and cheaper, and transport more accessible. Unfortunately, the chief modes of intraurban transport that were available for much of the period—the horse-car and the horse-drawn omnibus on rail lines—were not terribly satisfactory.

The horse-drawn omnibus was first used in North America in New York City about 1830. Shortly thereafter, the first system on street rail lines began regular operation in 1831. The economic success of this system led to the use of the horse-car in New Orleans, Chicago, Baltimore, St. Louis, Cincinnati, Pittsburgh, and Newark by 1860. By 1885, the peak of the horse-car era was reached for horse power was about to be replaced in public transport by the electric streetcar. At that time, 593 horse-cars were licensed to operate on 27 routes in New York City. Although the average speed of a horse-car (5 miles per hour) was not greatly in excess of normal walking speed, the innovation saved human energy and provided a way for those who could afford the fares to live farther away from their place of work. Urban areas were able to expand, but not that much, as can be seen from the case of Chicago in Figure 7.2. The mercantile

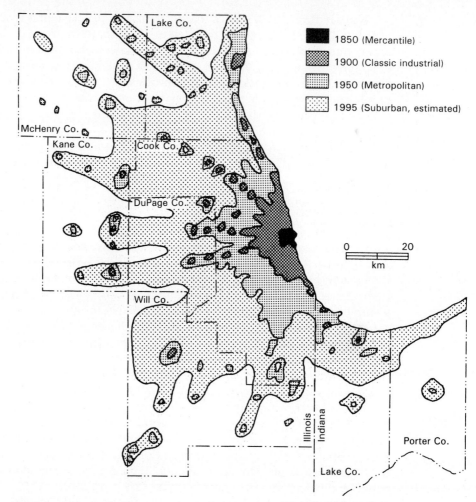

Figure 7.2 The spread of urban development in the Chicago region, 1850–1995.

(*Source:* Modified from Cutler, 1982; recent additions by author; the area demarcated is less than the entire CMA.)

City of Chicago in 1850 had a population of about 30,000, but in 1890 the population was 1.1 million and the urbanized area was extremely congested.

There were attempts to use the steam engine in intraurban transport, but it proved inappropriate for it was cumbersome, noisy, and dirty and, above all, tended to frighten horses, causing pandemonium and panic when emitting blasts of steam. Nevertheless, the steam engine did have a considerable influence on the form and growth of the city because land along the rail lines within the cities, and around classification yards, was developed by railroad companies as industrial sites. Furthermore, the rail lines were used for commuter railroads for those who could afford this mode of transport on a daily basis. For example, the prosperous merchants and new wealthy industrialists of Boston had generated sufficient demand to warrant 118 scheduled daily commuter trains as early as 1848. These commuters lived in *exurbs*, or dormitory settlements beyond and distinct from the immediate environs of the urban area but tied to it like beads on a string (Figure 7.2) by the railroad line.

The Spatial Separation of Economic Activities One of the outcomes of the greater specialization of economic activities that occurred with industrialization is a clearer spatial distinction of land uses within North American urban areas. There were, perhaps, two main reasons for this: transport improvements and the gradual shift from small workshops to larger-scale factory operations. Manufacturing industries congregated along transportation routes—first the waterways and later along the side of the railroad tracks.

For example, in 1865 a consortium of nine railroad companies and a number of meatpacking companies (such as Armour and Swift) consolidated numerous small stockyards that were distributed around Chicago into one huge operation—the Union Stock Yards—which eventually occupied one square mile south of Thirty-ninth Street and west of Halsted Street. This consolidation spawned a vast number of other industries at the periphery of the Yards which used as inputs by-products of the meatpacking industry. At its peak, in the early 1920s, the Yards employed more than 30,000 people and had given rise to large working-class districts, which often suffered from horrifying unsanitary and stench-filled conditions, epitomized in the neighborhood known as the "Back of the Yards" and immortalized in Upton Sinclair's classic urban novel *The Jungle* (1906; Jablonsky, 1993).

With manufacturing becoming increasingly concentrated along transport facilities, the commercial, retail, and financial center of the city gradually became more focused on service activities. Retail, financial, wholesale trade and entertainment facilities concentrated in a central business district (CBD), and began to cluster in subareas within the CBD. Also, as urban areas increased in population, neighborhoods of different social classes and ethnicity became much more noticeable.

The Spatial Separation of Social Classes The innovation of the horse-car and horse-drawn omnibus, the growing population, the increase in the number of people in different occupations and with quite different levels of income, and the establishment of distinct commercial and industrial areas, gave rise to a more distinct spatial separation of income-based social classes within urban areas. Low-income groups tended to locate or be located by industrialists close to their place of work or in less environmentally attractive, cheap, and shoddy housing. In contrast, some of the households in the rapidly growing middle-income segment of the population began to seek more attractive locations away from the muck, dirt, and squalor of the industrial districts, and noise of commercial areas. The income-based class divisions of society became replicated quite clearly in a spatial separation of social groups.

This is illustrated in a generalized diagram of land uses for Toronto in 1885 when the city had reached a population of about 150,000 (Figure 7.3). The first-class residence homes of the upper-income groups are located at the periphery

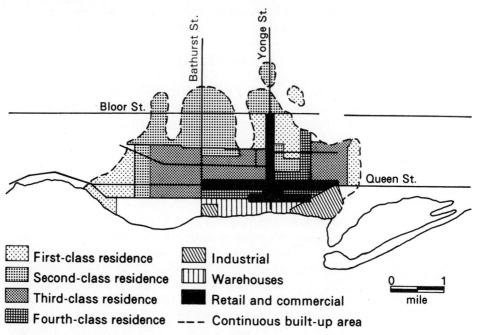

Figure 7.3 A generalized diagram of land uses in Toronto in 1885.
(*Source:* Redrawn from Baine and McMurray, 1977, p. 31.)

of the city, whereas the fourth-class residences of the lowest-income workers are located in the center of the city around the business, warehouse, and industrial districts. This concentric pattern of the location of classes and income groups is a classic form for industrial cities. Furthermore, it is noticeable that the first- and second-class residences, while on the periphery, are also adjacent to each other in a wedge, or sector pattern.

One feature of these neighborhoods that is particularly important in North American urban areas is that the class distinctions are frequently mixed in with ethnic, racial, or linguistic differences as well (Ward, 1971). For example, in the case of Toronto, the area of fourth-class residences in Figure 7.3 was predominantly Irish-Catholic, whereas the first- and second-class residential areas were predominantly English. The area around the Union Stock Yards in Chicago, and particularly the Back of the Yards, consisted of thousands of recent immigrants from eastern Europe, and each group often formed its own local cultural and religious community.

Metropolitanism, 1895–1945

The era from 1895 to 1945 was one in which a number of large metropolises emerged and epitomized urban organization in North America. During this period, the spatial separation of economic activities and socioeconomic groups becomes more distinct and more complex, urban areas become more spread out

(Figure 7.2), and suburbanization becomes more pervasive for the burgeoning middle class (Jackson, 1985). By 1890, twenty-six cities in North America had a population in excess of 100,000, of which New York City (2.5 million, including Manhattan, Brooklyn, Queens, Richmond, and the Bronx), Chicago (1.1 million), and Philadelphia (1 million) were the largest. However, at the same time as the major cities of North America were creating generally higher average levels of prosperity, they were also becoming appallingly congested and overcrowded, because rapid industrialization had exacerbated inequalities and spatial differentiation.

Population Decentralization   Adna Weber (1899; 1929), writing in the last decade of the nineteenth century, noted with alarm the high levels of population and economic congestion in urban areas in the Western World, particularly New York City. Seven wards in that city in 1880 had population densities in excess of 250 persons per acre (central city areas today average about 30 persons per acre or 19,000 persons per square mile), housed in walk-up tenements (Weber, 1899, p. 460). For most people it was still impossible to live too far away from their place of work. So home, community, and work were strongly interrelated (Hiebert, 1993). Most people resided in close proximity to the economic activities in and around the central business district (CBD), and around other major industrial centers within the city. Weber's solution to this unhealthy congestion was to advocate decongestion. However, this was not possible for the bulk of the population until decentralization (outward spread) became feasible with the advent of means of intraurban transportation that were faster and cheaper than the horse-car and omnibus—first, the electric streetcar and rapid transit, and then the automobile.

The Streetcar   The electric streetcar (or trolley) was first used successfully on the Richmond Union Passenger Railway in Virginia in 1888, and was subsequently adapted by Whitney in Boston in 1889. The immensely superior and efficient operation of the streetcar in Boston, where fares dropped from a dime on horse-cars to a nickel on streetcars, resulted in the building or ordering of more than 200 systems in North American cities in the following three years. The Metropolitan Street Railway transit company in New York City was immensely profitable, generating, between 1893 and 1902, fortunes totaling $100 million for its owners (Jackson, 1985, p.109). In 1901, there were 15,000 miles of electric streetcar rail lines in the United States. Although the systems required considerable capital outlays, they were relatively easy to construct as the rail lines were laid on the surface, and the electric wires could be slung overhead. There were no noise and visual-environmental impact studies required in those laissez-faire days! Thus, it is not surprising that the trolley became the first form of public transportation to be used on a large scale by employees going to and from work. Thus, the spatial patterns that developed became some of the major characteristics of North American cities in the twentieth century.

The electric streetcar made it possible for the middle-income population to spread out along major radials from the center of the city, and for residential densities to decrease. The spread that occurred was quite dramatic, as can be seen for Chicago between 1875 and 1925 when its population surpassed 3 million (Figure 7.3). For the population to spread out, however, land had to be provided and serviced and homes had to be built. Because the population of most urban areas in North America was increasing rapidly as a result of rural-to-urban migration, high rates of net natural increase, and massive immigration in the decades around the turn of the century (Ward, 1989), the innovation of the trolley is associated with great increases in land sales, values, and speculation.

Warner's (1962) seminal analysis of the interrelationship between the spread of the electric streetcar on the then suburban developments of Roxbury, West Roxbury, and Dorchester (in the southern sector of, and immediately adjacent to, the old central city of Boston) provides a detailed discussion of the effect of this transport innovation. Most of the land in these three independent towns, which became suburbs in the late nineteenth century, was beyond the zone of dense settlement in the pedestrian city of Boston prior to 1850. The horse-car prompted the beginnings of suburbanization in these towns, so that in 1870 their population had reached 60,000. By 1900, their combined population had boomed to 227,000. This increase in population was housed in 22,500 new buildings, of which 53.3 percent were single-family houses, 26.6 percent were two-family houses, and the remainder were multi-family dwellings. This extensive development of single- and two-family houses for families in the middle income range heralded the first of a number of phases of *mass* suburbanization in North America, which incrementally increased in penetration through nearly all income ranges in the twentieth century. Such developments could take place only through transport improvements. In this wave, the improvements involved the expansion of streetcar lines and associated land development.

Throughout much of urban North America the expansion of "streetcar suburbs" involved close cooperation between developer-speculators and transit companies, and lines were strategically placed and threaded through pre-planned subdivisions (Nelson, 1983, pp. 261–273). It was the developer, therefore, who determined the land use. If the developer thought an area of land was capable of supporting high-value housing, then the land was planned accordingly with large lots. The owners of the transit companies learned fairly quickly that the greatest profits could be achieved through "knowledgeable" speculation (which is no speculation at all). Farmland at the edge of cities about to be served by streetcar lines became instantly valuable for urban development. Hence emerged a close collaboration, often involving common ownership, between transit companies and real estate developers. Transit barons, such as Huntington in Los Angeles, who owned the Pacific Electric Railway Company with lines throughout the basin from Santa Monica to San Bernardino and from Pasadena to Balboa (Jackson, 1985, p.122), were really more interested in selling land. The streetcar lines were merely a means to that end.

Although the streetcar systems played an important role in facilitating the suburbanization of the population, it was the booming economy and rapid population growth along with the pressure for decongestion in North American cities at the turn of the century that stimulated the outward expansion. Speculators and investors were the agents providing the fixed capital enabling the deconcentration, whereas the streetcar network in cities, which often extended outwards beyond the old horse-car lines, further channeled and reinforced these preexisting patterns of growth. Given the enormous pressure for deconcentration, the transport and land developers could cloak themselves with an aura of moral purpose as they argued they were assisting the removal of people from unhealthy high density environments to low-density "garden" suburbs (Whitt and Yago, 1985).

Rapid Transit   Rapid transit is a cross between the streetcar and the railroad, involving the application of the electric motor to a system involving heavy rails placed in isolated structures on the surface, subsurface (subways), or above the ground (elevated). Thus, rapid transit facilities proceed at a faster pace (25–35 miles per hour) than streetcars (15–18 miles per hour) and can carry many more passengers. Because both the streetcar and rapid transit involve use of the electric motor, both modes were applied to urban transport at about the same time during the 1890s. However, as rapid transit involves much greater fixed costs in construction, its application was slower and more deliberate than the streetcar, with the first subway in North America being opened in Boston in 1897 and the first section of the New York subway in 1904. Nevertheless, the development of rapid transit had a great impact on the form and structure of urban areas—particularly with respect to radial (or sectoral) growth and through the creation of centers of economic activity around local transit stations.

Rapid transit is economically efficient only when it is moving large numbers of people daily. Consequently, rapid transit systems were constructed to connect the CBD, the chief place of employment and commercial activity, with the principal residential areas of urban population. In most cases, the direction of growth had already been defined by the horse-car/omnibus and the streetcar. For example, in Chicago rapid transit routes on elevated lines (the El) followed routes originally established by horse-drawn omnibuses and extended by the streetcar system. The expense of the systems was (and still is) such that only the most-traveled routes in certain sectors could be served. Hence, rapid transit accentuated the radial growth and "fingers-like" spread of the urban area (Figure 7.2).

Apart from further accentuating sectoral growth, rapid transit also added a new dimension to urban structure through its necessity for discrete passenger loading and unloading points spaced farther apart than ordinary streetcar stops. Thus, each station on a rapid transit facility has a unique tributary area from which it draws customers, although some of these areas are larger than others. Street transportation in the larger tributary areas focus on the rapid transit stations. This convergence of transportation and people results in the generation of business activity around rapid transit stations, and those stations with the largest tributaries may have very high densities of business and residential activity in the immediate vicinity. These foci, such as at Sixty-third and Halsted in Chicago, became significant business nucleations (outlying nucleations) beyond the CBD serving extensive local communities. Sixty-third and Halsted, for example, became an important focus for business, service, and, hence, employment activities for the extensive African American community that developed on southside Chicago during the era of metropolitanization.

One interesting aspect of rapid transit and streetcar systems is that after the great surge in construction of these systems in most large North American urban areas in the early part of the twentieth century, investments virtually ceased toward the end of the 1920s and many of them stopped operation for good during the Depression. By this time the automobile and truck had become the transport modes of choice, and roads the infrastructure imperative. The resurgence of interest in rapid transit and streetcar-like light-rail systems in North America since the 1950s will be discussed in Chapter 14.

Spatial Differentiation Within the Metropolis   The structure of urban areas that emerged by the end of the era (the early 1940s) is well summarized in three classical descriptive models. The first of these is Burgess's concentric zone model, which was formulated in the early 1920s and emphasizes the ring-like growth of urban areas (Park, Burgess, and McKenzie, 1925). The second is the sector model, which focuses on radial aspects of urban growth (Hoyt, 1939). Both models are derived from different types of processes and data, the former immigration and population redistribution, and the latter residential location as reflected in rents. The third is the multiple nuclei model, which metamorphoses the concentric and sector models into a cartogram of metropolitan structure (Harris and Ullman, 1945).

The Concentric Zone Model   The Burgess concentric zone model is based directly on the era of massive external immigration and internal migration to North American metropolitan areas and demonstrates the geographical outcome of assimilation of the migrants into an urban (and American) way of life. It envisages a situation in which the newcomers invariably flock to neighborhoods around the center of the original city, where there are cheap apartments, other persons and families from the same sociocultural background, and job opportunities. The process envisages that immigrants become settled, accumulate economic wealth, and begin (perhaps within one generation) to be absorbed into the American "melting pot." This increase in wealth and socio-cultural assimilation is accompanied by a spatial move outward into higher-income neighborhoods. The hypothetical consequence of this process of immigration, establishment, accumulation, and assimilation is an urban area consisting of six concentric zones (Figure 7.4a):

1. The *central business district* (CBD) is considered to be the focus of commercial, social, and civic life, and of transportation. This area contains the department stores, shops, tall office buildings, clubs, banks, hotels, theaters, and museums that are of importance to the whole urban area.

2. The second zone surrounds the central business district and is an area of wholesaling, truck, and railroad depots. It is called the *fringe* of the central business district.

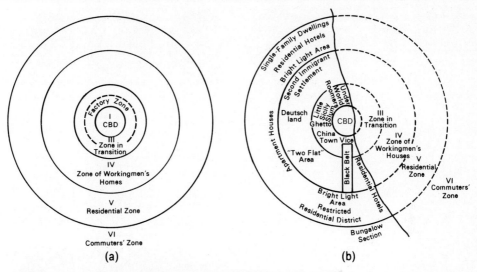

Figure 7.4   (a) The Burgess concentric zone diagram of urban structure; and (b) its practical application in 1920s Chicago.
(*Source:* After Park, Burgess, and McKenzie, 1925, Charts I and II.)

3. The *zone in transition* is a zone that used to contain some of the homes of the wealthy, but as the urban area expanded and immigration occurred from rural areas and overseas, the wealthy have been replaced with lower-income families and individuals. Consequently, this zone contains the slums and rooming houses that are envisaged to be common to the peripheral areas of the central business district. In the metropolitan era, business and light manufacturing encroach into this area because of their intensive demand for services and low-wage labor. So many immigrants who became entrepreneurs located their businesses in the neighborhood in which they first settled.

4. The *zone of independent workers' homes* consists primarily of industrial workers who have moved from the zone in transition. It might be regarded, therefore, as an area of second-generation immigrants and families who have accumulated sufficient wealth to purchase their own homes.

5. The *residential zone* is a zone of better residences containing single-family dwellings, exclusive restricted districts, and a few high-income apartment buildings.

6. The outermost zone is the *commuters' zone*. It encompasses a broad suburban commuting area containing satellite cities, with middle- and upper-class residences, located along rail or rapid transit lines.

The details of the practical application of this model to Chicago in 1920 are indicated in Figure 7.4b.

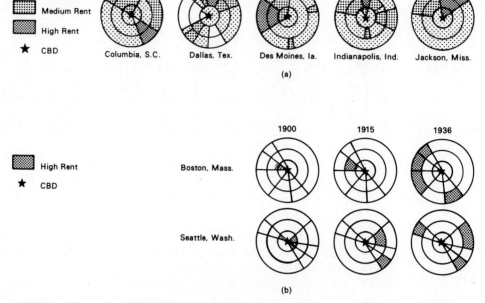

Figure 7.5 The Hoyt sector model of urban structure: (a) idealized pattern of rental districts in five sample cities in 1930; and (b) generalized change in location of high-rent districts in Boston and Seattle, 1900–1936.
(*Source:* Hoyt, 1939, Chapters I and II.)

The Sector Model Hoyt (1939) based his sector model on an intensive study of the internal residential structure of 142 North American cities. From an analysis of the average value of residences (by block) in each city, Hoyt presents a number of specific conclusions, the most important ones being (Figure 7.5a):

1. The highest rental (or price) area is located in one or more specific sectors on one side of the urban area. These high-rent areas are generally in peripheral locations, though some high-rent sectors extend continuously from the downtown area.

2. High-rent areas often take the form of wedges, extending in certain sectors along radial lines leading outward from the center to the periphery of the urban area.

3. Middle-range rental areas tend to be located on either side of the highest-priced rental areas.

4. Some cities contain large areas of middle-range rental units, which tend to be found on the peripheries of low- and high-rent areas.

5. All urban areas have low-rent areas, frequently found opposite high-rent areas and usually in the more central locations.

Thus, Hoyt hypothesized that as urban areas grow, the high-rent (or price) areas follow definite sectoral paths outward from the center of the city (high-rent

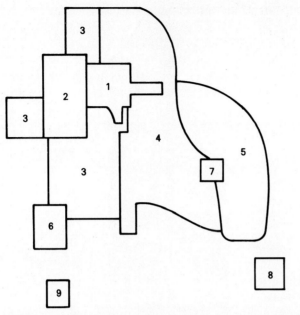

1. Central business district
2. Wholesale light manufacturing
3. Low-class residential
4. Medium-class residential
5. High-class residential
6. Heavy manufacturing
7. Outlying business district
8. Residential suburb
9. Industrial suburb

Figure 7.6 The multiple nuclei model for urban areas in America in the 1940s.
(*Source:* Harris and Ullman, 1945, p. 11.)

areas in Boston and Seattle provide examples in Figure 7.5b), as do the middle-rent and low-rent areas. Each sector, therefore, retains its character over fairly long distances once the initial rent differentiation is established.

The Multiple Nuclei Model Harris and Ullman (1945) recognized the particular, and hence conceptually restricted, nature of the concentric zone and sector models and proposed a descriptive model of the North American metropolis as it had emerged by the early 1940s that takes into account both concentric and sectoral features *and* the emergence of some outlying business and commercial districts (Figure 7.6). Called the *multiple nuclei model*, it suggests that urban areas have developed a number of areas that group around separate nuclei. These nuclei include the central business district, outlying business areas, and industrial districts around which the various parts of urban areas tend to be organized for their daily activities. The model indicates that transport improvements, particularly low-cost public transportation and the automobile, which had become a middle-class acquisition by the 1930s, had facilitated considerable population deconcentration and separation of home, community, and work by the end of the metropolitan era.

Land use information for Toronto in 1940 (Figure 7.7), when it had a population of about 900,000, provides some illustration of the general observations in the descriptive models. First, it is clear that industrial activities are generally transportation oriented (that is, railroad oriented). Second, the residences of the poor, which are labeled "fourth-class" in Figure 7.7, are centrally located

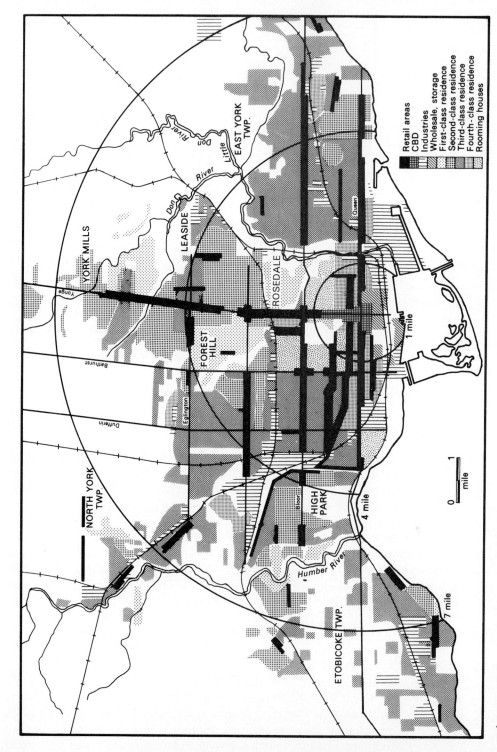

Figure 7.7 Generalized diagram of land uses in Toronto in 1940.
(*Source:* Redrawn from Baine and McMurray, 1977, p. 34.)

around the central business district and the industrial areas. Third, the first-class residences, presumably occupied by the wealthy, are located in the northern sector, whereas the second- and third-class residences of the middle class are in peripheral locations served by streetcars, commuter railroads, and other types of public transit. Fourth, outlying commercial (retail) areas have emerged at major intersections to serve the suburbanized population, and there are also some outlying industrial districts around classification yards.

THE CURRENT ERA: THE SUBURBAN REINVENTION OF URBAN NORTH AMERICA

The last half of the twentieth century was the period of greatest suburbanization of the North American urban population. The deconcentration that occurred prior to 1945 was but a prologue to the complete transformation of urban America that occurred after 1945 as metropolises were turned inside out as decentralization broadened and accelerated (Short et al., 1993). The proportion of the population of *metropolitan areas* in the United States residing in suburban (i.e., non-central city) locations increased from 23 percent in 1950 to 47 percent in 1990 (Survey, 1995, p. 4). This relentless centrifugal spread of population and economic activities has made the late-twentieth-century urban areas quite unrecognizable from the metropolises that emerged from the early and middle industrial eras.

This extensive suburbanization brought with it accentuated social separation, for a large proportion of the lower-income population resides in parts of older inner cities or pre–1945 suburbs, and much of the middle-income and wealthier population resides in more recent suburbs. In urban areas in the United States, this central city–suburban differentiation involves high levels of racial polarization as much of the African American and Hispanic American population in poverty (Table 7.1) is concentrated in quite well-defined parts of inner cities. These older inner cities also contain a large portion of the white population and other persons living in poverty. The inner cities often also include established ethnic neighborhoods, and clearly defined middle- and upper-income enclaves.

Forces Accelerating Deconcentration

The underlying forces that came to the fore to stimulate this massive deconcentration of population and economic activities in the second half of the twentieth century can be summarized as: (1) high rates of population growth and household formation, and fairly consistent high levels of per capita wealth; (2) technological innovations that facilitated decentralization; (3) governmental programs; (4) development and financial institutions; and (5) African American and Hispanic American in-migration and "white flight" from the central city. These forces acted in concert to restructure and fragment most North American metropolises.

Population Growth, Wealth, and Family Formation   Since 1945, the urban population (in places of more than 5,000 people) of North America has in-

Table 7.1

National Person Poverty Rates for Major Racial Groups and Selected Central Cities, United States, 1989

RACIAL GROUP/CITY	RATE	NUMBER OF PEOPLE
United States as a Whole	12.8	31.6 million
White (not of Hispanic origin)	8.3	15.5
Hispanic origin	26.2	5.4
African American	30.7	9.3
Other Races	16.4	1.4
Detroit	32.4	
New Orleans	31.6	
Atlanta	27.3	
Baton Rouge	26.2	
El Paso	25.3	
St. Louis	24.6	
Memphis	23.0	
San Antonio	22.6	
Milwaukee	22.2	
Baltimore	21.9	
Chicago	21.6	
Houston	20.7	
Philadelphia	20.3	
Tucson	20.2	
New York	19.3	

(*Sources:* Bureau of the Census (1993), p. 472; and Bureau of the Census (1992).)

creased by about 120 million, per capita income has doubled, and the rate of household formation has risen much higher (Figures 2.2, 3.1, 4.5). As a consequence, the demand for housing and hence the number of new housing units started was in general about three times that of the average level in the previous era. In both Canada and the United States there have been three building cycles: (1) an immediate post-war boom, catering to some of the pent-up demand from the depression and limited housing construction of the war years, which lasted until the mid-1950s; (2) a second wave, catering to the needs of the growing families of the "baby boom," which occurred during the 1960s; and (3) a third wave, catering for both new families and new households (the declining birthrate being offset by declining household size and an increase in number of one- and two-person household units), which occurred during the maturing of the "baby boom" cohort in the 1980s.

Furthermore, because of changing income and tastes, the housing units that were built were generally larger, with more rooms and bathrooms per unit, on new land at the periphery of metropolises (see the section on the influence of government). The volume of units built, and increase in average size of unit meant that urban areas expanded outward considerably in this era, particularly during the second and third of the postwar building waves when deconcentration was assisted greatly by new highway construction in the 1960s (see Figure 2.6) and the established auto-oriented transport infrastructure in the 1980s (Cervero, 1989). The diagram illustrating the outward expansion of metropolitan Chicago (Figure 7.3) indicates that the urban area more than tripled in size during this period, with the outer limits some 60 miles (100 kilometers) from the original "mercantile" core, while the population of the CMSA increased only 45 percent (from 5 million to 8 million). The decongestion advocated by Weber (1899) had now become a series of tidal waves of outward expansion.

Technological Innovation The technological innovation that had the greatest impact on shaping cities after 1945 was the automobile. In 1903, when Henry Ford sought $100,000 to establish a motor company, he was able to raise only $28,000. Five years later, when W. C. Durant, the founder of General Motors, told an investment banker that eventually a half-million automobiles would be produced per year, the banker was incredulous and showed Durant to the door of his office. The application of the streetcar to intraurban traffic and the immediate profits being made, its extension into interurban transport, and the use of rapid transit and the commuter railroad, led many persons to believe that the automobile had a limited future. However, the forces that these innovations had set in motion in reshaping and restructuring urban areas proved ideally suited to automobiles and trucks. All the facets of urban life that came to the fore along with the innovation of the streetcar—the suburbanization of the middle class, the desire for travel and recreation beyond the city, the emphasis on individual and family activities, the beginning of the outward migration of employment opportunities and longer-distance commuting—could be developed a great deal further with the use of cars and trucks (Bottles, 1987).

Until the mid–1930s, although the increase in automobile ownership was quite dramatic (Table 7.2), it was really used more for social and recreational purposes than for the journey to work. After 1945, when the ownership ratio dropped below four persons per automobile, they were used more and more for commuting purposes and, by 1970, more than three-quarters of all workers living in SMSAs used them for driving to work. By 1990, with about 40 percent of households in the United States owning two or more motor vehicles (BTS, 1994, p. 65), and the dependence on the automobile had become so great that it accounted for almost two-thirds of all inter- and intraurban passenger miles (BTS, 1994, p. 50). It is, however, important to realize that even with automobile ownership at what must be near saturation point (one car for every 1.7 persons in 1990), 10 million households in the United States are without vehicles, and a large number of these are households in poverty. Although about half of non-automobile-owning households in poverty are white, the highest concentrations are in the numerically smaller African American and Hispanic American groups. This lack of automobile ownership in households in poverty has a negative impact on the possibilities for access of some inner-city-located poor to jobs (see Chapter 9).

Table 7.2

Automobile Ownership in the United States, 1910–1990

YEAR	AUTO REGISTRATIONS (INCL. TAXIS, IN THOUSANDS)	AUTO OWNERSHIP RATIO
1900	8	9,511.3
1910	458	196.6
1920	8,131	13.1
1930	23,035	5.3
1940	27,466	4.8
1950	40,339	3.8
1960	61,682	2.9
1970	89,230	2.3
1980	121,600	1.8
1990	143,550	1.7

(*Source:* U.S. Bureau of the Census, *Statistical Abstract of the United States,* 1973, pp. 5, 547 (Washington, D.C., U.S. Government Printing Office, 1973); and, Bureau of the Census (1982), p. 579.)

The Impact of the Automobile on Public Transit   The impact of this high level of automobile use for commuting to work has not only permitted the spread of urban areas into fringe zones far from the old city core, but it has also led to the demise of public transportation (see Chapter 14). During the 1940s the bus began to replace the relatively inflexible fixed-rail electric streetcar, after 1950 bus transport within urban areas declined steadily (Table 7.3). Although public transport services exist in most North American cities—in the

Table 7.3

Passenger Traffic by Streetcar, Trolley Coach, and Motor Bus in Urban Areas in the United States, 1920–1990 (*In Billions of Total Passengers*)

YEAR	STREETCAR AND TROLLEY COACH	MOTOR BUS
1920	13.7	—
1930	10.5	2.5
1950	5.6	9.4
1970	0.4	5.0
1990	0.1	3.5

(*Sources:* Owen (1966), Table 16; 1970 data from American Transit Association, *Transit Fact Book, 1970–1971,* Table 2.2; 1990 data estimated from BTS (1994), p. 50.)

United States at least 293 of the 373 urbanized areas defined in the 1980 census provided some form of transit—passenger trips are few compared with private transport. There is some debate as to whether this decline in use of public transport has occurred as a result of user choice, or whether the decline in availability and quality of public transport left commuters with little choice. Whitt and Yago (1985) argue that the automobile industry helped to destroy a number of streetcar and bus systems in Los Angeles by purchasing and liquidating them.

The basic reason however, would seem to be the massive suburbanization of the population to low-density, single-family housing tracts that make the provision of public transport services uneconomic. As public transport requires high population densities and focused travel markets, the entire post-war North American urban experience of deconcentration runs counter to this requisite (Bottles, 1987). Whatever the reason—and one suspects that individual consumer choice and a lack of service provision have had a combined effect—the result is that in the United States by 1980 only 12 percent of all workers in SMSAs of 1 million or more population used public transportation for the journey to work, with exception of the New York SMSA (45 percent; Fulton, 1983). The commuter traffic using public transit is concentrated in a few metropolises: Only New York, Chicago, Boston, the San Francisco Bay Area, and Washington, D.C., have integrated rail, rapid transit, or bus transportation facilities which carry more journey-to-work trips than the national average. The Canadian situation is somewhat different for the proportion of total commuters using public transit in the three metropolises of 1 million or more is much higher (about 30 percent) due to greater public investment in such systems in recent years.

The Impact of the Truck   Not only people require mobility for the inputs and outputs from industries, goods and services have to be transported to all parts of urban areas as well. Some commodities, such as water, gas, and electricity, can be provided with their own specific transport networks, but others are not suitable for pipe or wire dispersion. Before the development of the internal combustion engine, very little commodity movement within cities was undertaken by anything other than horse-drawn vehicles. Horses were quickly superseded by 1900 for the movement of people but remained in service much longer for the movement of goods. The number of trucks in use increased sevenfold between 1950 (8 million) and 1990 (56 million) as the flexibility of these vehicles contributed to the decentralization of commercial and industrial activities from the center of the city. The dominance of the truck for the transport of goods both within and between North American urban areas was assured by national and state/provincial road transport policies during the suburban era (see below).

The Influence of Government on Decentralization   A number of policies at the local, state/provincial, and federal levels also influenced to an enormous extent the suburbanization of the population and economic activities. Perhaps three examples will illustrate this process, though there are many that could be cited, some general across the United States and Canada, and some quite specific to individual urban places. The examples relate to housing, transportation, and fiscal policies.

Housing Policies The tremendous increase in volume of post–World War II housing was made possible by the greater accessibility of new housing. Families could afford new homes in the United States because of policies put in place through various revisions to the Federal Housing Act (FHA). This act set mortgage guidelines and provided a mechanism for insuring mortgages that were made available through private-sector financial institutions. For example, in 1963 it was possible to obtain a twenty-five- (or even thirty-) year amortized FHA-insured mortgage at 6 percent interest with $1,000 down payment for a new, $20,000 single-family detached home ($20,000 is almost $120,000 in 1995 dollars). A veteran could obtain a similar loan but with no down payment under a Veterans Administration scheme. These government-insured mortgages were for new homes only, not for renovating older homes. As a consequence, there was an enormous expansion of housing subdivisions on new land at the periphery of cities. Homeownership in the United States is further encouraged through the allowance of the deduction of mortgage interest, on both primary and secondary homes, from income in the calculation of personal federal income tax. This, it can be argued, has further encouraged the purchase by middle-income households of larger homes on larger lots as well as excessive speculation in recreational properties.

Suburbanization in Canada after World War II was similar but not quite as dramatic as in the United States because the Central Mortgage and Housing Corporation (CMHC), established by the federal government, did not implement such universal schemes with similar low down payments (Harris, 1991). Furthermore, there is no allowance in Canadian income tax regulations for the deduction of mortgage interest from income in the calculation of personal income tax. Thus, the comparative rate of new construction, space consumption, and outward expansion, has been less than that in the United States.

The amount of suburbanization occurring during the second building boom era in the 1960s is reflected in the case of the Los Angeles SMSA—that most "twentieth century" of twentieth-century North American cities (Marchand, 1986; Davis, 1990). One-half of the housing stock existing in the Los Angeles SMSA in 1970 was built between 1960 and 1970 (Abler and Adams, 1976, p. 223), and the greatest concentration of the 1960s housing is in areas that were then suburban tracts at the periphery of the sprawling city in the San Fernando Valley (Figure 7.8). The older housing constructed prior to 1940 is mostly within the older city, the central business district, which is now demarcated by a circle bounded by the intersection of freeways.

Transportation Policies The movement of population to the suburbs was also accelerated by the construction of urban freeways (or expressways) through and around most large North American metropolises (Downs, 1992). Although some parkways (such as those in New York City) and limited-access superhigh-

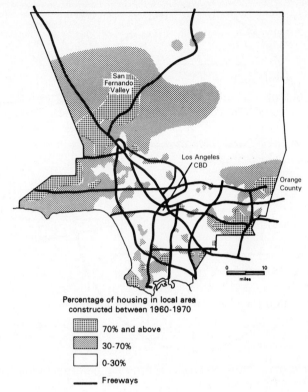

Figure 7.8 Housing built between 1960 and 1970, and freeways in Los Angeles.
(*Source:* Housing data from Abler and Adams, 1976, p. 22.)

ways (such as a few in Los Angeles) had been built before 1940, the tremendous construction of limited-access freeways occurred in the late 1950s and 1960s. The impetus in the United States was the 1956 Interstate and Defense Highways Act, through which the federal government, in the interest of national defense and the possible need for the rapid transcontinental deployment of forces personnel and ordnance (this was the height of the Cold War), created a trust fund that paid 90 percent of local construction costs if an expressway were tied into the national interstate system. Of the 42,000 miles comprising the planned interstate system, 5,000 miles have been located in metropolitan areas, and those roads have consumed about 90 percent of the trust fund budget (see Figure 14.1).

In Canada, the integration of the Windsor-Quebec corridor has also been enormously stimulated by the construction of limited-access highways commencing in the late 1950s with federal/provincial matching funds for the Macdonald-Cartier interprovincial highway, which connects into the United States interstate highway system, and in recent years with provincial road funds (see Figure 14.1). The extensive suburbanization around both Montreal and Toronto could not have

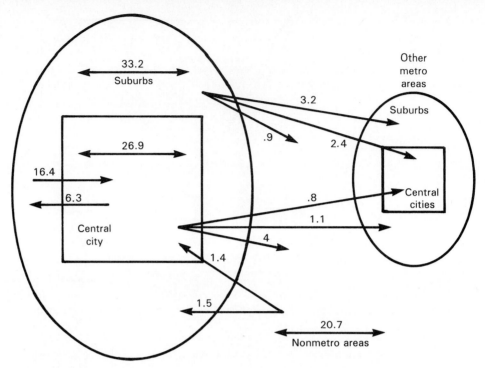

Figure 7.9 National commuting patterns in the United States, 1990.
(*Source:* BTS, 1994, Figure 3.4, p. 52.)

occurred in the 1960s without these highway and associated infrastructure developments, such as the Mercier and Champlain bridges, which opened up the south shore of the St. Lawrence River to Montreal suburban development.

These governmental programs have thus been in large part responsible for the decentralization of the middle- and lower-middle-class white population, and many types of commercial and industrial activities, from central cities in the post–World War II years. Large amounts of land, as far as 20 to 50 miles from the old central business districts, have been opened up for urban use. The suburbs are now served by their own system of circumferential, or beltline freeways, and the intersection of these radial highways has created locations of great accessibility to the central city, the suburbs, and the continent. Some, in the parlance of the 1990s, have become "edge cities" involving concentrations of integrated full-service malls incorporating business, commercial, retail, and entertainment activities (Garreau, 1991). Roughly two-thirds of the 19 million new jobs created in the United States in the 1980s were located in the suburbs (BTS, 1994, p. 52). The freeways make it easy for circumferential locations to be serviced by trucks and facilitate an increase in size, carrying capacity, and power of the vehicles.

Consequently, in most large metropolises, crosstown circumferential commuting is now more prevalent than the suburban–central city journey-to-work pattern that had been the basic feature of metropolitan commuting patterns

(Taaffe et al., 1963). In the 1990s, crosstown commuting or suburb-to-suburb daily work trips (33.2 million) in United States metropolitan areas are twice the volume of suburb–central city work trips (16.4 million), while the volume of within–central city daily work trips (26.9 million) remains quite large (Figure 7.9). Interestingly, in the context of the previous discussion concerning inner-city auto ownership, *reverse commuting* from the central city to the suburbs (6.3 million) has become quite significant. The deconcentration of people and employment opportunities to exurban locations in nonmetro counties adjacent to major metropolitan areas now generates 20.7 million commuting trips per day. Also to be noted are the significant commuting interactions between adjacent metropolitan areas.

Financial Policies   The third wave of construction during the suburban era in the 1980s appears to have been influenced to a degree, in the United States, by financial deregulation, particularly as it has been applied to the Savings and Loans companies. The raising of saving deposit limits insured by the Federal Deposit Insurance Corporation encouraged many dubious development activities, amongst which has been a great increase in the number of new office and retail malls, and master-planned residential communities at the edge of larger metropolitan areas (such as Denver and Washington, D.C.). When the hothouse construction boom of the 1980s came to a quick end in 1989, many ambitious developments went into receivership. This overspeculation (Mills, 1995), force-fed by the pyramiding of loans in a deregulated and badly monitored financial industry, undoubtedly helped to create greater real estate development and hence decentralization at the periphery of large metropolises than might otherwise have occurred.

The Impact of Developers and Financial Institutions on Decentralization   One of the major features of urban development since 1945 has been the increasing dominance of large corporations working closely with financial institutions and local governments in the development process. These groups have been able to take advantage of federal housing programs and the demand for housing, office, industrial and commercial space to completely reshape the North American urban areas after 1945. Prior to that time, most urban development was undertaken by relatively small construction companies. After 1945, the scale of building required made it possible for large corporations to develop continuous pouring and factory assembly methods of construction involving prefabricated materials for housing, shopping centers, and all types of commercial and industrial space.

Levittown   Probably the most dramatic example of this type of production is provided by Levitt and Sons on Long Island (Checkoway, 1984). The firm had gained experience with prefabricated building during World War II. Starting in 1947, they used assembly line and single-style ("Cape Cod") mass production techniques to construct 17,000 identical houses on 1,400 acres for more than 70,000 people, at a uniform price of $7,990. Each house came with all basic facilities (equipped kitchens, bathrooms, and central heating). Buyers knew what they were getting in advance from the displayed "model" home. So the purchase could be made easily with minimal legal costs in two half-hour steps—one to purchase and the other to clear title. This method of construction and marketing became, in effect, the model that was used across North America for most suburban tract new-home sales after that time. The system also has the beneficial effect of making resales relatively simple as well.

Roots of Suburban Sprawl Hise (1993) argues that the role of developers in the creation of the post–1945 multicentered suburban city is more powerful than is usually recognized, and that this role has its roots in the decentralization of industry that occurred in Los Angeles immediately prior to and during World War II. The suburban forms that were created provided the "model" for subsequent large-scale suburban developments across North America. The argument is as follows:

1. Immediately prior to and during World War II some of the major manufacturing companies, particularly in the aircraft industry, established huge production facilities in the Los Angeles region. These innovative firms included Lockheed, Douglas, Northrop, North American, General Motors, Vultee, and Vega (some of these have subsequently become merged with others). The reasons for selection of the Los Angeles region are generally summarized as topography (extensive flat desert land), a temperate climate, a local skilled nonunion labor force (which grew rapidly as people flooded in to the region in search of jobs, and became unionized!), and the local research and training environment provided by postsecondary educational institutions. Numerous manufacturing, service, and research activities linked directly or indirectly with these industries were established in proximity to them, creating decentralized employment complexes. Federal military and research expenditures associated with the war against Japan in the Pacific in the 1940s, North Korea in the 1950s, and North Vietnam in the late 1960s and early 1970s, and the Cold War with the Soviet Union until 1989, provided the high-octane fuel for this expansion. These major companies chose sites at the edge of the city along a 10-to-15 mile radius from the Los Angeles central business district—and in subsequent decades, as the metropolis burgeoned, many firms relocated outward much farther (50 to 100 miles).

2. Residential housing developers, imbued with the workplace-residence link of the classic-industrial and metropolitan eras, responded to the peripheral employment opportunities with new types of suburban communities similar to those exemplified by Levittown on the East Coast. For example, Panorama City in the center of the San Fernando Valley, constructed by Kaiser's Housing Division close to Lockheed and General Motors, consisted of 2,000 detached homes in a fully planned, self-contained community with its own schools, churches, and commercial center, also including facilities for recreation and health care. These were, in effect, the prototype for the master-planned communities associated with many "edge cities" of the 1990s (Knox, 1991).

3. These suburban communities, which provided affordable homes with two bathrooms, continuous-surface designed kitchens (with dishwashers and garbage disposal), and two-car garage or parking facilities, provided ease of purchasing (with FHA- and VA-underwritten mortgage schemes) and, potentially, ease of sale. Thus the residential communities interwove with what has become one of the central chracteristics of southern California suburban culture—ease of entry and ease of exit.

4. This wave of *urban restructuring* in Los Angeles, initiated by the rapid growth of decentralized industrial complexes, has provided an opportunity for the development industry to play a crucial role in reinventing the North American city. These new suburban forms have become so much a part of everyday experience for much of the North American population that they are now inextricably linked to late-twentieth-century lifestyles and culture.

The workplace-residence link ethos collapsed by the late 1950s, leaving a restructured sprawling suburban form that depended on near universal automobile ownership and continuous fixes of new freeways and highways. Thus, the crucial role of the development industry in the suburban era interweaves with the roles of transport innovations and government.

Migration of Visible Minorities and "White Flight"   During the late 1930s and 1940s, there was a massive movement in the United States of African Americans from the South to metropolises in the Middle West and Northeast. The migrants came in response to the availability of jobs generated by World War II and the growth of the consumer economy, particularly the wide range of industrial activities associated with the automobile and food processing industries, in the decades immediately following the cessation of hostilities. They settled in the parts of the inner cities that had been the destination of previous, but less voluminous African American migration from the South: for example, Harlem in Manhattan and southside Chicago.

In Chicago, earlier African American migrants from the South had managed to settle close to the bus and train stations south of the downtown around Twelfth Street and on the near westside (Cutler, 1982, p.124). But, even by 1940, the African American population (277,000 persons) in Chicago was only 8 percent of the city's population. These numbers increased rapidly during the war and immediate postwar period to 813,000, or 23 percent of the city's population, by 1960. This rate of increase of the African American population in Chicago reflects similar levels of migration to other midwestern, northeastern, and west-coast United States cities as people came from the South to what appeared to be a "promised land" of jobs, higher wages, and better social conditions (Lemann, 1991). As newcomers poured into the existing overcrowded African American communities, the established black middle class moved outwards. In the case of Chicago this was to the south and west into previously restricted (often by covenants) white neighborhoods. Processes associated with this type of movement, such as block-busting and red-lining, are discussed in Chapters 11 and 12.

Widespread automobile ownership, the opening up of relatively cheap sub-urban land, and mass-produced single-family housing, facilitated the move-ment of the white population that wished to relocate to suburban areas. Various financial and real estate restrictive practices had the combined effect of exclud-ing African Americans from parts of central cities and suburbs where the ma-jority white population did not want the black population to reside. This type of discrimination in housing location was not declared illegal until passage of the Fair Housing Act (1966) after deeply contentious hearings and debate, and the Civil Rights Act (1968). Even after passage of this landmark legislation, discrim-inatory practices remained in place in parts of many metropolises for years to come (see Chapter 12).

With a significant reduction of racial discrimination in the housing market and the continued improvement of the economic circumstances of some of the African American population, during the 1970s, the black population started dispersing into previously white areas of parts of central cities and some sub-urbs. The suburbanization that occurred was, however, primarily into older suburbs close to the central cities being vacated by the white population. By the mid-1990s still, the level of segregation between African American and white residential areas in the United States remains high (see Chapter 11). As a result of persistent economic decentralization, many African Americans remaining within central cities are in need and isolated from opportunity. Decentraliza-tion and restructuring have left people in these areas with limited access to em-ployment opportunities, and this, plus largely ineffective (Head Start being an exception) or outdated social programs, has resulted in widespread persistent social breakdown.

The increasing volume of Hispanic and Asian immigration to the United States from the 1970s has added to the racial diversity of urban areas and the problems being faced by segments of inner-city African American communi-ties. Many of these new immigrants have settled first in the central cities, where they have found it easier to integrate (or coexist) with white populations than in the suburbs (Frey, 1990, 34–35). Although the majority of Hispanic Americans reside in central cities, they are more numerous in suburban loca-tions than African Americans. (Note: In 1990 the Hispanic-American popula-tion in the United States totaled 22 million as compared with 30 million African Americans.)

Inner-City Decay Parts of the central cities of many metropolises in the United States are in a desperate state. One of the most desperate, for example, is the City of Newark (New Jersey), located within the extensive New York met-ropolitan region, which declined in population from 382,000 in 1970 to 275,000 in 1990, and is surrounded by middle- and upper-income suburbs. The Newark PMSA in 1990 comprised 1.8 million persons, 1.5 million of whom reside out-side the city. 58.5 percent of the population of the City of Newark is African American, and 26.1 percent Hispanic American. One-half the residents of Newark live in households headed by single women, and one-third are on wel-fare, with the highest rates of economic distress being experienced by the African American and Hispanic American groups. Unemployment is generally twice the average for the State of New Jersey. Among young male African Americans the rate is usually close to 70 percent. Since 1975, the city has lost about one-fifth of its property tax base as the number of stores has decreased by one-half. There is only one supermarket in the city, which is located on the boundary with the more prosperous suburb of Elizabeth.

Given this level of desperation, where does the extreme case, and others that are approaching it, have to go—farther down or back up? There are signs that the situation in Newark may be bottoming out because some companies are decentralizing part of their office and manufacturing functions from Manhattan and elsewhere in the New York PMSA. They have realized that Newark has land and property that is now cheap, and that it is also well connected by transport facilities to the rest of the New York-New Jersey metropolitan region and the world. The poor and unemployed of Newark have, however, become economically and socially marginalized. A factor contributing heavily to the perpetuation of this marginalization and the creation of a possible underclass (Rose, 1991) is the growth of single-parent households in poverty (Santiago and Wilder, 1991). The problems of parts of many United States central cities, which have become no-exit hells of abandonment for many inhabitants, therefore, arise "when high rates of joblessness trigger other problems in the neighborhood that adversely effect social organization, including drug trafficking, crime, and gang violence" (Wilson, 1996, p. 59).

Edge Cities

Sixty percent of the population of the forty-four *largest* metropolitan areas in North America, each containing more than 1 million people, resides in suburban, or non-central city, locations. These 85 million big-metropolis suburbanites inhabit an environment which is developing its own foci, usually at the intersection of major highways, such as Irvine situated in the general vicinity of the junction of I-405 and I-5 south of Los Angeles. Garreau (1991), in his study of suburban experiences in Boston, San Francisco, Houston, southern California, Detroit, and Washington, D.C., refers to these large agglomerations of office and retail activities as "edge cities." In general, they are described as having 5 million square feet of office space, 600,000 square feet of retail space in malls, and as being virtually new to the urban landscape since 1960. Some have names like Perimeter Center, Atlanta, which caters primarily to a suburban African American middle class, whereas others are identified as much by coordinates as anything else (e.g., "495 and Mass Pike").

There appears, in general, to be two locational types of "edge cities." Some edge cities are externally oriented, catering to markets within but particularly beyond the metropolis in which they are located. Large metropolitan areas serve not only national markets, but they are also increasingly driven to be internationally oriented as they become linked to the global economy (Ley et al., 1992). Activities with such external orientations tend to be "information" based, and they establish themselves in large campus-like office complexes and planned industrial technology parks along major transport routes connecting to other metropolises and close to international airports. Hoffman Estates, close to

O'Hare Airport (Chicago); Reston, near Dulles International Airport (Washington, D.C.); and Rockville and Shady Grove in a high-tech corridor extending along I-270 to the Washington beltway (Knox, 1991), are examples of this type of externally oriented development.

A second type of edge city involves activities that cater more directly to the needs of the metropolitan consumer market and tap into the suburbanized labor force. These activities include large sophisticated shopping malls and office buildings. Because of their need for maximum access to the metropolitan consumer and labor markets, they congregate at major transport intersections. Knox (1991, p. 190) describes one such example of an edge city at Tyson's Corner on the outer suburban side of the Washington beltway between the Dulles Airport toll road and I-66. There are two malls at its center, anchored by (in total) seven up-scale clothing and department stores, which generated in 1990 about 4 percent ($1 billion) of the retail sales in the metropolitan area of 4 million people. The retail sales volume of this edge city is 25 percent greater than that in the entire central city of Newark. These two malls are surrounded by buildings containing more than 20 million square feet of office space. The magnitude of the retail and office space and the annual consumer sales indicate the enormous impact that edge cities are having on metropolitan areas (Fleischmann, 1995).

Beauregard (1995) points out that the notion of "edge" cities raises once again the ambivalence that many people feel about cities—the city of light and the city of dark. On the one hand, middle- and upper-income North America has virtually contained central cities, regarding them as places to be avoided except for producer service employment, occasional recreational trips, and perhaps more *risqué* activities. On the other hand, people are seeking to recreate "cities" in other forms in the suburbs that have become the habitat for most people. He argues that edge cities do not offer an alternative to "the dangerous, destitute, and deteriorated central city" (Beauregard, 1995, p. 708) because edge cities cannot exist isolated from central cities or be immune from public ills that occur in society at large. The edge city concept may thus be viewed as escapist, ignoring the essential interrelated nature of urban areas.

Characteristics of Late-Twentieth-Century Suburbs

There are, perhaps, six main characteristics of late-twentieth-century suburbs that distinguish them from the streetcar suburbs and those built prior to 1945:

1. The postwar suburbs are built on the assumption that every family owns an automobile. So little thought is given to access to, or servicing by, public transit. As a consequence, *public transit is now virtually irrelevant* in all but a handful of metropolises. Those persons or households without access to a vehicle, particularly those in poverty, are in many ways locationally trapped. Adolescents in suburban areas are similarly confined, with hanging out in a local mall being a noticeable pastime.

2. The standard prewar house was land conserving for most were two stories high and built detached or semi-detached as part of row housing. The stan-

dard postwar house is a detached single-story bungalow or split-level, and in many suburbs they are built in subdivisions that have minimum lot-size zoning (i.e., requirements for half-acre, one acre, and so forth minimum lot sizes). Postwar housing, particularly that built in the booms of the 1960s and 1980s is, therefore, *land consuming*.

3. Suburbs are usually built by fairly large development corporations on large, preassembled tracts of land. These developments involve, in the aggregate, *a wide variety of housing types* (detached, duplex, low-rise, town-houses), *and* a mixture of retail, business, and industrial activities in malls or park-like settings. As a consequence, many large metropolises are ringed with corporate held *"land banks"* awaiting the appropriate financial and demand conditions for development (Pond and Yeates, 1994a; 1994b). Hence the difficulty of defining the outer limits of metropolises (Orok, 1993).

4. Although suburbs prior to 1945 were built mainly for the growing middle class, after 1945 nearly all suburban tracts are *income stratified* because (like Levittown) they tend to be price stratified. The labeling of subdivisions— Golf Course Estates, Pine Ridge Estates, Chateaux on the Green—reflect the shades of subtle pretensions. In some parts of North America, middle- as well as higher-income suburban enclaves are fenced in, with private security-controlled access, and twenty-four hour surveillance (e.g., the suburb of Rosemont, near O'Hare Airport, northwest Chicago).

5. During the last half of the twentieth century, much of the population and many economic activities that used to be located in the central cities has decanted to the suburbs. For example, Sears, Roebuck and Company, which used to occupy the world's tallest building, Sears Tower in downtown Chicago, has moved to Hoffman Estates, 37 miles into the outer suburbs. As this type of relocation has inexorably progressed outwards, *central city–suburban polarization* has, with deindustrialization, become more complex. Rings of apparently middle-class and wealthier white suburbs surrounding generally poorer inner cities can also be interpreted as rings of relatively prosperous well-educated professional elites surrounding poorer educated populations for whom job opportunities are becoming increasingly scarce. Some of these suburbanized economic activities and communities have been coalesced into "edge cities."

6. For the most part, development companies have ignored the need for redevelopment in the old central cities. Suburban growth has absorbed most of the available investment capital, leading to inner-city disinvestment and decay. The *suburbs in many metropolises have replaced the central cities*. Although there are some examples of private inner-city redevelopment, most that has occurred has been consequent to local government action. This has involved combining funds from federal and state programs, and local taxes, to build public housing or undertake central business district "revitalization" through the means of public-private partnerships. These partnerships are responsible for many of the hotel-convention-office-museum-recreational complexes visible in most downtown promotional material. The office and industrial parks associated with commercial/retail developments in many suburban areas

have created new business and employment foci that replace the traditional central business district. The CBD, even if it has retained some office and financial functions, has lost most commercial/retail activities.

Urban Structure in the Suburban Era

Thus, by the end of the twentieth century, North American urban areas are quite different from those at the end of the nineteenth century, or even those half-way through the period. They are, however, products of the accumulation of infrastructure, economic activity, and social change. It is, therefore, impossible to understand the internal structure of urban areas without a firm awareness of the layering of historical processes. Multicentered urban areas are frequently an agglomeration of the traditional and the new. For example, the Atlanta metropolitan region is organized around five large centers, each providing significant employment in retail and service activities (Fujii and Hartshorn, 1995, p. 688). These include: the traditional central business area of the City of Atlanta; the older mid-town area immediately to the north of the central business district; Buckhead/Lenox around I-400 in the near north inner suburbs; at the intersection of I-75 with the I-285 peripheral beltway; and at the intersection of GA-400 and I-285 (see Figure 10.1).

Taking many of the changes in internal structure into consideration, it is possible, albeit complicated, to develop a descriptive diagram of the internal structure of a typical North American metropolis by the 1990s (Figure 7.10).

1. The central business district frequently remains the area of maximum vertical development of buildings, with new expensive hotels and convention facilities, entertainment facilities, and renovated business areas. Some central business districts have remained alive, day and night, because they have retained real business functions and a resident population with disposable income (such as New York, San Francisco, Toronto, Vancouver); others appear to be alive because of concerted efforts to attract people downtown, often with new stadiums and other recreational facilities (Cleveland, Baltimore, Denver, Chattanooga, Atlanta, Phoenix, Miami, Cincinnati); while other downtowns appear to be in terminal decline (Barnett et al., 1995).

2. The rest of the inner city consists primarily of lower-income households as well as clusters of middle- and higher-income groups in well-defined areas. The lower-income areas contain much of the urban population in need, and are on a downward spiral of deteriorating infrastructure, declining services, and decreasing safety.

3. Beyond, or at the periphery of, the central city are older inner-suburban rings. They include previous streetcar suburbs, businesses and low-rise apartments around rapid transit stations, and some 1920s and 1950s suburban developments. These often quite handsome, tree-lined districts may exhibit vestiges of previous ethnic, racial, and income settlement, and are served by local retail nucleations and older shopping centers. They may also have a number of older planned industrial districts.

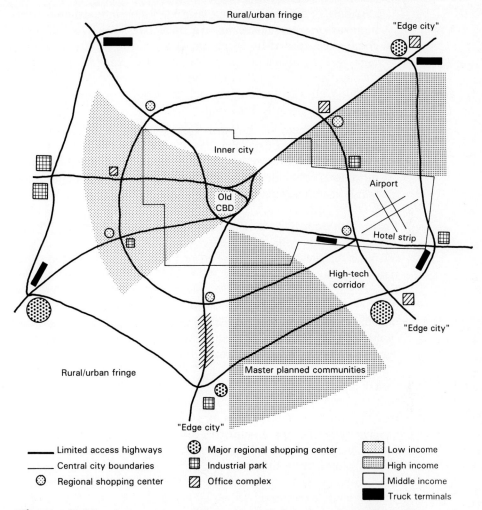

Figure 7.10 A cartogram of the reinvented metropolis of the 1990s.

4. The middle- and outer-suburban rings, the latter consisting of those suburbs built since 1965, have an internal structure that is related to and influenced by outer circumferential limited-access highways. Between the middle- and outer-suburban rings, indoor climate-controlled regional shopping centers, and industrial and business parks, prevail. The middle and outer rings are also subdivided into many income-differentiated communities which may occupy similar or same sectors as their counterparts in the inner-suburban ring and inner city. In large macro-urban regions, such as the New York-Northern New Jersey-Long Island CMSA, the Los Angeles-Riverside-Orange County CMSA, or the Washington-Baltimore CMSA, these extensive outer suburban rings mesh and "edge cities" may emerge.

5. Major limited-access highways, particularly those connecting the urban region to local international airport(s) or other large metropolises, give rise to

corridors of offices, research, and high-tech industrial activity. These major highway-oriented corridors are attractive locations because they provide good access internally to a large sector of the suburbanized metropolis, connect well to adjacent metropolitan areas and developing activity on the rural-urban fringe, and link the local region externally to the rest of the world. Well-traveled, high-speed, transport corridors are also regarded as a high-profile, prestige locations.

6. At the periphery of the sprawling metropolis is the rural-urban fringe, an amorphous zone of exurban developments, urbanizing rural service centers, farmland, scattered housing on severances, land held for speculation, and numerous other developments serving the urban region (see the rural-urban land conversion section in Chapter 8).

From Monocentric to Polycentric Urban Regions   The pre-1945 North American metropolis was clearly *monocentric*, with one preeminent central business district to which the rest of the urban area related. The predominant pattern of the journey to work was toward the downtown area. People looked to the central business district for specialized shopping and entertainment often on a weekly basis. The outcome of decentralization is that metropolises have become increasingly *polynucleated*, with many centers of commercial and economic activity (Berry, 1993). In a few large metropolises, downtown office *and* retail activities remain significant, but in most metropolitan areas the old central business area is one of many foci of activity, and in many cases its retail function, exemplified by large department stores, is quite diminished. In large metropolitan areas most retail employment is now located in climate-controlled suburban malls (Erickson, 1986).

The growth of these suburban retail and business malls and industrial/technology parks, and the relocation of activities to them, often involves considerable subsidies from different levels of government. Sears' move to Hoffman Estates was partly financed by $186 million in government subsidies paradoxically permitted by the extension to "greenfield" sites of a program originally designed to help central cities compete with suburbs for employment generating businesses. Similarly, the construction of the 4-million-square-foot Mall of America (on the pattern of the West Edmonton Mall in Alberta) in suburban Bloomington adjacent to Minneapolis-St. Paul, which has worked hard to retain two viable downtown business areas, is facilitated through the provision of $200 million in public funds for highway improvements and parking (Barnett et al., p. 56). Thus, governmental activities continue to foster decentralization.

Urban regions have become a galaxy of clusters of economic activities organized primarily around freeway systems. Crosstown commuting is far more voluminous than central city–suburban commuting, and people are living in a wide variety of types of racial, ethnic, and income-stratified enclaves. This type of galactic arrangement (Lewis, 1983) is illustrated in Figure 7.11 in the case of the Los Angeles-San Diego supermetropolis, which extends in an almost continuous built-up area from Oxnard 200 miles south to San Diego and the United States–Mexico border and contains, by the mid–1990s, about 18 million people.

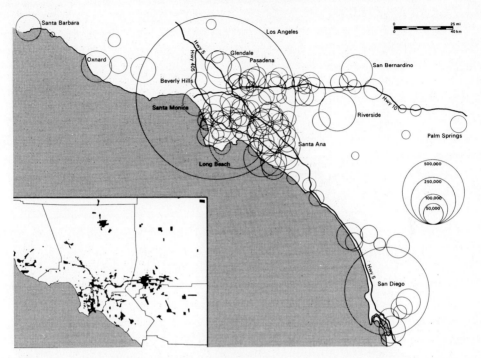

Figure 7.11 The distribution of population in the Los Angeles-San Diego region and the distribution of industrial land.

(*Sources:* Bureau of the Census, 1992, "Summary Characteristics for Governmental Units, 1990"; and Nelson, 1983, p. 201.)

The map (Figure 7.11) indicates an extensive spread of population and municipalities throughout this region of southern California. Industrial, commercial, and recreational activities are spread (in clusters) over this entire area as is witnessed from the inset map which illustrates the spatial spread of one type of land use—industrial land.

From Congestion to . . . Congestion Reference has been made previously to Weber's (1899) contention that overcrowding and high population densities were having a detrimental effect on health and the quality of life in parts of large cities at the end of the nineteenth century. The population decongestion that he advocated has occurred as a result of decentralization during the twentieth century. Improvements in transportation have permitted people to be enormously flexible in their residential location *vis a vis* their places of employment and socio-recreational activities. Much of the population of North America resides in homes that would be considered to be luxurious in many other parts of the world, and even households at the lower end of the income scale have housing that people in poorer countries would regard as an improvement on their present living conditions. The provision of adequate, safe housing does, however, remain a concern for much of the population in poverty. Nevertheless,

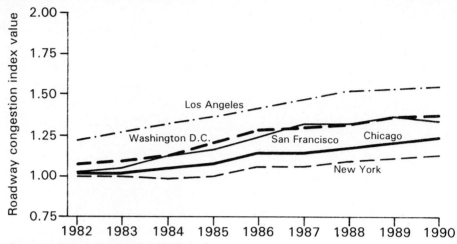

Figure 7.12 Increasing levels of congestion in the five most congested United States metropolises, 1982–1990.

(*Source:* BTS, 1994, Figure 4.20, p. 89. Note: The congestion index is based on estimates of delay hours on highways and major arterials modified by population size. The greater the delays, on a size-modified basis, the higher the value of the index.)

decentralization has allowed for the provision of acceptable (or better) housing conditions for the vast majority of North American households.

The spatial spread of cities, which has come to be almost entirely dependent on the automobile, has, however, created chronic periodic congestion on the highways of some of the larger cities (Hodge, 1992). Large metropolises simply cannot provide highway space sufficient to cater for the volume of motor vehicles at peak periods in central cities or suburbs (Carvero, 1986). As decentralization has progressed and more highways have been built, traffic congestion has, paradoxically, increased (BTS, 1994, p. 88), largely caused by incidents (accidents, breakdowns) rather than recurring high volumes (Downs, 1992). Figure 7.12 shows that road congestion has increased continuously since 1982 in the five urban areas that account for most of the "hours of delay" experienced in United States metropolises. The urban areas experiencing the greatest levels of congestion are Los Angeles, Washington, D.C., and San Francisco. Los Angeles and New York, because of their size, account for one-third of the hours of delay experienced in the fifty largest United States cities. Severe central-city overcowding in the largest urban areas at the end of the nineteenth century has, therefore, been replaced by suburban traffic congestion at the end of the twentieth century (see Chapter 14).

CONCLUSION

This overview of the internal structure of urban areas, therefore, centers on the experience of urban change in North America. It is an experience that involves a number of themes. The first is decentralization, which is leading to polynucleation in metropolises. The second is the effect of a large number of aspatial factors, such as technology, demography, the economy, government, and developers on spatial patterns within cities. The third is the wave-like impact of these various factors on the accumulation of urban infrastructure, and the consequent identification of a number of eras in which certain spatial features predominated. These eras have been defined as mercantile, classic industrial, metropolitan, and suburban. The fourth is the economic, social, and cultural interrelatedness of the various parts and peoples that make up modern urban areas.

Urban Land Use and Population Change

The basic themes of deconcentration of people and economic activities consequent to decentralization within urban areas may be analyzed within the context of three groups of interrelated models. The first, and most fundamental, concerns the value of urban land and location of land uses. Therefore, the discussion in this chapter provides an introduction to the extensive literature on urban land use theory, or concentric space–price equilibrium models (Mills and Hamilton, 1989; Shieh, 1992). The second restates the land use–land rent discussion in the context of population density models (Stern, 1993). The third examines public policy issues associated with the allocation of land to various urban uses—specifically, implications that may be derived from various models of rural-to-urban land conversion, zoning, and the taxation of property.

LAND USE, ECONOMIC RENT, AND LAND VALUES

The land uses of urban areas can be divided into six major categories: residential, industrial, commercial, roads and highways, public and quasi-public land, and vacant land. For cities in the 250,000-to-1.5 million population range in Canada and the United States the typical land use structure is about 30 percent residential, 9 percent general manufacturing, 4 percent commercial, 20 percent roads and highways, and 15 percent public use for government buildings and parks. The remaining 20 percent is either vacant or being held in some form for later development. A major issue concerning urban land is the ways in which the amounts in each particular use and the pattern of location of the different uses within urban areas are changing. Such information is useful because businesses and households require this type of information before making location decisions. Local governments as well want to know the ways in which urban areas are expanding and specific land uses are changing in order to forecast future expenditures and tax revenue growth (Mills and Lubuele, 1995). Urban land use theory provides a way of addressing this issue.

Land Use Theory

The work of von Thünen (1793–1850) represents the first serious attempt to provide a theoretical understanding of the location of land uses in geographic space (Hall, 1966). Von Thünen envisaged a country with no connections with the outside world. A metropolis (or marketplace) is located within an unbounded plain over which uniform soil and climatic characteristics prevail. Prices for agricultural goods are set externally at the marketplace: Since the price of labor and its productivity is uniform, market prices can be quoted *per unit area of production.* All producers have exactly the same production and living costs and the same-size farms (that is, they have the same costs of being in business). The *only* variable is the transport costs which are borne by the producers. Although transport possibilities are equal in any direction over the plain (providing a homogeneous transportation surface), the costs of transport increase in a linear fashion with distance and vary among commodities.

Economic Rent   Assume that the price at the marketplace per unit area of production (for example, one acre) for wheat is $4.00. There are farmers at locations A, B, and C distance (miles or kilometers) from the marketplace (Figure 8.1a). The transportation cost for shipping one unit area of wheat from A to the marketplace (MP) is $0.40, from B to MP is $1.20, and from C to MP is $2.00. The production and living costs for each farmer is $2.00 per unit area of production. With this scenario, the farmer at location A will make $1.60 more per unit area than basic production and living costs, the producer at B will make $0.80 more, and the one at C will break even by receiving for one acre of wheat exactly the same as the production and living costs per unit area of production. The farmer at C will, therefore, receive just enough income to remain in business.

With all farmers having equal standards of living, putting in the same effort, having the same costs of production, and wishing to maximize their financial returns, it would be natural economic behavior for them all to wish to locate at A or closer to MP than A. Consider, the farmers at C who are located at the margin of production for wheat (that is, where production and living costs per unit area equal net returns) will perceive that they can make $1.60 more for exactly the same effort at A than at C. This difference, in effect, represents the economic rent at A with respect to C. In general terms, *economic rent* can be defined as *the difference in return received for the use of a unit of land compared with*

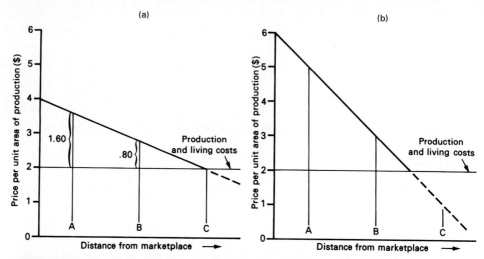

Figure 8.1 Von Thünen's concept of economic rent. (a) Wheat: the economic rent of wheat at location A is $1.60; (b) vegetables: the economic rent at location A is $3.00.

that received at the margin of production. In this particular case the difference is due solely to variations in transport costs for wheat, which are directly related to distance from the market place.

Assume now the existence of a second commodity, vegetables. Vegetables can be produced more intensively than wheat, but they are bulky and perishable, hence difficult to transport. Consequently, the transportation costs for the output of a unit of land in vegetables are higher than that for wheat—say $1.00 for A to MP, $3.00 from B to MP, and $5.00 from C to MP. However, the price for the output of one unit of land in vegetables at MP is higher than that for wheat—$6.00 as compared with $4.00. This situation is depicted in Figure 8.1b. In effect, the economic rent curve for vegetables is much steeper than that for wheat because of the difference in the transportation costs. Also, note that there is no way that farmers at C can produce vegetables because the transportation costs they would have to bear would not permit production and living costs to be met.

Land Use Zones Von Thünen envisaged a market for land that is structured in a particular way. A landowner owns all the land and places it to auction for rent each year. Under this situation, it is evident that the farmer practicing the type of production that is yielding the highest net return per unit of land at a particular location will be able to bid highest for that site. To illustrate, in Figure 8.1, it is clear that a farmer choosing to produce vegetables at A can outbid a person wishing to produce wheat at that same location by up to $1.40 per unit of land. Also, a farmer choosing to produce vegetables at B can just outbid one producing wheat. But, at C a farmer has to produce just wheat in order to survive.

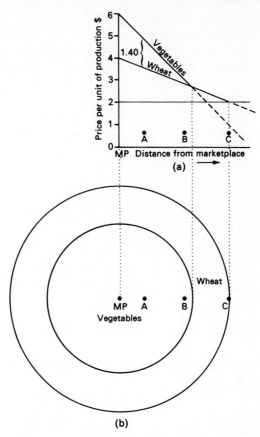

Figure 8.2 Von Thünen's concept of economic rent: the zonation of land uses.

Thus the "highest" (meaning highest price) and "best" (meaning the type of land use that is able to take the best economic advantage of a particular location) use of the land at A and B is vegetables, and at C is wheat. This results in concentric zones of land use around the marketplace, with the inner ring producing vegetables and the outer ring growing wheat (Figure 8.2). With many more crops and types of animal husbandry there will be many more rings. This model is in equilibrium when all the land is in its highest and best use, represented by concentric rings of different land uses around the marketplace. Hence, these types of models are known as concentric space–price equilibrium models. In this particular case, where the price at the marketplace is predetermined, it is a partial equilibrium model.

The Disposition of Economic Rent An important issue is who gets the economic rent, part of the classic "who gets what, where, and how" questions that follow as narrowly defined economic theories are examined in the broader contexts of welfare and social justice (Smith, 1994). In this case, the competition for the surplus is between the landowner and farmer. For example, farmers at A producing vegetables can afford to bid up to, but not exceed, $3.00 per unit of land; the actual contract rent (that is, the amount of rent paid to the landowner) will depend on the availability of land (supply) and the number of farmers competing for locations (demand). This kind of bidding process depends, of course, on the existence of a "market." It could be that the landowner exerts monopoly power and sets the contract rent equal to the economic rent, or, alternatively, the landowner may be benevolent and charge somewhat less.

Land Use Zones in the Urban Environment

The von Thünen concept of economic rent and land use zones can be applied to the urban scene with a few modifications to the assumptions (Alonso, 1964; Mills, 1967; and Muth, 1969). The flat, featureless plain with a homogeneous transportation cost surface and a central marketplace is replaced in the urban case by a *monocentric* (or single-centered) city on a flat plain over which transportation is equally easy in any direction. A bidding process is implied along with the absence of monopoly power. The users of land are assumed to have equal access to the land market, be equal in tastes, and so forth. All employment opportunities and the buying and selling of goods occur at the center of the city, so that people have to go there for economic purposes. Since transportation costs increase with distance, *aggregate transportation costs* (that is, the total cost of getting to a place from all other places in the city) are lowest at the center of the city.

Imagine that there are only three types of land use in the city: central business district functions (including commercial, finance, retail, and wholesale activities), manufacturing, and residential. Central business district activities must be central to their market and labor force (in the city). So their general economic rent curve is the steepest (Figure 8.3a). Manufacturers also wish to have maximum access to labor. Although they require centrality, the general economic rent curve for manufacturing is not as steep as that for central business district activities because the utility (or revenue received per unit of land) at central locations is greater for central business district activities than for manufacturing. Finally, residential activities have the shallowest economic rent curve because individual households cannot generate as much economic rent (or utility) at central locations as central business district and manufacturing activities. In such a situation, the zonation of land uses will be as represented in Figure 8.3a.

Land Use Zones in a Multicentered City   As has been indicated in Chapter 7, the type of city described above existed in general form until the 1940s, for at that time it became evident that multinucleated arrangements of land uses were emerging. The multicentered, or *polynucleated* form is not that much different conceptually from the monocentric city, just more complicated. The central business district may remain the locus of minimum aggregate travel costs for the entire urban region, or it may be just one among a number. The net result is that the arrangement of land uses (central business district, manu-

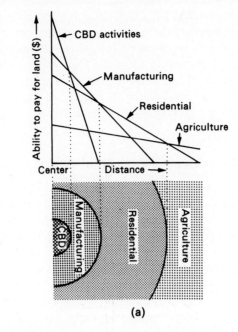

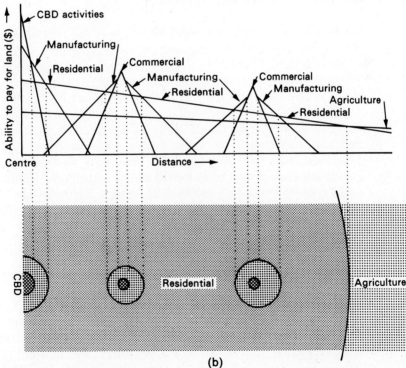

Figure 8.3 Hypothetical land use zones in: (a) a monocentric urban area; and (b) a polynucleated urban area.

facturing, residential) in the three-land-use monocentric model described earlier becomes restructured in a manner similar to that presented in Figure 5.3b. Each of the business-retail-office conglomerations (or edge cities), which are usually located in proximity to major highway intersections, serve as foci for large parts of the metropolis. So a pattern of economic rents and land uses becomes distributed around them in much the same way as the monocentric model envisages with respect to the central business district.

Empirical Examinations of the Changing Patterns of Economic Rent

The changing pattern of economic rent in a city can be examined through land values, for these are the outcome of the bidding process by users for sites. The land value is the value of the site excluding "improvements," that is, the buildings and infrastructure established on them. Of course, the envisaged bidding process is by no means perfect, for a particular site comes up for sale infrequently, but assessment offices and real estate appraisers do make estimates of the value of sites on the basis of sales of adjacent or similar lots elsewhere in the city. Thus, it is assessed land values (which are used for property tax purposes) that are frequently used as surrogates for the theoretical economic rent at particular locations.

Monocentric Land Value Models   Maps of the pattern of land values within the City of Chicago for 1910, 1930 (at the peak of land speculation prior to the 1930s crash), and 1960 (Figure 8.4) demonstrate visually the decreasing applicability of the monocentric economic rent model to urban areas. The land value maps for Chicago in 1910 and 1930 demonstrate that in the metropolitan era values in the city declined with distance from the central business district, with the peak value intersection (PVI) at the intersection of State Street and Madison Avenue, and a zone of higher values tonguing north and south along the lakeshore reflecting amenity values. The PVI is, theoretically, the locus of minimum aggregate travel costs for the entire metropolis. In the metropolitan era, rapid transit, commuter railroads, streetcars, bus lines, and urban arterials, all focusing on the downtown area defined by the "loop" of elevated transit lines surrounding the central business district. Foci of higher values around major outlying shopping nucleations become apparent by 1930 for these are business/retail areas just as well as essential collecting points in the network of public transport facilities.

By 1960, however, this simple structure is replaced by one that is more complex. Land values are still highest within the central business district and ridge northwards along the lakeshore, but they drop off rapidly in the middle part of the city and then rise toward the periphery. The distribution of land values looks much like one-half of a sombrero with a high core in the middle and a rim around the outside encircling a trough within. Edel and Sclar (1975) demonstrate that this "sombrero" pattern of real estate values existed for Boston up to 1940. After that date the pattern did not for by 1970 residential

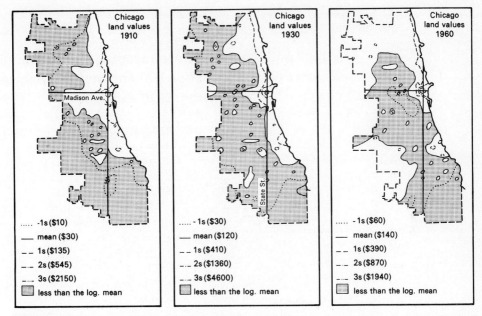

Figure 8.4 Land value (in dollars per front foot) surfaces for the City of Chicago in 1910, 1930, and 1960.
(*Source:* Yeates, 1965, Figures 1 and 2.)

land values *increased* with distance from the center of Boston. Similarly, McDonald (1981) shows for 1960 and 1970, bracketing the opening of the freeway system, that the commercial land value surface in Chicago was essentially flat—one more piece of evidence demonstrating that the advantages of access to the central business district are principally offset by access to the suburban circumferential highway pattern.

The decline in explanatory power of the monocentric model for land values between 1910 and 1960 for the City of Chicago can be demonstrated using multiple regression models. The research model, in which the *dependent* (land values) and *independent* (distance measures) *variables are in logarithms,* is expressed as follows:

$$V_i = a - b_1C_i - b_2M_i - b_3E_i - b_4S_i + e_i$$

where, V_i = land value at any ith location

C_i = distance of the ith location from the CBD

M_i = distance of the ith location from Lake Michigan

E_i = distance of the ith location to the nearest rapid transit station

S_i = distance of the ith location from the closest business nucleation within the city

i = 1, 2, 3, . . ., n; n being the sample size based on a systematic stratified random sample design of front-foot land values for

Table 8.1

Chicago Land Values: The Monocentric Model—Multiple Coefficients of Determination (R^2) and Multiple Regression Coefficients, 1910 and 1960

Year	R^2	MULTIPLE REGRESSION COEFFICIENTS			
		C	M	E	S
1910	76.2%	−.935	−.469	−.300	−.035*
1960	11.2%	−.250	−.120	+.029*	−.124

*Not significantly different from zero at t(.05).
(*Source:* Yeates (1965).)

484 blocks within the city extracted from Olcott (1911; 1961).

a = a constant (or intercept); in the case of the monocentric model it is the PVI.

b = parameters, or regession coefficients pertaining to the independent variables. Given the logarithmic form of the equation, the regression coefficients can be interpreted as elasticities. The larger the coefficient, the greater the impact (or weight) of a given proportionate increase in the magnitude of a variable on land values.

e = error term, or residual.

The signs of the regression coefficients in the research model are hypothesized to be negative in accordance with economic rent theory.

The applicability of the above model with respect to Chicago land value date is indicated in Table 8.1. The model in which three variables are significant describes the 1910 land value distribution extremely well (the multiple R^2 being 76.2 percent). Clearly, although the lake as an amenity attraction and the rapid transit system influence land values, the most important variable in 1910 is distance from the central business district ($b_1 = -.935$). The model for 1960 explains little of the variation in land values (the multiple R^2 being 11.2 percent), though the signs of the coefficients for the variables that test as significant are as hypothesized. Distance from the central business district remains the most important variable, but its influence is quite local. The monocentric model, therefore, turns out to be applicable to the metropolitan era but not to the rapidly expanding city of the suburban era, which, by 1960, in the case of a large metropolis like Chicago, was well on the way to being a multicentered urban region (Getis, 1983).

Polycentric Land Value Models   Even though there has been a great deal of empirical work undertaken on the determinants of site values for particular types of land uses (for example, Burnell, 1985), accounting for variations in prices in terms of site and improvement characteristics and neighborhood attributes, monocentric assumptions have remained either explicit or implicit in most studies (Berry and Kim, 1993). This is unfortunate because in a highly decentralized urban environment access is required to a multiplicity of places (nodes). In particular, single- and multiple-worker households (and their families) in highly mobile labor markets require access to multiple possible workplaces, a range of educational opportunities, and a variety of social-recreational facilities. It is, therefore, important to include multiple centers in site rent, or land value, models.

A Polycentric Model for Los Angeles County   The approach taken by Heikkila and colleagues (1989) examines how the existence of multiple centers affects *residential* property values in Los Angeles County in 1980 (population 7.5 million). The value of a house is considered to be related to three bundles of characteristics associated with dwellings: house features (these are essentially the characteristics used by assessors in valuing houses); neighborhood attributes (of secondary importance to assessors); and accessibility to employment, shopping, and amenity areas. Residential property values, expressed per unit lot area for a sample of almost 11,000 recent property sales, subsume residential land values.

The inclusion of house features, such as number of bathrooms, condition (five-point scale), living area, age of house, and month of sale, permits the property values to be interpreted in terms of land values. Neighborhood attributes are derived from the census tract in which a property is located and include median household income, percentage of professional jobs held by residents in employment, percentage of African American residents, and percentage of Hispanic American residents. Accessibility is estimated by determining the distance of each property to thirty-one possible centers in Los Angeles and adjacent Orange County, and the Pacific Ocean (an air quality and amenity value), and, through iterative stepwise regression procedures reducing these to the most important.

The model presented by Heikkila and colleagues (1989) explains almost all (the multiple R^2 is 93.3 percent) the variation in residential property values. Given the logarithmic transformation of the dependent and independent variables, the regression coefficients can again be interpreted as elasticities (Table 8.2). Living area and condition are obviously powerful determinants of the value of the residential structure, whereas neighborhood status determining variables (the income level and "professional" nature of the census tract) and accessibility measures (particularly to Santa Monica and the Ocean, and away from South Bay, which has a reputation for social disorganization) influence residential land values. The accessibility measures as a group add 10 percent to

Table 8.2

Los Angeles County Residential Values: A Polycentric Model, 1980

VARIABLE (LOGARITHMS)	REGRESSION COEFFICIENTS
House features	
Number of bathrooms	0.077
Condition	.225
Living area	.645
Age	.012
Month of Sale	.048
Neighborhood attributes	
Income	.293
Share of professional jobs	.212
Percent African American	−.009
Percent Hispanic American	.018
Distance to	
Pacific Ocean	−.099
Central business district	.022*
Santa Monica	−.224
South Bay	.110
Palos Verde	−.047
Glendale	−.029
Southwestern San Fernando Valley	.016
Malibu	−.028
West San Gabriel Valley	−.048
Burbank	−.029

*Not significantly different from zero at t(.05).
(*Source:* Heikkila et al. (1989), Table 1, p. 224.)

the explanatory power of the model. Most noticeable, a price–distance residential value gradient with respect to the central business district is not significant.

A Polycentric Model for Dallas County Waddell and colleagues (1993a; 1993b) refine and extend some of the issues raised by the Heikkila team (1989) by addressing more specifically the influence of accessibility to the central business district on the variation in housing prices (including additional types of spatial variables) and involving a larger housing database for a more extensive time period. The Dallas study focuses on house prices for 40,000 residential sales transactions over a five-year period (1985–1990) to capture seasonal and cyclical fluctuations. Using the same type of multiple regression format, the dependent variable is house price (logarithmic transformation), and the wide array of dependent variables are again grouped into those related to housing features, neighborhood attributes, and accessibility (Table 8.3). As with the Los Angeles study, the Dallas study is restricted to the analysis of the effect of a

Table 8.3

Dallas County Residential Values: Independent Variables by Category

ACCESSIBILITY MEASURES (AM)	HOUSE FEATURE MEASURES (HF)	NEIGHBORHOOD ATTRIBUTES (NA)
Highway	Living Area	Average living area
Central business district	Bathrooms	Neighborhood size
Las Colinas	Half-bathrooms	Floodplain
Galleria	Pool	Census tract land use in:
Preston/NW Highway	Fireplace	Single-family housing
Love Field Airport	Wetbar	Multifamily housing
Major retail	Sauna	Offices
Minor retail	Foundation Type	Industrial
University	Heating Type	Roadways
College/Junior College	Air-conditioning	Utilities
Hospital	Roof type	Parks
	Wall type	Percent African American
		Percent Hispanic American
		School district

(*Source:* Waddell et al. (1993a), Table 1, p. 7.)

polycentric environment on one type of land use. Values relating to other important uses, such as offices, commercial and business, and retail, are excluded.

The sequential construction of the models provides some useful nuances with respect to the monocentric-polycentric discussion.

1. The first model, relating accessibility measures (AM) to house prices, indicates that 47 percent of the variation in house prices can be related to accessibility measures (Figure 8.5).

2. The second and third models, relating, in one case, the accessibility measures and house features and, in the other case, accessibility measures and neighborhood attributes to variations in house prices, increase the explanatory power of the model to about 86 percent. This similar amount of increase indicates that the bundles of house features (HF) and neighborhood attributes (NA) variables are measuring similar things: prestigious "professional" neighborhoods tend to have large, high-quality houses, whereas poorer "working class" neighborhoods usually have smaller, less expensive housing.

3. Consequently, the fourth model, involving accessibility measures, housing features, and neighborhood attributes, increases the explanatory power only slightly to 88 percent.

Given the high level of explanation provided by the fourth model, it is possible to isolate the effect of various individual accessibility measures because the in-

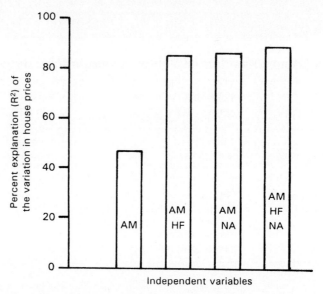

Figure 8.5 Dallas County residential values: explanation (R^2) of variation in house prices associated with accessibility measures (AM), house features (HF), and neighborhood attributes (NA).
(Source: Waddell et al., 1993a, p. 7.)

fluence of housing features and neighborhood quality on house prices are controlled for (or held constant) by their inclusion in the model.

Separated price–distance gradients derived from the fourth model—the "prices" essentially reflecting the values of location (that is, land values)—indicate the complex nature of spatial price patterns in a multicentered metropolis (Figure 8.6). The price–distance gradient, for example, with respect to Galleria (Figure 8.6a), an "edge city" complex involving an upscale shopping mall surrounded by a large concentration of offices on LBJ Freeway in north Dallas, is pervasive over a large distance. Proximity to Galleria (that is, within 1 mile), incurs a negative premium of 5 percent to what the house prices would have been if they were not located so close possibly because of traffic congestion in this area. The premium rises to a positive value of 10 percent between 1 and 2 miles from the complex, thereafter decreasing steadily to 12 miles where there is no discernible effect. Similar types of price–distance premium gradients exist with respect to the other commercial/retail complexes, Love Airfield, shopping nucleations, postsecondary facilities, and hospitals (Figure 8.6a, b, and d).

The gradient from the central business district to the north is quite different from that to the south—echoes, perhaps, of Hoyt's (1939) sector model (Figure 8.6c). To the north there is a positive premium as high as 30 percent, which increases to almost 35 percent for locations in the high-status communities of Highland Park (HP) and University Park (UP). To the south there is an immediate negative premium of 50 percent, associated with blight and social decay in this extensive sector. The negative premium extends as far as 3 miles from the central business district. Beyond this distance, on both the north and south, the

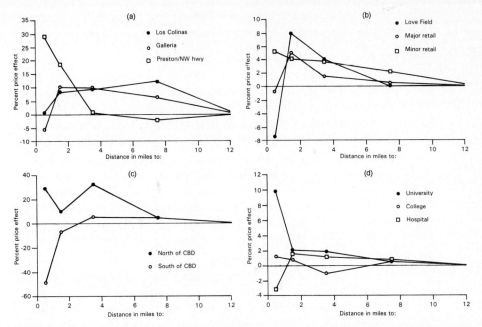

Figure 8.6 Dallas County residential values: price–distance gradients with respect to (a) major retail/office complexes; (b) airport and retail malls; (c) the central business district; and (d) post-secondary educational facilities and hospitals.

(*Source:* Waddell et al., 1993b, Fig. 5, p. 128; Figs. 7 and 8, p. 130; and Fig. 9, p. 131)

central business district has little influence on prices. The central business district has, therefore, a quite restricted effect on prices compared to, for example, an "edge city" such as Galleria.

The Los Angeles and Dallas studies, both using similar methodologies, demonstrate that the central business district-oriented monocentric model is no longer applicable—with respect at least to land values and prices—to the multicentered nature of highly suburbanized large metropolises. Modern urban areas are, however, becoming exceedingly polycentric as work, shopping, and recreational facilities are dispersed over large urban regions. Furthermore, both studies emphasize the close interrelationships that exist between accessibility, housing features, and neighborhood characteristics. Location, function, and form entwine to determine prices.

Sources of Economic Rent

The discussion thus far has focused on one type of rent, differential rent, derived from a simple space–price equilibrium model. It is, however, important at this juncture to recognize that there are a number of sources of rents or prices that have not been included in the detailed discussion because they do not have location (or accessibility) at their core. These various sources of economic rent, which are alluded to in parts of subsequent chapters (for example, Chapter 12), may be summarized for comparative purposes as follows:

1. *Differential rent* is defined here as the difference in return received for the use of a unit of land compared with that received at the margin of production. Hence, the discussion of price–distance gradients with respect to Dallas County is couched in terms of premiums. Ricardo does, of course, define economic rent in a similar way, but with respect to varying fertility of land rather than varying accessibility to a central marketplace.

2. *Monopoly rent* occurs whenever a monopoly comes into being in any part of production or consumption and the value of land is determined only by the price that the monopoly producer is willing to sell at, or the price a monopoly consumer is willing to pay (Scott, 1980). For example, Lorimer (1978, p. 100) claims that suburban land developers were able to add 20 to 30 percent to the price of new homes in many Canadian metropolises in the 1970s due to the monopoly control of land.

3. *Publicly induced rent* refers to land values in particular parts of metropolitan areas that are influenced by public spending—for example, a new bridge across a river that changes the pattern of accessibility. This is evident with public transit improvements, new highway interchanges, and many other types of government investments, such as office establishments. For example, the location of the LBJ Space Center south of Houston has drawn growth in the city in that direction and has acted as an enormous stimulant to land values. The issue of the ways in which the public might recoup the private gains from publicly induced rents is discussed in a later section.

4. *Amenity rent* is the term used to describe the high values some locations attain by virtue of their physical characteristics (e.g., on hills with views), or their location next to social or recreational amenities. The high land values along waterfronts in many North American cities are good examples. A public policy issue that arises in urban planning arises again concerning the degree to which individuals should be permitted to appropriate scarce amenity resources. National and provincial/state public park systems have been developed across North America in response to this issue. A market system for the allocation of land to certain uses both within and beyond urban areas does not, because of its nature and immediacy, cater to the needs for public access to amenities.

5. *Social, or class rent* arises when land values become affected by the incremental choice of households to locate in certain parts of cities and to exclude other households by various means from these areas (Gruen, 1984). The issue arises in many ways and has numerous complications, for the "rents" that are induced may be both positive and negative. For example, the increasing prevalence of master-planned communities at the periphery of cities involves, in their marketing, distinct elements of exclusion (through income, occupation, age, and so forth) and, consequently, the creation of social rent. On the other hand, the "containment" of particular households in parts of urban areas may induce negative social rents.

MODELS OF URBAN POPULATION REDISTRIBUTION

The clustering of people into urban areas is continuing unabated, though this clustering is far less concentrated than it was. This clustering occurs because: (1) urban areas provide a pool of labor which attracts industry and business, and people who wish to be part of that labor force to have access to the employment and services that are available; (2) the complex of business activity provides opportunities for people to use a wide variety of specialist skills or interests in a range of productive capacities; and (3) the intense diversity of activities and peoples generates creativity and new enterprises through the transference of ideas, practices, and social behavior (sometimes referred to as synergy). Given the underlying importance of the clustering of people to this urban dynamic, the changing patterns of population densities within urban areas has been a continuing concern in urban studies (McDonald, 1989). These changing patterns cast light on the ways in which urban areas reinvent themselves to maintain advantages of concentration.

Population Densities in Monocentric Urban Areas

From von Thünen-type arguments pertaining to economic rent it can be argued that population densities should be greatest at the center of the city and diminish toward the periphery, with some ridging of densities along major radials. If it is assumed that the center of the city is the primary focus of employment and economic activities and that locations close to the center are more expensive (high land values) than those at the periphery of an urban area, then the population will have to trade off space costs with transport costs (to work) at any particular location. If a household wishes to minimize both space and transport costs (including the cost of time spent commuting) it will have to live at higher densities close to the center of the city. If a household wishes more space, it will invariably have to move to cheaper locations farther from the center of an urban area and thus incur higher transportation costs. This individual household space cost–transport cost trade-off is central to population density models.

The Clark Model The relationship between population densities and distance from the center of an urban area was modeled by Clark (1951) and modified slightly by Newling (1966). Clark collected data pertaining to many urban areas around the world and concluded that population densities were related systematically to distance from, or accessibility to, the center of urban areas. Specifically, he suggests that urban population densities decrease in a negative exponential fashion (that is, decrease at a decreasing rate) with distance from the CBD. The formulation of this model is:

$$D_d = D_0 e^{-bd}$$

where, D_d = population density at distance d from the CBD
D_0 = a constant indicating the population density at the center of an urban area; that is, the *central density*:

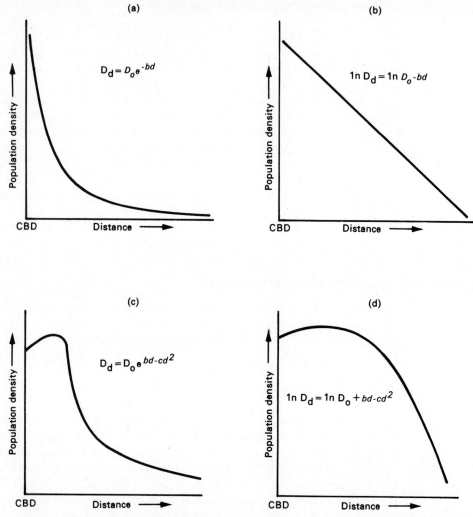

Figure 8.7 Population density models: (a) Clark's model, and (b) its logarithmic transformation; and (c) Newlings's modification of Clark's model, and (d) its logarithmic transformation.

b = a parameter relating to the distance variable, which, as it is negative, indicates the rate of decrease of population with distance; i.e., the *density gradient*.

d = the variable distance

e = base of the natural logarithms

In general, all that such an equation indicates is that population densities decrease rapidly at first with distance from the central business district and then tend to flatten out. This situation is expressed in Figure 8.7a, whereas the straight-line logarithmic transformation of the model is presented in Figure 8.7b.

One feature that the Clark model does not incorporate relates to densities at the center of a monocentric urban area. Few people actually reside in the central business district because nonresidential land uses, such as commercial, office, and business, tend to preempt other uses in the core. The result is a population "crater" at the center, which Newling (1966) incorporates in his second-degree polynomial modification of the Clark model. This recognition of the existence of a central business district "crater" is presented for comparative purposes as Figure 8.7c and its logarithmic transformation as Figure 8.7d. Outside North America, where urban areas tend to be less decentralized, the Clark and Newling monocentric negative exponential population density models are applicable in many urban areas (Stern, 1993).

Change in Central Densities and Density Gradients   Edmonston and colleagues (1985) use a particular form of statistical analysis (the two-point method) to estimate the parameters of Clark's model—the *central density* and the *density gradient*—for 204 United States SMSAs and 20 Canadian CMAs from 1950 to 1975. The results of the analysis are presented in Table 8.4 along with the graphic representation for United States cities in 1950 and 1975, and Canadian cities for 1951 and 1976 (Figure 8.8). The results demonstrate quite clearly that on a highly aggregated scale the Clark model is tenable for recent time periods, but that central densities and density gradients have been continually decreasing; that is, the population density curves are becoming flatter (Figure 8.8). Also, it is evident, for reasons indicated earlier (Chapter 4), that Canadian metropolitan areas have in general higher densities than urban areas in the United States.

The information in Table 8.5 shows how central densities and density gradients change with the population size of an urban area. It appears that in both countries central densities increase with size, but the density gradient decreases with size. Thus, when the densities of central cities collapse and the

Table 8.4

Density Gradients and Central Densities in Canadian and American Metropolitan Areas, 1950–1951 to 1975–1976 (CMAS and MSAS)

YEAR	DENSITY GRADIENTS		MEAN CENTRAL DENSITIES	
	CANADA	U.S.	CANADA	U.S.
1950–1951	0.93	0.76	50,000	24,000
1960–1961	0.67	0.60	33,000	17,000
1970–1971	0.45	0.50	22,000	13,000
1975–1976	0.42	0.45	20,000	11,000

(*Source:* Edmonston, Goldberg, and Mercer (1985), Table 1.)

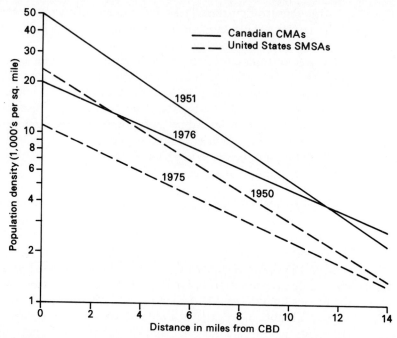

Figure 8.8 The relationship between population density and distance from the center of a metropolis, for United States and Canadian metropolitan areas, 1950–1951, and 1975–1976.

density gradients flatten, as they have done particularly in the United States for large metropolitan areas, the population literally floods outwards to consume vast quantities of space on the rural-urban fringe. This flooding outwards is exemplified in the case of Detroit (Figure 8.9). Anderson (1985) uses a Newling-type model in order to capture not only the density crater at the center of De-

Table 8.5

Density Gradients and Central Densities in Canadian and American Metropolitan Areas, by Population Size, in 1975–1976

POPULATION SIZE OF THE CENTRAL CITY	CENTRAL DENSITY		DENSITY GRADIENT	
	CANADA	U.S.	CANADA	U.S.
100,000–500,000	18,000	10,000	0.43	0.48
500,000–1,000,000	36,000	10,000	0.43	0.26
1,000,000 +	40,000	16,000	0.30	0.19

(*Source:* Edmonston, Goldberg, and Mercer (1985), extracted from Table 6.)

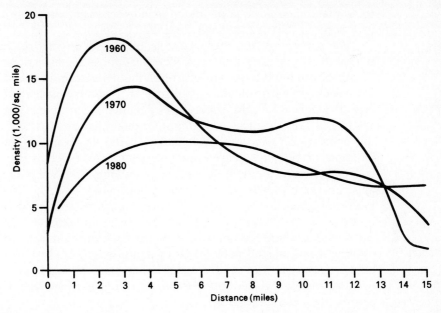

Figure 8.9   Changing density-distance relationships in Detroit in 1960, 1970, and 1980.
(*Source:* After Anderson, 1985, p. 418.)

troit, but also the dampening wave of population decentralization. The population density curves for 1960, 1970, and 1980 (Figure 8.9) demonstrate quite clearly the depopulation of the central city of Detroit that has occurred since the early 1960s. In this particular case, the general density gradient is now almost flat. It must be recognized, however, that Detroit (along with St. Louis and Cleveland) is one of North America's more extreme examples of the abandonment and depopulation of a central city.

Population Turnaround   The precipitous decline in central densities is symptomatic of the decrease in inner-city populations that has been occurring in many North American urban areas (as well as other urban areas in the industrial world) since 1945. There have, however, been some recent signs of a decrease in the long-term decline of inner-city populations in a few North American cities (Bourne, 1991). This apparent population turnaround in a few places has attracted great interest because a return of even a few households to the inner city might herald the beginnings of reversal of the negative image which prevails for most central cities.

With respect to *central-city* populations *per se,* most places that have incurred declining populations since 1970 have continued to do so. The only large metropolises in the United States that experienced a decrease in population between 1970 and 1980, followed by an increase between 1980 and 1990, are New York City (including the boroughs of Bronx, Brooklyn, Manhattan, and Queens), Boston, San Francisco, Seattle, and Oakland. Of the others, those that experienced a decrease between 1970 and 1980 have continued to do so, and those that increased in population (usually because their central cities are quite extensive) in that decade have continued to do so.

The perception of a turnaround is, of course, directly related to the definition of the "inner city" that is used because those regenerations that have been identified are quite local. For example, the population in the area immediately surrounding the central business district of Toronto (inside the central area cordon in Figure 14.8 and Table 14.4) declined from about 150,000 in 1951 to 114,000 in 1981, but has subsequently rebounded by 1991 to 140,000. Much of the population turnaround that has occurred has been in newly constructed buildings on recently cleared redevelopment sites (such as False Creek in Vancouver), often in condominiums. The population in this new housing stock usually consists of single- or two-person households with no children, in multiracial neighborhoods, with employed persons involved in professional occupations in the service, finance, and communications industries. In a few isolated areas there has been some "gentrification," that is, complete renewal of existing older housing stock occupied in a previous era by lower-income households. These more wealthy households exist cheek-by-jowl with the less wealthy, but in high-security complexes or enclaves (neighborhoods), thus accentuating social fault lines.

Population Densities in a Multicentered Urban Area

It is clear from the previous discussion that the monocentric model, theoretically useful though it may be, provides a poor representation of metropolitan areas in North America. The polycentric model is more representative. Crosstown commuting and highly dispersed employment opportunities are facts of life in the multicentered metropolis of the 1990s (Waddell and Shukla, 1993). For example, in the 11.5-million-population five-county area around Los Angeles (Los Angeles, Orange, San Bernardino, Riverside, and Ventura counties), Gordon and colleagues (1986) identified the existence of forty-five population peaks in 1970 and fifty-eight such peaks in 1980.

The complex model used for the five-county Los Angeles region, therefore, examines population densities with respect to the peaks existing in both 1970 and 1980. Table 8.6 lists the results for the twenty-six peaks that existed in both 1970 and 1980 in Los Angeles County *only*. The feature to notice from Table 8.6 is that there is no general tendency for the peak densities and gradients to change in the same way: Some areas are increasing in density while others are decreasing, and they can be adjacent to each other geographically. Thus, whereas the monocentric model has central densities and density gradients becoming flatter over time, the multicentered model demonstrates a far more complex pattern of change.

Monocentric Patterns, Polycentric Patterns, and Race   Although changes in the availability of transportation, government programs in housing and transportation, and the housing requirements in the three post–1945 building

Table 8.6

Los Angeles County Peak Densities and Associated Density Gradients, 1970 and 1980

PEAK REFERENCE NUMBER	1970		1980	
	PEAK DENSITY	GRADIENT	PEAK DENSITY	GRADIENT
1	19,300	−1.06	21,800	−1.05
2	5,800	−0.11	4,500	−0.42
7	9,860	−1.77	8,630	−1.48
8	7,640	−0.46	8,610	−0.55
9	13,000	−1.55	14,800	−1.11
11	5,930	−0.71	5,390	−1.89
14	15,500	−1.90	11,400	−1.08
15	11,100	−1.22	13,300	−1.63
16	13,300	−0.96	14,100	−1.72
17	17,500	−1.42	15,900	−1.44
18	22,300	−1.89	24,100	−1.70
19	31,000	−1.70	33,400	−2.10
20	11,500	−3.88	20,000	−10.50
21	6,410	−0.26	7,240	−0.48
22	6,310	−0.95	7,560	−0.90
23	23,700	−1.94	23,400	−1.56
24	14,600	−4.64	16,700	−4.85
25	19,000	−1.85	30,000	−2.10
26	9,450	−1.05	14,600	−1.49
27	21,600	−1.13	32,900	−1.10
28	11,200	−0.63	36,300	−1.66
29	20,800	−2.17	36,300	−1.66
30	6,070	−0.03	9,150	−2.73
31	20,500	−1.48	21,700	−1.72
33	5,300	−1.18	10,500	−0.61
34	5,140	−0.22	12,900	−1.86

(*Source:* Gordon, Richardson, and Wong (1984), p. 12.)

booms, have been primary influences on decentralization and the shift to the polycentric city, there is no doubt that racial factors have played an important part in the development of highly suburbanized metropolises. Reference has been made previously to the movement of the white population in the United States to the suburbs in the post–World War II period. This is documented quite clearly in population density curves calculated by Reid (1977) for the African American and white populations in a sample of thirty-three SMSAs for 1960 and 1970 in the United States (Figure 8.10). Whereas both the African American and white population density curves in 1960 were (in general) of the Clark type with the African American population highly centralized, by 1970 the white

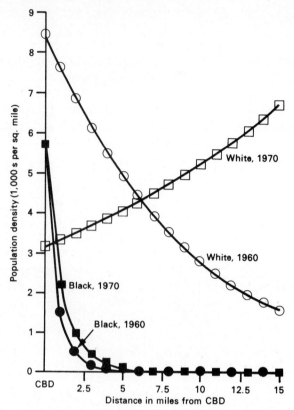

Figure 8.10 The generalized population density—distance relationship for the white and African American populations in a sample of thirty-three metropolitan areas, 1960 and 1970.
(*Source:* Reid, 1977, p. 356.)

population density curve had reversed with white population densities increasing toward the periphery of the cities (and then, presumably, decreasing toward the outer suburbs).

The stark gradient reversal in a ten-year period provides a clear demonstration of the precipitous nature and volume of "white flight" and infers segregation. Although a few of the distinctions between the location of African American and white households have something to do with income and affordability, a large part of the spatial separation is related to prejudice-induced discrimination. As a consequence, as has been mentioned previously in Chapter 7, toward the end of the 1960s the federal government increased its attempt to prevent discrimination in housing (such as through Title VIII of the Federal Fair Housing Act of 1968). Although evidence concerning the access to housing and the extent of integration that is being achieved will be discussed in greater detail in later chapters, segregation within central cities has increased. It appears that even where African American suburbanization in the United States has occurred high levels of spatial separation have been perpetuated (Farley, 1986).

PUBLIC POLICY AND LAND USE CHANGE

There is, then, a market for urban land in North America in which people and business activities compete for certain locations. In the market, "prices" of various kinds are established which, in effect, allocate land among the various types of uses. This market is not, however, allowed to run unfettered, for, if it did, the negative effects on society would be far greater than any positive effects (such as land being in its "highest and best use"). For example, in a free and unfettered land market it is highly unlikely that parks would be available for public use in prime locations, such as on waterfront land or in high-density areas. Government, and particularly local government, has to intervene in some way to influence the allocation of land to certain uses to achieve social objectives.

In this section the discussion will focus on: (1) the issue of the amount of land being consumed for urban purposes, particularly at the urban fringe where planning controls often appear to be difficult to implement; (2) a discussion of zoning as a method often used by planners to intervene in the allocation of land to certain uses; and (3) the use of the property tax to raise revenue and influence the use of land and the rate of land conversion.

The Increasing Consumption of Urban Land

The decentralization of people and economic activities, along with decreasing central densities and decreasing density gradients, imply increasing consumption of land for urban purposes and, hence, increasing amounts of land being transferred from rural to urban uses. This is a matter of international concern (Pacione, 1990) as well as a public policy issue in North America since the onset of large scale suburbanization in the 1950s (Clawson, 1971). The amount of space being consumed by urban areas is of particular environmental importance in certain parts of North America (for example, California, Florida, southwestern Ontario, and the lower Fraser Valley of British Columbia) where scarce, highly productive agricultural land is being consumed rapidly for urban purposes (Furuseth and Pierce, 1983; Platt, 1985; Gaylor, 1990).

The general reason for this concern can be indicated by using *land consumption rates* (LCRs) which measure the amount of urban land being consumed (per 1,000 persons) for urban purposes at a particular date. Land consumption rates are, therefore, the reciprocal of population densities. Figure 8.11 shows the change in the general relationship between land consumption rates, measured in hectares per thousand population, and city size for forty urban agglomerations in central Canada, and the conceptual relationship for cities in the United States (based on information in NALS, 1981). The graph emphasizes three major points.

First, land consumption rates are increasing through time. As the population becomes wealthier, transportation improves, and techniques of production and styles of consumption change, more land (space) is consumed by each individual. In general, the *increase* in urban consumption of rural land appears (on

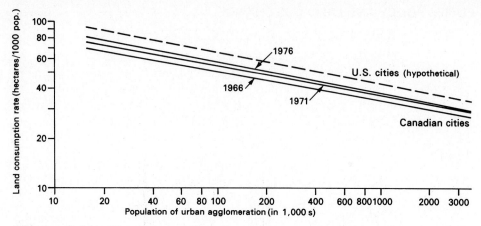

Figure 8.11 The relationship between land consumption rates and population size for a sample of urban areas in Canada and the United States.
(*Source:* Yeates, 1985a, p. 7.)

the basis of Figure 8.11) to be 10 hectares (or 24.7 acres) per thousand people per decade. A corollary to this is that cities that have grown the most in recent decades will tend to have higher average land consumption rates than those that grew most in earlier epochs because their eras of greatest development were when these rates were the highest.

Second, land consumption rates tend to decrease with city size. The graph shows that cities of about 100,000 population tend to have rates around 60 hectares per 1,000 persons, whereas cities of about 2 million persons tend to have rates around 30 hectares per 1,000 persons. An implication of this is that large cities tend to be more efficient users of land than smaller cities. Evidently, greater competition for locations in larger cities forces up land prices and hence intensity of use. This point needs to be reiterated with small-town environmental advocates—small urban areas are, on a size-modified basis, greater consumers of land than large urban places.

Third, United States cities tend, on the average, to have larger land consumption rates than Canadian cities. The inhabitants of United States cities consume, on the average, about 10 hectares per 1,000 persons more than those in Canadian cities. Although there are undoubtedly a large number of interlocking reasons for this difference, there does seem to be some evidence to indicate that public attempts to shape development increase land prices in Canadian cities to a higher general level than those in the United States and, therefore, development occurs generally on smaller lots in Canada.

For example, in a comparative study of four rapid-growth cities—Phoenix, Houston, Calgary, and Edmonton—Sands (1982) demonstrates that local governments in Calgary and Edmonton limit development to areas served by public water and sewer systems and require developers to pay most of these facility costs up front, whereas development in Houston and Phoenix is allowed to occur in *privately* established utility districts which are financed by tax-free bonds repaid by user fees. The net result is that Canadian suburban developments tend to be more compact than those in the United States.

The Direct and Indirect Impact of Urban Development on Rural Land

Unfortunately, the land consumption rates summarized in Figure 8.11 tell only part of the rural-urban land conversion story. They measure, as do all well-known conversion measures, the *direct,* that is, contiguous and visible impact of urban development on rural land. What is generally ignored or overlooked in such measures are the *indirect* impacts relating to conversions that are not necessarily contiguous or visible. These discontiguous indirect impacts include quarries, landfills, trailer parks, car pool lots, airports, "greenfield" factories, truck stops, and container yards. The land consumption rates as currently measured, therefore, seriously underestimate the amount of rural land consumed for urban purposes.

Measuring the Direct–Indirect Land Conversion Ratio   In a detailed study of land conversion in Oxford County, Ontario, and Laprairie County, Quebec, Pond and Yeates (1993) use the following simple accounting model to estimate the amount of land in urban use in each county:

$$LC_t = DL_t + IVL_t + ILVL_t$$

where, LC_t = the amount of land influenced directly or indirectly by urban development at time t

DL_t = the amount of land in direct urban use at time t

IVL_t = the amount of land in indirect visible use at time t

ILVL_t = the amount of land in indirect, less visible land use at time t

The amounts of land in each of the three categories were calibrated for a number of years using assessment rolls, aerial photography, ground surveys, household surveys to define exurbanites (Davies and Yeates, 1991), and a variety of statistical models to define transitional land (Pond and Yeates, 1994a; 1994b) on the basis of recent sales and landownership information.

The results for 1986 are as follows (land in hectares):

Laprairie   15388.5 = 6414.6 + 1361.9 + 7562.0
Oxford   14261.4 = 4419.4 + 5684.0 + 4158.0

Thus, the direct–indirect ratio for Laprairie County is 1:1.4, and for Oxford County 1:2.23; that is, for every one unit of rural land (hectares or acres) converted to direct urban use in Oxford County, 2.23 units are converted indirectly, and the corresponding ratio for Laprairie County is 1.14 units indirectly for every one unit directly. The difference between the two ratios is related to the difference in density of development between the two counties. Oxford County, located about 100 kilometers west of Toronto between Kitchener-Waterloo and London (Ontario), is quite rural with three country towns (Woodstock, Ingersoll, and Tillsonberg), whereas Laprairie County has been transformed since 1960 to become part of suburban Montreal south of the St. Lawrence River (Reid and Yeates, 1991).

Public Policy Implications There are, perhaps, two observations that should be made related to the rural-urban land conversion research described in the previous section, both of which imply a need for much larger-scale regional planning than often currently occurs. First, the amount of land usually measured as "urban," that is, land in direct contiguous urban use, greatly understates the impact of urban development. The calculated direct–indirect ratios suggest that the total urban impact is twice or three times that indicated by the amount in direct use. It appears that urbanism is far more pervasive at the urban-rural fringe than is often recognized. So sensitive lands may well be "urbanized" before actual physical development occurs. Regional planning relating to urban development and its impacts should, therefore, extend over an area far more extensive than that currently under direct or immediate urban use. Planning that does not involve such extensive urban regions fails to address more than half the story.

Second, the amounts of land being held in speculation at the rural-urban fringe in anticipation of possible future urban development are quite large—hence the comparatively large indirect as compared with direct land use measurements for Oxford County. Unfortunately, areas under intense land speculation are invariably in local government jurisdictions that are not well prepared to deal with the more egregious plans of development companies and the implications of their deals on the public provision of infrastructure (roads, drainage, storm sewers, waste disposal, schools, community facilities, and so forth). Furthermore, it is not unknown for local governments lying in the path of possible urban expansion to be hijacked by representatives of development interests. This again infers an important role for large-scale regional planning to harmonize the diverse interests of land owners, developers, home buyers, local government, and environmentalists.

Thus, the more inclusive land consumption accounting model described above provides a useful way of addressing urban land consumption issues in a planning context. The direct–indirect ratio, in effect, is a form of *leading indicator* of possible future development. The larger the ratio, the more the land that is being occupied by exurbanites (IVL) or by individuals or companies that may be presumed to be speculators (ILVL). The traditional land consumption rate is a *lagging indicator*, because, as it is based on direct contiguous urban land consumption, it involves land that is already developed for urban purposes.

Zoning

One of the most general methods used by planners in North American metropolises to control densities and the use to which land is put is through zoning (NLC, 1993). Comprehensive zoning first appeared in New York City in 1916 as an outgrowth of the British common law of nuisances, which has as its central principle the tenet that one must use his or her own rights so as not to infringe upon the rights of others. This idea is interpreted in the land use situation as meaning that one should not use ones own property in such a manner as to interfere with ones neighbor's use of their property. The concept has been extended over the years to include not only property but also the protection of life, health, comfort, and peace in the community.

It is, therefore, important to recognize that zoning was devised not as a planning tool for regulating the use or density of occupancy of land but as an extension of the common law of nuisances (Hason, 1977). This extension achieved a high level of legal validity as a result of a number of appeals of zoning decisions through the courts, the most celebrated case being a decision by the United States Supreme Court in favor of the enactment of a zoning bylaw by the suburb of Euclid, Cleveland, in 1926 (Weaver and Babcock, 1979). The need for zoning arose because the real estate industry and property owners perceived that the value of urban land and property could be seriously affected by what use was made of adjacent properties, or of other properties in the general vicinity. Thus, an expensive upper-middle-class house or row of housing could be affected negatively in value and marketability by the sudden establishment of an incongruent economic activity within the vicinity. Examples include the establishment of sleazy bars in residential communities and the location of noisy, noxious industries within the confines of a quiet neighborhood.

Therefore, the basic premise for zoning, as perceived by the owners of property and the real estate industry, was that the private land market could not be left to operate in an unrestricted manner. What the property industry had realized was that *externalities* (that is, factors not derived from the site itself) profoundly influence the market value of a property. So they wished to have some level of control over these externalities in order to protect (and even enhance) values. Ever since the implementation of zoning, there has been much debate about whether it does protect and enhance values. Chressanthis (1986) demonstrates, using a time-series model, that zoning has enhanced house prices. McMillen and McDonald (1993), on the other hand, argue that the introduction of zoning in Chicago in the 1920s probably did not increase land values or improve the allocation of land over and above that which could have been derived from an unrestricted land market. The effectiveness of zoning depends, however, on the type of zoning used and the vigor of its implementation.

Types of Zoning   Zoning, as it has developed over the years, has become an exceedingly complex political business, in part because it involves decisions at the *local* level, made by locally elected officials, that impinge directly on local individual households and businesses. The stakes may be quite high, and the impacts felt personally by voters who invariably have immediate avenues of access to local politicians and government officials. NIMBY ("not in my back yard") coalitions arise remarkably quickly, for example, with respect to the location of waste dumps, hospices, psychiatric care units, and half-way houses for parolees. Three types of zoning (that tend to overlap) are examined in the following sections: general zoning for the separation of uses, zoning for planning purposes, and zoning desired to achieve performance standards.

Zoning for Separation of Uses   Much of the zoning in cities is designed to designate certain parcels of land or whole parts of cities for particular uses, often with designated household unit or space densities. There are two types—*hierar-*

chical zoning and *specific zoning*. Hierarchical zoning, which still exists in many smaller communities, in effect ranks land uses according to their need for protection. In general, land for housing is at the top, commercial in the middle, and manufacturing at the bottom. Anything can be built on land zoned for manufacturing, only commercial and residential uses on land zoned commercial, and only residential on land zoned residential. Specific zoning, on the other hand, designates land for one use (and stated maximum coverage) and one use only—that is, land zoned for manufacturing can be used only for that purpose. Most communities have discovered hierarchical zoning to be dangerous in times of rapid growth. For example, abandoned industrial property on waterfront land may be purchased and redeveloped with high-rise condominiums before a local government has the wit and inclination to publicly think through the most appropriate re-use of the site. As a consequence, specific zoning is now used most often.

Zoning and Planning   City planners in North America are in a frustrating position: They may go through long procedures with local communities to develop goals and plans, but they find they have little power to actually implement much since it is the developers who actually build things. As a consequence, planners have devised many ways of trying to influence the course of events through zoning. The direction of growth of a city may be influenced by zoning land in advance of development; this is called *impact zoning*. Thus, the construction of new housing and an associated retail mall may be prezoned for one part of an urban area but not another. Impact zoning requires the preparation of a well-articulated regional plan with good forecasting procedures of the likely demand for different types of urban land.

Historic sites may be preserved by allowing the use (or density of use) that a developer wishes for that site to be transferred somewhere else; this is known as *transfer zoning*. In fact, transfer zoning, for all types of objectives such as building setbacks and child care centers, not just historic preservation, has become the biggest money-making game in many metropolises today as developers trade social or site-planning objectives for increases in floor space densities. *Percentage zoning* is often used by planners to encourage a mix of land use types—for example, high-, middle-, and low-income housing plus commercial uses in an area being redeveloped. Also, the outward spread of urban development at the urban-rural fringe may be controlled through the use of designated *green belts* involving agriculture and forest uses.

Zoning and Performance Standards   Zoning to achieve certain performance standards, such as the same type of service and industrial activity in a technological park, or with similar types of housing units (such as bungalow or split-level detached houses in one area separate from low-rises) in a suburban tract, is common in many North American metropolises. Restrictive residential land use and minimum-lot-size zoning are used by planners and developers as substitute ways of controlling the population density of future residential development (Bates and Santerre, 1994). In general, these procedures work quite well with respect to controlling the intensity of development, but it is often argued that the procedure has also been used for (or perverted into) *exclusionary zoning* in which certain income groups are excluded from certain areas by the regulation of lot sizes, minimum floor space for housing, and even the facilities within the house. It is a short step, of course, from such income discrimination to race and class segregation.

Thus, zoning activity in many communities is exceedingly complex involving a mix of planning offices, planning boards, zoning commissions, appeal boards, zoning officers, building inspectors, environmental protection agencies, and the local council itself. Given the complexity of the process, and the specific interests that are usually involved in any one case, it is not surprising that the process is dominated by the groups whose livelihoods are most affected—real estate agents, developers, and lawyers. It is an important administrative arena that shapes to a large extent the environment in which people reside. Hence, local government in general, and planning and zoning in particular, must not be the preserve of those with a direct (or indirect) financial stake.

Zoning and Land Use Change   In a wide survey of land use change in metropolitan areas, Wilder (1985) concludes that on the whole:

1. Land tends to remain in the same type of use even though there may be changes made to the buildings on the land.
2. The change that occurs usually results in more intense use of the land.
3. Certain uses invariably do not follow one another—for example, industrial and residential uses rarely follow one another.
4. Land use change seems to vary with location—succession within the same use is common toward the center of cities, whereas conversion of land use from one type to another is more common at the periphery of cities.

Given these general types of change, it is clear that zoning and its attendant constraints play an important role in influencing the types of changes that occur. The lesson from this is that if a situation demands an orderly land market with fairly predictable succession and land use change, then strict specific zoning is desirable. If, however, conversion is necessary to promote more radical change, then zoning will have to be less rigid and more flexible. However, major zoning changes can significantly affect prices (Chressanthis, 1986), thus providing gains for the owners of rezoned properties. There should be some mechanism that enables society to recoup all or part of the gains made by individuals or corporations through the act of rezoning (or other public investments in urban infrastructure). Such recoupment would maintain credibility in the zoning process and for those involved with it. One way of recouping such gains is through the property tax.

The Property Tax and Land Development

The property tax, which will be discussed in greater detail in Chapter 15, is one of the chief ways that local governments have for raising revenues. In 1990, in Canada, one-third of all municipal revenues came from taxes on property. In the United States, 18 percent of all local government revenue is derived from the property tax, but 52 percent of tax revenue raised by local governments themselves is derived from this source. The *property tax* (T) is generated on the basis of *assessed property values* (B), which, in most urban areas, are essentially based on housing features with some acknowledgment of neighborhood attributes, and the local *millage rate* (t), which is the amount to be collected for each dollar of assessed valuation. The property tax for an individual property is the assessed property value modified by the mill rate:

$$T_i = t_i \times B_i$$

where, i = 1, 2, 3, . . ., n, there being *n* properties in the city.

The mill rate is set locally and varies among municipalities. It also varies within municipalities according to type of land use. Some land uses (for example, office and commercial activities) frequently have a higher mill rate than residential uses. The sum of B_i is the *property tax base*. Given the magnitude of B_i, T_i is usually divided by a factor such as 1,000 for calculation of the actual property tax bill!

Although assessed values may originally have been close to market values, with the passage of time this is no longer the case, and reassessment only occurs if there has been an improvement to the features of the building structure (Barlowe, 1986, p. 446). As this can lead to enormous difficulties when actual prices in some areas incur a premium due to changes in the value of particular locations (recall the Dallas study discussed previously), or old buildings have been upgraded (for example, with asbestos removal, or new conduits for fibre optic cabling), in many jurisdictions assessed values have been replaced with estimated market prices as of a certain date (a "valuation" date). *Market value assessment* is difficult to implement because of the greatly increased property tax payments that might be required in some parts of the metropolis. For example, market value assessment invariably seems to imply higher property taxes for commercial and retail properties in sensitive areas, such as central business districts, where local governments maybe trying to retain employment opportunities.

The assessed property value, or market value, (B) is made up of two components: the value of the land (site value), known as the unimproved capital value (UCV); and the value of the improvements to the land (such as the buildings placed on the land), known as the improved capital value (ICV). As has been indicated previously, in North American cities the assessed property value is made up largely of the improved capital value (Bird and Slack, 1993, p. 73). The question, therefore, arises as to what would be the effect on urban land use and change if the property tax system were based on site values (UCVs) rather than improvements on the land (ICVs)?

In general, it would seem that property taxes based on improved capital values make the holding cost of land itself too low (Ohta, 1984). If land is not improved, that is, if it is without buildings, the property taxes are low. This creates a low cost for merely holding the land—other than the cost of foregone revenue that might be realized from the improvements. On the other hand, basing property taxes on improved capital values may well, in some cases, discourage owners from making additional improvements because of the fear of higher property taxes. Thus, it can be argued that a taxation system based on unimproved capital values (that is, land values):

1. Promotes rapid land use change because it discourages people from holding either unimproved land or improved land that is generating little revenue (or satisfaction).
2. Generates improvements closer to the "highest and best use" at particular locations.
3. Helps to reduce the extent of sprawl at the rural-urban fringe through the promotion of more efficient land use in general, and to foster more compact development.

A sure sign of a bad taxation system is one that encourages the holding of land by developers and speculators on the urban fringe, and the holding of land within cities in the form of surface parking lots. If land was "expensive" to hold but "cheap" to use, property developers would be discouraged from holding it too long before development and encouraged to use it more efficiently because of the high holding cost.

CONCLUSION

The discussions of land use theory, land value and population density models, and associated public policy implications, emphasize that the internal structure of North American cities is the product of a variety of forces that operate in different ways at different times. Undoubtedly, cyclical forces play a powerful role, though it must be recognized that the individual responses of urban areas to upturns or downturns in the economy varies according to the economic structure, location, and entrepreneurial dynamism of the city at that particular time period. The decentralization that these models document emphasizes the need for urban and regional planning that includes and extends far into the rural-urban fringe (Bryant et al., 1982) beyond the more densely built-up suburban areas. These models in effect crystallize the discussion of change and dispersal in Chapter 7, and also provide a basis for more detailed subsequent analysis of the changing locational aspects of employment opportunities, social groups, housing, transportation, local government provision, and sustainable urban development.

Chapter

9

The Intraurban Location of Manufacturing

*I*n Chapter 5 it has been noted that while total manufacturing output (in constant dollars) in North America has been increasing, manufacturing employment has been fairly static and declining relative to large increases in employment in consumer, producer, and governmental and public services. Nevertheless, the dynamics of the intraurban location of manufacturing remains a vital part of contemporary urban analysis because manufacturing is concentrated in urban areas. It is fundamental to the economic health of many metropolises. Also, the changing patterns of the intraurban location of manufacturing jobs have enormous social repercussions. Although manufacturing land occupies only about 9 percent of the developed land in most North American industrial cities, this land provides space for a vital aspect of the urban economy. Not only do the manufacturing activities on the land provide a large number of jobs, they also contribute a large proportion of the *basic* activity (that is, city supporting) in the urban region in which they are located.

In the United States, for example, nearly 80 percent of total manufacturing employment is located in the 281 metropolitan areas (MSAs or CMSAs) defined by the Bureau of the Census. Forty percent is located in the twenty largest metropolitan areas (Table 9.1). In Canada, 38 percent of the manufacturing employment in the country is located in the CMAs of Toronto and Montreal. However, most of the metropolises that are particularly dependent on manufacturing tend to be relatively small in population size. Figure 9.1 shows

Table 9.1

Manufacturing Employment and Contribution of Manufacturing to Total Metropolitan Employment and Earnings, United States, Twenty Largest Metropolitan Areas

METROPOLITAN AREA	MANUFACTURING EMPLOYMENT (THOUSANDS)	PERCENT OF	
		METRO EMPLOYMENT	METRO EARNINGS
New York-New Jersey-Long Island CMSA	1393.2*	15.5	15.1
Los Angeles-Anaheim-Riverside CMSA	1270.3	18.9	20.8
Chicago-Gary-Lake County CMSA	732.7	17.8	21.4
San Francisco-Oakland-San José CMSA	507.1*	15.8	19.8
Philadelphia-Wilmington-Trenton CMSA	472.8	20.3	21.2
Detroit-Ann Arbor CMSA	518.3	28.3	35.0
Washington, D.C. MSA	89.5	4.3	4.4
Dallas-Fort Worth CMSA	343.8	16.6	20.4
Boston-Lawrence-Salem-Laurel-Brockton CMSA	377.7*	19.0	19.5
Houston-Galveston CMSA	187.1	14.1	14.6
Miami-Fort Lauderdale CMSA	137.9	8.7	9.3
Atlanta CMSA	178.1	12.5	13.6
Cleveland-Akron CMSA	296.4*	27.2	30.1
Seattle-Tacoma CMSA	194.2	20.8	23.1
San Diego MSA	135.0	12.2	14.0
Minneapolis-St Paul MSA	263.9	23.2	25.4
St. Louis MSA	224.1	22.3	24.9
Baltimore MSA	130.5	11.6	14.4
Pittsburgh-Beaver Valley CMSA	132.6	16.2	19.7
Phoenix MSA	139.3	17.8	17.9

(*Source:* Bureau of the Census (1991), Table A.)
*1989 data unavailable; 1987 data used.

the locations of the sixty-six MSAs in the United States that have more than 30 percent of metropolitan earnings derived from manufacturing. Only five of these (Rochester, Detroit, Greensboro/Winston-Salem, Milwaukee, and Cleveland/Akron) have resident populations in excess of 1 million. Most of the manufacturing-dependent MSAs are located in the Northeast and Midwest.

Although the largest urban places the United States (Table 9.1) have quite diversified economies, a significant number have at least 20 percent of total

Figure 9.1 The sixty-six metropolitan areas in the United States most dependent on manufacturing.

earnings within their respective metropolitan areas derived from manufacturing employment. Detroit remains the large metropolis most dependent on manufacturing, with 28 percent of employment within the region, and 35 percent of total earnings derived from manufacturing. The difference between these two percentages is also indicative of the high hourly wage rates in the unionized automobile and related industries, for Detroit is ranked fourth highest in manufacturing earnings per employee in the United States. It is, therefore, extremely important to know something about the changing pattern of manufacturing location within urban areas, generalize these patterns to a typology of intraurban manufacturing, and examine some of the basic processes influencing growth or decline in employment among these locational types.

THE ACCUMULATION OF MANUFACTURING INFRASTRUCTURE WITHIN URBAN AREAS

Consistent with themes developed earlier (in Chapters 3, 4, and 5), the historical evolution of manufacturing in urban areas can be discussed in terms of developments in technology and organizational change in manufacturing. These two features are examined historically and spatially in terms of: (1) the pattern before the industrial revolution in the mercantile era ; (2) the early-industrial-revolution waterway/canal pre-rail era, which reached its peak by the early 1850s; (3) the middle-industrial-revolution railway era of the classic industrial and metropolitan eras, which extended for almost one hundred years to 1945; and (4) the mature industrial highway era, characterized by decentralization during the last half of the twentieth century.

In the preindustrial United States, manufacturing was usually related either to the shipbuilding and repair industry, or the fabrication of consumables (milling, clothing, leather, or horse-servicing products) for local domestic markets. These activities were located on the only good means of transportation—the sea, rivers, and lakes. Waterfronts were, therefore, prime locations, particularly those close to the center of towns which were also invariably located adjacent to waterfronts. Attracted by these favorable logistics, both industry and commerce launched intense competition for good waterfront locations that provided good access to supplies and were central with respect to the local population.

Later, during the early part of the industrial revolution, waterways became even more important for they were not only the mode of transportation for raw materials and supplies and hence industrial production, but in the case of rivers also a source of power and material input into the manufacturing process. As a mode of transportation, rivers were improved through dredging, straightening, and the construction of locks; canals were built to extend the reach of water transport and make more land serviceable for manufacturing and commercial development. The improvement of rivers also increased their use as a source of power, for along with canalization came the construction of small dams and weirs. These powerful access and production locational forces were mediated by a third, the relative immobility of labor located within the towns and cities. Thus, manufacturers either established both plants and industrial towns at suitable waterpower locations, for example, Waltham, Massachusetts, or Paterson, New Jersey, or located their plants along improved rivers or canals close to the central parts of existing cities.

The Middle-Industrial-Revolution Railway Era

The rapid development of the North American railroad network in the middle and late nineteenth century reinforced the advantages of central areas in most urban concentrations. This was particularly due to the twin effects of the focusing of lines on the downtown areas and the central concentration of labor. Thus, a melange of wholesale and manufacturing establishments became located on the fringe of central business districts and on the lands around the rail lines close to the downtown terminals. Furthermore, in the process of serving each other, they developed a complex network of ancillary industries. Thus an enormous congestion of industry, people, and jobs concurred in a relatively small area, which could only be relieved by making more land accessible for development.

Not only were the lands along the rail tracks attractive locations for industry seeking space, it was also in the best interests of the railroad companies to attract manufacturing to them. The railroad companies wanted to capture the traffic generated by the manufacturing plants, because they also owned much of the property adjacent to the tracks, an ownership that was invariably the result of government land grants. Thus, right from the outset, the railroad companies were also in the urban land development business. They used their land grants and purchased rights of way to lure factories to their facilities. This is well illustrated in the case of belt-line railroads which developed in almost every large metropolis during the latter quarter of the nineteenth century.

Belt-Line Railroads and the Decentralization of Manufacturing Belt-line railroads were constructed to allow interconnections and transfers of rolling stock, and the goods they contained, between the major intercity trunk lines that tended to focus radially on the center of major metropolises such as New York, Chicago, St. Louis, and Philadelphia. They facilitated interline transfer, and the bypassing of metropolitan areas for through traffic. Key to the operation of belt-lines are classification yards in which the incoming rolling stock is decoupled and reassembled for local or other destinations. The belt-lines and classification yards, therefore, serve as break-of-bulk places analogous to port facilities. Consequently, the land adjacent to the belt-lines and classification yards became prime locations for manufacturing with the railroad lines serving as the conduit for the supply of material inputs and the distribution of outputs.

The Belt Railroad Company (BRC) in Chicago provides a good example of a company which fostered industrial production adjacent to its lines from the time it was opened for traffic in the 1880s. The inducements were cheap land and direct rail access. However, the BRC inducements were greater than those offered by the major railroad companies because they offered freedom of access to all of Chicago's trunk-line railroads. The plentiful supply of cheap, transportation-serviced land was particularly important for those industries that could not obtain land for new plants or for expansion near the center of the city.

Up to the 1880s, manufacturing in Chicago had concentrated where the Chicago River enters Lake Michigan near the present central business district (the Loop). Soon this area became overcrowded. Consequently, the BRC became active in promoting the mouth of the Calumet River as an alternative industrial site, for the Calumet district was on BRC rails. The largest industrial district promoted and operated by BRC was, however, the Clearing Industrial District adjacent to Clearing Yard. This city of factories was developed by a land syndicate, organized by the BRC in the 1920s and 1930s as a planned industrial community with its own streets, water, and sewage services. Around those industrial areas clustered the homes of the people employed in the factories, the rhythm of the home being tied to the daily rhythm of the factory.

Thus in one sense, the decentralization of manufacturing began during the latter part of the nineteenth century and was related directly to the need for cheap space and the cost advantages that could be derived by manufacturers from more flexible forms of access. Around these new industrial areas grew working-class neighborhoods that, because of the limited availability of transport in the classic industrial city, were tied to the fortunes of the local factories. The rate of decentralization was, however, limited. For example, in 1910, 75 percent of the manufacturing employment in Manhattan was still in a small area south of Fourteenth Street (Pratt, 1911). The actual number of manufactur-

Table 9.2

Estimates of the Changing Geographical Distribution of Production Workers, New York Metropolitan Region, 1869–1987

YEAR	NUMBER OF EMPLOYEES (THOUSANDS)	NUMBER OF EMPLOYEES IN MANHATTAN	MANHATTAN (PERCENT)	METROPOLITAN REGION (PERCENT)
1869	240.3	131.4	54.7%	45.3%
1889	683.0	356.5	52.2	47.8
1919	1,158.6	384.7	33.2	66.8
1956	1,483.3	376.8	25.4	74.6
1970	1,889.9	332.6	17.6	82.4
1987	1,374.1	244.0	17.8	82.2

(*Sources:* Chinitz (1960), p.131; comparative estimates for 1970 calculated from 1970 census, Bureau of the Census [PHC(1)–145] (microfiche); and 1987 from Bureau of the Census (1991), Table C.)

ing employees in Manhattan did not begin to decrease until after 1945 (Table 9.2). This pattern of mainly inner-city manufacturing was not broken until the widespread use of truck for inter- and intraurban transport.

The Late-Industrial-Revolution Highway Era

The period of the late industrial revolution, starting in 1945, is characterized by the decentralization of manufacturing activity within metropolitan areas. The decentralization of manufacturing activity on a large scale was made possible by the truck, which allows goods to be moved much more easily around cities, and the growing use of the automobile for commuting to the decentralized workplaces. However, decentralization did not occur uniformly in all urban places, for large places experienced decentralization a great deal sooner than middle-sized cities.

One urban area that experienced decentralization at quite an early date is the New York metropolitan area. Although the actual number of manufacturing workers in Manhattan remained fairly steady up to the early 1950s, the proportion of total manufacturing employment located in Manhattan compared with the rest of the metropolitan region decreased particularly from 1890 onwards (Table 9.2). There has, therefore, been a comparable decrease in manufacturing employment in Manhattan (compared with the metropolitan region as a whole) for more than one hundred years, but an absolute decrease in manufacturing employment for forty years. This decrease in manufacturing employment in Manhattan is related to a mix of factors such as plant closures, downsizing, the movement of plants to suburban locations, and perhaps the replacement of old large plants by new smaller ones. Manufacturing decentralization, which is represented empirically in aggregate form in Table 9.2, involves many possible types of change.

A Birth-Death-Migration Model of Manufacturing Plant Location   Thus, aggregate central city–suburban figures relating to changes in the location of plants and manufacturing in a particular urban region do not reveal the complexity of the change. It is, therefore, necessary to develop a simple statement of change which incorporates an array of possible causes and shifts. Steed (1973) has developed such an accounting equation which is of significant use in analyzing plant location dynamics:

$$x_i = b_i - d_i + m_i - e_i$$

where, x_i = the net change in the number of plants in region *i*
   b_i = the number of plant births
   d_i = the number of plants closing down completely (plant deaths)
   m_i = the number of plants migrating to an area
   e_i = the number of plants emigrating from an area

This equation is used to analyze the change in the distribution of manufacturing in thirteen subareas within the greater Vancouver area for the two periods 1954 to 1957 and 1964 to 1967. The thirteen subareas were chosen so as to isolate the city core from several other foci (e.g., North Vancouver and Burnaby) and suburban areas around the city. In the aggregate, the innermost and historic industrial area exhibited by far the largest decrease in total number of plants between 1955 and 1965, whereas the greatest absolute increases were experienced in the outer zone, particularly along the north area of the Fraser River and in East Vancouver and Burnaby.

Application of the equation indicates that the absolute decline in the central part of the Vancouver region was related to heavy outmigration of plants during the early part of the 1955–1965 period. During the 1954–1957 period, the city core generated one in three of the region's migrant plants, and though a number stayed within the downtown area, a large number moved to other locations within the metropolitan region. By the end of the period, the situation was more stable, for although the heavy outmigration of plants from the core persisted, it was offset by a high plant birth rate and fewer mortalities. The areas of greatest net increase received their gains primarily through new plant births, and secondarily as a result of plant immigration. On the average, about one-third of the net increase in the suburban areas was due to immigrating plants, whereas two-thirds was the result of the establishment of entirely new plants in the region. Most of the immigrating plants were emigrants from the downtown area.

Thus, in Vancouver, the apparent decentralization of manufacturing plants in this first part of the truck/highway era was primarily the result of the entry of new plants. These new plants located either in the inner industrial core or in the suburban growth areas. The plant births in the inner core were, however, offset by a large number of plant deaths and emigrants, whereas the growth areas did not exhibit these symptoms. The inner core, therefore, continued in its role as a place in which new plants could *incubate,* that is, experiment and establish themselves with limited capital outlay. The growth areas gained in importance, primarily because it is here that space was available (though in the

Vancouver region industrial space is extremely expensive), and there was good access to less congested road transportation.

Impact of Manufacturing Restructuring on Intraurban Plant and Employment Location From about the mid–1970s on, manufacturing in North America has been subject to an enormous number of pressures as a result of major changes in technology, international competition, and methods of financing. These have been discussed in some detail in Chapter 5. The consequences of these changes have been felt at the intraurban level as well as the interurban. As the decentralization of manufacturing has accelerated, the most evident spatial outcome has been the establishment of whole new complexes of industrial activity in particular parts of the outer suburban ring.

There are, perhaps, three main features of this restructuring that had a major effect on the acceleration of the decentralization process. The first is the simple fact that in order to be competitive many existing firms have had to reinvest in new means of production. New capital investments invariably mean new plant locations on newly developed urban land. This is reinforced by the second element, which is that the amount of capital required for investments in new technologies is quite massive, and the machinery often environmentally sensitive. It pays, therefore, for a manufacturing plant to build new plants, perhaps in a "greenfield" location at the periphery of a city. The third element relates to the complex arrangements of suppliers and marketing activities in modern manufacturing. Because production runs tend to be shorter, the output has to be marketed quickly. Inventories have to be kept as low as possible because inventories cost money to purchase and store. So firms require maximum access to both suppliers and markets. These three features all combine to reinforce the tendency for modern manufacturing plants to seek peripheral locations.

The Growth of Manufacturing in Suburban Locations The growth of manufacturing in suburban areas within metropolitan regions is illustrated for the Toronto, Montreal, and Vancouver CMAs in Table 9.3 with respect to employment. Each metropolitan area is divided into three subareas: (1) the central city for each region; (2) the inner suburbs; and (3) the outer suburban periphery. The proportion of each CMA's population (in 1986) located in these subareas varies according to the definitions used, but in general the central cities have about one-third of the metropolitan populations and the suburbs two-thirds. The population of the Vancouver CMA has been increasing the fastest since 1975, followed by the Toronto CMA, with the Montreal CMA fairly stagnant (see Chapter 4, and Table 4.1).

It is interesting to note that the Toronto CMA was the only one of the three metropolises to experience an increase in the growth of manufacturing employment. Manufacturing employment was fairly static in Canada during the 1980s (see Table 4.2). Each of the three metropolises experienced a decrease in central-city manufacturing employment—in Montreal the decrease exceeded 25 percent. All three experienced an increase in suburban employment, but only in Toronto was this increase sufficient to offset the decrease in

Table 9.3

The Change in Manufacturing Employment in Vancouver, Toronto, and Montreal, 1975–1985

METROPOLIS AND SUBAREA	1975 EMPLOYMENT	1975 %	1985 EMPLOYMENT	1985 %	CHANGE %	POPULATION 1986 %
1. City of Montreal	146,861	52.1	108,941	43.4	-25.8	34.8
2. Montreal Island (excluding the City of Montreal)	91,929	32.6	90,393	36.0	- 1.7	22.1
3. Remainder of the Census Metropolitan Area	43,210	15.3	51,869	20.6	20.0	43.1
Total	282,000	100.0	251,203	100.0	-10.9	100.0
1. City of Toronto	79,747	24.3	66,388	18.0	-16.8	18.5
2. Metro Toronto (excluding the City of Toronto)	165,157	50.4	181,700	49.3	10.0	47.8
3. Remainder of the Census Metropolitan Area	83,040	25.3	120,662	32.7	45.3	33.7
Total	327,944	100.0	368,750	100.0	12.4	100.0
1. City of Vancouver	27,961	40.6	22,678	33.5	-18.9	29.4
2. Greater Vancouver Regional District (excluding the City of Vancouver)	36,925	53.6	37,452	55.4	1.4	63.1
3. Remainder of the Census Metropolitan Area	4,026	5.8	7,471	11.1	85.6	7.5
Total	68,912	100.0	67,601	100.0	- 1.9	100.0

(*Source:* Statistics Canada (1979), Table 6, and Statistics Canada, (1988b), Table 6.)

the central city. In Montreal the suburban increase was confined to the suburban periphery beyond Ile de Montreal (which defines subareas 1 and 2), which includes a number of older communities such as Laval to the north and Longueuil to the south. Metropolitan Toronto is experiencing a tremendous growth in employment in its outer suburban ring, which, given the steady-state situation of manufacturing employment in Canada, really implies a continuing concentration of manufacturing jobs to the suburban parts of the Toronto region.

A TYPOLOGY OF INTRAURBAN MANUFACTURING LOCATIONS

At this juncture it is useful to gather together some of the accumulated elements of locational structure that have developed through time in urban space into a general description of the intrametropolitan location of manufacturing. An interesting definition of locational types is suggested by Cameron (1973), who defines four major spatial groups of industry in an urban area: (1) central clusters, (2) decentralized clusters, (3) random spreads, and (4) peripheral concentrations. These four groups are defined in the context of classic locational choice behavior in which plant location is related to the cost of appropriately zoned land, availability and range of local servicing, access to supplies and inputs to the manufacturing process, access to labor, and access to markets.

Central Clusters

In the case of manufacturing firms located in central clusters, which are defined as groups of plants located close to the central business district (Figure 9.2), most of their market is within the central business district and the metropolis itself. Even though land costs or rents may be comparatively high compared with peripheral locations, easy access to the complex array of localization economies (Chapter 5) that are focused in the core area is one of the main features that attracts firms and permits them to be competitive and remain in business (Scott, 1983a). To these localization economies can be added an additional advantage of face-to-face contact with other firms or final purchasers within the cluster, the central business district, and the urban area in general (Ihlanfeldt, 1995).

Plants in central clusters tend to be characterized by some of the following features: they are involved with less standardized production; they are small in scale; the firm may be developing a new or slightly different product line; or they may be producing a particular product to satisfy a small market niche (Steed, 1976). Good examples of industries that exemplify a number of these characteristics are the high fashion portion of the garment district of Manhattan, and small printing shops and plants that are often located in defunct "fringe of the CBD" industrial buildings. Manufacturing firms focusing on these activities are involved in the production of less standardized products; the plants are small in scale; the product lines are continually changing in response to innovations in fashion, style, and design; and the firms cater to particular

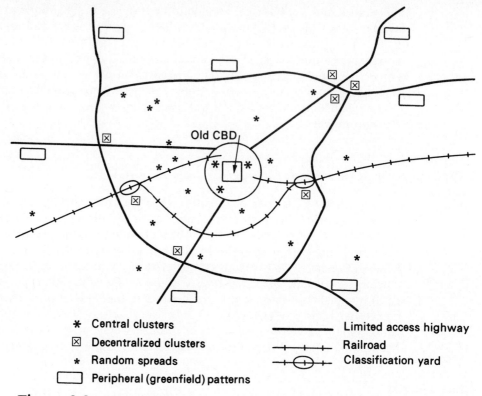

* **Central clusters**
⊠ **Decentralized clusters**
* **Random spreads**
▭ **Peripheral (greenfield) patterns**

────── **Limited access highway**
─┼─┼─┼─ **Railroad**
─┼─⬭─┼─ **Classification yard**

Figure 9.2   Manufacturing locations in centralized and decentralized clusters, random spreads, and peripheral plants in a hypothetical urban area.

market niches usually located within the central business district or specialty outlets within the metropolitan area.

The decentralization of manufacturing has left many of these central clusters vacant and in decay. The low-rise, often brick-veneer buildings, with their copious loft space, are neither in the best locations nor do they provide the most useful type of structures for modern plants which require one-floor flows rather than multi-floor patterns. However, in those few urban areas where the central business district has remained fairly vital (see Chapter 16), and the downtown area is an attractive residential location for some city workers, these old buildings can be dezoned for multiple-purpose uses, including residential uses such as loft condominiums. The success of such dezoning in hope of transition to new uses is directly related not only to the economic health of the downtown area, but also to neighborhood safety and the cleaning up of relic industrial wastes (Colten, 1990).

Decentralized Clusters

Plants locating in decentralized clusters purchase most of their material inputs from, and sell most of their product output to, customers located outside the metropolitan area. As a consequence, accessibility to markets and suppliers is greater and easier at locations adjacent to transport facilities that provide a high level of access to markets beyond the urban area. Today, when truck transportation and both road and rail container terminals are located in areas peripheral to the city, accessibility costs per unit of output will be lowest in these peripheral areas around major transport foci (Figure 9.2). Although locations close to transport facilities are of vital importance in creating the decentralized cluster, some external economy considerations enter to enhance the advantage of the clustered situations. For example, some plants may share common transport facilities, warehousing, and business services. So similar businesses can take advantage of pools of appropriately educated and experienced labor.

The Rise and Fall of the Route 128 Decentralized Cluster   A good example of the way in which firms and plants quickly perceived the advantage of a peripheral-belt highway as providing a high level of access to areas beyond the metropolis is Route 128, a circumferential expressway located about 12 miles radial distance from the center of Boston. During the 1970s, there was a rapid increase in the number of manufacturing plants clustered along Route 128 that collectively provided at the end of the decade more than 66,000 jobs. The firms locating in the decentralized cluster, such as Wang Laboratories, Digital Equipment Corporation, Prime Computer, and Data General, were involved primarily with the new minicomputer industry requiring the manufacture of computer hardware and the creation of software. There was also a number of firms concerned with the manufacture and packaging of drugs. The highest employment level was reached in 1984, but after that time, even though a considerable range of external economies had developed, particularly with respect to pools of appropriately trained and educated labor, the plants along Route 128 went into decline.

The basic reason why the firms in the Route 128 cluster went into decline is because they stayed, in general, with the production of minicomputers and failed to diversify into microcomputers (Norton, 1992). The microcomputer industry began in 1977 (Apple Corporation). Already by 1981 the available hardware and software was such to indicate that this was the area of future growth. The Route 128 minicomputer firms should have been in a good position to move into this rapidly developing new market. However, they did not develop new microcomputer product lines because they preferred to stay with the product and market that they had. Norton (1992, p. 164) refers to this as a "collective failure of management nerve." The Route 128 cluster did not compete with the Silicon Valley (adjacent to San José, California) microcomputer-oriented industrial cluster, which expanded with great rapidity in the 1980s (Saxenian, 1992; 1994). Contributing to this failure to move into the new product line was the momentum of great high-tech success in the 1970s, which led to high wages and high house prices. Thus, decentralized industrial clusters can form, grow, and decline, just like central clusters. Especially high-tech clusters involving fluid product innovation adapting to new demands with rapidity.

Industrial Districts   Whereas accessibility to suppliers and markets may influence the location of decentralized clusters, it may be argued that the actual formation of identifiable industrial districts involves three interrelated features that are vital for a firm to be competitive in a world economy (Markusen, 1996). One is that the firms involved make use of small-batch or *flexible production* systems (Gertler, 1992), which have as their prime characteristics output variety and rapid product and process innovation. The objective is not only quality production but also continuous product innovation and development. Thus, a particular type of product, such as microcomputer hardware and software, can be continually improved or renewed, with respect to both use and quality, through integrated research and development.

Flexible production systems, therefore, require labor forces that work in teams in a multi-tasked environment, because repairs, corrections, product variation, and the implementation of improved processes and product innovation occur to a degree as part of the ongoing production process (Schoenberger, 1988a). It is interesting to note that companies appear to have better success rates in the shift to *multi-tasking* if the job restructuring is negotiated within a unionized environment because unions can help protect and support employees who are worried about changes (Holmes and Kumar, 1993). The integration of flexible production with continuous improvement and innovation means that firms making use of these types of production systems tend to cluster in those parts of the city where persons involved with R&D, multi-tasked production, and administration can be brought together in one location. This invariably means decentralized clusters in middle-income suburban locations.

The second feature is that the firms operating with these flexible production systems can generate external economies through spatial linkages; that is, they form *industrial networks* through forward and backward linkages, some sharing of information, competitive technological innovation, and interfirm labor mobility. Thus, "Flexible production is marked by a decisive geographical re-concentration of production, and by the resurgence of the industrial district" (Walker and Storper, 1989, p. 152). In the cases of the 1970s Route 128 cluster, and the 1980s/1990s Silicon Valley cluster, these are "technology districts" (Storper, 1992), or "technopoles" (Scott, 1993), similar in some ways to the old garment district in lower Manhattan. These types of forward and backward linkages are enhanced through the formation of industrial concentrations in decentralized locations.

Just-in-time input and output requirements (see Chapter 5), which along with flexible production and networking comprise the components of modern (particularly high-tech) manufacturing systems, provide a third reason for the formation of industrial districts. Just-in-time systems are essentially about speed, or time compression, with respect to both the placement of orders through to the delivery of the final product and product development (Schoenberger, 1995). Thus, just-in-time production is a system of guarantees, for supplies and deliveries have to be made on schedule and to the right specifications as per contracts and orders. Firms that can meet such guarantees, and those that do have a huge competitive edge in the global economy, concentrate many aspects of their operation (such as sales, marketing, R&D, and production) in close proximity in order to achieve speed of implementation and rapid reaction. These industrial districts naturally situate within urban areas in locations that are highly accessible to external sources of supply and markets near to good road and air transport facilities. Districts of this type, therefore, invariably appear on the urban landscape as decentralized clusters.

Industrial Parks Some decentralized clusters and suburban industrial districts may be formed by or into industrial parks. These have been in existence in North American cities for many decades and have themselves been an important factor in decentralization. An *industrial park* is a tract of land that is subdivided and developed according to a comprehensive plan for the use of a community of manufacturing plants or offices. Basic infrastructure, such as streets, water, sewage, transport facilities, and electrical power, are provided by the company or city building the park, and in some cases buildings are provided for rent. These parks can serve as an incubator for new firms, and may also be associated with the development of new subdivisions. Although many industrial parks are privately developed, a large number are owned by local governments who use the facilities to attract industry to the local community.

Random Spread

In the case of plants that appear to be located in a random-spread fashion within metropolitan areas (Figure 9.2), one of the most significant influences on location is that their output is directed almost entirely toward the metropolis in which they are situated. The types of plants involved are usually those concerned with the assembly of consumer durables, such as washing machines, dryers, and refrigerators, though many are also involved with subcontracted work for larger companies. They use prefabricated, nonbulky parts as inputs, and assemble these into relatively bulky, space-consuming, final products. Thus, plants in this group require space for assembly and warehousing, and good access to the retail distributors throughout the metropolitan area. The transportation of the final product is invariably by truck, and, as a consequence, the plant can locate almost anywhere in the urban region where the cost of land is relatively low.

Random spreads of manufacturing activity are found most clearly in cities that have grown rapidly during the last fifty years, such as Los Angeles, Dallas-Fort Worth, and Denver. Nelson (1983) notes that manufacturing in Los Angeles is distributed in a dispersed fashion (see the inset map in Figure 7.11) throughout the extended megalopolis in much the same way as retailing, thus reinforcing the polycentric nature of the modern city. Frequently, the industrial sites occupied are adjacent to railway lines even though the rails themselves are not used for the transport of materials and goods. These are relic sites, where cheap older buildings on industrially zoned land can be used easily with little cost outlay by activities requiring fairly large amounts of space for both manufacturing and storage.

Peripheral Patterns

There are two critical characteristics of plants that are located in a peripheral pattern within metropolitan areas. The first is that they require large amounts of ground floor space for production, storage, and parking for their employees.

The second is that they comprise industries with nonlocal markets and are, therefore, oriented toward regions beyond the metropolitan area in which they are located. Firms with these characteristics usually establish themselves where there is plenty of available land at a reasonable cost in *greenfield* sites. Locations in peripheral areas adjacent to good highway facilities are preferred (Figure 9.2).

Examples of industries that are generally found in greenfield, or, in the case of California and Arizona, "dryfield," locations are those involving the production of aircraft, motor vehicles and parts, and farm machinery. Nelson (1983) observes the way in which the aircraft industry, which became established fairly early in the growth of Los Angeles (Hise, 1993), has periodically had to move farther out to obtain more space for larger hangars and longer runways as the metropolis expanded during the past thirty years. In some instances, the peripheral location of a new plant has led to the development of a new town on the periphery of a metropolis, such as a firm involving the production of transport equipment, which stimulated the growth of Glendale, Arizona, for example.

RESTRUCTURING PROCESSES AND THE CHANGING LOCATION OF MANUFACTURING

Although the discussion in the previous section provides a useful description of the major locational types of manufacturing within cities in the context of site location and changing accessibility considerations, the discussion would be richer if matters related to product development and capital-for-labor substitution were also included. "So on our heels a fresh perfection treads" (J. Keats, *Hyperion*). *The restructuring of manufacturing essentially occurs in response to product or price competition.* The term *product* includes ingredients of quality and service, whereas *price* subsumes marketing and delivery. Thus, both the spatial aspects of product development, which is touched on in the discussion of the formation of industrial districts, and the continuous drive by entrepreneurs to reduce costs through the substitution of capital (that is, machinery) for labor, require some elaboration.

Product Life Cycles

The product life cycle approach relates the stages in the development and marketing of the main output of a plant to the probable location of the production facility within a metropolis (Mack, 1993). The product cycle, in its simple form, is envisaged to consist of four stages (Figure 9.3): (1) an *innovation* stage, which entails relatively high development costs, short production runs, experimental marketing, and a high level of personal involvement by the developers of the product; (2) a *growth phase*, which occurs after the market has been sufficiently developed, and the product is found to have market appeal; (3) a *standardization phase*, when the market share can be clearly identified and the plant is concerned mainly with volume production; and (4) a final stage of *output decline* characterized by first decreased market share, followed by a decline in volume

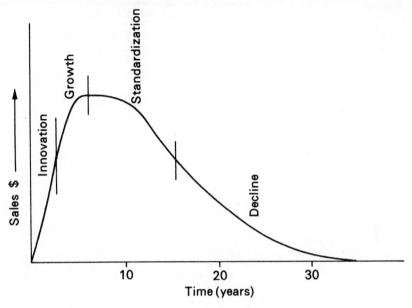

Figure 9.3   Stages in an industrial product life course.

produced, and declining profitability (Malecki, 1981; von Tunzelmann, 1995, Chapter 1).

Whereas the first phase is concerned mainly with product innovation, the second and third phases usually involve some process innovations. The third phase of product standardization typically involves *production dispersement* as firms shift their production to locations that provide lower labor costs or possibilities for greater economies of scale. The length of the product cycle varies according to the goods being produced. Oakey (1985) cites the case of medium-technology consumer products, such as washing machines and toasters, as having a possible cycle of twenty-five to thirty years, whereas modern high-technology computer products have a cycle of about five years.

Product Life Cycles and Location   In some instances, the change in location of an industrial plant within a city can be related to changes in the product cycle. It is not unusual for new firms with a new product to be established within central clusters, for the possibilities of short-term rentals and external economies generated by existing plants and the activities of the central business district creates an environment which serves as an incubator for the new plant. An alternative location could well be in a decentralized cluster, perhaps established as an industrial park, where low-cost facilities and other services are provided by the developers.

If the product becomes accepted and reaches a growth phase, the incubator facilities are no longer congruent with the production requirements. Frequently, a specially constructed plant is required in order to gain ready access to low-cost but productive labor. Such plants may be built in the decentralized clusters, but they are more likely to be established in peripheral locations. Often, plants at this stage seek capital or tax advantages from state/provincial or local governments as part of their profit-maximizing strategy. Older plants involved in commodities that have a long product cycle, such as refrigerators, may be found randomly distributed within the metropolitan area.

Flexible Production, Product Life Cycles, and Urban Agglomeration Economies Flexible production implies rapid product development, for it is a production process that facilitates continuous innovation. It is a means by which the spheres of process, product, and technical change in the modern firm may be drawn together more immediately and interactively than in the ("Fordist") assembly line when these spheres operate generally in separate environments. Thus, flexible production is essentially directed toward shortening the innovation stage of the production cycle, but elongating the growth and standardization phases (Norton, 1992). The just-in-time input-output flow processes and concentration on quality in the flexible production process reduce the need for volume production in the standardization phase to achieve economies of scale in order to generate the surplus needed for further investment and product development.

The locational implications are twofold and twin scaled. The requirement for immediacy of the spheres means that firms will increasingly tend to concentrate their production into urban regions that can cater to the range of labor, market, and access needs of a more complex concatenation of industrial circumstances. Thus, the first implication, at the macro-urban scale, is that flexible-production, just-in-time systems *revalue the economies of agglomeration* of large metropolises and extensive urban regions such as CMSAs. This revaluing occurs not because of the sheer size of the urban region, but because of the size and range of the *diversity* of labor, skills, transportation, forward and backward linkages, and educational and cultural opportunities that it contains. The second implication, which has been mentioned at the micro-scale in the context of decentralized clusters, is that such firms tend to concentrate into industrial districts.

The product life cycle approach, therefore, relates decentralization to the fact that firms are concerned with the development and marketing of something that at some time or another is superseded by another product. The various locational types tend to consist of plants at different stages in the life cycle sequence. Central clusters have an incubation function, but they also have plants at the tail end of the cycle that exist despite inertia. Decentralized clusters also serve an incubation function, but they consist also increasingly of firms constituting industrial districts with plants in the growth and standardization phases. Peripheral locations invariably have plants in the standardized phase, whereas random spreads frequently consist of plants in the late standardization or decline phase.

Capital-for-Labor Substitution

In a remarkable sustained accumulation of theoretical and empirical research, Scott adopts a political economy approach to examine the intraurban location of manufacturing (1983a; 1983b; 1984; 1986; 1988a; 1988b). Two fundamental questions concerning the intraurban location of manufacturing involve the rationale underlying (1) manufacturing decentralization, and (2) the clustering of firms in noncentral locations. In the previous discussion, decentralization and clustering have been examined in the context of access and agglomeration economies. Scott demonstrates that the incessant substitution made by capitalists of capital for labor provides a fruitful basis for examining manufacturing decentralization and the consequent intrametropolitan spatial division of labor; and he also reemphasizes that agglomeration economies, both localization and urbanization (see Chapter 5), provide a useful framework for analyzing clustering. The capital-labor aspect of his work has its roots in an earlier theoretical discussion by Moses and Williamson (1967). Departing from that paradigm, the recognition of agglomeration considerations allows Scott to analyze the degree to which different types of firms are able to integrate the various components of their production process, and the effect that this has on the creation of spatial clusters of interlinked economic activities.

A basic feature of Scott's work is the explicit recognition of spatial differences in the strengths of external and agglomeration economies, and the availability of land and labor within monocentric and polynucleated metropolitan areas. To summarize, it is argued that in the aggregate there are significant differences (that is, there is a positive or negative distance gradient) between the centers and peripheries of major business nucleations, in mono- and multicentered metropolises, in the strength of external and agglomeration economies, and in the costs of land and labor. The advantages to be derived from external and agglomeration economies are greatest at the center and least at the periphery (Figure 9.4a). Land costs, by comparison, are highest at the center and lowest at the periphery (Figure 9.4b); and labor costs are lowest at the center and highest at the periphery (Figure 9.4c).

Centrifugal/Centripetal Economic Forces   For various reasons, such as the drive by firms to lower costs in general and, in some instances, to control labor by reducing its importance to the production process, there has been an inexorable tendency for firms to substitute capital for labor. As land is cheaper and in greater supply at the periphery of nucleations, and, therefore, more attractive for new capital investments, this invariably implies decentralization of manufacturing. As a consequence, there is a spatial Heckscher-Ohlin effect (Blackley and Greytack, 1985), with plants having a lower capital–labor ratio in central locations, and plants with a higher capital–labor ratio in peripheral locations (Figure 9.4d). Thus, the general decentralization of manufacturing in North American cities can be related perhaps as much to the underlying long-term tendency for firms to substitute capital for labor as to changes in access due to transport improvements.

This centrifugal pattern is offset to a degree by the desire for some types of manufacturing to locate centripetally in clusters where they can gain either external economies, or agglomeration advantages with respect to the provision of inputs and the marketing of outputs. This applies particularly to those plants that either choose not to, or cannot, integrate (vertically or horizontally) many aspects of their production and marketing functions. Thus, the integrated plants can locate in free-standing peripheral patterns (or random spreads),

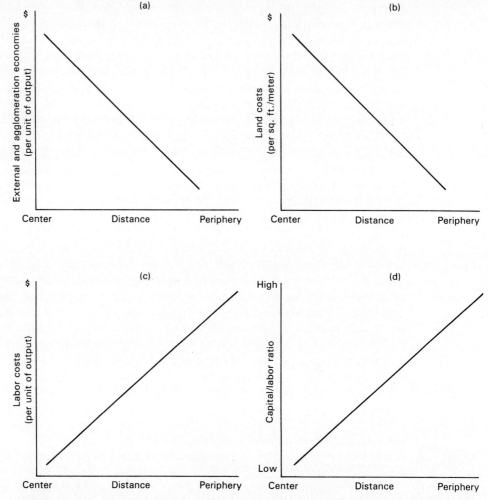

Figure 9.4 Theoretical relationships between (a) external and agglomeration economies, (b) land costs, (c) labor costs, and (d) capital–labor ratios, and distance from the center of business nucleations in metropolises.

whereas those types of manufacturing activity that participate in a great deal of subcontracting (Holmes, 1986), or have to be close to their markets, tend to locate in decentralized clusters within the vicinity of their associated activities. A number of these locational features are demonstrated in detailed case studies of the printed circuits, women's dress, and high-technology industries in the Los Angeles metropolitan area (Scott 1983b; 1984; 1986).

Los Angeles: The Printed Circuits Industry The locational patterns of various linked components—(a) suppliers, (b) customers, and (c) subcontractors—of

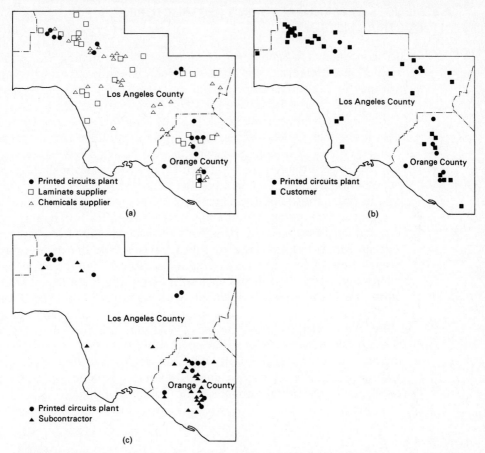

Figure 9.5 A sample of printed circuit plants and the location of their main (a) suppliers, (b) customers, and (c) sub-contractors in Los Angeles.
(*Source:* Redrawn from Scott, 1983b, pp. 357, 359, 361.)

the decentralized printed circuits industry in the Los Angeles metropolitan area demonstrates that the smaller and more labor-intensive plants tend to be more clustered than larger plants, which tend to be more integrated (Scott, 1983b). The data for a sample of printed circuits plants indicate that they tend to locate close to their main suppliers of laminates and chemicals, main customers, and main subcontractors (Figure 9.5). Thus, the size of the individual plants and organization of the industry influence the locational pattern of the firms in metropolitan space.

In a more general context, Blackley and Greytack (1985) provide evidence with respect to Cincinnati that industrial activities tend to be capital intensive in peripheral locations, and labor intensive in central locations. Thus, the inexorable substitution of capital for labor is expressed spatially in decentralization.

CONSEQUENCES OF MANUFACTURING RESTRUCTURING IN AN INDUSTRIAL METROPOLIS

In this concluding section the main concepts concerning the intraurban location of manufacturing are applied to the changing pattern of manufacturing within the Detroit metropolitan area. Detroit provides a useful example for summary purposes because among the large metropolitan areas it is most dependent on manufacturing (Table 9.1); its economy is contingent on a product, the automobile, which has been the central dynamic of North American industrial and urban development during the last half of the twentieth century; and the wealth generated from manufacturing provides relatively high-paying jobs, which rank the metropolitan area on a per capita basis as among the wealthiest on the continent (Pollard and Storper, 1996).

In a social context, it was a major destination for immigrants from eastern and southern Europe during the first few decades of the twentieth century, and African Americans from the southern states during the middle decades. Both groups were attracted by employment opportunities in the automobile and ancillary industries. Detroit has also been one of the main loci of strength of the North American union movement, with its post–World War II United Auto Worker and Teamster contracts setting the pace for collective agreements. Finally, the consequences of the post–1970 industrial restructuring, along with the suburbanization of the white population, have had starkly extreme consequences for racial polarization and inner-city decay. Thus, in many ways the Detroit urban region provides a good, albeit exaggerated, portrait of the impact of change in the intrametropolitan location of manufacturing on North American cities.

Growth of Detroit

Within the first few decades of the twentieth century, the automobile industry transformed Detroit from a city of 285,000 people with a diversified economy related to its situation at the mouth of the Detroit River on the interface of two of the Great Lakes, to a metropolis of 1 million in 1920. From the turn of the century, Detroit became the product of the requirements of the three giant corporations (General Motors, Ford, and Chrysler) that came to dominate the industry (Sinclair and Thompson, 1978). In geographic terms, this has meant that the growing automobile industry determined the spatial development of the city through a recurring need to establish large plants to achieve economies of scale in assembly line mass production ("Fordist"). Initially, the preferred sites were vacant land adjacent to radial railroad lines that focused on the center of the city and provided access to external sources of supplies and markets. These factories gave rise to local factory communities peopled by immigrants of particular ethnic backgrounds. World War II demands for armaments, and the postwar economic expansion, generated the need and market for even greater output, which could only be provided by new larger factories located on extensive tracts of vacant land farther from the city on the railroad lines.

This postwar outmigration of production facilities coincided with the suburbanization first of managerial, professional, and white-collar workers; and, second, of blue-collar workers who had achieved unusually high levels of affluence (for production workers) consequent to negotiated collective agreements. These new factories and housing subdivisions on new land were made accessible through the construction of a network of limited-access highways and improved arterials that catered to the transport needs of an enthusiastically home-owning auto-oriented populace. It will be recalled that limited-access highways proliferated during the 1960s and 1970s as a result of the Interstate Highway Program, and new housing became affordable to a wide range of income groups through, primarily, FHA-insured loans. Tragically, the persistence of direct and indirect discrimination practices effectively prevented racial minorities, particularly African Americans, from participating in this extended range of housing choice, thereby extending and intensifying racial segregation (see Chapters 11 and 12).

Economic and Social Impacts of Restructuring

The restructuring of the automobile industry after 1970, which was necessitated by price, quality, and style competition, primarily from Japanese and German producers, has had a profound impact on the Detroit urban region—as similar restructuring has had in many other metropolises (Wilson, 1996). The global integration of the industry has fostered foreign partnerships which helped accelerate the implementation of new robotic production technologies and the establishment of auto-related high-technology R&D firms, such as CAD/CAM (computer-aided design, computer-aided manufacturing), often in industrial districts in decentralized clusters. The large new assembly plants are established in highway-served peripheral, often greenfield, sites in the interstices between the old radial railroad transport corridors.

Post-Fordist production processes, involving flexible production and just-in-time delivery systems, have revolutionized and reinvigorated the automobile industry in the Detroit urban region and much of the American Midwest and southwest Ontario by making it more competitive (Florida, 1996). In social terms, the restructuring has been accompanied by "downsizing," which means job losses, and renegotiation of collective agreements to reduce fringe benefits and job demarcation and allow more multi-tasking. In spatial terms, the restructuring has reinforced abandonment of the central city by much of the white population and industry, which has been exacerbated by large-scale deindustrialization in the pre–1970 manufacturing areas during the recessions of the early 1980s and 1990s (Waddell and Shukla, 1993). Concomitantly, there has been the emergence of an extensive outer-suburban job crescent (Sinclair, 1994).

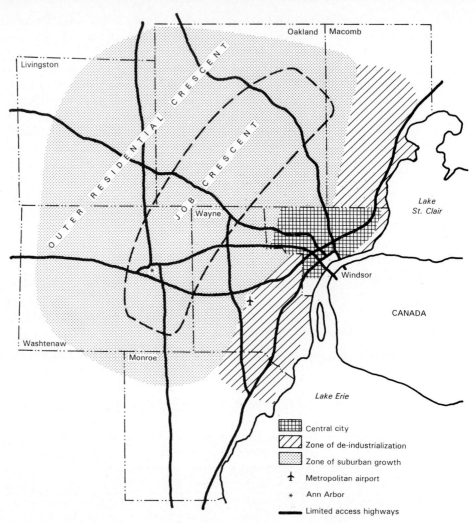

Figure 9.6 A structural model of the location of manufacturing in the Detroit urban region.

(*Source:* Redrawn and modified from Sinclair, 1994, p. 216.)

Spatial Consequences of Restructuring

The spatial consequences of this manufacturing restructuring are summarized in Figure 9.6 and Table 9.4. It should be noted that the Detroit urban region, defined in the table and illustration, refers to a six-county region that is more pertinent to this discussion than the large ten county Detroit-Ann Arbor-Flint CMSA (population 5.3 million) demarcated for the 1990 census. An extensive zone of deindustrialization occurs along lands marginal to Lake St. Clair, the St. Clair River, and Lake Erie, and the railway lines and lands that run parallel to the water frontage. The blighted Rust Belt appearance of these properties barely masks the insidious and hazardous decay, decomposition, and associated

Table 9.4

Detroit: Manufacturing Employment and Population (in Thousands), Population Change, and Income per capita for the Central City and Inner Suburban and Outer Suburban Crescents

DETROIT URBAN REGION	EMPLOYMENT		POPULATION				INCOME PER CAPITA
	NUMBER	% OF REGION	NUMBER	1970–90 SHIFT	% OF REGION	% AFRICAN AMERICAN	
1. Central City							
City of Detroit	98.5	19.9	1,027	−32.2	23.1	75.7	9,662
2. Inner Suburbs		46.2		+1.1	40.5	(3.1)	13,678
Wayne Co.**	116.9		1,085	−6.5			
Macomb Co.	112.2		717	+14.5			
3. Outer periphery	113.9	33.9	1,084	+22.6	36.4	(3.1)	15,943
Oakland Co.				+19.4			
Monroe Co.	7.8		134	+12.6			
Livingston Co.	6.4		116	+123.1			
Washtenaw Co.*	39.9		283	+20.1			
Total urban region	495.6		4,446	−3.5		20.5	13,578

* = Coterminous with the Ann Arbor PMSA.

** = Excluding City of Detroit.

( ) = Unable to partition inner and outer suburban figure.

(*Source:* Bureau of the Census (1991), Tables C and D. Data compiled by the author.)

leakage into the groundwater of chemicals, oxides, and metals generated by the defunct and relic manufacturing plants. The central city is the home of 73 percent of the African Americans in the urban region, and more than three-quarters of the city's population. Limited access to, and appropriate training for, jobs in the restructured peripherally located firms and plants is reflected in the income information (Table 9.4): The population of the central city appears to have, on a per capita basis, 70 percent the income of residents of the inner suburbs, and 60 percent the income of residents in the outer periphery.

Thus, economic restructuring has had profound consequences on the inner city of Detroit, turning many parts of the city into a wasteland of demeaning decay, despair, dependency, depopulation, and degradation. Many of the descendants of the African Americans who, between 1920 and 1960, came north to Detroit and other midwestern and northeastern cities, to work on the automobile assembly lines of the factories of this part of the promised land (Lemann, 1991) have, in effect, had the rug pulled out from under them. Jobs in automobile and associated industries have diminished, and those jobs that remain as part of the revitalized industry are located in a crescent some distance from the African American ghetto in the central city (see Chapter 11). The resulting breakdown of family and community life can be linked directly to massive job decentralization associated with spatial relocation of manufacturing and other associated activities (Rose and Deskins, 1991; Wilson, 1996).

The other side of "the spatial consequences of restructuring" coin is the outer crescent of employment and housing well beyond the central city (Figure 9.6). The bulk of the manufacturing employment in the Detroit urban region of the 1990s lies in a crescent occupying land left partly suburbanized in the 1960s and 1970s. Excluding the Macomb County employment located outside the perceived job crescent, probably as much as 57 percent of the manufacturing employment in the Detroit urban region is contained within this zone. Also, within the crescent is the university/college city of Ann Arbor which, with its research and educational resources, serves to concentrate many R&D activities into industrial districts, occasionally in the form of planned industrial parks. Beyond the job crescent, particularly in Oakland County, lies the outer residential crescent containing the homes of professional, white-collar and managerial workers who have gained in income, relative to blue-collar workers, since 1980.

CONCLUSION

The restructuring of manufacturing has, therefore, accelerated the decline of old manufacturing spaces, and fostered the development of new spaces in peripheral locations and decentralized clusters, frequently as part of "edge cities" (Fleischmann, 1995). These new industrial zones and districts appear, in the case of the

Detroit urban region, to have few linkages with the previous industrial spaces and the diminished central city. The relentless cost-reducing and workforce controlling drive by industrialists and entrepreneurs, which compels an inherent tendency to substitute capital for labor, continues to be reflected spatially in the decentralization of plants and the disappearance of manufacturing jobs in the Detroit urban region as it does in North American cities in general.

Chapter

10

The Intraurban Location of Employment in Service Activities

*I*n Chapter 6, it was observed that service activities provided nearly all of the new jobs created between 1970 and 1990 (Table 6.1). Furthermore, the three components of the service sector—consumer, producer, and governmental and public—each provide more employment than manufacturing. The largest component is the consumer services sector, consisting for the most part of retail activities along with more limited employment in wholesaling, whereas the fastest-growing sector involves producer services, particularly finance, insurance, and general business services. The purpose of this chapter is to describe contemporary patterns of the intraurban location of these three sectors of service employment, review various theories relating to the location of each component within urban areas, and discuss the extent to which these theories provide an understanding of locational dynamics and change.

A fundamental contrast in the intraurban location of the different components of service activities becomes obvious from Table 10.1, which provides a summary of the location of employment in retail activities (part of the consumer services sector) and finance, insurance and real estate (part of the producer services sector) in the Atlanta MSA. This large urban region is divided into four zones (Figure 10.1): the *central city* of the City of Atlanta with 14 per-

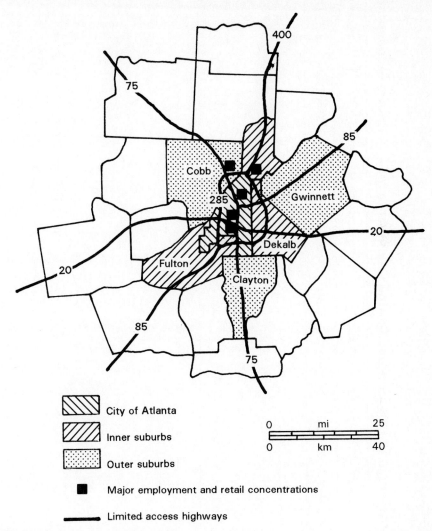

Figure 10.1 The Atlanta MSA: City of Atlanta, inner suburbs, outer suburbs, and extensive fringe. (Employment concentrations from Fujii and Hartshorn, 1996, p. 688.)

cent of the MSA population surrounded by the I-285 beltway; the *inner suburban areas* with 28 percent of the MSA population, which, together with Atlanta, comprise the public transport region served by MARTA (see Table 1.3) and contain well-established African American and white housing areas; the predominantly white *outer suburbs* not included within MARTA, which contain 35 percent of the MSA population embracing also a section of the burgeoning I-85 Atlanta-Greenville-Charlotte-Raleigh corridor; and an extensive *fringe area* with 23 percent of the MSA population involving new suburbs, exurban properties,

Table 10.1

Retail and Producer Service Employment in the Suburbs and Central City: Atlanta MSA (numbers in thousands)

Zone	MANUFACTURING		RETAIL		FIRE		POPULATION		% AFRICAN AMERICAN	INCOME PER CAPITA
	Number	% Share	Number	% Share	Number	% Share	Number	% Share		
1. Central City Atlanta	20.3	10.0	47.0	17.4	37.8	37.9	394	13.9	67.1	11,689
2. Inner Suburbs (MARTA)	79.4	39.1	100.8	37.3	38.7	38.8	801	28.3	31.8	15,814
3. Outer suburbs	62.4	30.8	90.6	33.5	17.4	17.5	983	34.7	5.1	14,595
4. Fringe	40.7	20.0	31.6	11.7	5.7	5.8	656	23.1	19.0	11,443
Total MSA	202.8	100.0	270.4	100.0	99.6	100.0	2,834	100.0	24.5	13,806

(*Source*: Bureau of the Census (1991), Tables C and D.)
Note: FIRE = finance, insurance, and real estate.

rural settlements (some with manufacturing plants), and agricultural activities. Whereas retail activities appear to be distributed in general accordance with the location of consumers, albeit focusing somewhat within the central city and inner suburbs, producer service employment is quite concentrated in the central city and the inner suburbs (as compared with population). The intraurban locational patterns of the major types of services are, therefore, being influenced by different processes, or similar processes that are having differing spatial outcomes.

THE INTRAURBAN LOCATION OF CONSUMER SERVICES

Although only about 3 to 4 percent of the developed land in an urban area is devoted to retail activities, these commercial activities are extremely important aspects of the urban environment. Not only are they large generators of employment, they also are the places where a large part of the income (about 60 percent) of the residents of the city is spent. Commercial areas are also the generators of a large share of the traffic flow. Suburban retail shopping malls, with their associated recreational facilities, have become the focus of social activity for the vast proportion of the population of North America, and a favorite hangout for some of the "teenage" population (Hopkins, 1991). As will be noted later, the public–private space controversy that private ownership of malls generates is but one of the interesting issues emanating from the "malling" of North America (Kowinski, 1985). The retail structure of urban areas is also being affected by a range of high-volume, tightly managed, income-geared, new retail formats which are competing effectively with malls and large department stores (Simmons et al., 1996, pp. 78–81).

Until about 1895, most of the retail activities in a city were located in the central business district, though there were a number of food and other convenience stores servicing local neighborhoods (Domosh, 1990). Between 1895 and 1945, as the population spread along the major transport arterials, so did a number of retail activities that served the day-to-day needs of the decentralizing population. This gave rise to unplanned strips of retail activities along the major streets, and nucleations at major traffic intersections. From 1945, the automobile and the demand for parking resulted in the widespread growth of suburban planned shopping centers, first open and then, from the 1960s, enclosed and climate controlled (Simmons, 1991). In the United States, shopping malls with more than 100,000 square feet of leasable space account for 65 percent of retail sales, excluding sales generated by automobile dealers, gasoline stations, and building suppliers (Bureau of the Census, 1993, p. 777). There has also been the development of specialized retail outlets involved with food, furniture, electronic goods, clothes, and general household goods as well as large discount department stores.

The literature on the geography of retailing (for example, Potter, 1982; Jones and Simmons, 1990; 1993) has involved two main components:

1. Spatial aspects of *retail location*, which is concerned with where stores or clusters of types of stores are located in a metropolitan area.

2. Spatial aspects of *consumer behavior and cognition*, which involves the factors influencing where and how far consumers are prepared to go to shop, and how they learn about the retail system. There is a particular focus on the ways in which different socioeconomic groups learn about the availability of different retail activities, goods, and prices.

In the next section the discussion focuses on the first of these interests, although there will be a brief review of the second.

Urban Retail Location

The geographic structure of retailing is based on three theoretical elements:

a) The idea of *multipurpose shopping behavior* by a population of consumers spread throughout the urban area (Eaton and Lipsey, 1982). Multipurpose shopping behavior results in the concentration of stores in retail nucleations. These nucleations are distributed within urban areas in a hierarchical fashion analogous to central places in the central place model (see Chapter 6). These retail nucleations may be either planned or unplanned. If they are planned, they may be either unenclosed or enclosed malls.

b) The concept of *external economies*, which argues that certain advantages of joint attraction can be gained by individual retail outlets locating together. This gives rise to a *clustering* of commercial activities.

c) The search by providers of consumer services for *profit-maximizing locations,* with those requiring the largest market area, such as department stores, vying for the most accessible locations. Retail activities are, therefore, found along radial and ring arterials and expressways, particularly at intersections. Those retail activities requiring the largest market area locate at major expressway interchanges, whereas those with local markets are established more ubiquitously around the urban area on neighborhood through streets. This desire for profit-maximizing locations generates the peaks and ridges in the urban land value surface (see Chapter 8).

These ideas form the basis of a typology of urban retail structure, defined first by Berry (1963), which has been demonstrated, with a number of refinements, to exist throughout North America as well as other parts of the world (Berry, 1967; Potter, 1982; Morrill, 1987; Jones and Simmons, 1990; 1993). This typology is defined geographically in terms of specialized areas, ribbons, and nucleations (Figure 10.2), and has proven to be remarkably robust even in the ever-sprawling North American urban environment where the suburban mall has become the dominant element of urban retail structure, and the central business district has been reduced for retailing purposes, if it exists noticeably in that role at all, to a specialized role often serving local specific socioeconomic or ethnic groups (Simmons, 1991).

The geographic pattern of consumer activities is strongly related to the intraurban location of market income, which is in turn strongly related to popula-

ASPATIAL PROCESSES	SPATIAL PATTERNS		
	Ribbons	Specialized	Nucleations
Multiple shopping			hierarchies
External economies		clustering	clustering
Profit maximising locations	freestanding	planned/unplanned	planned/unplanned

Figure 10.2 Connecting aspatial processes to the basic components of the intraurban pattern of retail location.

tion densities. In polynucleated metropolises the distribution of market income is quite dispersed, whereas in mononucleated cities market income is more concentrated. Figure 10.3 illustrates the contrast in distribution of market income between a highly dispersed polycentric metropolis such as Dallas-Fort Worth CMSA, and a less dispersed metropolis such as the Toronto CMA, both of similar population size. With each dot equaling 1 percent of each metropolis's market income, the pattern is much more concentrated in the Toronto CMA than the Dallas-Fort Worth CMSA. Hence, the location of commercial clusters (Figure 10.4a) in the Greater Toronto Area (GTA, slightly larger than the CMA) tends to be metro Toronto and downtown oriented, and the pattern of ribbons and nucleations is strongly related to the highway and rapid transit networks (Figure 10.4b).

Specialized Areas Specialized areas are recognizable clusters of commercial activity within urban areas. They have traditionally been well developed in the central business district where two types have historically been identified: (1) clusters of establishments of the same business type, for example entertainment districts; and (2) functional areas consisting of different commercial activities that are interrelated, for example, printing districts which include copying, advertising, magazine publishing, and computer layouts and graphics. Specialized areas basically serve *"niche markets,"* that is, the market for live entertainment, the shopping needs of particular ethnic or socioeconomic groups, the service requirements of the office sector, or the visual and sound needs of the home entertainment market.

Many of these specialized areas are so strongly developed that the districts or streets in which they are located have become household names, are known throughout the country or even the world, and have become "icon" identifiers in global marketing. In Manhattan, for example, Wall Street is synonymous with the financial community, Park Avenue with the headquarter offices of

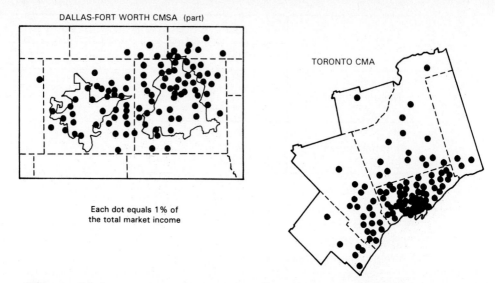

DALLAS-FORT WORTH CMSA (part)

TORONTO CMA

Each dot equals 1% of
the total market income

Figure 10.3 The spatial distribution of market income: Dallas-Fort Worth and Toronto.

(*Source:* Redrawn from Simmons, 1995, Figure 1, p. 7. Note that the Dallas-Fort Worth diagram involves only the central part of the extensive CMSA.)

many of North America's largest corporations, Madison Avenue with advertising, and Fifth Avenue with high-quality, high-price retailing. In Toronto, Yonge Street is well known for its cluster of popular music, video, and electronic equipment outlets as well as some of the more risqué aspects of life's rich pattern. In San Francisco, Union Square is synonymous with upscale shopping, Fisherman's Wharf with seafood restaurants and the "fun of the fair," and Chinatown with ethnically oriented theme retailing.

The dominant locational factor for this component of the urban commercial structure is either direct accessibility to the portion of the market served, or "magnet" locational identification as an attraction for occasional visits by tourists and the inhabitants of the urban region in which the attraction is located. Hence, these areas—be they planned or unplanned—are typically found in or adjacent to the old downtowns, or on major arterial roads within the urban area, or at the foci of public transport or freeway systems. Clustering occurs because the firms or stores involved find that there are considerable external economies to be gained by locating close to other similar activities, or because they want to share some type of common facility or joint power of attraction.

Examples of specialized areas encountered within cities include:

1. "Automobile rows," comprising contiguous strings of establishments selling new and used automobiles and providing facilities for parts, repairs, and servicing. These areas are strung out along the main highways and radial streets. Automotive dealers and related auto-servicing activities, such as gasoline sales, account for 27 percent of the total retail trade in the United

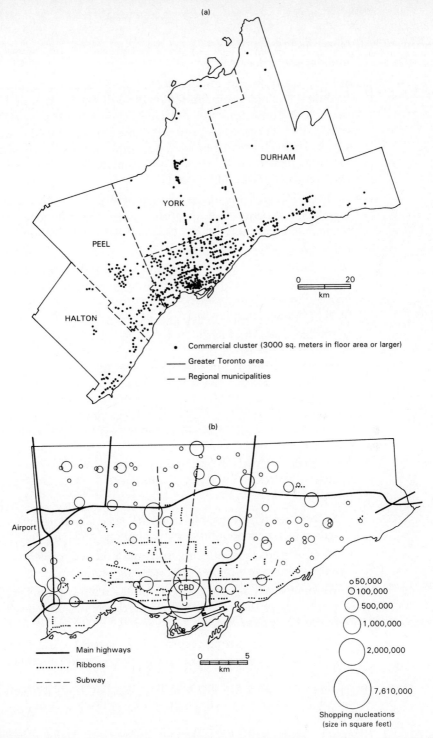

Figure 10.4 The location of (a) commercial clusters (greater than 30,000 square feet) in the greater Toronto area; and (b) ribbons and nucleations in Metro Toronto.

(*Source:* The GTA map is redrawn and modified from Simmons, 1994, Figure 8.5, p. 75.)

States and, therefore, comprise a highly significant part of the urban commercial landscape.

2. Medical districts, including offices of doctors, dentists, and related medical facilities. The older, unplanned medical districts are often located at the upper-floor level within the outlying retail nucleations, although more recently the tendency has been for these to be replaced or supplemented by planned office complexes separate from retail nucleations but at accessible locations near them, particularly in the suburbs. Frequently these medical districts cluster around hospitals.

3. Household-furnishing districts, comprising concentrations of warehouse-type facilities containing furniture, household appliances, home decorating, kitchen and bathroom outfitters, and other related stores.

4. Exotic markets, such as Maxwell Street in Chicago, Chinatown in San Francisco, Kensington Market in Toronto, and the many unplanned clusters of retail outlets serving particular ethnic groups within central cities, or the planned "ethnic" or lifestyle *theme malls* serving the needs of particular cultural or socioeconomic groups in the outer suburbs.

5. In the late 1980s and early 1990s, large specialized retail outlets have been transformed by the incorporation of new inventory and sales technologies (such as scanning), more efficient just-in-time types of distribution systems, and more effective income-targeted marketing, into *new retail formats* (Jones et al., 1994). As a result of the implementation of these more efficient types of business systems, these new retail formats are able to offer a wide range of goods under one roof, with 50,000 to 100,000 square feet of retail space, at lower prices than their competitors. However, as profit margins are slim, the continuing viability of these types of outlets depends on maintenance of a competitive edge. Examples of new retail formats in "big boxes" are:

    *   Superstores, which include groceries, specialty divisions (delicatessen, fish, hairdressing, video rentals, etc.), clothing, and bakeries, under one roof
    *   Membership clubs, offering discount food and a range of other household and automotive supplies to fee-paying members
    *   Warehouse clubs, for fee- or non-fee-paying members, operating in a wide variety of markets ranging from office supplies to furniture and household goods
    *   Home and design centers, providing a variety of interlinked home-related (e.g., garden, lighting, wall furnishings, carpets) specialty stores.

Increasingly two or more of these new box-like retail formats are locating in one unenclosed site connected together by parking facilities.

Ribbons   Ribbons, the second major component of the commercial structure, are noticeable throughout urban areas where their constituents form the backdrop to most well-traveled routes. Business types that have a tendency to locate in one or another of the various types of ribbons discussed in this section are those that need a certain minimum level of accessibility to the market served. They do not, however, require sites of maximum centrality within urban areas, and consequently can survive at locations along major arteries and urban highways (Figure 10.2). Although several different types of ribbons can be recognized, it is important to remember that the character of these is repeated from one part of the city to another.

Retail activities located on *highway-oriented ribbons* serve demands originating from traffic on the major highways. This traffic may be both local and nonlocal. Generally, the greater the volume of traffic on the highway, the more important it is as a generator of demand and, consequently, the greater the density of commercial development along it. In the older parts of the central cities, these ribbons are uncontrolled developments, but in the vast areas of post-1945 housing, and along newer intraurban highways running through the older central city, planned service plazas are the rule. Typical businesses on highway-oriented ribbons include gasoline and service stations, chain restaurants and drive-ins, ice cream parlors, and motels. These are essentially free-standing establishments among which there is very little functional linkage, and which cater, essentially, to one-stop service needs originating from the market channeled on to highways.

Commercial activities located on *urban-arterial ribbons,* invariably in *strip malls,* serve the demands of the local area, and they are, in general, large consumers of space or services that are needed only occasionally and involve special-purpose trips. The businesses located in these areas include appliance stores, self-assembly and mid-price-range furniture outlets, office equipment sales, funeral parlors, garden supplies, lumber yards, building and household supplies, and large supermarkets and new retail formats. These types of activities are free-standing in that they cater for one-stop shopping trips. Urban arterials tend to thrive during periods when the part of the urban area in which they are located is growing with new housing, young families, and hence high levels of local consumer demand. They decline as the wave of population growth passes and the local community settles into its life course (see Chapter 11).

The Nucleations The third feature of the commercial structure of cities are the so-called nucleations, which vary enormously in size but are distributed fairly evenly throughout urban areas in accordance with the distribution of population (Figures 10.3 and 10.4). In the older, built-up parts of the central city, shopping nucleations are usually found at the intersection of arterials, where high vehicular and pedestrian traffic intensity provide the rationale for their focused unplanned growth. Accordingly, they are associated with high degrees of centrality, which is reflected in the locational correlation between them and peaks of higher land values at street intersections throughout the urban area. Since 1945, planned nucleations have been added to, or have replaced, the pattern of unplanned nucleations in the older communities. As a result, there now exists a hierarchy of planned and unplanned nucleations in which the planned mall component dominates.

Table 10.2

The Size Distribution of Planned and Unplanned Retail Nucleations, Within the Hierarchy, in Edmonton

	NUMBER OF NUCLEATIONS	
NUMBER OF STORES	PLANNED	UNPLANNED
5–19	5	27
20–49	5	11
50+	12	4
Total	22	42

(*Source:* West (1993), Table 1, p. 34.)

Planned and Unplanned Retail Nucleations In their reformulation of the central place model, Eaton and Lipsey (1982) demonstrate that a hierarchy of central places, the intraurban equivalent of which is a hierarchy of commercial nucleations, can be generated in a model that incorporates two quite acceptable behavioral assumptions. One assumption is that consumers tend to prefer *multipurpose shopping* over single-purpose shopping, for journeys to multipurpose destinations maximize purchaser trip utility. Mulligan (1987) has demonstrated that the rate of multiple-purpose shopping behavior is high and provides a persuasive rationale for hierarchies of central places. A second assumption is that the firms which are the providers of the shopping facilities will, not unsurprisingly, seek to locate their stores at *profit-maximizing locations*. The economic behavior, therefore, of both producers and consumers of retail services (Ingene and Ghosh, 1990), tends to result in the clustering of stores in a hierarchy of central places, or nucleations, according to the frequency of the need by consumers for particular retail services, and the economies of service provision.

West (1993) shows that these two assumptions result in a combined hierarchy of unplanned and planned nucleations within cities, and also discusses why it is that small nucleations tend to be unplanned, whereas large nucleations tend to be planned (that is, malls). Table 10.2 indicates that in Edmonton, where the unplanned exceed the planned nucleations (within the hierarchy), there are three times as many large planned nucleations as large unplanned nucleations. This is partly related to the "age" of the city, for urban areas with extensive pre-1945 settlements may tend to have relatively more large unplanned nucleations, but partly also due to the search for profit-maximizing locations by firms.

In unplanned nucleations there is no restriction, other than possible shortage of appropriately zoned space, to the entry of stores. Thus, while a certain amount of business-type duplication may be useful to generate external economies of joint attraction and consumer comparative shopping, too much duplication may result in *profit dissipation* rather than profit maximization. The entry of stores into planned nucleations is, however, restricted by the mall owner. As the owner wishes to maximize mall rents, business-type duplication will be limited to the minimal amount sufficient for comparative shopping and consumer attraction. Thus, planned malls will tend to be the location of choice for business types characteristic of higher-order nucleations. Unplanned nucleations, allowing for relatively unrestricted entry of stores, can survive for low-threshold (that is, lower-order) business types.

The Intraurban Hierarchy   The hierarchy of shopping nucleations usually involves five to seven levels depending on the size and complexity of the metropolis. A fivefold hierarchy, typical of cities in the 0.5 million population range, is as follows:

1. Isolated store clusters at street corners

2. Neighborhood nucleations, including traditional shopping streets (with 50,000–200,000 square feet of retail space)

3. Community level nucleations (containing 200,000–500,000 square feet of space)

4. Regional nucleations (containing 500,000–1 million square feet of space)

5. Major regional nucleations (containing more than 1 million square feet of commercial space)

In a mononucleated city the central business district may well be the largest nucleation forming a sixth level, but in a polynucleated city it may be just one of a number of nucleations that cater to a particular geographic sector of the city.

This is illustrated in Table 10.3, where, as has been indicated previously, Toronto is regarded as being on the more "monocentric" end, and Dallas-Fort Worth on the more "polynucleated" end of a hypothetical mononucleated–polynucleated spectrum. The largest nucleation in the Toronto CMA is the central business district with 10 percent of metropolitan retail sales. In contrast, the two separate CBDs of Dallas-Fort Worth attract only 3 percent of metropolitan retail sales, involving sales mostly to the adjacent resident low-income African American population, and occasional sales to office workers. On the other hand, there are more regional and major regional nucleations in Dallas-Fort Worth than Toronto, accounting for significantly more of total metropolitan retail sales. There has been an attempt in Canadian cities to reinforce downtown areas by anchoring them to an enclosed mall of the regional or major regional size, such as Eaton Centre in Toronto (2.7 million square feet) or Pacific Centre in Vancouver (1 million square feet). Community-level nucleations, of both the planned and unplanned types, are more prevalent in the Toronto area, accounting for a greater share of metropolitan retail sales than in Dallas-Fort Worth.

It should be reemphasized that the number of levels in the hierarchy that may exist in a particular city depends on the size of the urban area. A large metropolis will have all the levels in a number of complex forms that reflect the socioeconomic structure of the part of the city to which they relate, whereas a

Table 10.3

Estimated Share of Metropolitan Retail Sales in Central Business Districts, Regional and Community Nucleations: Dallas-Fort Worth CMSA And Toronto CMA

	TORONTO CMA	DALLAS-FORT WORTH CMSA
Central Business Districts	Single binodal CBD	Two separate CBDs
	10%	3%
Regional and Major Regional Nucleations[a]	17 nucleations	23 nucleations
	12%	18%
Community Nucleations[b]	63 nucleations	55 nucleations
	18%	13%
Automotive	34%	37%
Other	26%	29%
CMA/CMSA Population	3.9 million	3.9 million
Population Density[c]	697	215
Income per capita ($US)	$17,900	$15,800

(*Source:* Extracted from Simmons (1995), Table 3, p. 9, Table 7, p. 17.)
[a] 500,000 square feet and more.
[b] 200,000–500,000 square feet.
[c] persons per square kilometers.

smaller and relatively homogeneous urban area may have only three or four. So it is evident in many metropolises that there is a hierarchy of retail locations relating primarily to a black community, and another relating to the white, predominantly suburban population. For example, Perimeter Center, at the intersection of the I-285 beltway around Atlanta and Route 400 is a major regional mall serving primarily the needs of a suburbanized African American population within the inner suburban region defined in Figure 10.1.

The Commercial Structure of Nucleations Figure 10.5 indicates some of the types of retail activities that enter at different levels in the hierarchy. The higher-order nucleations generally have the activities particular to their level and those of nucleations at lower levels in the hierarchy. Planned enclosed nucleations are generally the favored locations of retail outlets owned by chains, whereas unplanned nucleations involve a fairly large number of small independent retail stores. *Street corner* developments, or *minimarts,* constitute the most ubiquitous element in the pattern of nucleations and are scattered throughout residential neighborhoods to ensure maximum accessibility for basic goods to consumers residing within a short distance of the cluster. They normally comprise from one to four businesses of the lowest threshold types, with the grocer-drugstore-gasoline station combination most common in suburban areas.

Neighborhood nucleations are characterized by the addition of higher-threshold convenience types, which supply the major necessities to consumers living

Figure 10.5 A representative example of retail activities that are typical of different levels in the hierarchy of commercial nucleations.

in the vicinity. Grocery stores, supermarkets, delicatessens, barbers and beauty shops, taverns, and perhaps small restaurants are the most usual business types represented. In more suburban areas these are often located in unenclosed *strip malls*. Jones (1991), using information from a case study of suburban malls in San Antonio, Texas, argues that nucleations with about 100,000 to 150,000 square feet of leasable space, which are anchored by supermarkets and include a post office, are becoming an increasingly important part of the suburban retail landscape. His study demonstrates that patronage rates of stores in malls anchored by supermarkets is much greater than in those that are not anchored by such a magnet.

In the older parts of cities, neighborhood, community, and regional nucleations frequently occur in unplanned form as retail strips where the stores may reflect the various waves of immigration as well as the existing ethnic and socioeconomic structure of the local neighborhood. In some instances, the ethnic heritage of the street is preserved to become a form of specialized shopping area visited by suburbanized descendants of the previous inhabitants of the neighborhood long after the group that generated the character of the street has left. On the one hand, if the general level of purchasing power in a neighborhood decreases too close to the poverty level, the retail strip may become a "skid row." On the other hand, with gentrification or redevelopment, an older retail strip may begin to reflect the needs and tastes of the newer, wealthier market.

Nucleations at the higher levels in the hierarchy are distinguished by shopping facilities that are anchored or led by large supermarkets (greater than 50,000 square feet), which are added to the convenience functions of the lower-level centers. In addition, they provide a considerable range of personal, professional, and business services. *Community nucleations* provide less specialized shopping types, like variety and clothing stores, small furniture and appliance stores, florists, jewelers, and banks or other kinds of financial institutions. Supplementing these are a range of more ubiquitous services and entertainment facilities. Between 1986 and 1992, the number of malls in the community nucleation category increased at a greater rate (47 percent) than any of the other categories, including the neighborhood. This accelerated increase can be attributed to the invariable inclusion of large supermarkets at the community level, along with the convenience of a range of other business types (with little duplication), which together maximize the advantage of this size of facility to consumers for multipurpose shopping and store owners (particularly chains) with respect to sales.

Regional nucleations are set above the community nucleations by the addition of higher-order functions—for example, a branch of a large upscale department store chain—and by greater duplication of functions typical of the lower-level centers. Increased specialization of business types thus becomes the keynote. In addition, a far greater range of services is provided from offices on the upper floors of nucleations at this level, including a variety of medical services, professional services (architects, lawyers, and real estate agents), specialized training (dance and language), branches of educational facilities, and business services. Although the number of malls in the regional category in North America has increased at a relatively low rate in recent years—19 percent in the 1986–1992 period—planning applications suggest that in the less real estate–speculative age of the mid–1990s this size of mall is more economic to build and rent, primarily because it provides sufficient space for a variety of business types and some duplication to achieve external economies.

The *major regional nucleations* are much larger than the regional nucleations, and frequently have a number of upscale department stores that provide a full range of retail and service activities. Also, large enclosed malls contain not only numerous specialty stores, but also "exotic" forms of recreation. In the United States, enclosed nucleations in this category account for 10 percent of all leasable mall space, and 11 percent of all mall retail sales. They are, therefore, becoming an important feature of the intraurban retail landscape and, with the inclusion of extensive recreational, hotel, and convention facilities, are being promoted as foci for tourists and economic development.

In particular, *mega-malls*, with more than 2 million square feet of leasable space, are being promoted by development companies, such as Triple Five Corporation, as generators of urban growth (Jackson and Johnson, 1991). The first of Triple Five's developments, the West Edmonton Mall, has more than 4 million square feet of space, contains 827 stores and services, and provides about 15,000 mostly part-time jobs. It also has an ice rink, a wave tank for wave-riding year around, carnival-type entertainment facilities, and hotel facilities (Johnson, 1991). The mall has drained retail activities from downtown Edmonton, and the financial operation appears to be tenuous with struggles to refinance a $300 million mortgage and delays in payment of local taxes (Feschuk, 1995). Nevertheless, a similar Triple Five mega-mall, Mall of America, is now in operation in Bloomington, Minnesota (just south of Minneapolis), and one (the

"American Dream Mall") is being promoted as an antidote to the deteriorating downtown core of Silver Spring, Maryland, just north of Washington, D.C.

The Hierarchy of Retail Nucleations in Seattle Morrill (1987) has shown how a hierarchy of nucleations containing retail and service businesses exists in the Seattle PMSA, which has experienced one of the highest metropolitan growth rates in North America—25 percent, from 1.6 million in 1980 to 2.0 million in 1990. His study involves an analysis of the number of establishments, the general type of activity, and the sales volume in the thirty-four largest nucleations in the metropolitan area. The study is of particular importance because it covers much of the same geographic territory and research methodology as the earlier classic work of Berry and Garrison (1958). It is, therefore, possible to see the effect of suburbanization and the growth of planned shopping malls on the hierarchy of nucleations in this urban area over a thirty-year time period.

The Seattle metropolitan area is polynucleated, but not as much as Dallas-Fort Worth. The hierarchical structure is, in part, a complex outcome of the merging of a number of cities. Thus, at the highest level, the structure involves the Seattle central business district (Table 10.4), with the Bellevue central business district at the second level, and the Everett central business district among five nucleations at a third level described as "large regional." The information in Table 10.4 demonstrates that the higher the level of the nucleation, the greater the average number of retail and service establishments per center, the greater the proportion of those establishments that are high level, and the larger the average volume of retail trade per nucleation. With reference to the number of levels in the hierarchy, it is evident that the Seattle metropolitan area must have at least seven or eight levels as the data presented by Morrill concern only the larger nucleations. Interestingly, the Seattle central business district has a similar share of total metropolitan retail sales (10 percent) as downtown Toronto has of its metropolitan area.

The structure of this hierarchy of nucleations has changed considerably since the 1950s. In particular, the nucleations in levels III and IV (see Table 10.4) consist primarily of suburban planned regional shopping malls: It is these that have been growing fastest in sales volume. In this case, the impact of newer shopping malls along with suburbanization is evidenced most clearly at the regional level in the hierarchy of nucleations, with the Seattle central business district declining in relative importance. For example, downtown Seattle accounted for more than 30 percent of total metropolitan retail trade in 1954, but only 19 percent in 1982, whereas the five at level III increased their share from 7.2 percent to 9 percent, and those at level IV increased theirs from 14 percent to 17 percent. This relative decline of downtown Seattle compared with the rest of the metropolitan area, and increase in share of the level III and level IV nucleations, has continued into the 1990s.

Impacts of Shopping Malls in North America Given the pervasiveness of shopping malls in North America, some discussion of their impact on urban ar-

Table 10.4

The Hierarchy of Retail Nucleations in the Seattle PMSA

LEVEL	NUMBER OF NUCLEATIONS	AVERAGE NUMBER OF RETAIL ESTABLISHMENTS PER NUCLEATION	AVERAGE NUMBER OF SERVICE ESTABLISHMENTS PER NUCLEATION	% HIGH-LEVEL ESTABLISHMENTS	WEIGHTED AVERAGE OF RETAIL SALES PER NUCLEATION (MILLIONS)
I	1 Seattle CBD	1,420	2,640	43%	$925
II	1 Bellevue CBD	500	615	39	470
III	5 Large regional	363	347	33	316
IV	9 Regional	253	227	30	214
V	18 Community	115	120	30	77
Metropolitan area totals		12,500	15,100	28%	$9,140

(*Source:* Data retabulated from Morrill (1987), Table 2, pp. 109–110.)

eas and society in general is required. The mall is finely attuned to an automobile-oriented society that demands consumer safety and ease—that is, consumers wish to be safe and have ease of travel to, parking at, movement around, purchasing in, and exit from the facility. Furthermore, the stores within the mall must provide an array of goods that is well displayed, marketed at competitive prices, and purveyed by competent salespersons aided by instantaneous purchasing technology. For the most part, the planned shopping mall with its centralized control is better able to combine and assure such high levels of safety, ease, and competitiveness, than unplanned facilities. These advantages, which are evidently overwhelming as planned malls now dominate the consumer service structure of North American cities, are, however, provided at a price. The price often involves the demise of retail functions in downtown areas and other large unplanned nucleations, and the privatization of public space.

The Hollowing of Central Business Areas   Most central business districts and unplanned regional nucleations within the older central cities of North America are no longer leading locations for consumer services. A combination of decades of population decentralization, preference by consumers for multipurpose shopping and store owners for profit-maximizing locations, have led to the hollowing of unplanned nucleations (Schneider, 1992). Among the larger North American metropolises, only a few downtowns, such as New York (Manhattan), Philadelphia, San Francisco, Pittsburgh, Seattle, and the major Canadian cities, have retained a significant proportion of metropolitan area sales. Central business districts in Canadian cities have managed to retain significant shares through the implantation of enclosed malls downtown (Simmons, 1991).

The business districts of nonmetropolitan urban areas throughout North America have been similarly hollowed, with boarded store fronts prevailing along many a "Main Street." Lloyd and Beesley (1991) show that the demise of the Peterborough, Ontario (population 70,000), central business district is related to the establishment of two suburban malls, Landsdowne Place and Portland Place, which provide one-stop indoor shopping and greater ease of parking than the downtown. Similarly, in a study of five smaller urban areas in Ontario, Yeates and colleagues (1996) demonstrate that the traditional "Main Streets" in each of the places are extremely vulnerable to new retail formats located in strip malls and new shopping centers located at the periphery of the urban areas or in adjacent metropolises.

The Privatization of Public Space   One of the major legal difficulties with respect to malls is that although they contrive to appear to be public places, they are in fact private and run for a profit (Goss, 1993). For the most part, the public is not aware, or not bothered, by this *privatization of public space*, but there are at least two reasons for concern (Sorkin, 1992). First, public spaces have traditionally been places where free speech and assembly rights have been, and been seen to be, exercised (Goheen, 1993). With the demise of downtowns, there are now few public spaces where these rights may be seen to be employed (Mitchell, 1996). Mall owners, on the other hand, also have rights—the right to regulate use of private property, which, the owners would argue, includes the "common areas" of shopping malls. These two rights, therefore, collide. The real worry would appear to be in those cases where mall owners make choices and allow groups of which they approve to promote their cause in common areas, but disallow groups of which they do not approve. Six states in the United States have passed legislation permitting "limited" free speech in the common areas of shopping malls.

A second issue arising from the private ownership of malls concerns monitoring and social control. The purpose of the mall is to encourage people to make purchases. It is a space produced to promote consumption (Lefebvre, 1991). People will do this if they feel safe and are in the mood to purchase. The assurance of safety is both overt and covert. Overt safety is provided by mall security guards and the visible presence of video cameras. As malls are private property, owners frequently feel that they have the right to not only protect shoppers in malls, but also to expel from malls individuals who appear to be "undesirable" and not using the mall for its designated functions, shopping, or the purchasing of entertainment. The exercising of covert exclusion is fraught with possibilities for social injustice. Similar injustices may occur through technological surveillance: Individuals, couples, families, and groups may be monitored and the action of monitoring used as a means of behavioral control.

Spatial Aspects of Consumer Behavior and Cognition

The basic concepts underlying the notion of the hierarchy of nucleations are that individuals have perfect information about the alternative shopping facilities available to them in the urban environment, and that they act rationally to minimize the total costs (travel and purchase costs) of buying a bundle of goods. The *behavioral approach* relaxes these assumptions by recognizing that individuals have different values, desires, levels of mobility and cultural traits, and perhaps racial, ethnic, or cultural constraints, and that these characteristics also influence or limit spatial choice (Golledge and Timmermans, 1988). Individuals have learned different amounts about the availability of goods and services within urban areas. So these various *cognitive images*, or mental maps of retail structure, change in accordance with external stimuli such as advertising, and events particular to each individual such as life cycle changes or changing income levels.

As a result, studies of consumer behavior focus on the external and particular factors that influence the cognitive (learned) images that people have about the retail structure of urban areas. These two groups of factors are identified for illustrative purposes in Figure 10.6 where the possible range of retail alternatives is considered to be influenced by two groups of factors. One group consists of the particular characteristics of the consumer as represented by such factors as ethnicity, income, and age. These factors interact with each other to generate different ranges over which an individual consumer may be willing or able to travel for particular goods. Studies of *designative cognition* focus on how individuals with particular characteristics, such as the elderly, learn about retail opportu-

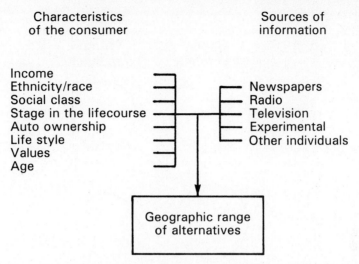

Figure 10.6   The external and particular factors that influence the range of retail alternatives considered by consumers.
(*Source:* Modified from Huff, 1960.)

nities, and how these are evaluated in terms of their life conditions. For example, Smith (1985) shows that the elderly have a more restricted geographic knowledge of retail opportunities than younger folk. Coshall (1985) further suggests that, perhaps due to restrictions on personal mobility, the elderly tend to overestimate distances to retail outlets and thus restrict their geographic consumer range. Furthermore, the geographic consumer range is related to income level (Potter, 1982) with lower-income households being more restricted than higher-income households. The influence of this combination of consumer characteristics and stage in the life course on consumer range is illustrated in Figure 10.7—the older the person and the lower the income, the more restricted the geographical area over which a person will travel to make purchases.

The second group of characteristics identified in Figure 10.6 is composed of the external factors that influence how much a person knows about the retail structure of an urban area and the possible places in which purchases of particular types of goods may be made. These factors fall into two basic groups: (1) sources of information, and (2) available experiential information, which is derived from an evaluation of previous information and practice. Individuals with limited sources of information will tend to have more restricted knowledge of the retail opportunities in an urban area. On the other hand, those with a greater variety of information sources will tend to have a much wider range of retail opportunities. In this context, Smith (1992) demonstrates, on the basis of sample surveys of the ambulatory elderly in both central city and suburban locations that their knowledge of the range of retail opportunities is quite restricted, with distance from home, prices, and the existence of a grocery supermarket being the prime influences on choice of retail outlet.

These information factors become particularly important when viewed in the context of the distinctive location of racial and different income groups in

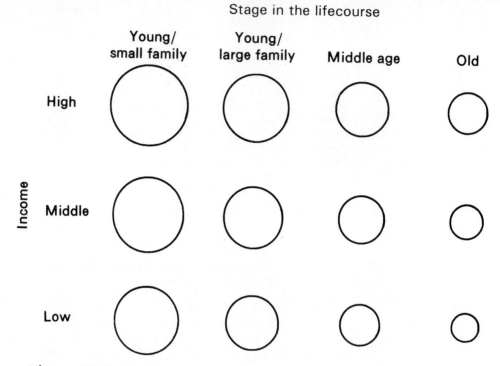

Figure 10.7 The way in which stage in the life course and income influence consumer range.
(*Source:* Potter, 1982, Figure 5.9, p. 159.)

metropolitan areas. The amount of information that people have about different retail opportunities is greatly determined by the marketing strategies of stores and shopping malls. Quite often, the retail opportunities in suburban malls are not relayed to those living in the inner city, even if inner-city residents have the desire and means to shop in suburban locations. Likewise, the variety of specialized retail facilities in ethnic areas within the older sections of large metropolises are frequently unknown to those residing in suburban locations.

THE INTRAURBAN LOCATION OF PRODUCER SERVICES

In Chapter 6 it has been suggested that the range of producer services, defined as activities that provide inputs to or that are linked with other economic activities, are quite extensive. They include transportation, communications, and public utilities; finance, insurance, and real estate; and business services. Wholesale activities comprise a separate category that have, for census purposes, often been included as adjunct to the retail sector, but these activities may just as reasonably be examined as part of the producer goods sector. Given the extensive and diffuse nature of producer services, in this section the focus will be on theories and principles related to the intraurban location of (1) wholesale and distribution activities, and (2) office activities, which embrace finance, insurance, real estate, and business services.

The Location of Wholesale and Distribution Activities

Wholesale and distribution (W&D) activities, involving the trade of goods in large lots from one place to another for resale purposes, are linked to both production and consumption (Glasmeier, 1990). As a consequence, the buildings that they occupy may be owned by, or directly serve, manufacturers, wholesalers, or retailers. It is estimated that manufacturers own, contract, or lease 73 percent of W&D facilities in the United States, wholesalers 22 percent, and retailers 5 percent (WERC, 1994). Thus, manufacturers in particular have a vested interest in selecting locations for W&D activities that minimize costs of storage and distribution and facilitate just-in-time deliveries. In this regard, it is estimated that costs associated with the transport of manufactured products to consumers account for 51 percent of total manufacturing W&D costs, and warehousing (including labor) 31.2 percent (La Londe and Delaney, 1993). Accessibility would, therefore, appear to be a prime influence on the location of W&D facilities. Accessibility, however, with respect to what?

Sivitanidou (1994) addresses this question, in the context of fifty industrial districts within the extensive Los Angeles urban region, by examining the relationship between single-tenant W&D land and property rents (the dependent variable) and a number of locational attributes with respect to the W&D facilities (the independent variables). It is hypothesized that four groups of attributes affect rents (Table 10.5): aggregate access to markets and manufacturing production; access to an array of transport facilities; access to blue-collar labor; and local land market conditions as indicated by a surrogate measure of the quality of the local environment, and the amount of available industrially zoned land. The results indicate that some of the specified locational variables account for, at best, half the variation in W&D land and property rents. Particularly noticeable in this Los Angeles case study is the insignificance of access to port and rail terminals.

There are intriguing differences in the significance of the various attributes for large and small single tenant facilities. The most significant group of attributes associated with the location of large facilities are access to freeways and air transport. The locations of the smaller facilities are, however, related to a wider array of variables, particularly local retail markets and manufacturing production centers, local land market conditions, and access to freeways and air transport. Sivitanidou (1994) suggests that the reason for this difference may be related to general orientation of trade in the two groups. Small W&D operations are oriented toward local retail markets and production centers, whereas larger W&D facilities are more oriented to markets and consumers external to the Los Angeles region. This internal-external distinction is reflected, therefore, in the need by larger facilities to have good transport links to the markets both within and beyond the metropolis. As Glasmeier (1990) observes, with respect to the

Table 10.5

Access and Site Characteristics Influencing the Magnitude of Large and Small Single-Tenant Wholesale and Distribution Property Rents in the Los Angeles Urban Region

Variables	SIZE OF SINGLE-TENANT FACILITIES	
	Large	Small
A. Spatial Linkages to Markets and Production		
Access to intraurban:		
Retail markets (market density)		$+\sqrt{}$
Production centers (manufacturing employment density)		$+\sqrt{}$
B. Transport Linkages		
Access to:		
Freeways (mileage within district)	$+\sqrt{}$	$+\sqrt{}$
Major freeway junction (1,0)	$+\sqrt{}$	$+\sqrt{}$
Airport facilities (distance from)	$-\sqrt{}$	$-\sqrt{}$
Port facilities (distance from)		
Rail terminals (distance from)		
C. Labor Linkages		
Access to:		
Blue-collar workers (distance from)	$-\sqrt{}$	$-\sqrt{}$
D. Land Market Conditions		
Amount of:		
Quality of environment (PCI)	$+\sqrt{}$	$+\sqrt{}$
Industrially zoned land		$-\sqrt{}$
Number of Properties	252	560
Multiple R^2	43%	51%

(*Source:* Extracted from Sivitanidou (1994), Table 3, p. 17.)
$\sqrt{}$ = significant at the .05 level.
+/- are the expected signs of the regression coefficients.
PCI = per capita income.

development of high-tech merchant wholesaling in Austin, Texas, it is the linkages with external markets that stimulate new industrial complex formation and strengthen the economic base of cities. The importance of ease of access to both internal and external producers and markets may well also underlie the decentralization of W&D facilities.

Intraurban Office Location

Office activities are not only concentrated among the large metropolitan areas of North America, they are also concentrated within urban areas as well, especially in central business districts within central cities and suburban nucleations. New

York, Los Angeles, Chicago, San Francisco, and Toronto are the metropolitan areas with the largest amounts of office space and office employment in the United States and Canada (Sui and Wheeler, 1993; Gad, 1991). There has been some spread of office activity to smaller metropolitan centers, as has been indicated in Table 2.1, which illustrates the change in the distribution of headquarter offices among the major metropolises of the United States. Even though New York has experienced a considerable decrease in the number of headquarter offices located within its environs, the headquarters that it has retained are responsible for assets in excess of the next ten largest metropolises combined (Holloway and Wheeler, 1991), and the amount of office space in the urban area also exceeds that of the next largest metropolises by a wide margin (Figure 6.6).

Information on the changing location of the headquarter offices of the 750 largest corporations in the United States provide a nationwide overview of trends in the location of office activities between central cities, suburbs, and nonmetropolitan areas (Table 10.6). In general, headquarter offices are becoming less central city located and more suburban and nonmetropolitan area located. Nevertheless, at the beginning of the 1990s, more than two-thirds of headquarter offices were still central city located. There is, however, a distinction between banking and insurance and all other types of headquarter offices (primarily manufacturing): The financial institutions remain much more central city oriented than all other types of companies (Shilton and Webb, 1995). This, as will be noted later, raises interesting questions concerning the importance for the command and control functions of large organizations of the agglomeration economies (such as forward and backward linkages) of central cities (Schwartz, 1992, p. 4). Some types of companies have headquarter functions that are more central city oriented than others.

The gritty persistence of the central city, and particularly the central business district, in the face of inexorable decentralization is indicated with respect to the changing location of occupied office space in the extensive New York urban region (Table 10.7). Although this region includes a large number of central cities, such as Newark, Jersey City, and Bridgeport, New York City (particularly Manhattan) is regarded as the traditional center for office activities. Whereas office space in the CMSA nearly doubled between 1970 and 1988, the greatest increases were beyond New York City, particularly in the suburban areas of northern New Jersey. Nevertheless, with the rapid globalization of the world's financial industry in the 1980s, New York City continued to demonstrate resilience as the leading office center of North America by adding more than 50 million square feet of space during the 1980s. This rapid growth of office space did lead to a cyclical market crisis with high vacancy rates between 1989 and 1993. However, there has been a noticeable recovery of occupancy rates in the mid–1990s (Mills, 1995).

Financial Districts and Offices Within A Multiplicity of Downtowns
The clustering of financial districts in the heart of downtown areas is illustrated in Figure 10.8a with respect to the central district of Toronto in 1990. Here the sky-scrapering financial district is clustered around Bay and King, adjacent to two other aspects of the broad service sector: consumer services in the form of

Table 10.6

The Distribution of Corporate Headquarters among Central Cities, Suburbs, and Nonmetropolitan Areas, United States, 1969–1989

| TYPE OF CORPORATION | METROPOLITAN STATUS | | | NUMBER OF CORPORATE HEADQUARTERS |
	CENTRAL CITY %	SUBURBS %	NONMETROPOLITAN %	
Banking and Insurance				
1969	96	3	1	100
1989	94	6	0	100
All Other Types				
1969	82	15	3	650
1989	64	29	7	650
Total, all Types				
1969	84	13	3	750
1989	68	26	6	750

(*Source:* Reaggregated from Ward (1994), Table 2, p. 472.)

Table 10.7

The Location of Office Space (in millions of square feet) within the New York-Northern New Jersey-Long Island CMSA

SUBAREA	1990 POPULATION	1970	1980	1988
New York City	7.3	239 (84)	259 (65)	309 (59)
Long Island (Nassau/Suffolk)	2.6	14 (5)	28 (7)	42 (8)
Northern New Jersey	4.6	26 (9)	76 (19)	110 (21)
Northern Suburbs	3.5	5 (2)	36 (9)	63 (12)
Total (millions)	18.0	284 (100)	399 (100)	524 (100)

(*Source:* Numbers calculated from Schwartz (1992), Table 1, p. 7; Bureau of the Census (1991), Tables C and D. Population figures in millions.
Note: New York City forms the large part of the "core" area demarcated in Figure 10.3 which includes the Jersey City PMSA.)

an extensive retail business district, and public services as represented by educational, government, and health care facilities. Associated with these producer and public service activities are residential areas—commonly a mixture of high-rise condominiums and older gentrified row-housing—which accommodate much of the younger and single labor force employed in this wide range of service activities. Public transport, in the form of rapid transit, commuter trains, buses and streetcars, and private transit as exemplified by the expressways, focus on this downtown financial district (Figure 14.8a). In metropolitan areas in which retail activities have largely decentralized, this aspect of land use is much less extensive than depicted in Figure 10.8a, the remains being directed toward serving the daily needs of office employees, or the local needs of the downtown population.

The general nature of the 1975–1990 change in the patterns of these land uses in the downtown core is suggested in Figure 10.8b. The office district, which incorporates the financial area, expanded eastwards; central core retailing expanded northwards beyond Bloor Street; and old industrial districts contracted with some of the land being redeveloped for public and private housing. Following redevelopment around a new City Hall in the late 1950s, China Town relocated to the Spadina/Dundas area, replacing a declining textile manufacturing district; and a more trendy retail/entertainment district recently spread westwards along Queen Street beyond Spadina Avenue. An entertainment district has become more identifiable with the renovation of some Art Deco theaters and the construction of some new ones, and the establishment of multiplex cinemas associated with retail center redevelopments such as the Eaton Centre. The central area consists, therefore, of multiple interconnected downtowns in which office and financial districts play a crucial role because of their employment and business generation.

Principles of Intraurban Office Location   A review of the literature (Lichtenberg, 1960; Goddard, 1975; Clapp 1980; Gad, 1991; Archer and Smith, 1993)

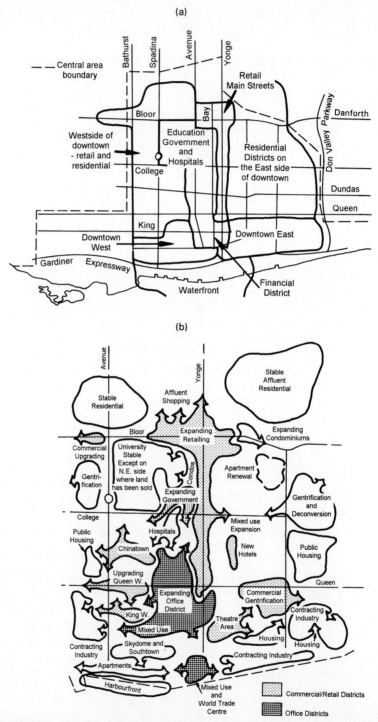

Figure 10.8 The multiplicity of Toronto downtowns: (a) districts within the central core; and (b) directions of change within the central core, 1975–1990.
(*Source:* Redrawn from Relph, 1990, pp. 78 and 79.)

indicates that four basic themes underlie the theory of office location. These principles are: contact intensity and variety arising from urban agglomeration; employee accessibility; location rent and taxes; and decentralization and suburban clustering. Some of these principles act in concert, whereas others counteract each other to a degree. The first two of these principles help to explain why the central business areas of major metropolises remain a strong focus of office activities, while the third and fourth may act together to give rise to the suburbanization of some office functions.

The Principle of Contact Intensity and Variety A major theme underlying many office location studies is that certain types of office occupations tend to cluster in central areas that have the greatest accessibility to contacts with respect to both intensity and variety (OhUallacháin and Reid, 1992). This is not meant to imply that all workers in the office need large numbers of contacts, but the managers and officers who bring in the business need them. Thus, the greater the contact needs of an occupation, the greater the attraction toward centers of intensive contact activity. Gad (1985) demonstrates that the businesses with the greatest need for contacts with others in downtown locations are: law firms, public relations consultants, banks, trust companies, insurance agents, and investment dealers. Most of these activities have to do with financial/securities functions, hence the continuing strong central city orientation of financial activities exhibited in Table 10.6.

The executives and professional managers in law firms, consulting firms, public relations businesses, real estate development companies, computer systems and network management consultants, and insurance agencies have a high need for contact variety as well as contact intensity. For example, the need for face-to-face contact is the main reason why 51 percent of the lawyers, 40 percent of the employment service firms, and 35 percent of the management and public relations consultants in the Edmonton CMA (population 840,000) are located in the central business district (Michalak and Fairbairn, 1993). In this context but at an entirely different population scale, Schwartz (1992) demonstrates that the core area of the New York CMSA (Figure 10.9 and Table 10.7) experienced the greatest growth in employment not only in securities and banking (contact intensity) but also business services, legal services, accounting and auditing, and advertising (contact variety).

The social structure of the central business district has become oriented toward facilitating face-to-face contact, particularly over extensive lunch times. Social, sports, and luncheon clubs, as well as restaurants in downtown areas are geared for this type of social/business activity during the working week. Also, professional sports events (at downtown or suburban locations) are increasingly geared for business entertainment—hence the continuous push by owners of sports franchises for stadiums and arenas with more high-priced "boxes." Many women quite rightly suggest that in this type of male-structured social business environment they suffer certain business disadvantages. As a consequence, there has been a strong effort in recent years to change this male business lifestyle to accommodate women business colleagues, and also for women to establish social networks of their own.

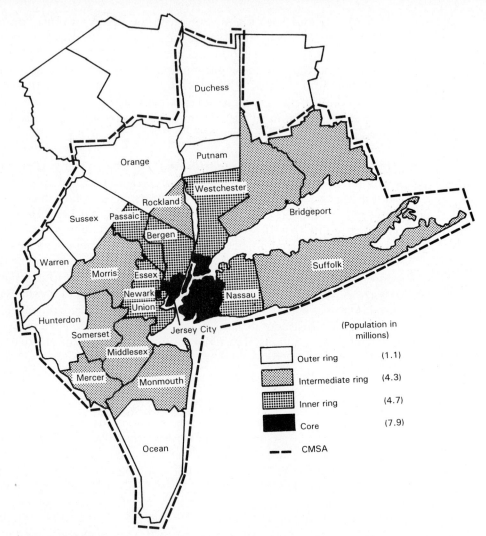

Figure 10.9 The New York metropolitan region.
(*Source:* Redrawn from Schwartz, 1992, Figure 1, p. 5.)

The Principle of Employee Accessibility Whereas the principle of contact intensity and variety applies to executives and highly trained professionals, the principle of employee accessibility includes the vast proportion of office employment that works at average and below average, wage levels. Given the numbers of individuals required, firms tend to locate where they have good access to skilled and reasonable priced labor. In this context, turnover costs are particularly important. Therefore, firms try to reduce turnover costs either by good labor policies and/or by having good access to easily substitutable labor. Central locations in urban areas are therefore preferred, especially if these locations are well served by public transit facilities. Since a large proportion of the office labor force is female, and women tend to have less access to automobiles than males, it is not surprising that those few urban areas with adequate public transit systems tend to have a large proportion of office employment in central locations.

Nelson (1986) demonstrates that as office activities have become functionally more complex and computer networks more universal, a number of "back-office" activities, mainly involving clerical and administrative support employees, have become decentralized in order to gain access to lower-wage female labor located in the suburbs. Her study pertains to the Walnut Creek-Concord area, which is essentially suburban with respect to San Francisco. A number of functions related to customer accounts, billing, inventory control, and routine management are located in office buildings constructed adjacent to freeways in order to gain the advantages of lower unit rents compared with the central business district and the large supply of well-educated female labor that is located in such rapidly growing outer suburban areas.

Location Rent and Taxes Location rents, or the price of land, are a product of accessibility with respect to a wide range of urban activities. If the forces of agglomeration in the downtown area are strong, particularly with respect to the availability of contacts, access to labor, and other advantages pertaining to centrality (discussed in Chapter 8), location rents will be high because of the competition for space. High location rents reinforce the clustering of office activities, often in skyscrapers. Firms will continue to pay high location rents as long as centrality remains important to their business activities. However, when accessibility at other places within metropolitan areas increases, the utility of central locations decreases, and firms may decide to move all or part of their office functions to noncentral sites where location rents are less. The ensuing changing competition for central business district and noncentral locations leads, in time, to readjustments in the pattern of location rents which may result in diminishing rent differentials between central and noncentral sites. Thus location rents, as has been noted in Chapter 8, sort use, and those activities that gain the greatest advantage from higher-rent downtown locations, such as financial activities, will remain in central locations, whereas those that do not realize commensurate competitive advantages may decentralize all or part of their office activities.

Office or business taxes, which are set by individual local governments within metropolitan areas, may also serve to influence office location. Although the matter of city expenditure and revenue raising in a highly fragmented local government environment is discussed later (Chapter 15), at this juncture it need be noted that office or business taxes are one way in which local governments raise revenue. Thus, in many jurisdictions central cities charge higher office taxes because of their great range of social, economic, and infrastructure needs, than suburban municipalities. Such high taxes do not necessarily drive offices out of downtown areas, for they can provide valuable facilities and quality-of-business-life incentives that may otherwise have to be purchased privately. High central city taxes become a liability when they are much greater than suburban taxes, where the suburban locations enjoy a "free-ride" on central city amenities. The central cities of Toronto ($9.36 to $13.80 per square feet), New York ($7.60–$11.88), Montreal ($8.14–$11.10), and Chicago ($7.00–$8.60) have the highest office taxes in North America (G&M, 1995).

Principles of Decentralization and Suburban Clustering   Suburban locations have evolved rapidly in the last half of the twentieth century as places with the greatest share of office floor space and employment within North American cities (Matthew, 1993b). Whereas in the mid–1950s approximately 95 percent of office space was clustered in the central business districts of United States cities, by 1986 only 43 percent was in central business locations (Fulton, 1986). In a similar vein, in 1988 in the Toronto CMA, 45 percent of office employment was located in the central city, and 55 percent in the suburbs (Gad, 1991, p. 434). Although many factors may give rise to decentralization, the most important are: the range of changes in intraurban accessibility associated with the automobile, the construction of intraurban highways, and the development of polynucleated metropolises; and improvements in communications technology, which have expanded ways of disseminating information, undertaking transactions, and cultivating contacts (Clapp et al., 1992). From the mid–1980s, office operations and methods of communicating have been revolutionized through continuous innovations in microchip technology, associated software developments, and widespread implementation of computer networking (Mills, 1995).

An interesting feature of this decentralization is that suburban offices tend to cluster in the same fashion as those in the downtown area. This clustering occurs in *office parks, suburban downtowns,* and in *corridors* along specific major highways (Hartshorn and Muller, 1989; Pivo, 1993; Matthew, 1993a). On the face of it, the suburbanization of offices would appear to suggest that agglomeration economies are less of a requisite in office location and clustering would not be so prevalent. The question, therefore, is: To what extent do the components of agglomeration economies that are traditionally associated with clustering, such as requirements for contact intensity and variety, demands for access to labor, and economies of scale in the provision of space necessitated by high location rents, underlie the phenomenon of suburban office clustering?

Archer and Smith (1993) pose, and address this question in the context of suburban office clusters in the sprawling Orlando MSA (population 1.2 million), home of *theme parks,* and a "second tier" office location with few headquarters but a large number of general purpose offices involving many clerical and administrative support employees. In this urban area, 50 percent of metropolitan office space is located in twelve suburban clusters, a "cluster" being defined as a "group of office buildings that are not separated by unrelated land use" (p. 55). In general, the authors find that the traditional factors provide only a partial explanation for the clustering.

Location rents in the Orlando area are low. So clustering does not occur to achieve economies of scale with respect to land use. Given the general purpose nature of the offices, there is little need for face-to-face business contacts—a characteristic also noted by Michalak and Fairbairn (1991) with respect to suburban office parks in Edmonton. The highways within the Orlando region provide reasonably good access to labor, but the fairly new and ubiquitous nature of the road network would appear to favor free-standing locations as much as clusters. Economies of scale in the provision of employee amenities, such as health clubs, day care centers, courier pickups, and car-pooling do appear to reinforce clustering. The major attractions of clustered locations, however, appear to be the *external economies of association* realized when a number of firms, some representative of identifiable "brand" companies, visibly group together with attractive buildings on a quality site to reinforce each other's corporate image.

THE INTRAURBAN LOCATION OF PUBLIC SERVICES

With employment in governmental and public services 40 percent greater than that in manufacturing, and almost as large as that in consumer services (see Table 6.1), components of urban infrastructure that facilitate the provision of public services form a dominant part of the urban landscape. The environment influencing *intraurban public facility location* is, however, more complex than that which theoretically influences the location of consumer and producer services. In a seminal article sketching the need for a theory of public facility location, Teitz (1968) argues that the fundamental difference between the location of activities in the public and private sectors relates to the role of the "market," be it consumers or other businesses.

In the study of the location of businesses in the private sector, issues concerning the size and location of the market are generally taken as given because the existence of the facility as well as its size and quality characteristics, are regarded as the *product* of the market. In public facility location, the size of the market is of crucial concern because the market and level of service provision are determined by the government, and the area to be served has to embrace the entire jurisdiction (Koontz, 1991). For example, a local government decision to provide publicly funded fire protection for all households necessitates decisions concerning the spatial distribution of fire stations throughout the city as a whole, the number of fire halls being determined by the schedule of available funds. The spatial distribution of fire stations thus comprises a fire prevention and control system for the urban jurisdiction as a whole, the level and quality of service of which is subject to budgetary constraints. The Teitz model, therefore, seeks to maximize consumption of a zero-priced service, such as fire services, supplied from a given spatial distribution of service facilities.

Characteristics of Public Facilities and Location

Given that the array of public facilities provided by governmental or quasi-governmental organizations is quite large, it is useful to discuss them in terms of a number of distinguishing characteristics:

1. Although there may be hybrids, public facilities may be characterized as *point* or *network* systems.

a. A point system is one in which the services are provided from a particular location, such as a hospital, employment office, post office, or court house. Services provided from point-located public facilities are generally used intermittently. Those used regularly by a large segment of the population, such as post offices, have to be located more ubiquitously with respect to the population than those used less frequently, such as court houses.

b. Facilities that require continuous connections in space, such as water, sewerage, electric power, telephone, and highway services, are distributed through networks. Figure 10.10 provides an example of a network of trunk sewer lines in the greater Toronto area which are required to drain toward pollution control plants close to the shore of Lake Ontario. Some of these networks, such as trunk sewer lines, waste refining facilities, and highways, involve considerable initial capital investment and, in consequence, their location becomes a vital determinant of urban growth. For example, the Las Vegas MSA, which is experiencing one of the largest population growth rates in North America (from 740,000 in 1990 to 1 million in 1995) spent $1 billion on new sewer, highway, and water systems in 1995, paid for by local sales taxes and new bond issues.

2. With most public facilities, decisions have to be made concerning the trade-off between *economies of scale and the advantages of dispersion* in service provision. Generally, the more frequently a service has to be used by individuals, the more dispersed its provision. The economies of scale/dispersion trade-off may well generate different locational decisions for various aspects of a particular service type. For example, postal services involve collection, sorting, and distribution. Collection, involving payment for the service, requires a dispersed pattern of customer-oriented offices. Sorting can be concentrated in large plants which permit economies of scale and technical innovation. Distribution involves the establishment of a relatively impersonal delivery network.

3. Many public facilities exhibit a *hierarchical pattern* of public service provision.
    a. A hierarchical pattern is inherent in networks because the flow tends to be cumulative with the capacity of the trunk lines being proportionately larger than the capacities of the branches.

    b. With point systems, a hierarchy of service provision occurs to take advantage, where possible, of economies of scale with its advantages of specialization. Some types of activity requirements for proximity and face-to-face contact also foster clustering. Three examples, which together accounted for 19 percent of total United States employment in 1991, illustrate the prevalence of hierarchical systems:

        (i). Schools, which are located in response to the residences of children and adolescents, are distributed hierarchically because of the age of children and pedagogy. Typically, primary schools are distributed ubiquitously within neighborhoods, junior-inter-

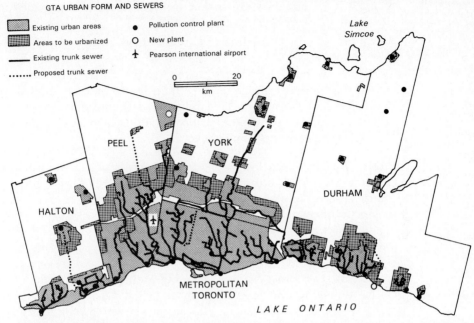

GTA URBAN FORM AND SEWERS

Figure 10.10   The influence of trunk sewer lines on urban form in the greater Toronto area.
(*Source:* Redrawn and modified from a map produced by the Greater Toronto Area Task Force, 1996.)

mediate schools are more dispersed serving wider communities, while high schools are large, with numerous specialized options and facilities, serving a wider region.

(ii) Hospital and community health facilities also tend to be distributed in a hierarchical fashion similar to schools, on sites on highways and urban arterials easily accessible by the population. There is a tendency for some of the largest hospitals, which provide an extensive range of services, and specialist institutions, to cluster in central locations, often around major medical teaching schools. They do this to gain certain external economies, particularly with shared high-tech equipment, specialist personnel, and collaborative research.

(iii) Local government services emanate from a local political system. Hence, the pattern of much local government employment reflects the command and control characteristics of the political process. The political process feeds on interpersonal contact, because the political machinery of government and the politicians and managers that head its departments require proximity. Thus, major governmental departments (such as those of the attorney general or housing), and the services the employees of the departments provide, are often located centrally, with certain services that are in greatest need provided in decentralized branch offices.

4. Public facilities in urban areas are provided from multiple levels of government. For example, some services in the City of Chicago are provided by the city, some by Cook County, a few by the wider multi-county (Cook, Du Page, McHenry) Chicago region, others by the State of Illinois, and others by the federal government. This multiple level nature of service provision generates a layered pattern of location facilities each of which reflects the locational characteristics cited above.

Public Facility Location-Allocation   A basic objective of public facility location is, therefore, to distribute offices or networks for the provision and distribution of services to a defined market within a particular jurisdiction subject to constraints concerning the availability of resources (Yang, 1990). One way of addressing this type of problem, in which a fixed number of locations are either known or to be selected, and "customers" or service provision resources to be allocated to them, is through the general methodology of mathematical location-allocation programming (Lea and Menger, 1991). Although mathematical location-allocation programming models have been applied most often in cases involving point systems (Brown et al., 1974; Sennott, 1991), there are also applications for network systems as well (Miller et al., 1992).

In general, most of the public facility location-allocation models have focused on the maximizing of "social utility" with respect to a given set of service locations, or the selection of x out of y sites as service locations (Dear, 1978). Social utility is frequently defined operationally in terms of accessibility, or the minimization of aggregate distance traveled by clients. The simplest form of such a model is the *transportation problem*, in which the "social utility" objective is defined as the minimization of aggregate distances traveled by clients, such as high school students, to a fixed number of service locations, such as high schools (Yeates, 1974, Chapter 10). The optimal solution, when aggregate distances are at a minimum, establishes optimum service areas. In the case of high schools and students served, the optimization of service areas implies high school districts with the greatest social utility. This school districting problem (Schoepfle and Church, 1991) can become more interesting with the extension of "social utility" to include equity goal constraints that, for example, minimize disparities in minority enrollment, student–teacher ratios, and overall enrollment (November et al., 1996). When such equity goals are included, they appear to be in conflict with crude distance-minimizing objectives.

The inclusion of equity goals raises a number of issues concerning the use of mathematical location-allocation programming models in the social sciences. The abstract numeric nature of the models and their computer-based GIS solutions can conceal matters related to community social values (Hodgson and Oppong, 1989). For example, many urban areas in the 1990s in consequence of massive public debt are "downsizing" public services. This often implies closures of some schools, hospitals, and other social services, and redefinitions of client eligibility for certain benefits or services (such as unemployment benefit requirements). Location-allocation models may appear to provide "objectivity" to such rationalization, particularly facility closures. It should, however, be clear from the school districting and equity examples cited earlier that such models merely provide a way of assessing only those aspects of the service impact of political decisions that are specified within the structure and assumptions of the models.

Public Facility Location and Public Policy

Accessibility is only one aspect of social utility. A more comprehensive theory of public facility location must also take into account *political* and *equity* dimensions of location. These have proven to be difficult dimensions to include because: (1) the multi-layered political dimension is likely to be less interested in locational efficiency, as expressed in maximizing accessibility, than in voter satisfaction with locational decisions and the reach-of-service provision; and (2) equity, or social justice criteria, involve not only considerations of equity in service provision (Newbold et al., 1994), but also understanding of the ways in which the nature of the North American economy generates wide economic and social disparities that require collective action as well as private response (Danielson and Wolpert, 1992).

Although the political and equity dimensions have proven conceptually difficult to include in public facility location theory, Dear (1978) argues that the theory may be enriched and made more relevant through an understanding of three additional concepts.

1. *Location*, to a considerable degree, *influences access*. Hence, many public services are decentralized not only to "meet the market" but also to "realize the market." This latter aspect is important because it is in society's interest to support care in situations where an absence of local care could lead to dire consequences not only for individuals and families, but also for society as a whole. An example of where location strongly influences the possibility of use is with child care services where the facilities, whether they be in local neighborhoods or adjacent to places of employment, have to be in highly accessible locations. The necessity for parents in most families, whether they be single-parent or two-parent, to participate in the wage economy generates a situation in which government is having to wrestle with the extent to which it is willing to provide subsidized child care spaces for those who cannot afford the full costs but who wish to remain in paid employment (Mackenzie and Truelove, 1993).

2. The delivery of a public service at a particular location generates *user-associated and neighborhood-associated externalities*.

   a. User-associated externalities develop when clients cluster around areas of service provision and are able, through their agglomerative presence, to influence the quality and range of service. There is a tendency for such agglomerations to occur in central cities where appropriate support facilities are available. For example, the homeless, living in parks or vacant lots, or moving diurnally between streets and shelters, may publish their own newspaper to disseminate information and address matters of concern to the homeless (Wolch et al., 1987).

b. All public facilities generate neighborhood-associated externalities. These may be positive, as with a new primary school which local residents may perceive as supportive of property values, negative, as with half-way houses for correctional service parolees which local residents may perceive as threatening, or neutral, as with a public health clinic. Negative neighborhood-associated externalities are often an example of the NIMBY (not in my back yard) syndrome which is most prevalent in neighborhoods dominated by single-family dwellings (Dear and Taylor, 1982; Wolsink, 1994).

3. The availability and location of particular public services is invariably influenced by the *changing public policy context* within which they are embedded. For example, persons requiring mental health care used to be incarcerated in hospitals (known colloquially as asylums). During the mid–1960s, an alliance between libertarians who believed that asylums infringed on civil liberties, fiscal conservatives who wished to reduce the costs of mental health care, and the medical profession which now had various chemical forms of therapeutic treatment available, resulted in the closing of asylums and the relocation of patients to communities where they could be treated as outpatients from community mental health facilities (Johnson, 1990). The result was a shift of patients from generally rural located hospitals to urban areas, particularly central cities. Therefore, in the 1990s, quite a few persons in need of more closely monitored mental health care are homeless inhabitants of downtown streets (Dear and Wolch, 1987; Isaac and Armat, 1991).

It is, therefore, impossible to separate the consideration of public facility location from social policy and the political environment that generates that policy.

CONCLUSION

The location of employment opportunities in consumer, producer, and public services has largely decentralized during the past few decades. In general, the location of employment in these different types of service activities reflects the polynucleated structure of North American cities. This is particularly true of retail activities, in which consumer demand and purchasing power virtually determine the location of shopping centers and malls. The only type of service employment that remains to a considerable extent in central locations is that derived from office activities in the producer and public services sectors. In these cases, there are strong forces fostering both centralization and decentralization. The modern North American metropolis has, therefore, become an environment in which employment opportunities, including those in manufacturing, are widely dispersed. An important consideration resulting from this employment dispersion concerns the relationship between the spatial distribution of different types of employment opportunities and the socioeconomic structure of urban areas. The next chapter will, therefore, address the nature of the changing socioeconomic structure of urban areas.

Chapter

11

Social Spaces Within Urban Areas

*T*he objectives of this chapter are to gain an understanding of the variety of demographic socioeconomic patterns that exist within urban areas and to appreciate the range of forces that generate different social spaces. Discussion of these forces is rooted in two general theoretical approaches. The first, *human ecology,* emphasizes the changes in technology and the functional roles of individuals and households in the formation of social spaces within urban areas. The second is referred to collectively as *new urban sociology* (Smith, 1995). Whereas the human ecology approach can be presented as a unitary theory, the new urban sociology represents the confluence of a number of conceptual themes in social theory such as behavioral analysis, political economy, and structural perspectives. As used in urban geography, both approaches focus on how patterns of dominance and subordination develop in urban space. Human ecology "explains" these patterns in terms of technological and functional imperatives; the new urban sociology "explains" them in terms of hierarchical power relationships between social groups.

THE HUMAN ECOLOGY APPROACH AND URBAN SOCIAL SPACE

The concentric zone model of city growth and social spaces has been introduced and discussed in Chapter 7. This zonal model was a product of the Chicago School of Human Ecology of the 1920s when the primary concern of urbanists was the ways in which the massive waves of immigration from overseas in the decades around the turn of the century, increasing volumes of migration of the domestic population from rural to urban areas, and technological innovations in transportation, communications, and production, were affecting the social stratification of society and urban social space (Park et al., 1925; Hawley, 1986). In this section we will examine the wider manifestations of this approach as it pertains directly to patterns of socioeconomic change, and discuss a derivative of this approach, social area analysis, and its extension into factorial ecology.

The Human Ecology Approach

The basic assumption of the human ecology approach to the study of urban social space is that the processes recognized by plant and animal ecologists as operating in the natural environment can be translated to the urban social sphere. Thus, the human ecology approach developed in the 1920s as a social science that is "fundamentally interested in the effect of position, in both time and space, upon human institutions and human behavior" (Park et al., 1925, p. 64). The immediate utility of this approach was twofold. First, it focused on the characteristics of people in certain places and the tensions that occurred at the boundaries separating people of different characteristics. In this way the human ecology approach provided a conceptual structure and array of processes that are intrinsically spatial.

Second, the human ecology approach involves processes that relate to *groups* rather than individuals and can, therefore, be applied to aggregations of people and households that are collected by the census and published in census units (such as census tracts, blocks, and enumeration areas) of one kind or another. In fact, the approach, with its emphasis on the definition of communities of perhaps 3,000 to 9,000 people, gave rise to the use of census tracts within cities. These units are supposed to represent collections of people and households of roughly similar characteristics. The ease of obtaining and using census tract data may well explain the continued relative popularity of the ecological approach in urban studies.

Invasion-Succession-Dominance   Human ecology theory involves a dynamic process of invasion-succession-dominance which suggests how change in dominance-subordination occurs in social space. The process envisages an initial stage in which a particular group of people, or land use, occupies a given space or territory. The people, or the land use, have sufficient characteristics in common to appear relatively homogeneous. Then, an incursion of another group with different characteristics occurs. This new group may be assimilated into the original resident group and adopt its characteristics, or it may be repulsed, or it may prove stronger and more able to live in that particular environment than the original resident group. In the last case, the new group may succeed the older group and, as in plant ecology, a new climax stage is reached in which the new species replaces the one that had occupied that niche before it. There is, therefore, an inherent assumption that stable homogeneous social groups or land uses are the norm, and that the invasion-succession stages will lead to a new norm.

It is not at all clear that stability is the norm, for most parts of cities are changing continuously. Furthermore, there is also an underlying supposition in the invasion-succession-dominance process that homogeneity is the natural end product, and that heterogeneity (the mixing of different types of peoples, households, and land uses) is a transition state. The process may, therefore, incorrectly provide subliminal support for preexisting racial and class prejudices in some people. We will observe in this chapter, and Chapters 12 and 13, that although there may be groupings of people in parts of urban areas that exhibit fairly similar characteristics, this cannot of itself be assumed to be a "natural" outcome, for it is invariably created by public and private actions in some way.

The invasion-succession-dominance process may be applied in a variety of ways to the analysis of the impact of class and culture on neighborhood transformation in which change rather than stability is the norm. Godfrey (1988), in his study of the evolution of the Mission and Haight-Ashbury neighborhoods in San Francisco, suggests two variants of the invasion-dominance-succession process, *ethnic* and *nonconformist,* based on the different experiences of two communities. The ethnic transition model is characterized by the experience of the Mission district, which, because of outmigration of low-income persons and households, had been an area in decline in the 1950s and 1960s. The district changed during the 1970s and 1980s to predominantly working-class Latino households. The stages of change involved: *migrant penetration,* in which a few upwardly mobile minority households entered the district; *minority invasion,* as the new minority households replaced the existing population which was moving away; and *ethnic consolidation,* when the new incoming Latino population began to predominate.

The second model, characterized by Godfrey (1988) as initiated by "nonconformist" groups, describes a situation of neighborhood transition following its *discovery* by artists and gays in the 1960s. As characterized by the Haight-Ashbury district, this involved the gradual replacement of lower-income families and individuals, often elderly, by artists, gays, and other "nonconformist" persons. These individuals were attracted by the short-lived hippie culture and the attractive but often decayed housing stock, which provided an appropriate ambience for creativity and nonconformist lifestyles. This new ambience then generated a *middle-class transition* with individuals and private developers gentrifying much of the existing housing stock for a larger population of inhabitants who wished to be associated with the risqué image. The last stage in this particular type of nonconformist-initiated transition, *bourgeois consolidation,* occurs when median housing prices and rents in the neighborhood reach a level that can be afforded only by wealthier households. This avant-garde–generated transition of older run-down neighborhoods to middle- and upper-class habitation occurs in other older central cities, such as Greenwich Village and East Village in Manhattan, and Cabbagetown in Toronto. In these examples, the bohemian or the hippie is the vanguard of the middle and upper class (Ley, 1994).

Social Distance   One of the ways in which neighborhood homogeneity may be created is through the process of social distance. It is, in fact, a process suggestive of the extent of racial or ethnic prejudice among people. Social distance indicates the degree of social separation between individuals, families, ethnic groups, neighborhoods, occupations, or any other grouping of individuals. As a social concept, it may be measured behaviorally in terms of who marries whom, residential proximity, and degree of social mixing; or it may be measured more abstractly by asking individuals to rank others according to a scale of hypothetical social relationships.

For example, Bogardus (1926) measured how native-born white Americans perceived other racial or ethnic groups by asking whether they would (1) admit ethnic group X to a close kinship by marriage, (2) have X as friends, (3) have X as neighbors on the same street, (4) have X in the same part of the city, (5) have X in the same city, and (6) have X in the same country! Numerous studies have demonstrated that "the less social distance there is between individuals, the greater the probability of interaction of some kind" (Knox, 1982, p. 41). The corollary to this concept is that the greater the social distance between individuals and groups, the greater the likelihood of a large physical distance between them as well. The Bogardus study also viewed *social distance* from the perspective of the dominant (i.e., white) group rather than regarding it as a possible *reciprocal relationship* in which diverse groups of people, perhaps differentiated by race, ethnicity, and income, exhibit varying degrees of *locational avoidance.*

A study by Waddell (1992) suggests the varying reciprocal strength of social distance between three racial groups—non-Hispanic whites, African Americans, and persons of Hispanic origin—on residential and workplace location in the Dallas-Fort Worth CMSA. The study, based on 2,000 worker records for each racial sample, concludes that "race influences urban structure by segregating not only residential neighborhoods but also workplaces" (p. 135). Although there is a high degree of avoidance between whites and Hispanics and African Americans in residence and workplace locations, the degree of avoidance is much greater between whites and African Americans than between whites and Hispanics. Furthermore, there is also a tendency for Hispanics and African Americans to exhibit residence and workplace avoidance. Although the level of avoidance decreases with income and education, it still exists at each level of attainment. It would, therefore, appear that a high degree of reciprocal social distance in both residence and workplace location exists between whites and African Americans, and lower degrees of reciprocal social distance between whites and Hispanics, and African Americans and Hispanics.

Defining Social Space

One of the issues concerning human ecologists is the way in which social space may be defined. It is possible to envisage a number of definitions based on a variety of socioeconomic characteristics. What is needed is either a theoretical understanding of the impact of urban development on the formation of social groups in the city, which would permit the selection of definitional criteria, or an analytical means of determining which, among a large group of possible cri-

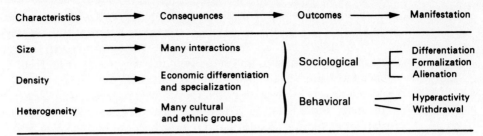

Figure 11.1 Wirth's (1938) view of the outcome of increasing urbanization.

teria, are the most relevant. Ideally, the two approaches, one theoretical and the other highly empirical, would support each other. Social area analysis, rooted in the classical theoretical work of Wirth (1938), provides a conceptual framework for selecting criteria to define social spaces. Factorial urban social ecology, an application of factor analysis (Rummel, 1970), provides a data-based approach.

The Social Area Concept and Social Space During the 1920s and 1930s, North American society was changing from one that was predominantly rural to one that was urban. Many people at that time thought that this was unfortunate as urban areas were regarded as generally iniquitous, if not downright sinful, as they provided an environment with greater opportunities and less constraints than rural areas in general and family farms in particular. This attitude underlies Wirth's (1938) analysis of the impact of society's transformation from a rural to an urban state. His interpretation is based on three indisputable characteristics of urban areas: They are generally large in size, have a reasonably high density of population and economic activities, and are reasonably heterogeneous in terms of people and occupations.

Wirth hypothesized that *size* created a large number of possible interactions. As a result, we are led to assume that most interpersonal contacts in the city are impersonal and secondary and that there are few close interrelationships. On the one hand, the individual has more freedom and less personal and emotional control from others; on the other hand, unsatisfying transitory contacts lead to a lack of meaningful interaction and consequently to feelings of alienation (Figure 11.1). *Density* reinforces the effect of size, for it results in differentiation and in specialization with respect to economic activities, residential location, and workplaces. As the complexity of society is thus increased, people are compartmentalized into specialized roles that predominate in particular subareas of the city. The result is the segregation of the city into a mosaic of social and economic worlds that are too numerous and complex for anyone to fully comprehend. Finally, cities are *heterogeneous* because they are the product of migration of people of diverse origins. They are also socially diverse. This characteristic is compounded by the differentiation and specialization of occupations. This heterogeneity results in no common set of ethical values.

Eventually money tends to become the measure of all things, with certain objects, such as a house or automobile, providing external status symbols.

There are, according to Wirth (1938), two main outcomes (Figure 11.1) of increasing size, density, and heterogeneity as society shifts from a rural to an urban state—one sociological, and the other behavioral (Berry, 1981). On the sociological level, urbanization is hypothesized as leading to differentiation, formalization, and alienation. The differentiation effect can be both spatial and aspatial, for economic activities and people become highly specialized in function and role, and both types of specialization can occur in particular locations. The development of numerous formal institutions is largely the result of excessive differentiation and the need for a means of communication. Alienation can be expressed in withdrawal. At the behavioral level, the stresses of urbanism, from such factors as high urban densities or excessive stimulation, result in a variety of "abnormal" behaviors, such as hyperactivity. The hypothesized result is that urbanism leads to greater levels of social disorientation and a faster breakdown of family life compared to rural settings.

Wirth's (1938) negative view of the outcome of urbanization on family life and social relationships is overdrawn. There is no doubt that as size, density, and heterogeneity increase, people are faced with excessive stimuli and a constant preoccupation with surviving the rat race of corporate competitiveness, commuting, and cacophonic environments. Cases of people exhibiting conditions of alienation, or some form of social deviation, do exist in cities, but they also exist in smaller towns and rural areas. However, do these conditions become more numerous in cities because of increasing scale, density, and heterogeneity? Abundant evidence suggests that when these conditions exist, they invariably occur for other reasons; there is little evidence that urbanization itself creates social disorientation and a breakdown of family life. Furthermore, large urban areas not only have agglomerative effects in economic terms, size and density also stimulate social interaction and cultural creativity (Gans, 1962; Wilson, 1991, Chapter 9; Sherlock, 1991). Nevertheless, high population densities, size, and heterogeneity persist in many parts of North America, such as Orange County (California), as perceived negative influences in quality-of-life ratings (Baldassare and Wilson, 1995).

Social Area Analysis Although Wirth's model can be criticized on the grounds that it is too negative, his view of the effect of urbanization on people stimulated an enormous number of questions. One question concerned how *the change from a rural to urban state was manifested in the social structure of the city*. Shevky and Bell (1955) hypothesized that the changes Wirth discussed could be related to ecological groups, which are defined statistically and geographically by census tracts. Furthermore, they suggested that the changes could be summarized in terms of three constructs relating to the groups: social rank, family status, and segregation. The latter can relate to the formation of distinct groups by characteristics other than race, such as age and language. It is hypothesized that increasing levels of urbanization result in clearer differentiation of social groups by social rank, family status, and segregation.

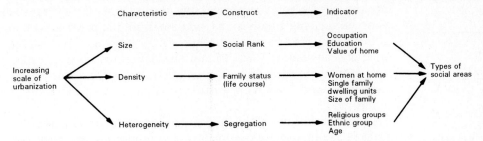

Figure 11.2 Steps in the development of social area constructs and indices.
(*Source:* Based on Shevky and Bell, 1955, p. 4.)

1. *Social rank* relates to size because large urban places, with their opportunities for many interactions, generate a wide variety of economic activities (Figure 11.2). Consequently, the indicators relating to this trend are the proportion of the labor force in different occupations, years of education, and various measures indicating the wealth of the group.

2. *Family status*, which is an outcome of increasing density, reflects the effect of increasing specialization on families. The members of urban families are involved with a number of different occupations with wide separation between work and home—quite different from traditional family-run farms. This diversification of occupations within families crystallizes most clearly with respect to the movement of women into waged employment. Thus, the family status construct can be measured by the proportion of women employed outside the home, average family size, and the proportion of single-family dwelling units in a census tract.

3. *Segregation* reflects the increased level of heterogeneity as society changes from a rural to urban state. A census tract is considered highly segregated if a high proportion of its population consists of individuals of a particular ethnic background, race, age, or gender.

It should be noted that the social area analysts define the word *segregation* numerically. Later in this chapter the meaning of segregation will be discussed in terms of process.

Social Area Constructs and Social Space in the City of Wichita   The original Shevky and Bell (1955) application of social area analysis result from studies undertaken in the cities of San Francisco and Los Angeles. In this section the constructs are defined on the basis of census tract data for the City of Wichita (Kansas) for 1990 in accordance with the indices suggested by the theoretical argument and listed in Figure 11.2. Although the Wichita MSA (1990 population 485,270) spreads over the counties of Butler, Harvey, and Sedgwick, the city of Wichita (1990 population 304,011), located within Sedgwick County, provides the bulk of the population. The population of the MSA outside the city is spread over a large territory on farms, exurban properties, scattered housing subdivisions, and small municipalities. Although the population of the MSA is mostly white (423,784), there is a significant African American population (36,979) and population of Hispanic origin (19,793). Almost 60 percent of the white population, 93 percent of the African American population, and 77 percent of the population of Hispanic origin, are located within the city.

As suggested in Figure 11.2, the social rank construct has been defined in terms of a composite of occupation—the proportion of a tract's employed labor force in executive, administrative, managerial, and professional occupations; education—the proportion of a tract's population 25 years of age and older with at least a bachelors degree; and median value of home. A composite index is achieved through a summation of the ranks of census tracts according to these three variables. The census tracts are then reranked with respect to the composite index and divided into three groups—the highest-scoring 25 percent tracts being labeled "high," the lowest-scoring 25 percent being labeled "low," and the 50 percent in between labeled "middle." The map of these tracts (Figure 11.3A) suggests the existence of two sectors of high-social-status-areas, one to the west and the other the northeast, and a band of low-social-status-tracts extending north and south in the central part of the city. The high- and low-status areas are largely separated from each other by wedges of tracts in the "middle" category that extend to the southwest and southeast.

The family status construct is constructed similarly, but is based on two of the variables in Figure 11.2—the aggregate number of persons per family in a census tract, and the proportion of housing units in a tract with three or more bedrooms (as a surrogate for single-family dwelling units). The variable "women at home" is not included because by 1990 the vast majority of women in most census tracts (female employment participation rates ranging from 50 to 70 percent) were involved in paid employment outside the home. The high-family-status/low-family-status distinction in Figure 11.3A exhibits a concentric pattern with a low-family-status area consisting of tracts with a high proportion of one- and two-bedroom housing units and few persons per family forming a crescent through the center of the city to the southeast, high-family-status tracts with larger housing units and more persons per family at the periphery, and middle-family-status tracts predominating in areas in between.

The third construct is defined in terms of the degree of concentration of African Americans and persons of Hispanic origin as these are the most clearly defined minority groups in the Wichita MSA. These two minority groups concentrate remarkably in very few, highly clustered census tracts (Figure 11.3C). Fifty-six percent of all African Americans in the Wichita MSA are concentrated in the ten census tracts demarcated in Figure 11.3, with most of the rest of the African American population located in tracts adjacent to this cluster. Although the Hispanic origin population is more dispersed, two census tracts in the central part of the city exhibit a significant concentration. The processes that give rise to these types of concentrations and clustering of minority groups are discussed later in this and the following chapter.

Patterns of Social Space   The spatial patterns of the social area constructs in Wichita in 1990, therefore, reinforce Berry's (1965) observation, which was based on studies by social area analysts in the 1950s, that three general patterns of social space exist in North American cities. The first is the axial, or sectoral, variation of neighborhoods by social status, the second is the concentric variation of neighborhoods according to family status, and the third is the localized concentration of particular minority populations. These hypothetical spatial patterns are presented in Figure 11.4. The social rank diagram obviously echoes Hoyt (1939). The concentric distribution of family status is the product of the greater availability of new single-family homes, requiring low initial mortgage down payments, in outer suburban areas. Minority groups, whether African American, Puerto Rican, Mexican American, Chinese American, or the low-income single elderly, usually concentrate in distinct parts of the city for a variety of reasons.

Factorial Urban Ecology   The human ecology approach to the definition of social space within urban areas can be criticized not only because of the antiquated "rural-to-urban" theoretical base, but because of the methodology used to define the constructs and social areas. The methodological objections pertain to the use of census tracts and the subjective selection of measures to define the constructs. The use of census tracts is based on the assumption that they represent fairly homogeneous groups of population. This gives rise to the "ecological fallacy" in urban analysis, for it is incorrect to assume that census tracts contain people with similar socioeconomic characteristics. An additional aspect of this fallacy relates to the treatment of census tracts in statistical analysis and mapping as if they are equal in population and physical size, whereas they are not. Nevertheless, given that some spatial level of aggregation is required to achieve confidentiality and that block data do not provide full coverage, census tracts often provide the most useful units of information for spatially distributed data.

   The second methodological objection to social area analysis, that the indicators are chosen subjectively to support the constructs, also cannot be denied. Consequently, it is interesting to know if the constructs are basic patterns of social structure that can be produced if all the variables pertaining to census tracts within metropolitan areas are used. The factor analysis model tries to answer this question by using all the information available (Davies and Murdie, 1993, pp. 54–56). Factor analysis involves a set of procedures that can be applied to summarize in a few dimensions a large number of characteristics that relate to particular items or observations. In this case, if the items are census tracts, the number of variables pertaining to census tracts is usually greater than thirty. Factor analysis helps, in effect, to find the underlying dimensions, or common threads, represented by these many variables.

Factor Analysis and Social Area Constructs   Although there have been many factorial urban ecology studies related to *particular* cities (for example, Rees,

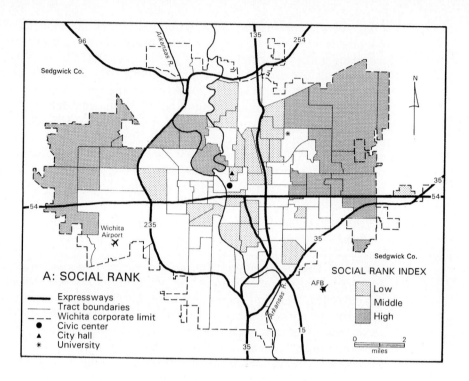

A: SOCIAL RANK

Expressways
Tract boundaries
Wichita corporate limit
● Civic center
▲ City hall
✳ University

SOCIAL RANK INDEX

Low
Middle
High

0   2
miles

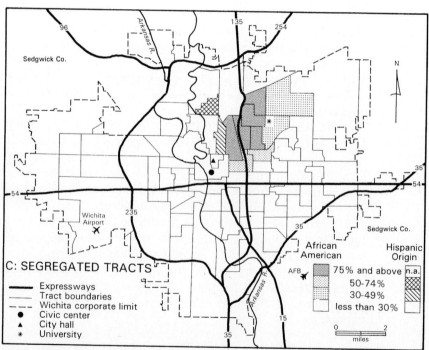

C: SEGREGATED TRACTS

Expressways
Tract boundaries
Wichita corporate limit
● Civic center
▲ City hall
✳ University

African
American   Hispanic
Origin

75% and above   n.a.
50-74%
30-49%
less than 30%

0   2
miles

Figure 11.3 The social areas of Wichita, Kansas, 1990.

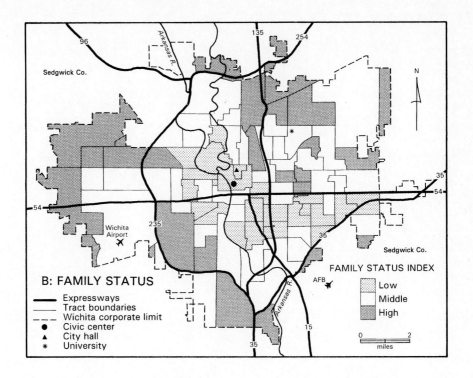

B: FAMILY STATUS

— Expressways
— Tract boundaries
- - - Wichita corporate limit
● Civic center
▲ City hall
＊ University

FAMILY STATUS INDEX

Low
Middle
High

0   2
miles

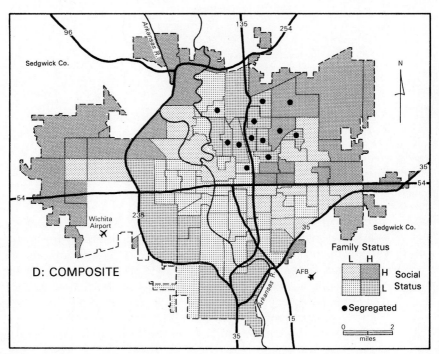

D: COMPOSITE

Family Status

L   H

H Social
  Status
L

● Segregated

0   2
miles

Figure 11.3 (continued)

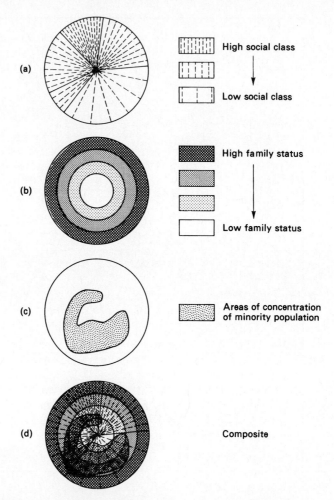

Figure 11.4 An idealized spatial arrangement of constructs and indices of (a) social rank, (b) family status, (c) segregation, and (d) a composite view of the spatial variation of social spaces, as defined by social area constructs.

1970; Perle, 1982/3; Murdie, 1988), Davies and Murdie (1991) have undertaken a study for a *system* of cities. The analysis involves the twenty-four CMAs in Canada and thirty-five socioeconomic variables pertaining to the entire set of census tracts (2,981, having an average population of 4,576) located within the metropolitan areas. The factor analysis of the thirty-five variables related to these census tracts indicates that 85.8 percent of the total variation inherent in these measures for all the census tract can be summarized by nine dimensions. These nine dimensions in effect summarize almost all of the "story" told by the thirty-five variables.

Table 11.1 provides a summary of the nine dimensions of the internal socioeconomic structure of Canadian metropolitan areas as defined by Davies and Murdie (1991; 1993). The qualitative descriptors summarize the meaning of each dimension in terms of the elements of the variables that are involved. For example, the "impoverishment" descriptor identifies a dimension that distinguishes census tracts that embrace, among probably a number of other characteristics, a conflation of female-parent families, low-income families, many dwellings that are rented, and above-average male unemployment, with a below-average representation of two-parent families. Again, the "ethnicity" descriptor identifies a dimension that distinguishes census tracts that have an above-average proportion of recent immigrants born outside of Canada and an above-average proportion of the population that is not of British or French ethnic origin and a below-average male unemployment rate.

These nine dimensions, in effect, provide a more subtle interpretation of the three social area constructs (grouped by the author in Table 11.1). For example, the family status construct consists of dimensions relating to "families," "non-families," "early/late families," "young adult" households, and "housing." The family status construct may, therefore, be more appropriately designated as a *life-course* group of dimensions (the concept of life-course is discussed in greater detail in the behavioral section of the chapter). Given the high proportion of women working outside the home in the paid labor force in the late twentieth century, it is interesting to note that higher-than-average percentages of women in the labor force are identified most specifically as part of the "young adult" dimension. The continuing importance of social status attainment, race or ethnicity, and life-course as factors differentiating population in urban space is attested also at the disaggregated level of individual workers, as well as the aggregated census tract level, in the Dallas-Fort Worth residential and employment location study (Waddell, 1992).

Factorial Urban Ecology and Patterns in Social Space   One of the earliest and most comprehensive analyses of urban social space using the procedures of factorial urban ecology was undertaken by Murdie (1969) for metropolitan Toronto in 1951 and 1961, which, by this latter date had not sprawled much into the outer suburban ring beyond the metro boundaries (see Table 9.3). Murdie attempted to determine objectively whether (1) the constructs recognized by the social area analysts could be defined for metropolitan Toronto, and (2) if they could be defined, whether they were distributed in the manner hypothesized by Berry (1965). The study involved eighty-six variables concerning occupation, income, dwelling units, ethnicity, language, religion, sex, and education, relating to the population aggregated to census tracts. The results emphasized the underlying importance of the social area constructs. In both time periods the constructs were distributed as hypothesized in Figure 11.4.

More recent studies of metro Toronto emphasize a remarkable consistency in the constructs and patterns since the 1950s (Fabbro, 1986; Murdie, 1988). For example, Fabbro (1986) establishes the existence of a social status factor based

Table 11.1

Variables Constituting the Nine Underlying Dimensions of Social Space in Canadian Metropolitan Areas

---

AFFINITIES WITH SOCIAL STATUS CONSTRUCT

VARIABLES (PRESENCE OF)	DESCRIPTORS	VARIABLES (ABSENCE OF)
University educated Males in managerial occupations High-income families (> $40,000 p.a.)	Economic Status ←→	Education (< grade 9) Males in manufacturing, construction
Female-parent families Low-income families Dwellings rented Male unemployment	Impoverishment ←→	Two-parent families

---

AFFINITIES WITH FAMILY STATUS CONSTRUCT

Persons per family ratio Children (0–14 yrs) Adults (25–34 yrs)	Families ←→	Middle-aged (55–64 yrs) Old age (>65 yrs) Children left home
Non-family/family ratio Singles (never married) Divorced/married adult ratio Childless	Non-families ←→	Children (0–14 yrs)
Young children (under 6 yrs) Adults (24–34 yrs) Moves (in last 5 yrs) New housing (under 10 yrs old)	Early/Late family ←→	Singles (never married) Middle-aged (55–64 yrs) Young adult (18–24 yrs) Early middle-aged (45 – 54 yrs)
Adults (18–24 yrs) Females in labor force	Young adult (pre-family)	Persons per family ratio
New housing (under 10 yrs)	Housing ←→	Poor housing Median length of occupancy

---

APPARENT SEGREGATION

Immigrants (born outside Canada) Non-British/Non-French ethnic origin	Ethnicity ←→	French ethnic origin Male unemployment rate
Local moves (within province) French ethnic origin Apartment dwellers	Migrant ←→	Distant migrants (out of province)

(*Source:* Retabulated from Davies and Murdie (1993), Table 3.1, p. 58; (1991), Table 2, pp. 63–7.)

Social status

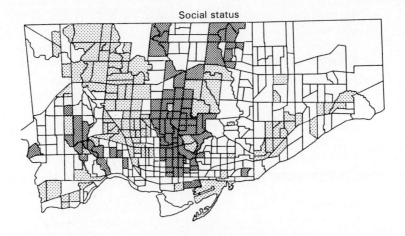

Family status (life-course)

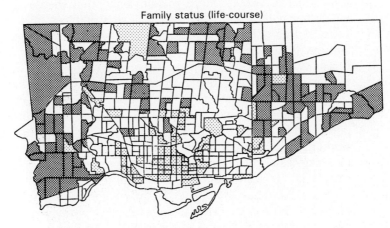

Ethnic or racial minority

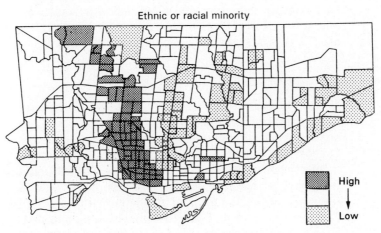

High

↓

Low

Figure 11.5 The social space of metropolitan Toronto, 1981, as defined by factorial ecology procedures.

(*Source:* Redrawn from Fabbro, 1986, Maps 4, 5, 6.)

on education, occupation, and house values that is sectorally distributed; a life-course factor (echoing family status) based on age of homeowner and residence in owner-occupied dwellings and an absence of young nonfamily persons and childless couples that appears to be concentrically distributed; and an ethnic status factor that identifies those parts of the metropolis with a concentration of individuals and families that may be defined generally as "non-Anglo-Saxon" (Figure 11.5). These include individuals who self-identify in the census as of Italian or Portuguese ethnicity, or African-West Indian or Chinese racial origin.

BEHAVIORAL PROCESSES, SOCIETAL STRUCTURES AND URBAN SOCIAL SPACE

Whereas the human ecology approach focuses on the ways in which immigration, rural-to-urban migration, and changes in technology influence the formation of social spaces, more recent conceptual themes focus on the ways in which behavioral processes and societal structures influence the formation of social spaces. Although both human ecology and the newer conceptual themes address issues pertaining to the spatial generation of inequalities (such as through segregation), the more recent approaches do so more specifically. Seminal works stimulating these more recent approaches were provided by Rossi (1955; 1980), who focused on behavioral processes affecting residential mobility; Weber (1922; 1968), whose writings provide important insights to contemporary racial and ethnic divisions in society (Stone, 1995); Harvey (1973; 1985), who argued that urban inequalities in housing, employment, and access to social services occur consequent to the operation of a market-oriented land and real estate economy; and Mollenkopf and Castells (1991), who emphasize the ways in which the globalization of the economy amidst the increasing emphasis on world cities is generating dual urban economies which have profound impacts on the restructuring of social space. In this section, we will examine issues relating to dominance-subordination in social space from behavioral, race and ethnicity, and competitive capitalism perspectives. Perspectives based on aspects of gender theory will be discussed in Chapter 13.

Behavioral Perspectives on Change in Social Space

North American society is highly mobile. Consequently, the social space of urban areas is continually reformed by this high level of mobility. Among a number of developed countries for which residential mobility studies are available, Canada, Australia, New Zealand, and the United States have the highest rates of mobility (Long, 1988; 1991). In the United States, about one-fifth of the population changes residence annually; 46.7 percent of the population alone changed residence between 1985 and 1990 (Bureau of the Census, 1993, p. 27). Some 54.6 percent of these 1985–1990 moves were short-distance moves within the same county, implying that most moves are intrametropolitan. These high rates of mobility suggest an average of thirteen moves in a person's lifetime, of

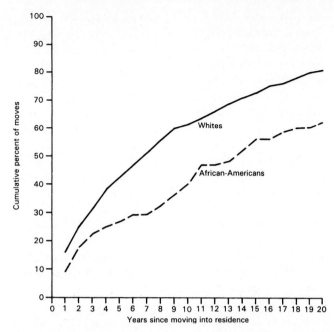

Figure 11.6 Comparison of residential mobility rates for whites and African Americans, Oklahoma City, 1992.
(*Source:* Graph constructed by author from data in St. John et al., 1995, Table 1, p. 721.)

which about ten usually occur after an individual's seventeenth birthday. These are, of course, averages because some people are much more mobile than others (Moore, 1986)!

At the outset it is important to recognize that there are clear differences in mobility levels between white and African American populations in the United States. Figure 11.6 is based on residential history survey interviews with a random sample of 376 residents in Oklahoma City and includes only those moving within the MSA. The data suggest a large difference in residential mobility rates between whites and African Americans. In one year, 16.2 percent of whites had moved from one residence to another within the MSA compared with only 9.3 percent of African Americans (St. John et al., 1995). The residential mobility gap widens until year seven, by which time 51.1 percent of whites and 29.3 percent of African Americans had changed residence once. This gap of about twenty percentage points is thereafter quite persistent. Among the implications of this white/African American residential mobility gap are either that African Americans exhibit greater residential stability and community attachment during their life-course than whites, or the range of choice of residential location for African Americans in Oklahoma City is restricted by a constrained housing market that makes improvement in housing conditions resulting from a move unlikely.

Mobility and Life-Course Considerations   If it is accepted that an average person changes residence about thirteen times in a lifetime, the residential situations that may be related to these moves must be examined. Because the thirteen moves include those undertaken within urban areas, those between rural and urban areas, and those from outside North America, the reasons for moving are quite diverse (Clark, 1986; 1992). Just over one-third—or four to five—of a person's lifetime moves are interurban and are related to job opportunities, differences in income levels between cities, and various amenity factors. The remaining two-thirds of the moves are intraurban. All moves can be related to life-course considerations, whether they be related to changes in employment or family requirements. The issue is not, however, just *why households move, but why households move where they do* (Rossi and Shlay, 1982). Thus, in examining the reasons for changes in location we must also examine factors influencing residential location. This second half of this issue is discussed in Chapter 12.

In a classic study related to family mobility, Rossi (1955; 1980) challenged Wirth's (1933) negative views on the effects of residential mobility. He provided evidence to show that the relatively high rates of mobility that were occurring among a rapidly urbanizing population were not due to rootlessness and aberrant social behavior but changing housing needs that accompanied changes in household and family structure, composition, and wealth. This new paradigm, after some years in gestation, gave birth to a whole new approach to analyses of residential mobility and public policy (Clark and Moore, 1982). Rossi demonstrated that the majority of intraurban moves is related to changing requirements for living space as families change in size, age, and income.

The notion of "life-course" is proving more useful in behavioral studies of residential mobility than "life-cycle" (Warnes, 1992; Gober, 1992). The concept of the life-cycle implies that an individual passes through a predetermined sequence of stages—childhood, school age, young adulthood, leaving parental home, marriage, childrearing, childlaunching, retirement, widowhood—that may not be applicable to all individuals. Thus, the idea of life course—childhood, adolescence, adulthood, middle age, young-old, older-old—is more inclusive with respect to the diversity of lifestyles, households, and family and nonfamily structures (McHugh et al., 1995). This diversity is reflected to a degree in the five family status affinity dimensions in Table 11.1. About three of the thirteen lifetime moves made by average North Americans occur before or during adolescence. The average of ten moves made by an individual after adolescence are related to a multiplicity of possible changes during the life-course.

Life-Course Perspectives and Social Space   Given the large number of variations in living arrangements of individuals in North America during the life-course, and hence the wide range in types of social spaces in which people may reside, the array of possible life-course combinations presented in Table 11.2 is merely suggestive. Although the inevitable consequences of age provide some organizational linearity to the table, the suggested life-course elements are not deterministic. For example, a person may choose to live in a one- or two-person household throughout their life-course. In general, however, as an individual

Table 11.2

Life-Course and Residential Mobility

AGE	0–14	14–20	20–35	35–55	55–75	75 +
Course	Childhood	Adolescence	Adulthood	Middle age	Young-old	Older-old
	School age family home primary/secondary/post-secondary apprentice reconfigured families	apartment house	one or two parent young children lifelong learning employment long-distance commuting commuting couples marriage/divorce/re-marriage	older children (second home/cottage)	grandparent empty nest later-life learning job changes retirement	retirement home retirement widowhood
Household size	2.35		3.06	3.26 2.89	2.34	1.57
Moves	11		1111	111	1	1

(*Source:* This table is prepared by the author from an idea in St. John et al. (1995), Figure 1, p. 253; McCarthy (1976), Table 2; and Bureau of the Census (1993), Table 69, p. 57.)

passes through the life-course, the type of house and the physical location that are optimally required vary.

As most of the residential location decisions are made by heads of households, in accordance with their perception of the household's requirements, there are possibilities for tension between decisionmakers and dependents. For example, the location and space requirements of an adolescent differ considerably from those of a child. Furthermore, as the heads of households age, they may become less responsive to needs for housing adjustments. This pattern is frequently evidenced by resistance to life-course residential adjustment in later years due to sentimental attachment to a house and location that are no longer congruent with the household's requirements.

There are, however, some general features of Table 11.1 that have implications with respect to the formation of social spaces:

1. Residential mobility is greatest during young adulthood when individuals are struggling to establish themselves. This "struggle" involves job searches, various types of postsecondary education and training, and often various forms of living arrangements. The family home may still provide a base, but the individual is establishing independence. Apartments and townhouses, either downtown or in planned complexes in suburban areas, often provide the locations for these residential arrangements. McCann (1995) describes a number of neotraditional high-security suburban developments, such as Kentlands in Gaithersburg (Maryland) outside Washington, D.C., that cater to the upscale end of this type of market because aggregate income in multi-person young-adult households may be quite high.

2. The three moves associated with the lengthy middle-age period are usually associated with family formation and requirements for more space. The initial move may well be to a low-down payment, highly mortgaged, single-family home in the outer suburbs, but with increasing family age and job mobility, the household may relocate to a larger home in a more exclusive "gated" suburban subdivision. Separation and divorce may, however, result in family impoverishment and shifts to "townhouse" and other types of high-density residential suburban developments.

3. With an increasing proportion of the population of North America over the age of 55, residential communities catering specifically to individuals in this part of the life-course are becoming more prevalent outside the traditional "retirement" states of Florida, Arizona, and parts of California and British Columbia. These range from large self-contained retirement communities, such as Paradise in the Sierra foothills outside Chico in the Sacramento Valley, to residence complexes for the mobile elderly and full-care retirement homes.

Social space is, therefore, inextricably entwined with life-course as mediated by income and, to a certain degree, race.

Social Space and Lifestyle Considerations Decreases in family size in North America are accentuating the importance of lifestyle preferences in the creation of social spaces within metropolitan areas. Lifestyle refers to the way a person wants to live, and the people with whom he or she wishes to associate or feels most comfortable being associated with. One general way of illustrating the effect of lifestyle on location is in the context of the values an individual or family may choose to emphasize. Although there may be many types of value sets, in an urban context these may be summarized as familism, careerism, localism, and cosmopolitanism.

Familism describes a situation in which an individual places a high value on the unity of the family and its function as a mechanism for propagation and socialization of the young. *Careerism* refers to individuals who place high values on the status gained from employment, particularly prestige professional activities, and who emphasize upward mobility, professional responsibility, the gaining of material benefits, and elite consumerism. *Localism* describes a situation in which one's interests are limited to people residing in a well-defined local area or community, and one's attitudes and behaviors are, in many respects, subservient to the norm accepted within the community. *Cosmopolitanism*, the opposite of localism, implies a value system that emphasizes the absence of control and the freedom to experience ideas and behaviors from anywhere.

These different values may result in quite different decisions concerning where to locate in a metropolis, and even in which metropolis to locate. Familism is widely accepted by the broad cross-section of North American society. Up to about 1975, the suburban housing developments were marketed almost exclusively as the apotheosis of this lifestyle (Castells, 1977, p. 109). Careerism is strongly followed in exclusive suburbs and affluent apartment complexes where privilege is maintained and social interaction structured through private schools, social and business clubs, and other types of social arrangements.

By contrast, people living in working-class neighborhoods and various ethnic areas seem to place a high value on localism, for example, neighborhood shopping areas, and may also be evidenced more dramatically by local gangs and boss-dominated precincts. Localism is emphasized out of a need to stress the identity of a place and to provide a supportive environment for established and recent immigrants to the community (Lai, 1988; Iacovetta, 1992). Cosmopolitan lifestyles are prevalent in large metropolitan areas, particularly downtown areas as they involve people who place a high value on particular cultural experiences or range of contacts. To an extent, possibilities for the realization of certain of these values are directly related to income. Across the spectrum, it is entirely possible for individuals to exhibit a number of the value traits and place greater or less emphasis on various traits during the life-course.

An example of the way in which a lifestyle can create social space and influence urban growth and redevelopment, relates to the creation of gay communities in a number of large metropolitan areas. Gay communities form for a number of reasons, ranging from the need by individuals for support to the exclusionary (or oppressive) response of an often nonunderstanding and antagonistic society at large. Such communities are invariably found in parts of downtown areas that were originally low cost, but where the vibrancy of the communities has frequently led to redevelopment (Lauria and Knopp, 1985). In some ways, the struggle by gay communities for legitimacy as an existing part of social space is similar to the struggles made by ethnic minorities in previous decades for acceptance within urban areas.

Ethnicity and Race

In the human ecology and behavioral models, differences in residential location patterns between majority and minority racial and ethnic groups are hypothesized to occur because of socioeconomic differences between the populations. Particular racial or ethnic groups locate in various parts of the city because of housing affordability and access to jobs. Human ecology theory suggests that as socioeconomic differences between groups decrease, levels of assimilation should increase. More recently, Massey (1985) argues similarly that the residential location of recent immigrants reflects their degree of cultural and economic assimilation with the rest of the population: As the cultural and economic characteristics of new immigrant populations become similar to the average characteristics for the population as a whole, so does residential proximity. But, although assimilation appears to have occurred with respect to many ethnic and some racial groups, there remain degrees of spatial separation between various racial and ethnic groups as has been described previously in the Dallas-Fort Worth study (Waddell, 1992). Race and ethnicity has had a profound influence on the shaping of North American cities (Castells, 1983).

Although Weber (1922/1968) regarded race and ethnicity as complex and somewhat elusive concepts, he provided a definition that is useful in the discussion of the formation of urban social spaces in North America. Weber suggests that racial and ethnic groups are "human groups (other than kinship groups) which cherish a belief in their common origins of such a kind that it provides a basis for the creation of a community" (Runciman, 1978, p. 364). Thus, ethnicity is not really based so much on some shared cultural characteristics, such as nationality, language, and religion—though these may help in demarcating the concentration of ethnic groups in urban areas—and race is not based so much on certain shared biological traits, such as color, but on a *belief in some common ancestry* (Stone, 1995, p. 396). The twentieth century is replete with examples of the ways in which dormant ethnic identification can be awakened by political activism, and how racial attributes may be assumed by or applied to individuals. A variety of individual and societal processes are involved in racial or ethnic group formation whether they involve "the figment of pigment" (Horowitz, 1971, p. 244), or the "invention of tradition" (Hobsbawm and Ranger, 1983).

Racial and ethnic identities, therefore, shift over time and place. The United States and Canadian national censuses are barometers of changing views of ethnicity and race and the rise of ethnic consciousness (Kelly, 1995). In general, various "panethnic" groups, such as Asian Americans or Asian Canadians, are requesting more culture-based possibilities for self-identification rather than less. In this context, Omi and Winant (1986) note that race is not so much an entity as a process in which racial categories acquire and lose meaning over time. The United States census, and most white Americans, used to view "blacks" as one racial group, but with increasing levels of education and economic status, and continued immigration from the Caribbean, African Americans are now seen as involving a complexity of racial and ethnic identities. For example, immigrants from Trinidad or Jamaica may prefer national or regional panethnic West Indian identities to the rather generic designation "African American" (Waters, 1990).

The strength that people derive from racial or ethnic identities is exhibited by the increasing importance of *transnational social systems* in the late-twentieth-century global economy (Patterson, 1991). A number of racial/ethnic groups, particularly those involved in extensive recent migrations to North American cities, such as persons of Chinese, West Indian, or Mexican origin, have developed well-integrated transnational economic and social linkages that are based on the human capital, organizational skills, and confidences generated through community cohesion. Although transnational social systems have been powerful forces in international finance and business in the past, the extension of global economic and financial integration to culturally based integration within a wider-variety transnational human communities, often based on extended kinship patterns, is a phenomenon of increasing importance to urban development.

Comparisons of Racial and Residential Proximity in the United States and Canada   In a highly generalized comparative analysis of neigborhood racial residential proximity, Fong (1994) demonstrates that race is more of a determinant of social space in urban areas in the United States than in Canada. This is not to suggest that race has not been, and is not, a determinant of social space in Canada. Anderson (1991), for example, shows that Vancouver's Chinatown is largely a creation of the white community over a one-hundred-year period. The Fong (1994) study, however, is more broad based as it involves samples of whites, blacks, and Asians from 404 urban areas in the United States and 41 in Canada. A neighborhood is defined as a census tract, which has similar average populations (4,000 persons) and population size ranges in both countries. The graphs in Figure 11.7 indicate that:

1. African Americans are *hypersegregated* from whites in United States urban areas as compared with the black population in Canadian cities (Figure 11.7a). Whereas 75 percent of African Americans in the 404 urban areas in the United States reside in census tracts that have populations less than 30 percent white, 70 percent of the blacks residing in the 41 urban areas in Canada reside in census tracts that are between 40 percent and 70 percent white. It should be noted that the 60 percent of the black population in Canada is comprised of recent immigrants from the Caribbean area.

2. The Asian population is far less segregated from the white population in both the United States and Canada, though the Asian population is more segregated from the white population in the United States than Canada (Figure 11.7b). Nearly 80 percent of the Asian population in the 41 Canadian urban areas reside in census tracts that have populations that are between 40 to 70 percent white, whereas 65 percent of the Asian population in the 404 urban

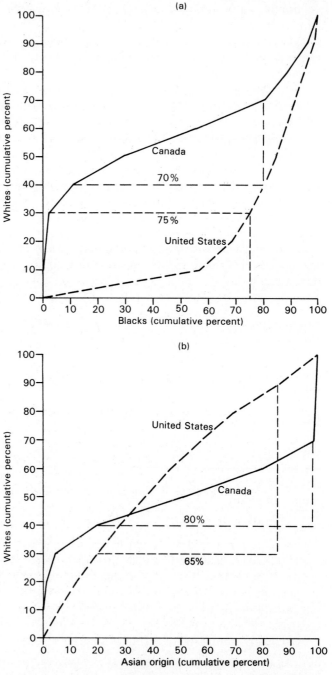

Figure 11.7 Comparative segregation: graphs of residential proximity percentages for (a) African-origin, and (b) Asian-origin populations, with European origin populations in census tracts in Canada and the United States.

(*Source:* Graphs prepared by author from data in Wong, 1994, Table 1, p. 288).

areas in the United States are located in urban areas that have populations between 30 percent and 90 percent white.

3. Socioeconomic gains in the black population in the United States are not necessarily translated into the greater levels of assimilation envisaged by human ecology theory. This observation is supported more directly by Farley (1995) in a study of housing segregation in St. Louis in which he demonstrates that, at all income levels, whites and African Americans with similar incomes experience similar levels of separation. Furthermore, he shows that residential segregation in St. Louis in 1990 remains close to its historic level, and that little of the separation can be accounted for by differences in income and housing affordability.

The important question concerns how these segregated areas come about. Are they the result of self-interested individual behavior, or are they the consequence of premeditated actions by particular groups and individuals?

Ethnic and Racial Residential Separation   The formation of high levels of racial and ethnic residential separation within urban areas at different time periods, including the present, may be related to a number of interrelated causes.

1. The human ecologists, as has been observed, generally regarded residential separation as the outcome of differences in income and access to capital. Thus, the choice of location for low-income families and individuals is restricted by the places where low-priced accommodation may be found. If certain racial and ethnic groups are concentrated at the lower level of the income spectrum, then these groups will most likely dominate within the lower-priced residential areas. Consequently, segregated patterns may form as the outcome of individual residential location decisions based on affordability. Hacker (1992), for example, argues that because African Americans are three times as likely to be poor and unemployed as whites they will concentrate in neighborhoods that whites avoid. Similarly, Fong (1995), while documenting the high levels of segregation between African Americans and whites in the United States, nevertheless reveals greater levels of racial proximity between Asian Americans and whites, which suggests that the human ecology view may prevail more for some racial and ethnic groups than others. The information cited above with respect St. Louis (Farley, 1986; 1995), however, is strongly suggestive of continuing high levels of residential separation between African Americans and whites at all income levels.

2. In a classic article on the location of ethnic groups in cities, Firey (1945) suggests that sentiment and symbolism may act as culturally based factors leading to the clustering of particular ethnic or social groups in certain areas. Some ethnic and religious groups may well concentrate in certain parts of cities because of the religious, shopping, recreational, and other attributes that may be available to serve the needs of a particular community. Some Sephardim Jews or residual Italian and Portuguese communities, may remain where they are for basically cultural reasons. Allen and Turner (1996) document the way in which twelve different new ethnic immigrant commu-

nities concentrate primarily close to the central part of the greater Los Angeles area in response to a need for cultural and relational support. Likewise, Johnson and Roseman (1990) show that extended family and kinship relationships play a strong role in influencing the direction and location of African American outmigrants from Los Angeles.

3. The rather gentle idea of "sentiment and symbolism" may be reformulated at the end of the century for many ethnic and racial communities as mutual avoidance, implying that various communities may choose to maintain and expand communities in particular parts of metropolitan areas because individuals and families feel more comfortable, or less threatened, in such environments. The spread of majority African American middle-income suburbs from south Atlanta to adjacent areas of Fulton County (see Figure 10.1) may be attributed to a long history of segregation, which is now transformed into mutual avoidance between African Americans and whites and is reflected in a noticeable north–south imbalance in jobs (Fujii and Hartshorn, 1995). The need to control interaction and maintain avoidance between people is taking extreme proportions with the proliferation of security systems and gated communities (Davis, 1992).

Segregated areas, which are created by the premeditated actions of others, are common features of American urban life. In particular, a majority of the 30 million African Americans, 14 million Mexican Americans, and 3 million persons of Puerto Rican origin, enumerated in the 1990 United States census, are constrained by a variety of external circumstances to live in restricted areas, as are a large number of Asian Americans. Almost every urban area in North America, whatever the size, that has a significant number of nonwhite inhabitants and persons of Hispanic origins has at least one segregated area of these minority populations. These segregated areas echo the former ghettos of central and eastern Europe where the Jewish community was not only maintained by the majority population as distinct from the rest of the city, but in many cases ultimately contained by a wall beyond which the residents were not allowed to travel without a pass.

African American Ghettos Segregated social spaces involving African Americans and Mexican Americans are particularly pervasive numerically in the largest United States metropolitan areas. Prior to 1910 most of the African American population was located in the South and the lands bordering the Mississippi River south of St. Louis. The phenomenal growth of African American–population ghettos in the largest metropolitan areas outside the south occurred between 1910 and 1970. Whereas in 1910 twelve of eighteen urban places in the United States with an African American population of 25,000 or more were in the South, by 1970 more than one-half of the seventy or more urban places that had at least 25,000 African Americans were outside the South (Rose, 1971). Given that the proportion of the African American population living below the poverty level increased from 30.7 percent in 1989 to 33.3 percent in 1992, a large proportion of the population of these urban ghettos is in economic and social distress (Bureau of the Census, 1992).

This shift in relative location and composition of the African American black urban population is important because in the South the form and structure of the urban ghetto differs quite widely from that in urban centers beyond the South. Southern African American urban ghettos are located in parts of cities that have been set aside by whites as black areas for generations. African Americans in southern urban ghettoes, then, have a clearly prescribed territory in which they have developed their own institutions and stratified residential structure. There is no question as to which race lives where, who is serviced where, and where the location of the African American boundary. This spatially segregated society is almost as strong today as it was in the 1960s before equal rights and fair housing legislation—hence, the previous reference with respect to African American middle-class suburbanization into Fulton County outside the City of Atlanta. In these types of situations, long-standing dominance-subordination has metamorphosed into mutual avoidance between the majority-minority racial groups (Rose, 1976; Galster and Hill, 1992).

Although a number of African American ghettos outside the South have been in existence for a century or more—just as long as Harlem in Manhattan, for example—they have been growing in number and size following the economic and social upheavals of the Depression, the economic opportunities of World War II, and the growth decades after 1945. However, African Americans came to northern and western cities to find a situation quite different from that in the South. Although there was no clearly prescribed territory, in general whites would not accept African Americans into their neighborhoods and were extraordinarily reluctant to accept them into their schools and community facilities. As a result, African Americans frequently occupied the most deteriorated or undesirable part of the urban area. Simultaneously, a plethora of formal and informal restrictions and controls were established by the white majority to constrain the population. As inmigration and growth caused pressure on land and housing, the ghetto expanded outward at the periphery, following lines of least resistance organized by the real estate and mortgage finance industries. Much African American suburbanization is thus viewed by Rose (1976) as being a "spillover" from adjacent central city ghettos.

The Expansion of the Milwaukee Ghetto   The pattern of growth of an African American–population ghetto in an American city is well illustrated by the case of the Milwaukee PMSA, which is the predominant part of the 1990 Milwaukee-Racine CMSA (Figure 11.8). The number of African Americans in the City of Milwaukee increased quite rapidly from 20,000 in 1950 to 192,000 in 1990, and nearly all of this population at each time period was located in one of the three parts of the contiguous ghetto demarcated in Figure 11.8. The continuing hypersegregation of the African American population within the central city is quite remarkable given the population growth that has occurred in the suburban parts of the PMSA since 1970 (Table 11.3). Evidently, while the population of the central city has declined by almost 90,000 persons since 1970, the African American population has increased by 90,000, implying a net loss of persons other than African Americans of 180,000. It is also important to note, as has

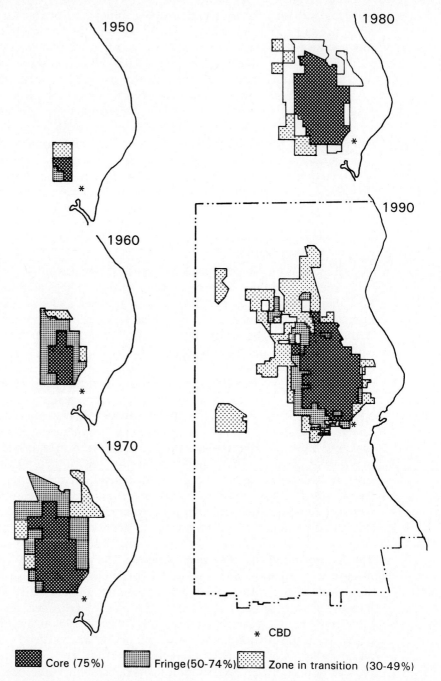

1950

1960

1970

1980

1990

* CBD

Core (75%)   **Fringe (50-74%)**   **Zone in transition (30-49%)**

Figure 11.8 The spatial spread of the African American ghetto in Milwaukee, 1950–1990.

(*Source:* 1950, 1960, and 1970 maps redrawn from Rose, 1971, p. 34; 1980 and 1990 maps prepared by author from census tract data.)

Table 11.3

Milwaukee PMSA : Central City–Suburban Comparisons in the Location of the African American Population, 1970–1990 (numbers in thousands)

	1970			1990			
		AFRICAN AMERICAN			AFRICAN AMERICAN		PER CAPITA
SUBAREA	POPULATION	NUMBER	%	POPULATION	NUMBER	%	INCOME*
City of Milwaukee	717	105	14.6	628	192	30.5	$10,593
Suburbs	680	2	0.3	804	4	0.5	$14,892
Total	1,397	107	7.7	1,432	196	13.7	$12,992

(*Source:* Bureau of the Census (1993), Tables 46, 43; (1991), Tables C and D.)
*1987 U.S. dollars.

been documented previously for other metropolitan areas, that per capita incomes in the central city are about 70 percent of those in the suburbs.

The African American ghetto was established to the immediate northwest of the downtown area and to the north of the original location of much of Milwaukee's industrial activity at the junction of the Milwaukee River and Lake Michigan. Rose (1971) defines a number of territories within the ghetto as follows: There is a core area of census tracts in which the population is 75 percent or more African American; a fringe area in which the population is at least half African American; and a zone in transition in which 30 to 49 percent of the population is African American. The interesting features demonstrated in Figure 11.8 are that the African American population is highly concentrated, and that an area changes in just a few years from one with few African Americans to one almost completely occupied by African Americans. The zone in transition barely lasts "in transition" for more than a few years! Furthermore, although the number of African Americans located in the suburban areas doubled between 1970 and 1990, the numbers involved are negligible compared with the population as a whole (Table 11.3). This is one of six extreme cases of hypersegregation in metropolitan areas in the United States (Massey and Denton, 1989).

A number of implications may be derived from the maps and table. First, forces generating interracial mutual avoidance must be compounding. These mutual avoidance forces are in large part the product of a continuing long history of impositions designed to perpetuate white/African American dominance-subordination. Second, given the rapidity of population change that occurs at the margin of the segregated area, there must be some set of mechanisms in place that facilitates maintenance of the ghetto and promotes rapid transition when change is inevitable. These mechanisms must continue to exist despite the discontinuance of legal sanctions supporting racial separation and racial inequalities. Clark (1984) suggests that real estate agents, including bankers, mortgage lenders, savings and loan institutions, and real estate brokers, play a powerful role in maintaining the ghetto and promoting rapid change when needed in fringe and transition areas. Federal, state, and locally funded public housing and renewal schemes also perpetuate the concentration of persons in need within central cities.

The Mismatch Hypothesis The relatively lower per capita incomes in the central city compared with the suburbs are related to a lack of sufficient appropriate employment opportunities for poorer central city residents, particularly minority groups such as African Americans residing in inner-city ghettos. With jobs in manufacturing, retailing, and offices more available in the suburban rings than inner cities, it is hypothesized that a *spatial mismatch* in employment opportunities has developed in the United States (Kain, 1968). Much of the white population has relatively good access to suburban employment opportunities and downtown office jobs, whereas minority populations have far less access to suburban jobs and insufficient training for many jobs in the producer services sector. When inner-city residents do gain employment in suburban locations, they often have to travel longer distances, thus incurring greater costs.

The spatial mismatch hypothesis has been debated and tested in a number of situations using a variety of methodologies. However, the results tend to be somewhat contradictory. Recent reviews of the literature suggest that polynucleated metropolises and continuing racial segregation are producing more evidence of mismatch (Holzer, 1991). There is, however, evidence to indicate that for those white and minority workers, such as African Americans and Hispanics, who are in full-time employment, commuting patterns and distances traveled to work are becoming similar over time (Taylor and Ong, 1995). This convergence in commuting patterns of employed persons of different racial backgrounds is explained primarily in terms of the increasing levels of suburbanization of the minority groups.

A micro-study of the imminent movement of a food-processing plant from the Milwaukee central business district to a new and more efficient suburban location indicates the different ways in which minority persons are affected by the decentralization of job opportunities and, hence, provides a basis for this contrasting evidence. Fernandez (1994) demonstrates that even with the firm selecting a site to minimize worker inconvenience, and every attempt being made by the company to retain its multiracial workforce, minority males and white and African American women were more likely to suffer negative travel cost and access consequences than white males. The study also demonstrated that there would be significant class impacts, with hourly paid workers suffering more negative consequences from the move than salaried employees.

Thus, plant relocation from the central city to the suburbs, in a highly segregated residential environment such as that in Milwaukee, is likely to have much greater negative impacts on minorities, women, and hourly paid (blue-collar) employees than white, salaried employees. Given the attempts by the firm to minimize the possible negative consequences of the move, the case demonstrates the underlying inequities that arise from spatial mismatch and the diminishment of human capital that results (Logan and Alba, 1993).

Mexican American Barrios The *enumerated* Hispanic population in the United States has grown from fewer than 2 million in 1940 to 22 million in 1990. Of this panethnic group, almost 3 million are of Puerto Rican origin residing primarily in the inner cities of the New York CMSA (for example, Spanish Harlem in Manhattan); 1 million are Cuban Americans who have been prime agents in revitalizing the economic life of Miami, and dominate it politically (Portes and Stepick, 1993); 14 million are Mexican Americans located primarily in metropolitan areas in the Southwest; and the rest are from a number of other Spanish- or Portuguese-speaking countries, mainly in Central and South America, who tend to locate in the central cities of the Northeast or southern California. Although the proportion of the Hispanic population of Mexican American origin has declined from 70 percent in 1950 to 60 percent in 1990, it is numerically by far the largest. The Mexican American population in the United States ranges from the wealthy to the poor, and resides in a variety of social spaces, from the "tony" Mexican American–dominated middle-class suburbs of San Antonio to the *barrios* (neighborhoods) of the poor and recent immigrants from Latin America in that same city.

Although some poor barrios in large metropolises, such as in south Bronx in New York City or east Los Angeles, were founded in a central city context, most *barrios* or *colonias,* as rural concentrations of Mexicans (and other immigrants from Central and South America) are often called in the Southwest, originally began as agricultural labor communities (Fuchs, 1990). They were often founded on the periphery of cities as colonies of migrant workers, or grew out of farm, mining, or railroad labor camps. The workers, often entire extended families of workers (for every member of a family sought paid employment), frequently moved from one barrio to another in search of work according to the cycle of agricultural production, or the availability of industrial or service jobs. The barrios were located at the convenience of the white majority population whose enterprises and homes were greatly advantaged by the availability of low-wage unregulated labor. The *barrios,* being tolerated parts of the landscape, had inadequate sewage, education, and other basic services. As a result of the rapid growth of urban areas in the Southwest since 1970, many of the *colonias* have been absorbed into the expanding metropolitan areas and basic living conditions considerably improved. Their populations are now oriented more toward urban rather than rural-based employment.

The proportion of the Hispanic American and African American populations living in poverty in the United States is the highest of any of the defined panethnic groups—between 30 and 40 percent of the population of each group (Bureau of the Census, 1992, Table B). Although the white population has the smallest proportion living in poverty, it has the greatest number of persons existing below the poverty level. A large number of each of these populations in poverty, which includes in all groups a significant number of the elderly, resides in metropolitan areas, predominantly in central cities, and in the case of minority populations concentrated often in public housing projects. It is in these social spaces of deprivation that the population is particularly subject to high rates of unemployment, poor education, limited provision of health care services, high poverty levels, high incidences of drug and alcohol abuse, high rates of teenage pregnancies, many single-parent households, and high levels of crime inflicted largely by local inhabitants on local inhabitants. These are places of formidable stress where a young man between the ages of 15 and 25 is more likely to be killed on home turf than an American soldier was in Vietnam (Kotlowitz, 1991).

Structural Perspectives on the Formation of Social Space

Structural perspectives in urban analysis pursue in a more focused manner the suggestions in the previous section with respect to the formation of ghettos and barrios: Social spaces within urban areas can be viewed as the product of forms of collusion between local and transnational economic elites and various levels of government. In particular, structural approaches place the conflict, tensions, and interactions between the major sources of production—land, labor, capital, and the state—at the core of urban analysis (Tulloss, 1995). For example, Edel and colleagues (1984) examine the way in which the owners of capital, and some entrepreneurs, have interacted to develop suburban environments in

Boston that, although they appear to give more residential choices, simply retain the same type of choices within spatially contiguous environments; and Hirsch (1983) shows how the leading economic and educational (e.g., the University of Chicago) institutions in Chicago managed to protect their interests during the expansion of the African American working-class ghetto that occurred between 1940 and 1960. Thus, by focusing on the interplay between the sources of production in the formation of social space, structural perspectives do not allow analysts to evade issues of social class.

Dual Economies, Dual Social Spaces In previous chapters we have discussed a number of characteristics of advanced capitalism: globalization of competition, the international integration of capital markets, the rise in importance of employment in producer services, the refocusing of control and command to transnational corporations headquartered in a number of world cities, and the restructuring of manufacturing and service industries. These changes are having a profound effect on the restructuring of urban social space in much the same way as the innovation of the factory system in the second half of the nineteenth century reformed social space in nineteenth-century industrial cities. For example, since 1970 almost the whole of Manhattan south of Ninety-sixth Street (Figure 11.9) has incurred commercial redevelopment, gentrification, and conversion of old manufacturing and warehouse buildings into up-market residential and boutique retail and restaurant districts. Previous low-income African American and white-ethnic groups have moved to other city boroughs or the outer suburbs, leaving only the African American ghetto centered on Harlem north of Central Park, adjacent Spanish Harlem, and an old tenement district and Chinatown on the Lower East Side as social spaces occupied by lower-income individuals and families (Lin, 1995).

It is argued that these changes in Manhattan are a reflection in social space of the polarization of development that has occurred in New York City as it has asserted itself as a national and administrative command center in the world's economy (Mollenkopf and Castells, 1991). This polarization involves increasing disparity between an affluent upper circuit and a more deprived lower circuit of economic and residential life, and perhaps underlies the "first world–third world" appearance of many North American cities (Lin, 1995). The upper circuit involves those individuals involved with management, the producer services, and associated information, communications, and political sectors. The lower circuit includes those involved with low-wage industrial production and most aspects of consumer services. In between, the proportion of the population that may be categorized as middle-income is gradually decreasing. This polarization of development is, therefore, generating clearer distinctions of social class. Eventually, these more polarized distinctions are becoming more clearly represented in not only downtown social space but suburban social spaces as well.

Social Class and Homeownership One aspect of social space that is related to social class is homeownership. Certain parts of cities are in effect controlled by others and rented out, whereas other parts of cities consist of homeowners who reside primarily in their own property. Inner cities contain a large amount of

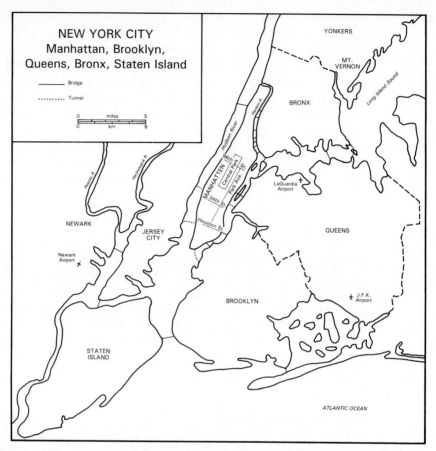

Figure 11.9 The boroughs of New York City.

rental accommodation; suburban areas contain by comparison much less (Harris and Pratt, 1993, p. 289). There are, as a consequence, different sets of interests between areas that have a high proportion of rental accommodation and those that do not, giving a distinct spatial dimension to such issues as rent control (Goetz, 1995). Areas with high levels of homeownership include people who profit from increases in equity due to economic growth and the resulting appreciating in home values, though, in the short term, fluctuations in the economic cycle may generate negative equity. Renters, on the other hand, do not have the opportunity to accrue wealth in this way, and neither do they have security of tenure or costs beyond the agreement of the lease, unless they are renting property that is subject to some kind of lease continuance and rent control provisions.

Information in Harris (1986) for samples of residential units in Toronto indicates a considerable variation in homeownership rates among different social classes (Table 11.4). Even in a country in which about 63 percent of all households own their home (except in the Province of Quebec where renting is more

Table 11.4

Home Ownership Rates by Class in Toronto, 1931 and 1979

	PERCENT HOMEOWNERS	
Class	1931	1975
Owners and managers	61%	82%
Middle class	54	65
Working class	42	49
Owners of small businesses	50	49

(*Source:* Selected data from Harris (1986), p. 80.)

general), homeownership rates among working-class Torontonians in 1975 was only about 50 percent. The homeownership rates among owners and managers and the middle class are quite high and increased much faster between 1931 and 1975 than those for the working class in recent decades. Thus, about one-half of the working-class population in the city at that date was living in residential units that were owned (controlled) by others.

The Underclass Debate In George Eliot's England, the period immediately prior to industrialization, there existed a large rural underclass, "almost indentured in their dependence on their masters" (Karl, 1995, p. 8). This dependence of people who had neither property nor capital on those that had, fostered a society based on custom and tradition that was essentially immobile. The underclass, for the most part, lived and died in the locale in which they were born. It was from this type of environment that many of the immigrants to North America in the eighteenth and nineteenth centuries were escaping.

Wilson (1987; 1996) has argued that just as the structural forces of feudal Europe created a predominantly rural-based underclass, so have the postindustrial economic, political, and social forces of the latter half of the nineteenth century created an inner-city underclass. The creation of this underclass is related to the main economic and social forces that have been discussed previously—the suburbanization of jobs and economic activities; the restructuring of manufacturing, which has reduced the number of good blue-collar employment opportunities, particularly for males; the decanting of much of the middle-class white population to the suburbs; underlying racism in housing markets which effectively contained much of the African American population in inner cities; and gross disparities in social and educational opportunities between poorer and richer neighborhoods.

The basic economic indicators illustrating the structural changes in the economy are telling. In 1979, the richest 1 percent of Americans held 22 percent of the wealth of the United States—by 1992 this 1 percent held 42 percent. In the same period, the proportion of *workers* classified as poor rose from 5.9

percent to 7.4 percent, and the proportion of families classified as poor reached 12 percent. More generally, the incomes of the poorest 5 percent of Americans dropped by 34 percent from 1969 to 1993, whereas the incomes of the richest 5 percent rose by 43 percent. Thus, structural changes in the United States economy have created strong forces for income polarization and joblessness among the already poor, situations to which governments are reacting largely with policies related to a bygone age. Nevertheless, because of high rates of joblessness, and the emergence since 1975 of a high proportion of single-parent households within the poorer parts of inner cities, this underclass is largely dependent on the state.

The underclass debate focuses on a number of issues (Wilson, 1991; Lee, 1994). One is the appropriateness of the use of such a distasteful term as "underclass" to pertain to *some* ghetto situations where the population consists largely of poor and unemployed persons who have been left behind after working- and middle-class ghetto dwellers have moved to other neighborhoods. What we are really talking about here are the problems, and possible public policy responses, of individuals and households in pockets of isolated deprivation in particular areas (Hughes, 1989).

A second issue concerns Wilson's thesis that "aspects of underclassness are primarily an outcome of macrostructural economic forces impacting vulnerable communities" (Rose, 1991, p. 492). Studies by Kasarda (1990) and Johnson and Oliver (1991), linking the social and economic outcomes of recent industrial restructuring to African American male unemployment, and Rose and Deskins (1991) linking changes in Detroit's automobile economy to high levels of illegitimacy among African American adolescents, support this view.

A third issue involves the attempt by some researchers to ascribe certain behavioral characteristics, such as single parents, long-term unemployment, and criminal tendencies, to the underclass (Murray, 1990). The allusion of behavioral characteristics to a class or group is at best misguided and supercilious, and at worst as wrong as it was of the landed gentry to ascribe general behavioral characteristics, such as shiftlessness and servility, to the rural poor as a class in eighteenth- and early-nineteenth-century England. There are pockets of poverty and deprivation within many North American cities "but despite overwhelming joblessness and poverty, black residents in inner-city ghetto neighborhoods actually verbally endorse, rather than undermine, the basic American values concerning individual initiative" (Wilson, 1996, p. 179).

A fourth issue concerns the extent to which United States federal and state governmental policies may have exacerbated the concentration of pockets of chronically and continually poor persons in the inner city (Gans, 1995). Macro-economic governmental policies that fail to address the income distribution question, and social policies that fail to respond to the education and training needs of a part of the population, clearly provide some support for this view. Even more troublesome is the contention that some antipoverty programs, such as Aid to Families with Dependent Children (AFDC), food stamp programs, and subsidized housing initiatives, may have unwittingly encouraged the proliferation of single-parent households. Furthermore, the concentration of poor persons in subsidized housing in housing projects has created easily identifiable areas of avoidance in inner cities and places where the drug trade, which has flourished since 1980 with the spread of crack, can incubate.

The concentration of pockets of poor folk in inner cities does not, however, imply the creation of a self-perpetuating class. Instead, it implies a need to break and prevent a cycle of deprivation that is shared to varying degrees by other people for varying periods in their life-course (Davis, 1995). In this context, Wilson argues for a political coalition advocating policies that would benefit a wide range of groups, not just pockets of inner-city poverty and deprivation. Such policies would "promote race-neutral programs such as job creation, further expansion of the earned income tax credit, public school reform, child care programs, and universal health insurance" (Wilson, 1996, p. 235). The important issue as far as urban areas are concerned is that the polarizing consequences of economic and social restructuring have become, at this "underclass" extreme, spatially fixed within many inner cities. And, urban areas can at best provide only a palliative response when more fundamental national policy responses are required to address the symptoms.

Class and Conflict   Given the increasing polarization of the distribution of political power, wealth, and opportunities among social classes in North America, there are occasions when conflict occurs over the sharing of resources. Studies of conflict have, in general, emphasized four general themes (Cox, 1986):

1. Conflict arising from positive or negative externalities that change the market or use value of a location. Highway improvements, for example, may provide positive externalities for some by improving accessibility to the urban environment, and negative externalities for others if the freeway cuts through a neighborhood and destroys its cohesion.

2. Conflict arising from monopolistic control. Urban land is not a liquid asset because it is fixed in a particular location. Each site is in this sense unique. The trouble is, some sites are more "unique" than others—for example, sites adjacent to parks, with views, or in waterfront locations. Conflicts between the need of the public for access to such locations, and the "rights" of the property owners to exclusive use, is one of the crucial issues in planning in market-oriented economies (Dear and Scott, 1981).

3. Conflicts arising with respect to the level of governmental intervention in the form of planning, zoning, taxation, and licensing. The public may well argue that its interests are best served by greater governmental intervention to assure more equal distribution of benefits, whereas owners frequently want less governmental intervention, arguing that the less the intervention, the more rapid the pace of development. There are frequent struggles over the extent of governmental intervention, and then over the control of government itself. For example, in 1981 a tenants' movement in Santa Monica (Los Angeles PMSA) obtained a majority on the local town council and subsequently enacted a "people oriented" program (Shearer, 1982) which emphasized tenants' rights, accessibility to amenity locations, and social programs for the elderly, unemployed, and disadvantaged.

4. Conflicts arising as a result of pressure from particular interest groups. Some interest groups such as chambers of commerce and real estate boards, have become so much part of the fabric of local government that their views are firmly meshed within the decision-making process. Others, such as tenants' associations and neighborhood preservation committees, represent interests that are often not entrenched within most government organizations and decision-making processes. As a consequence, pressure from the unrepresented groups often has to be public- and media-oriented, whereas influence from the more established groups that are well represented within the political process can be far less visible.

The establishment–interest group dichotomy lies at the heart of the "People's Park" issue in Berkeley, California (Mitchell, 1995). Between 1969, when a parking lot was envisaged for an empty lot owned by the University of California close to its Berkeley campus, and 1991 when development of the site into a planned well-lit open recreational area was announced, the property has been a touchstone for local conflict between "the public" and its owners. Berkeley City Council has generally been in favor of orderly development.

In this case the fundamental issue is—what is public space for? Is it a representational environment that may be used for democratic action, or is the prime concern the provision of an orderly and safe environment? Mitchell (1996) argues that there is a long history in the United States of local governments creating regulations and practices to control dissent in public places, particularly the dissent of unionizing and striking workers. The assumption is that dissent leads to violence that might harm or restrict property rights. This assumption essentially restricts the rights of marginalized groups to publicly present their views on the nature of a just society. Ruddick (1996) demonstrates that there is a complex variation by class and gender concerning conceptions of public space, and the need by different groups for an orderly and safe environment in private spaces, such as coffee houses, which by the nature of their use are deemed quasi-public spaces (recall the discussion in Chapter 10 concerning malls).

Conflict Over the Redevelopment of Times Square   The Times Square redevelopment project illustrates the importance of a number of these themes, in particular the consequences of monopolistic control of land, governmental intervention, and the reaction from particular pressure groups that are not represented within the existing political process, exemplified in this case by the Clinton Coalition of Concern (CCC). Earlier in the chapter it was noted that since 1970, primarily in response to the structural changes in the North American and world economy and its expanding role as a leading transaction center, Manhattan has largely been reformed. There has been an enormous decrease in manufacturing employment, except in the specialist and fashion garment industry, which has partially shifted to an expanded Chinatown production zone, and an increase in producer service employment. The explosive growth of jobs located in the financial and business districts of mid-town and lower Manhattan has created employment for the white professionals and executives who reside in suburban communities in Westchester, Rockland, Putnam, and Bergen counties and commute to Manhattan, and young professionals residing within the city. Furthermore, the increase in office activity created a demand for more office space at the fringe of the mid-town business area. Thus, the underlying cause of the issue relates to the restructuring of the city's economy in response to globalization.

The conflict arose over the plan for the conversion of the highly mixed-use Times Square area, around Forty-second Street, to a massive new office, wholesale, and hotel complex, and over the renovation of nine theaters (Fainstein and Fainstein, 1985). The vast variety of uses that existed included rooming houses, tenements, a large number of reputable small retail and commercial businesses, and a comparatively small but highly visible number of sex-related businesses that provided the image for the area. Getting rid of, or dissipating, this sex and sleaze image provided the "high ground" catalyst around which a wide range of interests could gather to promote wholesale physical redevelopment that would, in reality, cater to the changing business environment.

The actors involved were a *consortium of developers* as part of the New York State Urban Development Corporation (NYSUDC), which does not have to follow zoning and citizen participation requirements although the city does, the *owners of small businesses* who thought they would be driven out by predicted higher rents, and many local *residents* who also feared that the project would stimulate upscale apartment and condominium redevelopment and much higher rents. The strongest opposition came from the low-income residents of the Clinton neighborhood, who not only attempted to resist the intimidatory tactics of landlords who were trying to vacate buildings for redevelopment, but also attempted to negotiate some form of neighborhood compensation.

The result is that the Times Square redevelopment is reaching fruition, because the interests of the developers, the business community, and government were too much in harmony for the scheme to be stopped or even modified very much. However, the CCC did manage to gain from the NYSUDC project managers $25 million for the preservation of space for low-income housing and small-scale businesses. The recession of the early 1990s and the glut in office space slowed down the pace of redevelopment, but by the mid–1990s the project was well underway. For us, this example demonstrates that an approach that relates the restructuring of physical space to structural changes in the economy, which focuses on the politics of hierarchical power relationships as they exist between social groups and classes, provides a fruitful framework for an analysis of the dynamics of change in social space.

CONCLUSION

The various approaches to the analysis of urban social space that have been described and discussed in this chapter illustrate that (1) the types of social space that are defined depend on the concepts and data used in their definition; (2) the theoretical approaches that are used influence the types of issues that are engaged; and (3) whatever the theoretical approach used, there are an enormous number of different forms of social space that have arisen in North American urban areas as a result of a complex array of forces. Interestingly, it would seem that whereas at the beginning of the suburban era central cities appeared to be highly diverse with a wide array of different types of social spaces and employment opportunities, and the suburbs much less diverse, at the end of the twentieth century the situation in most urban areas, particularly those in the United States, seems to be reversed (Frey and Speare, 1995). That is, the urbanized areas outside many central cities embrace a complex variety of social and economic spaces, while the central cities exhibit less diversity.

Most of the social spaces within urban areas arise as a consequence of the wide variety of cultural and economic attributes of individuals and households. In particular, ethnicity and race have been profound influences shaping North American cities. What this means with respect to the structure of social spaces within urban areas is that cultural attributes in general have been a profound shaping force. These cultural factors are entwined with equally important economic and class differences that are expressed primarily in terms of variations in purchasing power and social group hierarchies. The diversity of social spaces within urban areas is the product of a multitude of residential location decisions that are mediated through a variety of filters. In Chapter 12 we will explore in greater depth some of the filters that influence residential location decisions.

Chapter

12

Residential Location and Urban Revitalization

*T*he spatial distribution of population and of various social, economic, ethnic, and racial groups within urban areas occurs as a result of the location of households in housing units. The factors that influence and shape residential location decisions are, therefore, of prime importance in urban analysis. Housing is a *complex good* because any given housing unit involves many attributes ranging from space, quality, number of rooms, number of bathrooms, type of house (detached, duplex, row house, low-rise, high-rise), amount of exterior space, location, to such features as atmosphere and design. Table 8.3 lists a number of these features that may be used in establishing assessments for property tax purposes. The basic premise of residential location analysis is that households choose different bundles of housing attributes according to their preferences for different types of housing, but that these choices are subject to a range of financial, social, and governmental constraints. In other words, the choice of a residential unit and its location (the spatial decision) is influenced to varying degrees by a host of aspatial factors.

MACRO INFLUENCES ON THE DEMAND FOR AND SUPPLY OF HOUSING

Housing, whether rented or owned, is an extremely important element of urban and human life because it is where most people spend a large proportion of their day. A house also consumes a large share of total income. Housing provides a physical space for the creation of a home, which, ideally, is regarded by the inhabitants of the unit as a protective and supportive environment (Harris and Pratt, 1993). Housing also conveys a statement about the socioeconomic status of the household unit, with a change in status frequently given a public face by a change in residence. In North America, the idea of an adequate provision of housing is often linked to another idea, that of homeownership. Homeownership provides not only a financial asset from which all members of a household unit may benefit, but it also provides the unit with a clear financial stake in the economic and political health of society. Homeownership has been the chief source of equity growth for the average household. Thus, an adequate provision of housing is important to any society because it is one of the bedrocks for personal growth, an expression of socioeconomic status, and a foundation for sociopolitical stability. Housing is, therefore, a great concern of governments. Thus, a large number of public policies have accumulated that are designed to influence the demand and supply for both rented and owned accommodation.

The Demand for Housing

Perhaps the simplest long-run influence on the aggregate demand for housing is the rate of formation of family and nonfamily households (Gober, 1986). The distinction between family and nonfamily households is important—family households involve children, and nonfamily households do not. Since 1980 the rate of formation of new family household units has been decreasing, while that for nonfamily household units has been increasing. This has changed the demand for different types of housing. There has been a leveling in demand for single-family units, and an increase in demand for nonfamily-oriented housing such as condominiums. Thus, the aggregate demand for housing (DH) at some time in the future may be estimated from:

$$DH = EFNF + DFNF + IMMFNF - DFS$$

where, EFNF = Existing number of family and nonfamily household units
DFNF = Forecast increase (or decrease) in the domestically generated demand for family and nonfamily units
IMMFNF = Number of new family and nonfamily units created as a result of immigration
DFS = Number of household units lost as a result of death or alterations in family status

As long as EFNF is known, the other elements have to be estimated taking into account the changing demographic structure of the population, change in family income, replacement demand, and availability of credit.

The demographic structure is comprised of the rate of natural population increase, growth from immigration, and rate of "aging," as discussed in Chapters 3 and 4, which directly influence the rate of formation of family and nonfamily households. As average household size decreases, and the population ages, there is a tendency for a shift from large single-family homes to smaller

Table 12.1

Historical Trend of Housing Unit Types, United States, 1940–1990
(in Thousands)

	1940		1960		1990	
	NUMBER	PERCENT	NUMBER	PERCENT	NUMBER	PERCENT
Single Detached	23,731	63.6	40,103	68.8	60,383	59.1
Multi-Unit (5 or more)	3,928	10.5	6,238	10.7	18,105	17.7
Mobile Homes	NA		767	1.3	7,400	7.2
Other	9,666	25.9	11,207	19.2	16,376	16.0
Total	37,325	100.0	58,315	100.0	102,264	100.0
Units per 100 Persons	28		32		41	

(*Source:* Bureau of the Census (1993), Table 1234, p. 724.)
Note: The "other" category includes duplexes, townhouses, row houses, and multi-unit buildings with four units or less.

units such as townhouses or apartments in multi-unit buildings. The peak in single-detached ownership in the United States occurred in the 1960s. Since that time, the number of units in multi-unit buildings and mobile homes has increased dramatically (Table 12.1). In Canadian CMAs the proportion of households residing in single-detached houses has decreased from 67 percent in 1951 to 57 percent in 1991 (Silver and Van Diepen, 1995). The large rise in real incomes that has occurred in North America since World War II has been an important factor influencing the increase in housing demand. Replacement demand is also important, for even in areas where there is no growth in the number of households, a demand for new housing may arise as a result of changing tastes and rising aspirations.

The availability of credit and associated interest rates also exert a strong influence on the demand for both rental and owned accommodation. Rental prices are influenced to a degree by the availability of credit and interest rates, for both affect the rate of construction and the end costs. Credit and interest rates also have a direct effect on homeownership because nearly all homes are purchased with the assistance of a mortgage. In these cases, the actual purchase price is often of less immediate importance to a purchaser than the size of the down payment and the monthly "carrying charges" to retire the principal and pay the interest on the mortgage. To these "carrying charges" must also be added nonmortgage items, such as the monthly property taxes and the costs of lighting, heating/air-conditioning, and other services. In 1992 the proportion of recent home buyers' household income devoted to carrying charges in the United States was 33 percent as compared to 24 percent in 1976 (Bureau of the Census, 1993, p. 734). Thus, it is not surprising that individual homeowners are very much interested in trends concerning interest rates, for a rise in rate of 1 percent results in an increase of about 10 percent in the total cost of a house over the normal lifetime of a mortgage.

The Supply of Housing

A housing unit is defined as a dwelling that has a separate entry or an individual entry off a hallway or corridor that does not pass through another dwelling. The total housing stock is defined as the existing stock, plus new construction, less deletions. Although annual new construction is usually between 1 percent and 2 percent of the total housing stock, these relatively few new additions exert a disproportionate effect because they constitute between 30 percent and 40 percent of the stock available for purchase or rent at any one time. Furthermore, this new housing stock influences living styles by adopting new designs and fittings, which in turn influences the rate of obsolescence of the existing stock.

One of the most important influences on the number of new housing starts is the *vacancy rate*, defined as the proportion of unoccupied dwellings as compared to the housing stock as a whole. Vacancy rates provide the most readily available source of information for developers and builders and give them a good indication of the state of the market and its ability to absorb new units at acceptable prices. Of course, vacancy rates depend on other conditions, such as income levels and the availability of credit, but as a direct yardstick influencing supply, they stand by themselves. Vacancy rates for rental units in MSAs are usually between 5 percent and 8 percent, and for homeowner units between 1 percent and 2 percent. When vacancy rates are low, as they were in Canadian CMAs for apartments between 1973 and 1990, both rental and purchase prices increase and developers are encouraged to increase the supply through new construction or conversions (Silver and Van Diepen, 1995). High vacancy rates are obviously advantageous to consumers because in such situations prices stabilize, or decrease, and the range of choice increases.

Builders and developers are just as influenced in their ability to build by the availability of credit and low interest rates as buyers are in their ability to buy. Thus, if a government wishes to induce capital investment into the housing sector to not only increase supply but also create jobs, and increase consumer demand, it often changes interest rates, provides various kinds of lender-insurance schemes, or offers targeted subsidies, to attempt to achieve these ends. Policies such as these have contributed to the tremendous growth of capital available for new housing construction in the United States after 1945, as reflected in the increase in number of residential units per 100 persons in Table 12.1. Interest rates in particular affect the shadow market in housing, a process that provides housing through redevelopment of existing structures, such as old office buildings or lofts transferred to residential use (Baer, 1986).

The Influence of Government on Housing Supply   The supply of housing in North America has been influenced enormously by governments of the United States and Canada. In the United States many schemes have been stimulated by federal initiatives; in Canada, programs have involved the federal and provincial levels of government. A major *indirect* influence on construction is the basic discount rate (in the United States) or the bank rate (Bank of Canada), which determine the lending rates offered by commercial banks and lenders of mortgages. But governments have also been involved *directly* in housing through a large number of different programs. In general, these programs can be grouped into four phases in which different types of intervention predominate: the slum clearance and urban renewal phase; the supply-side subsidies phase; the demand-side subsidies phase; and a recent phase of cutbacks and attempts to address unintended consequences.

Urban Renewal   In the United States, before 1960, urban renewal and low-rent public housing were the principal instruments of direct government activity, the primary goal of public policy being the improvement of housing standards. These programs, such as Title I of the 1949 Federal Housing Act, essentially consisted of subsidies linked to new construction, which in turn was linked to slum clearance. Subsidies were provided by the Department of Housing and Urban Development (HUD) to the suppliers of housing rather than to the consumers, and resulted in the stimulation of new housing construction in the periphery and the demolition of low-income areas in the inner city. In fact, slum clearance and urban renewal under Title I had the effect of actually reducing the total supply of housing generated by the program. Between 1949 and 1968, 439,000 mainly low-income housing units were demolished in inner cities in the United States and only 124,000 mainly middle-income housing units were constructed under the program. Similar programs in Canada were directed towards renewal in deteriorating parts of downtown areas (Bettison, 1975).

Supply-Side Subsidies   A second phase of direct government intervention in the supply of housing came to the fore in the 1960s with the growing disenchantment with urban renewal and public housing schemes (Bourne, 1981, pp. 191–214). By and large, these programs expanded *lender-insurance* schemes aimed at reducing homeownership mortgage costs so that moderate or low-income households could afford to live in new housing of reasonable quality. The previously mentioned FHA and VA schemes are prominent examples of these types of programs in the United States. In Canada, the Central Mortgage and Housing Corporation (CMHC) provided similar lender-insurance schemes. Thus, between 1957 and 1970, roughly one-fifth of new construction in the country was undertaken under the aegis of CMHC programs.

In both countries these programs, which in effect subsidize the cost of supplying homes by providing fully insured mortgages, received widespread support from both the property industry and the purchasing public. However, by the early 1970s it was becoming widely recognized that these supply-side subsidy programs had benefited primarily white families in the middle-income range by lowering marginally interest rates on mortgages, and had also helped to foster massive suburbanization and sprawl as a result of an emphasis on *new* homes on new land. Furthermore, the allowance in the United States taxation system for the deductability of mortgage interest from income taxes added to the view that lender-insurance schemes were a form of subsidy for homeowners.

Demand-Side Subsidies Thus, during the 1970s, there was a gradual shift in emphasis in both Canada and the United States toward demand-side subsidies and the establishment of housing programs for particularly disadvantaged groups, such as families or the elderly close to or below the poverty level (Sewell, 1994, Chapter 8). The primary goal of public policy was the improvement of housing conditions for the poor. To achieve this purpose, Section 8 of the United States Housing and Community Development Act of 1974 provided certificates guaranteeing landlords the difference between a tenant's rent and 30 percent of the household income. Although this act was dressed up as a demand-side subsidy, the way that it was administered made it a supply-side subsidy, as it was the developers, not the poor, who received the certificates for delivering units at an affordable price for low-income households. The scheme led to an enormous number of abuses as well as further concentration of the poor into large public housing projects. In addition to such housing support schemes, a plethora of particular supply-side programs continued to exist, such as the FHA and CMHC lender-insurance schemes, as well as others in partnership with local governments for the provision of housing for the poor, the elderly, and the physically handicapped.

Cutbacks and Addressing Unintended Consequences During the 1980s, a conservative political phase in both the United States and Canada led to a reduction in funds allocated to demand- and supply-side subsidies, an expansion of block grants to the states to support housing-related programs, and an attempt by a number of cities, such as Cleveland, Baltimore, St. Louis, and Montreal, to lure middle-class households back to homeownership in the inner city (Varedy, 1994). At the same time, a deficit-driven overheated economy, with a construction industry also fueled in part by the raising of federal insurance limits on deposits in savings-and-loan financial institutions, generated a boom and price inflation in the private property industry (Ball, 1994).

Despite growing waiting lists for public housing there were severe cutbacks in federal low-income housing programs and a reliance on federal-state block grants to fund local housing and community development initiatives. Between 1980 and 1988, the United States federal government cut budget authority for all federal housing assistance programs by 82 percent, from $32 billion to $5.7 billion. Furthermore, Section 8 of the HCD Act was virtually eliminated with the number of new certificates issued cut from 110,000 in 1981 to 10,000 in 1988, being partially replaced by a five-year rent supplement voucher system. The limited number of vouchers available (202,000 between 1984 and 1988) were allocated directly to eligible poor households. The result of these cuts in authorization, and the general increase in house prices due to the property boom and inflation, was that by the end of the 1980s housing *affordability issues* had risen to the top of the United States housing policy agenda as middle-income households began to find it difficult to afford homes in an overheated economy, and lower-income households were being squeezed into owner and rental markets (Linneman and Megbolugbe, 1992). The National Affordability Housing Act of 1990 reflects this public concern, but was *passe* for middle-income groups in many respects when enacted as a property market "crash" and declining prices were already underway (Mills, 1995).

The restructuring of manufacturing and financial institutions in the 1980s had, however, generated greater income disparity between social groups. So housing affordability problems among the poor remained the most important issue. Local and state provincial governments are, in the 1990s, under pressure to provide demand-side support for low-income residential accommodation during a time when public deficits in many jurisdictions have to be severely controlled. Furthermore, the *concentration of public housing* (or "projects"), constructed under the aegis of supply-side policies mainly in inner cities, such as Yonkers in New York, have now come to the forefront of concern. It is argued that many of these concentrations have stimulated neighborhood racial change and minority ghetto formation (Galster and Keeney, 1993), and may well have a negative effect on urban regeneration, particularly if projects are located close to revitalizing producer service–oriented central business districts (McGregor and McConnochie, 1995), leading to local *spatial containment* strategies (Cook and Lauria, 1995, p. 553; Goetz, 1992).

In consequence, HUD, which in 1995 had more than 200 housing and community renewal programs on its "books," is attempting to implement a plan called Moving to Opportunity, which was approved by the United States Congress in 1992. The program would give about 1,500 public housing families in six cities rent support and help in finding accomodation in low-poverty suburban areas. This program emanates from the apparent success of Chicago's Gautreaux program, which began in 1976 and has dispersed 5,000 poor African-American families from the Gautraux district of the inner city to apartments elsewhere, mainly in inner suburban locations. Follow-up studies have indicated that "dispersed" household heads are more likely to be employed, and children more likely to go to college and gain full-time employment with benefits, than those households remaining in the inner city (Ramos, 1994). The Moving to Opportunity program is plainly small, but it does indicate the direction of public policy with respect to the public housing and the subsidization of population dispersement that is being overlain on a vast number of existing programs.

Filtering and Housing Policy

The mobility of households from one dwelling unit to another involves a *filtering* of dwellings (Lowry, 1960). The filtering process has been of crucial ideological importance to the formulation of housing policies in North America (Ohls, 1975; Sewell, 1994, p. 8). In general, the filtering process describes the way in which a dwelling unit changes from being occupied by households or individuals from one income group is later occupied by people from another income group (Baer and Williamson, 1988). Thus, it is possible for a house to change in ownership from a lower-income household to a higher-income household, which is described as *upward filtering*. More commonly, as the price of a house or apartment relative to all other house prices at a given time decreases (given

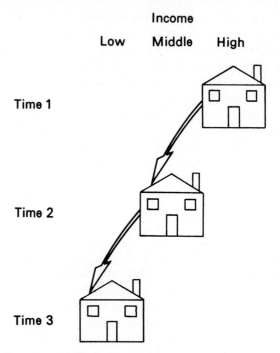

Figure 12.1 Downward filtering: A house passes, through time, from being inhabited by a higher-income household to progressively lower-income households.

no significant improvements) due to aging, deterioration, and creeping obsolescence, the house will through time pass to progressively lower-income families or individuals (Figure 12.1). This sequential occupance of property of declining quality to lower-income households is known as *downward filtering*. Filtering, therefore, lies at the heart of neighborhood change (Smith, 1964).

Five factors appear to influence, in different ways, the upward or downward filtering of dwelling units: the supply of housing, the rate of household formation, changes in real income levels, changes in housing quality, and changes in the lifestyles of households. The first of these has been particularly important with respect to the establishment by governments of housing policies in North America (Braid, 1984). The filtering concept implies that if the rate of new housing construction is slightly in excess of the rate of new household formation, then it will be possible for the relative value of older housing to be decreased slightly and hence become available to lower-income households due to higher vacancy rates. Thus, the filtering concept provides a plausible excuse for a public policy directed at providing well-insured mortgage loans for middle- and upper-income groups, for such policies increase the supply of housing. Therefore, all income groups could move into better-quality housing as the relative prices of dwelling units decline (Galster and Rothenberg, 1991).

There are, however, at least three objections to policies that have adopted the filtering principle. The first is that the policy appears to accept the idea that new housing should be provided, on the whole, for middle- and upper-income groups and that it is all right for the poorer segments of society to occupy older housing. As new housing is built primarily on the metropolitan fringe, and older housing exists in the inner city, the results with respect to spatial separation are obvious. Thus, the filtering concept can be used as an excuse for spatial inequality (Boddy and Gray, 1979). The second objection relates to the necessity of keeping the supply of housing in excess of demand. If the supply of housing is not kept ahead of demand, upward filtering of dwelling units may occur, and the general level of housing for all but the highest-income groups may deteriorate.

The third objection to acceptance of the filtering principle in housing policies is that it assumes that household units can move freely into dwelling units as they become available. Studies of filtering, however, indicate that different parts of metropolitan areas experience different rates of filtering and that price levels are affected not only by the supply of dwelling units and the demand in these different areas but also by competition from other land uses (Maher, 1974; Weicher and Thibodeau, 1988). It is, therefore, impossible to assume that the increase of housing supplied for one particular income group will affect the supply of dwelling units available for others. The housing market in a metropolitan area, as will be discussed later in the chapter, is not a single entity but consists of several submarkets. While there may be sufficient or excess supply in one submarket, there may be insufficient supply and excess demand in another.

MARKET-ORIENTED PERSPECTIVES ON RESIDENTIAL LOCATION

A vast body of theory relating to housing consumption and location has stemmed from the neoclassical economic approach, which emphasizes the decision-making processes of individual households in a market framework (Muth, 1961, 1978; Mills and Hamilton, 1989). In general, the central theme of the theory is the contention that households are seeking to maximize their utility by purchasing the bundle of goods—recall the definition of housing as a complex good—which jointly offers them maximum satisfaction, subject to individual household budget constraints. In terms of location in urban space, this theme can be rephrased: Households seek to maximize their place utility by seeking a location and housing unit which offers them the greatest satisfaction subject to the constraint of income.

The Residential Location Model

The residential location model assumes that the bundle of goods for which a household is spending income (Y) is housing (H), transportation (T), and a remaining category of all other goods (G), that is: $Y = H + T + G$. The "all other goods" category includes food, clothes, and entertainment. In the simple model, those components are assumed to consume a constant proportion of a household's income regardless of its level of wealth. Under these conditions, the residential location decision for a household is regarded as involving a trade-off between housing costs, and transportation costs (which consume about 17 percent of household income). The transportation cost component is represented by journey-to-work costs because work trips account for 40 to 45 percent of total trips. The model was developed with respect to a monocentric city situation with the greatest employment concentration assumed to be in and around the central business district, but it is also applicable (as we shall see) to multinodal polycentric situations.

In the model, the location of maximum place utility for a residential household involves a trade-off between the costs of a house and the costs of traveling to and from work sites in the urban area. The line *H-J* in Figure 12.2a represents the average budget line for an income group and defines the possible combination of purchases of the two goods, housing and commuting. Now, what will be the trade-off location (or maximum place utility) for an individual household? This question is answered by introducing household indifference curves, and sets of indifference curves, for three households A, B, and C with similar incomes that are mapped on Figure 12.2a. These indifference curves represent the consumption of housing and travel costs that give equal satisfaction at particular combinations of expenditures. The highest curves represent the largest combination of expenditures. Hence, a household will seek to locate where an indifference curve just touches, or is tangential to, the budget line *H-J* for the income group. The three sets of indifference curves (A, B, and C) represent the various preferences of households for these particular goods. In other words, household C weights housing costs higher and travel costs lower than the others. The location of an individual household, therefore, depends on the preferences of individual households *vis a vis* housing and journey-to-work costs.

The average budget line of another, perhaps lower, income group may be represented by the line *R-L* in Figure 12.2b. A number of low-income households, each with their own preferences indicated by maps of indifference curves, can also be positioned along this line, with perhaps the greatest number coinciding around the indifference map for household X. Likewise, it is assumed that the greatest number of higher-income households have indifference maps that coincide around B. There are cases where lower-income households are prepared to pay more for housing than higher-income households, but their journey-to-work costs have to be exceedingly low; and there are also cases where some lower-income households are willing to pay more than the higher-income households in travel costs, but their housing costs have to be very low. Nevertheless, the lower-income group may be able to exercise some of their preferences in locations that have not been appropriated by the higher-income group. Furthermore, the great heterogeneity of the housing stock in urban areas may make it possible for the different income groups to achieve satisfactory place utility, given the different budget constraints, in quite similar geographic locations within the city.

An important point to recognize concerning the residential location model is that it has provided the basis for the development of an elaborate theory

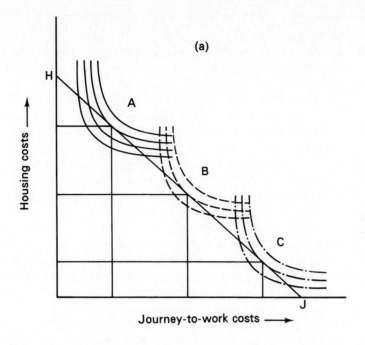

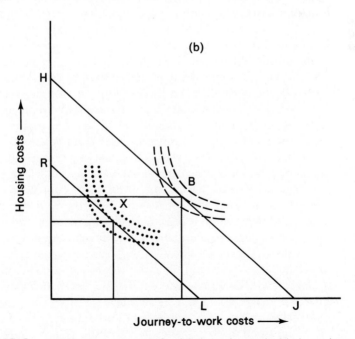

Figure 12.2 Indifference curves for (a) three households in an income group related to an average budget line; and (b) a comparison between the trade-offs that may be made by high-income and low-income groups.

(Wheaton, 1983). This theory has been incredibly influential in the analysis of housing and housing markets and the formation of public policy. The major criticisms of the approach focus on two basic issues. The first is the assumption that all households do have choices with respect to house type and location, within budget constraints. It is evident that in many situations, particularly those relating to the poor, elderly, and racial minorities, choices are limited. Secondly, the model, however complicated its construction, invariably assumes a large proportion of a household budget to be fixed in "all other goods." This is not the case, for often families make trade-offs among all aspects of their budget. For example, it has been demonstrated that when poor families receive demand-side housing subsidies, they often allocate their increased income to food and other goods rather than more or better quality housing because they have already been forced to spend more than they would like on accommodation.

Application of the Residential Location Model   A model based on the central premise of traditional residential location theory, which is that households substitute transport costs for housing prices and space in choosing residential locations, has been examined by Quigley (1978) using the actual choices made by a sample of 3,000 renter households in the Pittsburgh metropolitan area. The model also assumes that the locating households have known and permanent workplaces in a multicentered metropolitan area. The results indicate that households at different income levels do, in effect, substitute housing costs for transport costs by consuming different house types and greater or lesser space. As family size increases, households are more likely to locate in detached homes or duplexes, and for the same family size high-income households occupy considerably more space than low-income households.

The specific applicability of the residential location model in density gradient format in the polynucleated Los Angeles environment has been investigated by Song (1994), and on a national basis in the highly urbanized country of the Netherlands (Borgers and Timmermans, 1993). The Los Angeles study investigated the relationship between land rents, isolated from house and rental prices and property characteristics, and distance from six major employment nodes in Los Angeles County, including the central business district which provides only 3 percent of employment in the urban area. Song (1994) concluded that land rents decrease and space occupied increases with distance from the six major employment nodes, and that overall accessibility to employment opportunities is thus the primary influence on residential location choice even in a multicentered urban area.

In the Netherlands study it is hypothesized that residential choice is related to: characteristics pertaining to the residence, such as type, cost, neighborhood; transportation facilities available in the neighborhood, such as public transit service and access to highways; and travel time from the residential location to the workplace by public transportation and automobile. The heads of the sample of households were asked to rank the relative importance of the various characteristics in their choice of residence. The results indicate that the group of characteristics related to the residence itself and travel time to the workplace are important in the choice of residence, whereas those concerning the availability of transport facilities are of little importance. Both male and female respondents replied similarly. The results of these studies thus affirm the fundamental importance of house characteristics and access to workplaces in residential location decision making.

BEHAVIORAL PERSPECTIVES ON RESIDENTIAL LOCATION

The behavioral perspective enriches the neoclassical approach to residential site selection by including an array of noneconomic characteristics that are ignored in the traditional housing–workplace transportation cost trade-off model. The behavioral approach to residential location theory focuses on the ways in which individuals and households go about making location decisions, and it emphasizes the roles that various external agencies, such as real estate agents, play in influencing decisions. The approach accepts the crucial importance of budget constraints, and the trade-offs that households have to make between the housing, transportation, and "other" components of their budgets, but the thrust of the approach is that the locational decisions are *mediated* in some way by the actual process of making a first decision to locate or a subsequent decision to relocate.

The Brown-Moore Model

The mediation process is explained best in terms of the Brown-Moore model (Brown and Moore, 1970), which has had a considerable influence on the development of research in residential location (Popp, 1976; Huff, 1986). The Brown-Moore model focuses on the general way in which a household goes about making a decision to relocate as a consequence of "stress," which could arise as a result of dissatisfaction with the current residence or its location, employment change, or, in the case of new households, initial selection. The model envisages two stages in the decision process, first the decision to move (or not to move), and second the relocation decision itself. The various aspects of these two components are presented in flow-sequence form in Figure 12.3.

The Decision to Move   The decision to move is the end result of internal and external factors that create "stress" in the household. Internal stress arises from the changing needs of the household for tangible features, such as space and facilities, and for less tangible requirements arising from changing expectations. These stresses can be regarded as resulting for the most part from life-course and lifestyle changes surrounding the individuals comprising the household. For example, the addition of children invariably leads to requirements for more space, access to schools, and reasonable levels of access to employment. At the other end of the life-course, the "empty-nest" syndrome, loss of spouse or partner, and aging itself, frequently lead to a search for a home with less space in which workplace location is not a concern, but access to amenities or family are important (Warnes and Ford, 1995).

External stress occurs from changes in the dwelling itself, most commonly due to deterioration with age, and the neighborhood. With respect to the neighborhood:

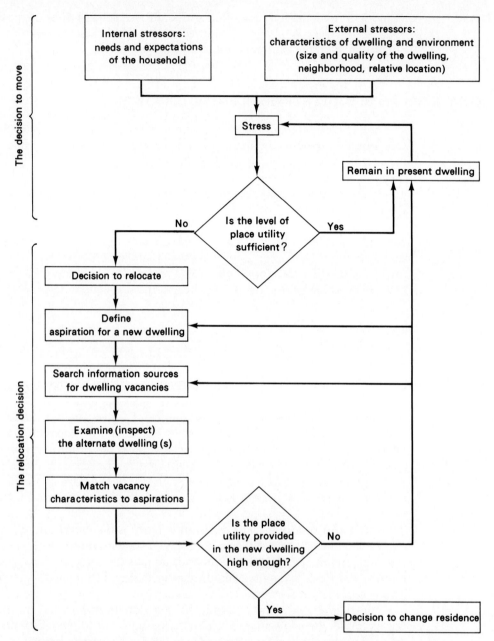

Figure 12.3 A behavioral approach to the residential location decision.
(*Source:* Redrawn from Popp, 1976, p. 301.)

1. Stress may occur as a result of changes in the immediate environment, such as with an influx of a different racial or ethnic group. For example, Taub and colleagues (1984) examined the way in which homeowners in neighborhoods that are undergoing rapid racial change, and those incurring increasing rates of crime, change their household investment decisions and decide to move elsewhere.

2. Stress may also be a result of changes in what a particular household desires in (and is prepared to put into) the immediate neighborhood. For example, young families make more use of certain features and facilities, such as schools and recreational parks, than older families.

These internal and external factors may give rise in a particular household to a level of stress that causes the household to consider whether the utility of that particular residential unit in that particular location is sufficient. This consideration will, of course, be influenced strongly by the magnitude of the household budget and access to employment as well as a variety of other services, families, and friends. If the level of utility of a particular place is sufficient, the household will remain in its present dwelling. If the level of place utility is insufficient ("No" in Figure 12.3), the household will begin to consider relocating. For example, with the elderly there is a retirement age peak in relocation propensity from large metropolitan areas among married persons in middle- and higher-income groups; and another increase in propensity to change dwelling for persons around 75 years of age, particularly among widows. Most of these are short-distance moves (Morrow-Jones, 1986; Longino, 1990).

The Relocation Decision   When the household decides to relocate, it has to define its aspirations; that is, the type of house and neighborhood to which it would like to move. The household has to decide on the amount of space and design characteristics it would like for the dwelling unit, and the locational attributes with respect to employment and services that are to be emphasized. These are complex decisions, because they often involve trade-offs between competing interests if a number of members of the household have jobs or educational facilities to attend. Furthermore, the type of housing unit and the most desirable location are also influenced strongly by life-course considerations. Households with two or more children put a greater premium on space and location with respect to schools than those without children.

Once the aspiration region of the household has been defined, members of the household then search through the housing market, as expressed in advertisements and real estate listings, to identify a short list of possible alternatives. The search is often not thorough, and is mediated strongly by real estate agents who have a powerful influence on the geographic and social spaces in which the household conducts the search (Palm, 1976). White middle-class families are conditioned by peer group pressure, and encouraged by real estate agents, to focus their search on particular suburban communities (Cutter, 1982). Similarly, various ethnic groups are also guided by agents to "appropriate" parts of the city (Palm, 1985).

When a vacancy is found that conforms with the joint aspirations of the household, the question then arises as to whether a sufficiently high level of utility might be realized at that location (in that particular unit) to warrant moving. If the utility is high enough, the household unit will move to the new residence. If it is not high enough, the household unit may decide to remain in its present location, redefine its aspirations, or look for other vacancies. In the case of households that own their own home, the decision to move is also predicated on the strength of the resale market for the current home, and the *financial equity* level of the homeowner. Relocations from high-vacancy-rate areas to low-vacancy-rate areas may be extremely difficult because of resale difficulties for the existing home. Similarly, owners burdened with *negative equity,* that is, the market value of the current home is less than the outstanding mortgage capital, may find it extremely difficult to move for financial reasons (Ball, 1994).

A Sovereign Consumer?

The neoclassical approach to residential location, and its behavioral modifications, are, therefore, useful ways of examining housing location decisions. The basic issue with the use of models of this type is, however, that they concentrate exclusively on demand, and assume that the consumer is sovereign in making household location decisions. The approach gives us only a partial explanation of the residential location process, for there are many instances in which the consumer is not sovereign and choice is limited by supply. There may, for example, be considerable dissatisfaction in certain areas with multiple-family housing, but the lack of any other affordable choice within the same area severely limits the possibility of alleviating stress. Studies within this perspective may, therefore, ignore important restrictions on the decision-making process and may conclude that the residential structure of cities is the way it is because that is what people want.

Housing affordability issues relate primarily to households with low incomes that rely on rental accommodation, though 40 percent of households with affordability problems are homeowners. Affordability issues become prominent when lower-income households pay more than 30 percent of their gross income on housing. In Canada, as in the United States, about 35 percent of all households renting accommodation paid 30 percent or more of their income on housing in 1990–1991 (Lo and Gauthier, 1995). The vast majority of households vulnerable to housing affordability problems involve the elderly and young adults, single-parent families, and people living alone. A large proportion of households with affordability problems also involves visible minorities. The prominence of the affordability issue is in many ways related to the vacancy rate for lower-cost rental accommodation, a rate which is in turn influenced by supply-side policies with respect to the provision of public housing, the adequacy of demand-side voucher or income-support subsidies, and the availability of low-rent private accommodation. The availability of private low-rent accommodation is in large part determined by the market, as measured by the vacancy rate, and to a debatable degree by local housing policies, such as *rent control* (Linneman and Megbolugbe, 1992; Goetz, 1995).

HOUSING SUBMARKETS AND RESIDENTIAL LOCATION

Previous references to differential vacancy and filtering rates and the mediating effects of real estate salespersons, have pointed to the existence of a number of housing submarkets within any urban area. Although there are a number of ways in which housing submarkets may be defined, there are, in general, two main approaches involving variables that may be used either singly or in combination with others. A definition of submarkets in terms of *housing stock* includes such factors as: type of housing unit (such as apartments or condominiums, row houses, and single-detached houses); form of tenancy (owner, rental, public sector); and price or rent (from high to low), which also represents quality. Figure 12.4a presents these three characteristics as variables that could be used singly, for example, in Melingrana's (1993) study of the condominium submarket, or in combination to define a large number of different submarkets. Most studies of housing submarkets define them in terms of one or more of these characteristics, particularly quality, as measured by prices or rents paid, and form of tenure (Rothenberg et al., 1991).

Housing submarkets may also be defined in terms of the occupants of the housing, that is, by *household type*. Figure 12.4b presents these submarkets in terms of age and family status, economic status, and race and ethnic origin. Perhaps a fourth dimension, lifestyle, should be added, though it is difficult to present this in diagrammatic form. These three characteristics could either singly, or in combination, and depending on the number of categories used, define a wide array of different submarkets (Cook and Bruin, 1993). It is the intersection of these various household types with the household stock, each unit of which has a fixed location in urban space, that gives rise to the great variety of social spaces. The important question is how these different submarkets are created, for when this is known it might be possible to correct some of the inequalities with respect to housing access that occur.

The Property Industry

The numerous factors that contribute to the creation of housing submarkets are often difficult to unravel because they arise from a complex industry dealing with property. A model of the property industry, which contains a sketch of some of the more important interconnections is presented in Figure 12.5. The property industry is divided into two components, the existing buildings sector and the building industry sector, which make common purpose with various aspects of local government, a range of public utilities and cable companies, the finance industry, and a variety of producer services.

At the heart of the *existing buildings sector* are the property investors, or owners, who range from large corporations, which might own many multimillion-dollar properties, to the landlord who operates a few apartment buildings, to the individual homeowners whose homes are their only big investment. Although the existing buildings sector is dominated numerically by the large number of individual homeowners, the dominant voices from the sector influencing the property industry come from the few large property investors. All

(a) Housing types

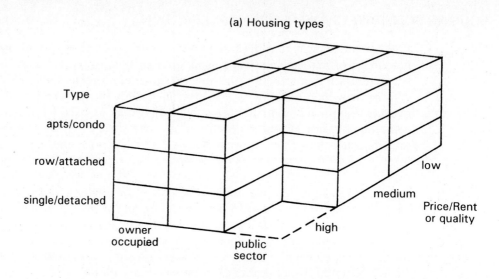

(b) Household types

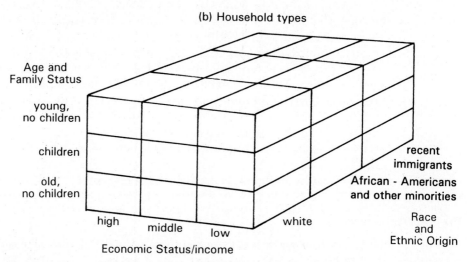

Figure 12.4 General types of housing submarkets classified by (a) housing, and (b) household types.
(*Source:* Redrawn from Bourne, 1981, p. 89.)

the property owners are serviced to a greater or lesser degree by property insurance companies, insurance agents, lawyers, real estate agents, mortgage lenders, and utility companies. Thus, the policies and attitudes of these property service industries influence the property investors. Their policies, in turn, are largely in accordance with the few large property-owning companies that have considerable political and economic clout.

The attitudes and policies of the service industries are also influenced by the groups operating in the *building industry sector*, at the center of which are

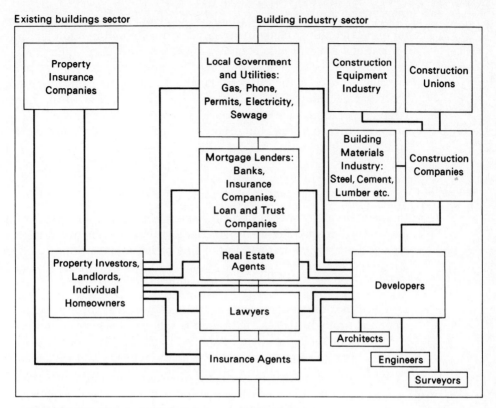

Existing buildings sector   **Building industry sector**

Figure 12.5 A model of the property industry.
(*Source:* Redrawn from Lorimer, 1972, p. 19.)

the developers. This sector is concerned with adding to the existing stock of buildings in the urban area. To do so, the developers have to obtain funds and insurance, deal with local governments to influence planning and public investment in infrastructure (such as trunk sewage lines and highways), obtain building permits, and operate through the construction companies with the various elements of the construction industry. Although the property industry has traditionally been characterized by a large number of small construction companies and individual developers, the complexity of the industry is such that a few large development corporations now dominate most of the new construction that takes place in metropolitan areas. Thus, on both the property investor and development sides, a few large companies have a major influence on the entire industry.

Given that property investment and development has consolidated into the hands of a few large corporations and that there is a complex, interlocked industry serving them, it is not surprising to note that many of the activities outlined in separate boxes in Figure 12.4 have become integrated both vertically and horizontally and many are operating multinationally. A single firm may assume many different functions. For example, many life insurance companies are now developers, property owners, and mortgage lenders, just as many banks have now gone far beyond the business of lending into property development partnerships. Thus, many large vertically and horizontally integrated organizations are now much more concerned with using their funds to finance their own projects, rather than meet the demands of small independent property investors or home builders. Furthermore, the concentration of capital, investment decision making, and property development in the hands of a few large corporations can lead to quite dramatic changes in the location and type of capital investment within urban areas, as was observed in Chapter 7 with respect to the burgeoning of edge cities.

The Role of Financial and Governmental Institutions

In a path-breaking study of housing submarket formation, Harvey and Chatterjee (1974) demonstrated that financial institutions and governmental programs created housing submarkets in the City of Baltimore primarily on the basis of the sources of funds that were available to different social groups in various parts of the city. They identified five general submarkets:

1. The inner city, which consists primarily of African-Americans residing in rental units. In those dwelling units that were sold cash and other forms of private financing predominate. The purchase price of the homes is by far the lowest in the city, and there is an absence of institutional financing.

2. White ethnic areas have a high proportion of dwelling units that are owner occupied and a fairly low proportion of rental units. Incomes are below average, but a large proportion of the houses sold are financed by local state savings-and-loan companies and cash transactions.

3. Middle-income African-American areas in which homeownership is relatively high and the value of houses about average for the city, are serviced mainly by mortgage banks operating with federal FHA loan guarantees.

4. White middle-income areas have housing financed primarily through the state savings-and-loan companies with FHA mortgage guarantees. More than half of the dwelling units in these areas are owner occupied. The value of the homes is about average for the city.

5. An upper-income area, located on the northern periphery of the city, receives loans directly from savings banks and federal savings-and-loan companies and makes relatively little use of FHA guarantees.

Thus the study illustrates the way in which federally insured loans are used to finance middle-income white and African-American homeownership but are rarely used to finance housing in low-income African-American and low middle-income white-ethnic areas.

Lending Discrimination   It is evident that the ability of household units to obtain at-market mortgages determines whether and where they are able to purchase a home. Creditworthiness has not been, and is not now, simply a decision made by financial institutions on the basis of a household's or person's income, but a decision structured by the information and decision-making procedures of private lenders and governmental institutions. Murdie (1986), for example, demonstrates that as a consequence of Canadian federal government mortgage-lending insurance policies in the 1960s and 1970s, the large financial institutions have tended to concentrate their loans in suburban parts of Toronto. This has tended to work in favor of established Anglo middle-income families. Immigrants during that period, particularly those from southern Europe, had to resort to less formal and often higher-priced lending operations in order to purchase older homes in the more central part of the city. As the immigrant community became more established, this pattern of using community-based financial resources continued along with increasing use of the large financial institutions by the second generation from immigrant families.

Lending discrimination is an extremely sensitive issue because it is illegal. In the United States, the possibility of discrimination is monitored by the Departments of Justice and HUD, and various public interest groups, with information obtained annually through such means as the Home Mortgage Disclosure Act (HMDA, 1975). This statute requires lenders to make publicly available information about mortgage applications and approvals sufficient to observe differences, if any, in acceptance rates by gender, race, income, loan amount, neighborhood, and other characteristics (Leven and Sykuta, 1994). The data obtained through HMDA have revealed higher rates of loan denial in the approval process for minority populations, particularly African-Americans, and high rates of denial in particular neighborhoods (Munnell et al., 1992). The issue of redlining, the denial of loans in particular neighborhoods, has been a matter of concern with respect to some parts of inner cities because such practices mitigate against homeownership and property improvements, and promote the formation of low-income minority ghettos (Dingemans, 1979).

Redlining and Lending Discrimination In the spring of 1975, the United States Senate Committee on Banking received data which showed a consistent pattern of refusals to make loans to creditworthy persons who wished to purchase property in older neighborhoods of Los Angeles, St. Louis, New York City, and Washington, D.C. One particular savings-and-loan association in Chicago had deposits of $10.4 million from residents in redlined neighborhoods but had not made a single loan in these areas. In cases where loans had been made to purchasers within redlined areas, the mortgages often carried interest rates that were much higher than normal and were to be reassessed at frequent intervals during the amortizing period (Brown and Bennington, 1993).

The data that were collected indicated that redlined neighborhoods usually have one or more of the following characteristics: (1) older housing, (2) blue-collar workers, (3) white-ethnic, African-American, and Hispanic-American populations, (4) racially integrated, (5) adjacent to low-income neighborhoods, and (6) many poor single individuals residing in rental units. Neighborhoods exhibiting one or more of these characteristics are usually located in inner cities of metropolitan areas, though there are examples of redlining in the downtown sections of commuter railroad exurbs. The result of redlining is that properties become hard to sell, maintenance and rehabilitation are discouraged, and ownership of the various properties falls into the hands of "slumlords," speculators who try to exact as much rent as they can for minimal investment (Long and Nakamura, 1993).

The reasons given for refusal by mortgage lenders to make loans in redlined areas generally revolve around the financial institution's notion of risk. The loan manager has a perception of the area as being inhabited by people with unfavorable credit histories, irregular employment, and unstable households. This image is often based on hearsay rather than substantiated facts and can, quite frequently, be a misinterpretation of the way in which minority groups cope with their situation. In some cases there is a real communication problem between the loan managers of the financial institutions and the would-be borrowers, for the managers may never have been near the area and have no first-hand knowledge of the condition of the area involved. But whatever the reasons for refusal, loans should be made or not made to people, not areas. Therefore, when a loan at normal interest rates is denied to an individual, the reasons for the denial should be made known in writing to the applicant.

The Community Reinvestment Movement Armed with information obtained through HMDA, and the Community Reinvestment Act (1977), which requires federally regulated lenders to serve the credit needs of their communities, neighborhoods have been able to seek financial compensation from institutions that have failed to live up to their statutory obligations. Activists' attempts to document discrimination against minority persons within minority neighborhoods and to obtain compensation for past redlining are known collectively as the Community Reinvestment Movement (CRM). The key to the effectiveness of CRM in such cities as Boston, Detroit, Chicago, Milwaukee, Atlanta, and a number of cities in California, has been the requirement that financial institutions filing for deposit insurance, branch office establishment, and many other federally regulated activities, must document the ways in which the credit needs of the local community being served by the financial institution and its branches have and will be met.

This documentation can be challenged by third parties, thus allowing community reinvestment organizations leverage to negotiate compensation or greater local investment packages. During the 1980s, community reinvestment campaigns generated more than $18 billion in financial commitments in more than seventy cities (Squires, 1992). This movement is interesting not only because it demonstrates the way in which federal laws can help individuals and local communities resist discrimination and seek redress, it is interesting because it also indicates the kinds of things that society has to do to achieve greater equality between the races. Smith (1994) emphasizes that social justice is manifest not by programs and good intentions but by measurable reductions in inequality, such as reductions in housing inequalities between African-American and white communities in United States cities.

Continuance of a Lending Discrimination Environment Although HMDA data permit the monitoring of possible discrimination with respect to denials in individual mortgage applications, there are a number of procedures and practices occurring prior to submission of a mortgage application that may create a discriminatory environment. Goldstein and Squires (1995) suggest that these include:

1. Use of appraisers who are unfamiliar with inner cities and are thus more likely to undervalue properties in minority communities.

2. A limited number of minority employees in lending institutions, which creates an unwelcoming environment for potential minority applicants, particularly those for whom English is not a primary language.

3. Advertising campaigns that rarely reach minority communities.

4. Minimum loan criteria, which have a *disparate impact* on low-income applicants who are generally seeking smaller loans than higher-income households.

Thus, there may not only be a continuance of discrimination in the approval process for applications, there may also be systemic discrimination in preapproval practices as well.

The Role of the Real Estate Industry

Houses can be sold in two ways, either through real estate agents or by the owners of the property themselves. About 85 percent of single-family homes on sale are sold with the assistance of real estate agents. This is normally done on the basis of a commission that is a fixed percentage (about 6 percent) of the sales price. Thus, although the seller pays the real estate agent, it is the buyer who actually provides the money. It is clearly to the advantage of the agent to get the highest price possible in a sale, or package of sales. Jud and Frew (1986) demonstrate that real estate agents, as the salespersons for the housing industry, generate effects similar to that of advertising. That is, they increase the level of demand among buyers and attempt to stimulate higher prices.

In terms of the creation of housing submarkets, real estate agents act as (Barresi, 1968; Palm, 1985):

1. Screening agents who filter would-be buyers and direct them to areas of possible purchase. This screening process is vital, for many movers, as shown in the discussion of the Brown-Moore search model, have limited information and may be only marginally interested in relocating.

2. Specialists who frequently have information only about certain types of housing or particular parts of urban areas. Thus, if potential buyers feel constrained to use only one or two agents, they will automatically be limited in the range and location of their choices.

3. Supporters of the *status quo*, for as part of the property industry, the agent is generally conservative in nature and therefore averse to social change. Consequently, the policies of many of the agencies support those of the lending institutions, development industry, and so forth. Real estate agents,

however, are in favor of any social trends or policies that promote high rates of homeownership and residential mobility.

As indicated previously, Palm (1985) has examined the degree to which the real estate industry in Denver is divided into submarkets along racial or ethnic lines. The specific hypotheses that she examines are that: (1) sellers choose real estate agents who are of their own (or similar) ethnic and racial group; (2) the sales area in which an agent operates is influenced by the race of ethnicity of the agent; and, (3) the network of business contracts with which the agent works and, therefore, obtains cross-listings and other types of cooperation from, is largely within the same general racial or ethnic group.

Using sample information and personal interviews, Palm demonstrates that each of these hypotheses is tenable. Furthermore, she argues that "sellers are not listing with an agency simply because it is located within the local neighborhood; they are expressing an ethnic preference in their behavior" (Palm, 1985, p. 64). In the case of sales area limitations, African-American agents far more than Hispanic agents recognize that their color severely limits the parts of the metropolis in which they may do business. Finally, the contact network within which agents work and receive information is strongly structured in terms of race and ethnic background. Anglo agents deal with other Anglo agents, African American agents deal with other African-Americans, and Hispanics with Hispanics. Given this racially biased and segmented structure, it is no wonder that submarkets remain firm and that social spaces persist as fairly homogeneous.

Landlords and Housing Submarkets

Landlords seek to obtain the highest returns from their capital investment. Harvey and Chatterjee (1974) have shown how the *principle of leverage* and the possibility for loans greatly influence the potential rate of return on a landlord's actual monetary investment. If landlords can borrow a large proportion of the needed capital investment, then the leverage on their own investment is great, and they can, in effect, charge fairly reasonable rents to receive an acceptable rate of return on their own monetary investment. However, if landlords have to put up a lot of their own money to purchase a property, as in older urban areas, the rate of return on their own investment is quite low, and the landlords will charge higher rents and perhaps skimp on maintenance in order to achieve an acceptable rate of return. Landlords may, consequently, create their own submarket of people, such as the elderly poor who will not complain and have no other choice than to live in highly subdivided multifamily housing. The principle of leverage thus adds a special *frisson* to the previous discussion of lending discrimination.

Landlords also make decisions as to which market they want to rent to. These decisions are based on effective rates of return to capital investment, the amount of maintenance required, and means by which the depreciation of the building may be controlled. For example, some apartment complexes with many amenities are available only for middle- and upper-income elderly, singles, and couples without children. The creation by landlords of submarkets of this type is obviously based on a perception of the increased noise and maintenance problems that may occur from families with children. Furthermore, in such situations the attractive facilities (pools, sauna, and games rooms) are rarely used. So depreciation on these lifestyle amenities is quite low. Thus, there may be an abundance of apartments available for middle- to upper-income elderly or no-children submarkets, but very few rental units available for families with children.

Students frequently ask how it is possible for landlords to discriminate in this way and create submarkets that are favorable to their particular interests. Certainly, a whole host of national and local acts and bylaws are in place to prevent discrimination. But in situations of high demand, landlords can do so by judicious use of the filters through which potential tenants have to pass in order to obtain rental accommodation: the application, the building superintendent, and the rental agency itself. For example, one maneuver adopted by landlords is to have standby "renters" (friends) who are willing to testify that they have just rented a unit if the landlord wishes to avoid renting to a particular applicant.

The Abandoned Housing Submarket

The abandoned housing submarket or nonmarket consists of housing units that have been withdrawn from the property market, and in which it appears that the owner is unable to return the dwelling unit to that same use. The problem of abandoned housing and business activities is focused in inner cities in many metropolitan areas, but is greatest in such cities as New York, Baltimore, Boston, Cleveland, St. Louis, Detroit, Newark, Chicago, and Washington, D.C. Although the number of abandoned dwelling units is difficult to measure, a visible sign is often boarded-up windows, particularly at lower floor levels. One of the spatial characteristics of abandoned housing and business activities is that the vacated properties are found in clusters in particular parts of central cities. This means that although in a city as a whole the actual number of abandoned properties might be quite low, in certain districts the concentration can be quite high. For example, in the City of St. Louis the number of boarded-up vacant housing units in 1990 was only 2 percent of the total housing stock, but 50 percent of these units were concentrated in twelve census tracts north of the downtown area and on the western edge of the city (Figure 12.6).

Causes of Abandonment   The question that arises is how and why abandonment occurs. Dear (1976) presents a persuasive explanation, which incorporates many of the concepts and trends we have discussed or referred to in this and preceding chapters. The explanation, presented in model form in Figure

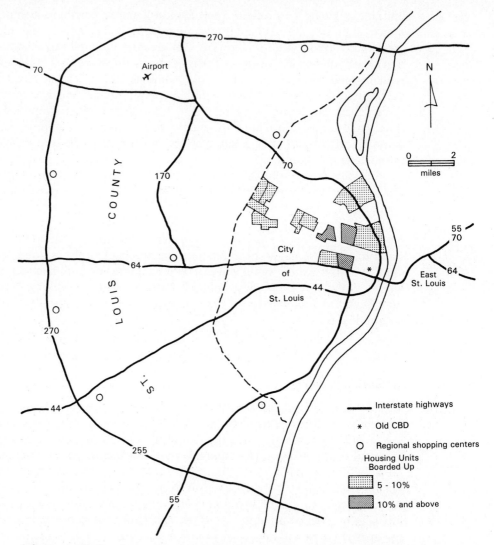

Figure 12.6 Concentrations of boarded-up housing units in the City of St. Louis.

(*Source:* Data from the Bureau of the Census, 1994, *1990 Census of Population and Housing, Census Tracts, MSA Reports.*)

12.7, suggests that driving mechanisms behind abandonment are the suburbanization of the population and economic activities, the concomitant decline in population of parts of many inner cities, and the decimation of the economic base with deindustrialization and restructuring in the public and private sectors. When suburbanization and population decline are coupled with high rates of unemployment in certain areas, individuals and households can no longer provide sufficient aggregate demand for the rents that landlords require to maintain the space that is available. In the case of the City of St. Louis, the pop-

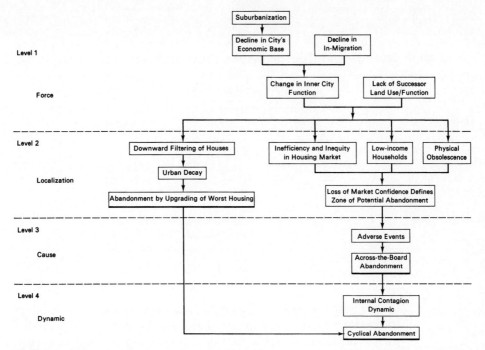

Figure 12.7   A diagrammatic representation of an explanation for abandoned housing and businesses.
(*Source:* Modified from Dear, 1976, p. 91.)

ulation declined by 36 percent between 1970 and 1990, and large income differences exist between the central city and the suburbs (Table 12.2).

It is suggested that there are two elements to the localization of abandonment in specific neighborhoods. First, the abandoned housing may be the lowest-quality housing in the oldest and least preserved part of the urban environment and, as such, forms the residue of the house-filtering chain. Second, the particular housing submarket for the poor is a factor, for the lowest-income households and individuals have insufficient income to purchase, improve, or pay rents that are adequate to encourage landlords to maintain buildings and prevent the spread of obsolescence. This lack of purchasing or renting power is accentuated by the kind of institutional practices discussed previously. The consequence is a decline in property values, which further discourages reinvestment. When values sink low enough, a zone of potential abandonment develops and provides an environment in which abandonment can occur and spread.

The actual cause of abandonment of a specific property is an adverse event occurring within the zone of potential abandonment that leads an individual or corporate owner to cease having any interest in the building even if it is in reasonable condition. This adverse event could be a fire or vandalism, but such adverse events outside this zone would not lead to abandonment. Once begun,

Table 12.2

Population Change and Central City–Suburban Income Disparities, St. Louis MSA

| | POPULATION | | PERCENT AFRICAN | PER CAPITA |
SUBAREAS	1990	1970	AMERICAN	INCOME
Central Cities				
St. Louis	396,685	622,236	47.5	9,718
E. St. Louis	40,944	64,500	98.0	5,835
Suburbs				
Suburban St. Louis Co.	993,529	951,671	13.1	15,654
Rest of MSA	1,012,941	790,969	23.2	11,130
Total MSA	2,444,099	2,429,376	17.5	12,655

(*Source:* Bureau of the Census (1991), Tables B and C.)

abandonment proceeds in a contagious manner, and further adverse events are no longer required, for the process has developed an internal dynamic of its own. This dynamic involves the property owners' negative perception of the housing market in the area, which may well be reinforced by the swift manner in which vacant units are soon stripped of all saleable items. Thus, the neighborhood can slip into a downward spiral of abandonment and decay. Property owners do not lose, however, for usually the original investment has been covered many times over during the years when it was rented. So it is not unknown for the insurance value to be realized consequent to the last rites of an arsonist.

Addressing Abandonment Some of the policy approaches recommended by Dear (1976) to address the issue of abandonment involve:

1. The establishment of a public or quasi-public corporation that is given the specific task of replacing the vacating landlords and property owners.

2. The provision by government of the necessary level of funds that are required for redevelopment and rehabilitation.

3. The implementation of a scheme of housing allowances for the poor in order to maintain some purchasing power in the marketplace for housing, so that low-income individuals and households can afford homes.

4. Preventing the spread of abandonment by moving swiftly to avoid the new abandonment of properties in particular areas (a so-called abandonment watch).

5. The development of community-oriented plans to rehabilitate areas of extensive abandonment, which may well imply a program for changing the economic structure and land use of much of the area.

This array of suggestions appears to have been followed, undoubtedly unknowingly, to various degrees in attempts to halt and reverse abandonment in the South Bronx.

Abandonment in the South Bronx In the 1940s, the South Bronx, located just across the Harlem River from Manhattan, was a middle-income mainly white-ethnic high-population-density area linked by excellent public transit facilities to Manhattan and the other boroughs of New York City. During the 1950s and 1960s, a considerable proportion of the population, particularly those households involving younger families, or in the family formation stage of the life-course, moved to the suburbs. They were replaced, in part, by a lower-income population representing a diversity of racial and ethnic groups. By the mid–1970s, the pace of change and decentralization had been such that abandonment had become widespread—1,500 buildings were abandoned or torn down, 500 acres of former industrial, commercial, and residential land lay waste, and 30 percent of the residents were registered as unemployed. Gang wars, drug wars, arson, decay, and homelessness were endemic (Carey, 1976).

The first reaction was for the federal government to propose a massive injection of public funds for public housing. However, this was not accepted by many local politicians, planners, and social activists on the grounds that such projects exacerbated the problem by creating further concentrations of the poor. About $2 billion in public funds have been spent on renovating and building new infrastructure and public facilities, such as several new schools, a two-year college (Hostos Community College), courthouses, an art museum, and a proposed New York City Police Academy. The major change has, however, involved pockets of grassroots neighborhood-based housing redevelopment schemes, some part of the community reinvestment movement described previously. The location of these schemes is indicated in Figure 12.8. These groups, such as the MBD (Mid-Bronx Desperadoes) Community Housing Corporation with the Charlotte Gardens project, have by the mid–1990s built 4,000 units of affordable new housing and reclaimed 22,000 units of old but structurally sound abandoned tenement housing (McKee, 1995).

Homelessness

Residents and workers in downtown areas of North America's largest metropolises encounter the homeless every day. The homeless are often seeking financial assistance at crossings or doorways, squatting by vents, under bridges, eating and sleeping in the basements of local churches, or the Salvation Army home, dodging in and out of the warm malls, often following well-defined daily mobility patterns (Wolch et al., 1993). Although nobody knows the exact number of homeless persons in the United States or Canada on any one night, because the population is too transient to estimate with any degree of acceptable accuracy, the reality is that the number is large, apparently increasing, with needs that cannot be ignored (Baum and Barnes, 1993; ABS, 1994). Most congregate in the downtowns of large metropolitan areas. Rossi (1989) estimates that

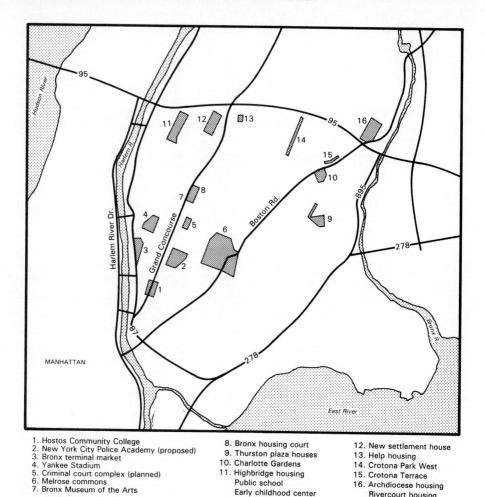

1. Hostos Community College
2. New York City Police Academy (proposed)
3. Bronx terminal market
4. Yankee Stadium
5. Criminal court complex (planned)
6. Melrose commons
7. Bronx Museum of the Arts

8. Bronx housing court
9. Thurston plaza houses
10. Charlotte Gardens
11. Highbridge housing
    Public school
    Early childhood center

12. New settlement house
13. Help housing
14. Crotona Park West
15. Crotona Terrace
16. Archdiocese housing
    Rivercourt housing

Figure 12.8 The location of pockets of redevelopment and abandoned housing and new institutional public infrastructure in the South Bronx, New York City. (*Source:* Redrawn from McKee, 1995, p. 87.)

each night in the United States as many as 400,000 to 500,000 people may be homeless, that is, do not have regular access to a normal dwelling, and another 4 to 7 million persons are so poor that their housing is so transitory that they may well suffer periodic homelessness. The number of homeless and precariously housed in Canada are proportionately similar (Dear and Wolch, 1993).

Given the caveats concerning the difficulty of surveying the homeless population, a United States Conference of Mayors (1990) report estimates that one-half of the homeless are single men, one-third families with children, one-eighth single women, and a small proportion (3 percent to 5 percent) unaccompanied youth. Almost one-half are African American, one-third white, one-sixth Hispanic, and a number Native Indians particularly in some western cities such as Winnipeg (CCSD, 1987). About one-third have some form of diagnosed mental instability, and a similar proportion are substance abusers. Dear and Wolch (1993) note that today's homeless are in many respects different from those of thirty years ago: There are more young people. Whereas thirty years ago homeless men were on average 36 to 44 years of age, today they are more likely to be 18 to 24 years old. There are also more women and children, which are the fastest-growing group among the homeless (Klagge, 1994). Finally, there are also more mentally disabled and more substance abusers. Given the wide range of individuals involved, it is not surprising that the reasons for homelessness are quite complex.

Reasons for Homelessness Although there are many intertwined reasons for homelessness, three may well account for most of the persons involved.

1. In Chapter 10 we noted that changing public policy with respect to mental health care, associated with newer forms of chemical therapeutic treatment, a clearer legal perception that institutionalizing people against their will infringed on civil liberties, and a desire to reduce costs, had resulted in large-scale deinstitutionalization and the closing of many mental care hospitals. Unfortunately, many of the touted chemical therapies have proven inadequate, and few communities had, or wished to develop, facilities adequate to cater for those requiring treatment and support. By the 1990s, a significant proportion of individuals on the streets are the *uninstitutionalized,* those that would have been placed in hospitals previously. Isaac and Armat (1991) recommend making involuntary institutionalization of the mentally ill easier in order to save them from further exposure to life on the streets.

2. A large number of the homeless are unemployed, young single males whose lack of education and training makes them totally unsuited for today's restructured, high-tech, postindustrial economy in which *menial jobs are scarce.* Wright (1991) comments that years ago such men would have hung around downtown skid-row flophouses, grabbing day labor jobs unloading cargo, or helping in construction. In the 1990s, these types of jobs are in scarce supply. The observation that liberal free-market economies, with their cyclical employment swings and episodic restructuring, do not do well catering for those at the extreme edge of the labor force suggests that more conscious public efforts at appropriate job creation are required (Lang, 1989).

3. The argument by Rossi (1989) that there are too many poor chasing too few housing units highlights how a shortage of supply of appropriate *affordable housing* for the poor, and those precariously hovering around poverty, particularly affects homeless families. This shortage of affordable housing for the homeless is related to the inadequacy of rent subsidies, cutbacks in the supply of public housing, and the difficulty that private (slum) landlords face in meeting the market (despite offers of local government subsidies) due to local housing regulations which have raised building standards, and therefore costs of renovation, and prohibited excessive room subdivision and cubicle hotels (Tucker, 1990; Jencks, 1994).

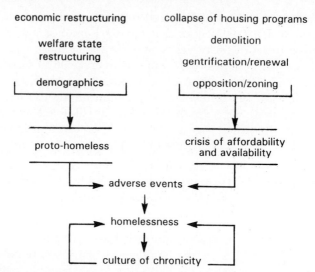

Figure 12.9 A path to homelessness.
(*Source:* Dear and Wolch, 1993, Fig. 16.1, p. 301.)

A special problem occurs with the unaccompanied young, many of whom are teenage runaways from dysfunctional family situations, who are often easily lured into lives involving drugs and prostitution (Ruddick, 1995). Shelters for the young, with counseling and health services, are a high priority. So is swift action by counselors to get to them early before the pimps, for example, as they arrive at bus stations.

Ameliorating Homelessness Wolch and Dear (1993) suggest that in the path to homelessness (Figure 12.9) two crucial stages are particularly sensitive and require special attention. The first is the *adverse event* which drives an individual or family into homelessness. This can be an eviction from the current residence; domestic violence, which, in the absence of knowledge about the location of interval houses or other support services, can drive women (and children) into a homeless condition; and loss of job or welfare support. An "early warning" help-line, providing vital information and help that people need in these types of stressful situations, and a system of temporary shelters can stall a slip into homelessness. The second is prevention of the development of a *culture of chronicity* through maintenance of simple social support networks and emergency shelters that provide food and washing facilities, mail-drop services, welfare information, and telephones (Laws, 1992).

One feature to be avoided in this provision of support services is the concentration of the services in one area so that it becomes a contained homeless ghetto, as happened in Los Angeles in an area adjacent to the central business district (Figure 12.10). This containment occurred as a result of a compromise between the city's Community Rehabilitation Agency, which wished to keep the homeless out of the CBD, and homeless activists who wanted more homeless services, transitional housing, health care, mental health care, alcohol and

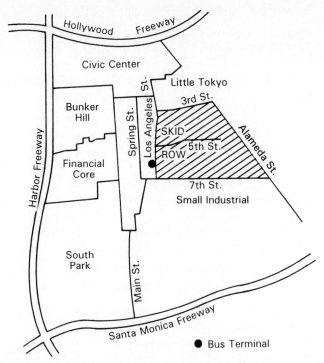

Figure 12.10 An attempt at spatial containment of the homelessness in downtown Los Angeles.
(*Source:* Redrawn from Goetz, 1992, Figure 1, p. 543.)

drug rehabilitation clinics, a less hassled environment, education and job training, and employment services (Goetz, 1992). Such containment by local governments may be tempting because it provides a focus for scarce public assistance resources, but it also has the effect of creating a self-perpetuating community which, like the experience of other kinds of ghettos elsewhere, has insufficient space and resources to cater for the population that gravitates toward it.

APPROACHES TO URBAN REVITALIZATION

This concluding section commences with a general discussion of urban redevelopment, and then focuses on gentrification and incumbent upgrading as examples of specific types of urban revitalization. In a classic study of urban growth in the New York metropolitan area, Hoover and Vernon (1959) hypothesized the existence of a five-stage evolutionary model of urban redevelopment based on the experience of older metropolises such as New York City. This model has proved to be useful for the examination of redevelopment in a number of North American cities (McCann and Smith, 1981). The evolutionary model of redevelopment can be summarized as follows:

Stage 1 is *residential development* in single-family houses.

Stage 2 is a *transition stage* in which there is substantial new construction and population growth, but in which a high and increasing proportion of the new housing is in apartments, so that average density is increasing. Much of the apartment construction replaces older single-family houses.

Stage 3 is a *down-grading stage,* in which old housing (both multi-family and single) is being adapted to greater-density use than it was originally designed for. In this stage, there is usually little actual new construction, but there is some population and density growth through conversion and crowding of existing structures.

Stage 4, the *thinning-out stage,* is the phase in which density and dwelling occupancy are gradually reduced. Most of the shrinkage comes about through a decline in household size. . . . But the shrinkage may also reflect merging of dwelling units, vacancy, abandonment, and demolition. This stage is characterized by little or no residential construction and by a decline in population.

Stage 5 is the *renewal stage,* in which obsolete areas of housing are being replaced by new multi-family housing (Hoover and Vernon, 1959, pp. 185–198).

Redevelopment will occur, therefore, either as a consequence of the pressures of demand (in Stage 2) or as a result of deterioration (in Stage 5). The South Bronx would appear to have experienced all these stages. Similarly, the Seattle Commons project, a grassroots scheme to redevelop a run-down inner-city district to create a 25-acre park, around which would be developed a new neighborhood, is a visionary scheme for wholesale community revitalization.

Redevelopment

Redevelopment, as differentiated from incumbent upgrading and gentrification, is frequently characterized by a combination of a number of changes (Table 12.3). These changes can include land use, structures, types of residential units, per capita income, social status, and stage in the life course of the population. The land use may change from residential units to commercial or multifamily structures. Areas previously commercial or in manufacturing may be converted to higher-density commercial uses or high-density residential activities. The succeeding population frequently consists of individuals or two-person households in either the young adult or late mature stages of the life-course. There is, therefore, a complete change in the neighborhoods involved, in which the original inhabitants are displaced. These previous inhabitants are invariably at the lower end of the income spectrum. The benefits have accrued to the wealthier succeeding populations, whereas the burdens of funding new locations in which to live and the costs of new housing have fallen on the poorer segments of society. But redevelopment may well have created additional employment opportunities while in process.

Table 12.3

Comparison of Change Characteristics among Neighborhoods Undergoing Redevelopment, Incumbent Upgrading, and Gentrification

CHARACTERISTICS OF CHANGE	REDEVELOPMENT	INCUMBENT UPGRADING	GENTRIFICATION
Social status	X		X
Per capita income	X		X
Land use	X		
Structures	X		
Housing type	X		
Population density			X
Persons per room			X
Stage in life-course	X	X	

Table 12.4 provides a conceptual framework summarizing various features of inner-city redevelopment as they have occurred in North American inner cities since 1950. The 1950–1970 phase, devoted primarily to *urban renewal* usually involved federal or state/provincial partnerships with the housing authorities of local governments for primarily residential renewal. Some of the worst features of these projects—blocks of high-rise subsidized housing— have been, or are being, removed. A *neighborhood redevelopment* phase, concerned primarily with transferring vacated inner-city industrial and transport property to residential and mixed uses, has lasted from about 1965 to date. These types of redevelopment invariably involve redevelopment agencies and public sector funds.

Commencing in about 1970, redevelopment in waterfront and downtown areas began to occur with *public-private partnerships* in which the redevelopment agency, in conjunction with the private developer, takes the lead. These public-private partnerships may take the form of *universal schemes* or *particular projects*. Examples of universal schemes are successive housing acts in the United States after 1970 that made the redevelopment of decaying parts of cities more attractive to private developers because of the involvement of public funds. The outcome, however, has invariably been displacement of the poor, and construction of some middle- and upper-income housing units, and many more sterile civic centers, convention halls, office buildings, and high-priced hotels. Specific schemes, such as those relating to Quincy Market (Boston) and phase II of Battery Park (New York) have been more successful because of their focused nature.

During the 1990s, much redevelopment of old railway lands has been occurring. As much of this land is owned by railway companies, the leadership has come from the private sector, with developers and local planning agencies

Table 12.4

Phases of Redevelopment in North American Inner Cities

	URBAN RENEWAL	NEIGHBORHOOD DEVELOPMENT	PUBLIC-PRIVATE PARTNERSHIP	PRIVATE-PUBLIC PARTNERSHIP
Time Period	1950–1970	1965–1990s	1970–1990s	1990+
Project Type	Slum clearance	New precinct	Waterfront/CBD	Railyards
Typical Projects	Regent Park [Toronto]	St. Lawrence [Toronto]	Harbourfront [Toronto]	Railway Lands [Toronto]
	West End [Boston]	False Creek [Van.]	Quincy Market [Boston]	False Creek N [Vancouver]
	Newark [NJ]	Battery Park City I [NY]	Battery Park City II [NY]	Mission Bay [San Fran.]
				Riverside South [NYC]
Land Use Change	Residential to residential and institutional	Industrial/Transport to residential/mixed	Industrial/Transport/ Commercial to mixed	Transport/Industrial to mixed use
Leadership	Housing authority	Redevelopment agency	Redevelopment agency & private sector	Private sector
Early Capital	Public sector	Public sector	Public & private	Private sector
Change in Ownership	Private to public	Public to public	Public to private	Private to private
Plan Responsibility	Authority & city	Agency	Agency & developer	Developer & city
Planning Mode	Master plan	Site plan & urban design guidelines	Design guidelines & site plan	Site plan & urban design guidelines
Key Success Factors	Public grants	Public grants	Private investment	Land ownership
	Expropriation power	Expropriation power	Expropriation power	Private investment
			Public borrowing	Political approval
Problems	Displacement	Limited public funds	Uncoordinated development	Loss of public control
	No market response		Few social benefits	Few social benefits

(*Source:* Table provided by Professor David L. A. Gordon, School of Urban & Regional Planning, Queen's University, Kingston, Canada.)

working together on site and design plans in a *private-public partnership*. Even though the leadership has come from the private sector, local planning agencies play a considerable role because many of the land use changes and surrounding infrastructure developments that are required need political approval. Furthermore, although the early capital is often derived from the private sector, later capital may be derived partly from the public sector for particular elements of the overall design, such as transit or public parks.

Waterfront Redevelopment   Waterfront lands have been a particularly important area for redevelopment because, since the 1960s, when technological changes such as containerization made old port facilities obsolete, these lands have often sat idle. Furthermore, in North America, these lands have also been adjacent to old downtown areas that have often been in need of redevelopment. In a discussion of waterfront redevelopment projects in New York (Battery Park City), Boston (Charlestown Navy Yard), and Toronto (Harbourfront), Gordon (1996) notes that although political aspects of these redevelopments may be extremely important, their success depends a great deal on an incremental urban design practice within the overall envelope of an approved master plan (Savitch, 1988).

This is because these developments take a long period of time during which tastes and needs can change over the course of large-scale projects. Gordon notes that Battery Park City has been generally considered a success because, although it was a large project, it was completed in clearly designed phases within an overall internal transport framework. This avoided the construction camp appearance of many projects, allowed existing buildings to be reused where possible, and integrated existing uses at street level with the newer ones. Although much of Battery Park City is in "prestige" uses, such as the World Financial Center offices, and middle- and high-income residences, general public uses are assured through the provision of parks, sites for a winter garden, Stuyvesant High School, a Holocaust museum, and access to the waterfront.

Distributing the Benefits of Redevelopment   The issue with redevelopment, therefore, is not so much that it occurs, but who benefits when it does occur. The evidence is that too often lower-income individuals and families are displaced and have to relocate in housing situations that represent no improvement on those they have left, whereas higher-income households and commercial developers benefit (Fainstein et al., 1983; Fainstein, 1990). However, Schill and Nathan (1983), in a study of nine revitalizing neighborhoods in five cities, contend that when forced to leave and find new accommodation, lower-income households on the whole improve the quality of their housing. The basic point, therefore, is that although redevelopment is obviously highly necessary in certain parts of most North American cities, it must be undertaken in such a way as to provide benefits for all segments of the urban community, particularly those households that are directly affected.

Gentrification

Although the term *gentrification* seems to have been coined by Glass (1964) with respect to a type of revitalization in London involving the upper-income takeover and redevelopment of Victorian mews, it is also commonly used to describe a process that came to widespread public attention with respect to North American cities during the 1980s (Palen and London, 1984). *Gentrification* refers to a process involving an influx of upper- and middle-class households into an area of old homes that were previously occupied by lower-middle and low-income individuals and households. The Victorian mews had previously been occupied by horses and stable personnel.

The gentrification process thus generally occurs in a small physical area, usually involving old homes that when renovated have charm, happen to be close to a place of historic (e.g., Society Hill, Philadelphia) or amenity interest, usually have good access to employment opportunities in a traditional central business district, and are invariably adjacent to high-socioeconomic-status districts (Beauregard, 1990; Ley 1992). The reason why gentrification has attracted a great deal of interest is that many people concerned about the health of old inner cities hope that gentrification heralds a larger-scale movement of people back to the inner city (Bourne, 1993a; Badcock, 1993).

Theoretical Explanations of Gentrification   The many case studies of gentrification emphasize that there may well be a number of different types of explanations of the gentrification phenomenon that vary in importance according to the stage in the economic and building cycles (Ley, 1992; Lees and Bondi, 1995). These explanations may be grouped into five categories:

1. *Economic causes*. A number of studies have emphasized that at the peak of building cycles the cost of new housing at the periphery of the city becomes so high in many North American suburbs that equivalent inner-city housing space seems cheap by comparison (Smith and Schaeffer, 1986). Thus, a number of middle-income households may respond to this price, or rent gap by moving into gentrified properties. Actually, this underlies one of the reasons why Canadian inner cities contain more middle-income families than those in the United States, for the cost of new housing in Canadian suburbs has invariably been higher than similar new housing in United States suburbs.

2. *Demographic explanations*. One observation from a number of case studies is that gentrifiers appear to consist of younger, higher-income professionals in less conventional households than the nuclear family (Ley, 1994a). The numbers in this group have swelled as the "baby boomers" are now reaching middle age, thus creating a significant demand for well-located residential units.

3. *Economic base explanations*. The cities that have experienced the greatest amount of gentrification appear to be those that have a high percentage of their labor force in financial, business, and associated commercial activities (Lees, 1994). Such cities invariably have strong financial and commercial

cores and a large, well-paid labor force, a proportion of whom may seek downtown residential locations.

4. *Lifestyle explanations.* A larger number of people in North America may be becoming more pro-urban and pro-downtown, consciously choosing the cultural, social, and recreational amenities of the city over those of the suburb and exurbia (Zukin, 1989).

5. *Political economy explanations.* These emphasize the interplay between capital, labor, and the state. Smith (1979) argues that landowners and other financial interests follow a policy of neglecting the inner city (disinvesting) until such time as local governments may change local zoning, housing by-laws, and regulations sufficiently to stimulate reinvestments that yield tremendous profits.

London and colleagues (1986) examined these five explanations for gentrification in forty-eight cities in the United States and concluded that while all, to various degrees, were quite tenable, the lifestyle explanation was particularly persuasive.

The Impact of Gentrification The impact of gentrification on North American cities has not been that great. The few examples that have occurred have created enormous interest, but the numbers of people involved are quite small except in a few cities such as Vancouver, Philadelphia, New York, San Francisco, and Toronto (Bourne, 1993b). The evidence indicates that a back-to-the-city movement in the United States has not really occurred, and the income gap between the suburbs and the old central cities still exists (Ledebur and Barnes, 1993). Many of the examples of gentrification discussed in the literature appear to involve either redevelopment (the tearing down of housing and the construction of different types of residential units), or incumbent upgrading, which is discussed in the next section. A few Canadian cities, such as Vancouver and Toronto, have experienced some inner-city redevelopment and gentrification. In general, the focus on gentrification may divert research away from the analysis of the economic, social, and political bases of poverty which persists as the most serious policy problem (Bourne, 1994).

An important issue, however, relates to the benefits and costs of gentrification. The conclusion seems to be that with gentrification city government may well benefit as a consequence of revitalization and a regeneration of the tax base; the property industry usually always benefits as a consequence of capital gains and tax breaks; and the gentrifiers may benefit through capital gains and locational attributes, but also bear financial costs and locational risks (Ley, 1994b). In contrast to that, the displaced bear the burden of seeking new residences which may, in some cases, result in an improvement in housing conditions but in others lead to no improvement but only to higher rents. Robinson (1995), for example, documents the lengthy struggle of community activists in San Francisco against the wholesale gentrification and redevelopment of the

Tenderloin district on the southern edge of the central business district in an attempt to persuade the local redevelopment agency to include a more neighborhood-sensitive vision. This vision embraces a greater mix of uses and structures to maintain some affordable housing and social services for lower-income households.

Incumbent Upgrading   According to Clay (1979), incumbent upgrading occurs primarily in lower-income neighborhoods where considerable renovation occurs to the existing housing stock, but the residential units remain occupied by the existing community. There may be some change in the age structure of the neighborhood as younger households replace older ones, or families return to the "old neighborhood." The types of neighborhoods in which these types of changes generally occur are invariably those that have a strong cultural or ethnic base, with established families who wish to reside close to traditional churches, social facilities, and shopping. As can be seen in Table 12.3, incumbent upgrading does not include the kind of social, income, and residential unit changes that are usually involved in either gentrification or redevelopment.

One of the major differences between this type of revitalization and gentrification or public-private redevelopment is the capital base for the renewal. The big disadvantage faced by lower-income neighborhoods is that capital frequently flows from the area, and new capital often avoids the area. There tends to be disinvestment as the houses and general infrastructure are allowed to deteriorate. The task, therefore, is to promote the reinvestment that is needed for individual homeowners, and renters, to upgrade their immediate environment. This can often only be achieved through the lobbying efforts of local groups, such as community reinvestment organizations, to persuade local governments to improve public services and security and to generate a flow of public and private funds for renovations. These districts often face a form of quasi-redlining. So governmental action is frequently required to ensure a flow of funds (Holcomb and Beauregard, 1981, pp. 46–50). The study by Murdie (1986) demonstrates quite clearly the way in which new immigrants to Toronto have had to use private funds to upgrade the houses in the neighborhoods in which they have settled.

CONCLUSION

Thus, the reasons why people and households locate where they do are extraordinarily varied and quite complex. At a general level, the price–space commuting cost trade-off envisaged in residential location theory appears to be tenable with respect to polycentric as well as multicentered cities. In this context, supply and affordability issues are paramount, as access by households to adequate housing is vitally important for maintenance of a healthy population and nurturing of the young. The structuring of housing submarkets by race, ethnicity, and governmental, financial, and property-related institutions, however, suggests that the choices exercised by households are restricted by many circumstances peripheral to affordability. Furthermore, even with the vast array of public supply- and demand-side programs implemented during the past fifty years, the housing of the poor and those in poverty remains the issue of central concern. This chapter has indicated the many things that society has to do to ameliorate housing problems associated with poverty and inequality.

13

Women and Urban Form and Structure

The previous discussions of the waves of urban development, and the ways that these have been translated into urban form and structure, have occurred largely within a gender-blind framework (MacKenzie, 1989). Though some discussion has focused on the effect of dominance-subordination relationships between social groups as manifest, for example, in redlining, it has not embraced the fact that North American urban development, as elsewhere in the world, has occurred within the structure of a patriarchal, or male-dominated society (Walby, 1990). Thus, while it has been emphasized that North American urban areas arise out of, and reflect, the changing economic, technological, and population bases that underpin them, the impact on urban development of the essential control by men of these levers has not been considered (Bondi, 1990). The issues to be addressed in this section include, therefore, how urban development can be interpreted as reflecting the patriarchal structure of society, the ways in which urban form and structure may help perpetuate gender dominance-subordination, and some of the changes that are needed to make North American urban areas more liveable for both women and men.

It is important to understand at the outset that the concept of patriarchy is complex because it involves socially constructed boundaries that evolve unevenly in time and operate differently at the same time in different places and among different classes (Halfacree, 1995). These socially constructed boundaries include those that separate men from women spatially, require them to behave differently, attach different meanings to similar behaviors, and cause appearances, personality traits, and attitudes to be interpreted in different ways. Gender can, therefore, be seen "as a lifelong process of situated behavior that both reflects and reproduces a structure of differentiation and control in which men have material and ideological advantages" (Ferree, 1990, p. 870). In times past, these material advantages were enshrined in patrilineal inheritance practices. In North American culture, however, unlike some others, patriliny has been a declining custom (Moser and Peake, 1994). The key loci of situated behaviors are the workplace and the home, with sex roles traditionally defined by each. Indeed, Walby (1990) argues that the workplace and domestic spheres are the chief places in which male control is exercised. Hence, it is not at all surprising that these are the spheres in which contemporary social changes are most visible, *and* contentious!

CHANGE IN HOME AND WORK

Traditional social area analysis and factorial ecology provide clear examples of the way in which dominance-subordination, and sex-role stereotyping, were perpetuated in the social sciences. Family status was originally defined in terms of female participation in the waged labor force and size of family, the assumption being that low female labor participation rates were associated with greater numbers of persons in the family. The social rank or economic status of the family was defined primarily in terms of male employment characteristics (Pratt and Hanson, 1988). Similarly, factorial ecology studies do not differentiate the data on the basis of gender and, as a consequence, infer female social spaces from male-oriented data (Davies and Murdie, 1991). Randall and Viaud (1994), on the other hand, using gender-specific information for Saskatoon, suggest the corollary to this: Male social spaces may be inferred from female-oriented data sets. They also note the presence of nonstereotypical gender relations in households where women are affluent career and white-collar workers, or single parents. The analysis of urban social space is, therefore, being influenced by the increasing involvement of women in the paid labor force and the changing nature of families.

Women in the Paid Workplace

One of the chief ways in which female dependency has been maintained is through the denial or the limitation of employment to women in the paid labor force. Such denial or limitation not only increases dependency on a (usually male) wage earner, it also curtails the range of benefits, such as private pension plans and work-related social contacts available to women. Employment changes or interruptions due to childbearing and child raising have invariably been means by which this dependency has been reinforced. Thus, women have historically had lower participation rates in the paid labor force than men, and earned less. Since 1975 this situation in both Canada and the United States has changed, and would appear, at first glance, to have improved significantly (Ghalam, 1993; Steinbruner and Medoff, 1994; Best, 1995).

For example, there are now proportionately more women in the United States and Canada in waged employment, and fewer men. In 1990, about 45 percent of all waged workers in both countries were women. The data for Canada indicate that the share of women in the labor force has been increasing, in spite of the sharp downturn in employment participation rates exhibited in Figure 13.1 (related to the 1989–1992 recession followed by an economic recovery that has generally been weak with respect to job creation). Increases in employment were particularly pronounced among women of ages 25 to 54. The decrease in male employment was related primarily to downsizing in the goods-producing sectors and induced early retirement among men 55 to 64 years of age.

But the waters are still muddy. In spite of the increase in the numbers of women in the labor force, women are still more likely than men to work in lower-wage, less secure, part-time employment. In both Canada and the United States, women have fairly consistently accounted for 70 percent of all part-time workers. Furthermore, the number of part-time jobs has been increasing at a greater rate than full-time employment. About one-third of both women and men working part-time are doing so because they can only find part-time jobs (Figure 13.2). Only 32 percent of women working part-time in Canada are doing so because they do not want to work full-time, and 11 percent are doing so because of family responsibilities.

More than 70 percent of women employed have jobs in traditionally female-dominated occupations—teaching, nursing or related health care occupations, clerical, sales, and personal service (Figure 13.3). Similar levels of *occupational segregation* exist in the United States where historic data suggest this has been a remarkably persistent phenomenon throughout the twentieth century. Nevertheless, there have been evident gains in such areas as management and administration, where womens' share of employment increased from 29 percent in 1982 to 42 percent in 1993 (Statistics Canada, 1994, Table 2.8).

Most women work in the service industries, and in 1993 predominated with 53.3 percent of service sector jobs (Figure 13.4) as compared with only 23.9 percent of goods producing jobs. While the proportion of women employed in the generally higher-paying goods producing sector as compared with the service sector has decreased (from 18.4 percent in 1975 to 14.0 percent in 1993), their share of total employment in that sector as compared with men has increased (from 19.6 percent in 1975 to 23.9 percent in 1993).

Although more married women are working—there has been a rapid growth in employment of mothers since 1975—single mothers are far less likely to be in paid employment. Although Figure 13.5 indicates that the proportion of women with children younger than 16 years in paid employment in Canada increased from 50 percent in 1981 to 63 percent in 1993 (with similar increases in participation rates for mothers with young children), only 52 percent of single mothers with children of 16 years and younger were employed as compared with 65 percent of mothers in two-parent households (Ghalam, 1993, p. 4).

Thus, although the general increasing participation of women in the paid

Figure 13.1 Paid labor force participation rates by gender, Canada, 1975–1993.
(*Source:* Redrawn from Best, 1995, p. 31.)

labor force indicates decreasing dependence, the perpetuation of a wage gap provides some evidence of its stubborn persistence. Despite the increase in women in the paid labor force, the earnings of women employed full-time on a full-year basis are, in general, 70 percent those of comparable men. This proportion has been increasing only slowly since 1975 (Figure 13.6a). In some professional areas, such as teaching, the wage gap between men and women is quite narrow, whereas in others it is quite large (Figure 13.6b). There is evidence to indicate that for specific subsets of the population, for example, college-educated women between the ages of 25 and 34, earnings have narrowed to 90 percent of those of men in comparable positions, but for other age/educational achievement groups the differential remains large (Steinbruner and Medoff, 1994). The large gap in managerial, administrative, and personal service occupations indicates not only that some women may be receiving less income for comparable work, but also that men in these areas continue to occupy the more senior positions. Although this managerial wage gap may be interpreted as age related, with the male group being of greater average age and therefore positioned higher in salary profiles, it is more likely reflective of the entrenched position of men in senior management.

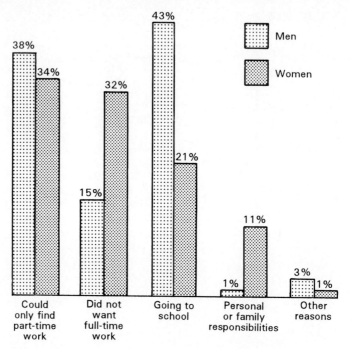

Figure 13.2 Reasons for working part-time by gender, Canada.
(*Source:* Redrawn from Best, 1995, p. 31.)

Families and Households

Female dependency in the home has been maintained largely through the traditional structure of the family, which has historically involved a domestic role for women dependent on male income. This historic situation, however, has changed considerably during the past few decades with greater levels of female economic independence, making it possible for more women and men to pursue a greater variety of lifestyles. The nature of these changes can be inferred from the information on life-course events that are summarized in Figure 13.7. The data in Figure 13.7 are derived from a 1990 questionnaire survey of major life-course events experienced by a random sample of 13,495 women and men ages 15 and above in Canada (Rajulton and Ravanera, 1995). The survey specifically requested information concerning dates of *possible* life-course events that could be experienced by an individual. There was no prejudgment concerning the particular events which a person might experience.

This range of possible events include: date of the respondent leaving home (HL); respondent's first or only marriage (FM); respondent's first cohabitation (FC); respondent's first separation/divorce (FS); death of the respondent's first spouse, or first widowhood (FW); respondent's second marriage (SM); birth of the respondent's first child (FB); birth of the respondent's last child (LB); respondent's first child leaving home (FL); and respondent's last child leaving home (LL). The occurrence of these events is tabulated for females and males

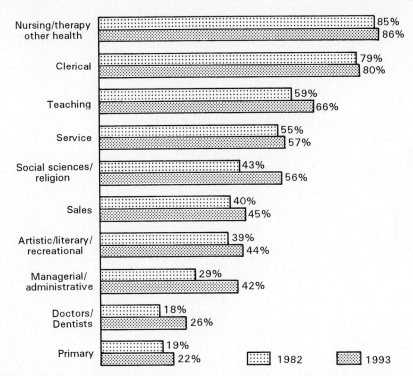

Figure 13.3 Women as a percent of persons employed in selected occupations, 1982 and 1993.
(*Source:* Redrawn from Best, 1995, p. 32.)

grouped by ten-year age cohorts by the respondent's date of birth. Two of these cohorts, one consisting of those born in the 1930s and the other for those born in the 1950s, have been selected for comparison. The older cohort represents the population that was making social, economic, and residential location decisions, and hence affecting urban form in the 1950s and 1960s; the younger cohort may be viewed as the first set of cohorts' children (the "baby boomers") who have been making socioeconomic and locational decisions and affecting urban form in the 1980s and early 1990s.

The cumulative proportion graphs suggest a number of changes in the family (and, by inference, household) structures and life-course events that may well be related to the changing economic position of women *vis-a-vis* men:

1. A larger proportion of women and men married by the age of 35 in the older group than in the younger group. At age 35, 92 percent of women born in the 1930s had married, whereas only 81 percent of the 1950s cohort had married by that age (Table 13.1). Furthermore, whereas the marriage curve for the male 1930s cohort continues to increase so that by age 45 about 90 percent of men are married, the curves for both males and females in the 1950s cohort tend to level off after age 35. Thus, it would appear that while a

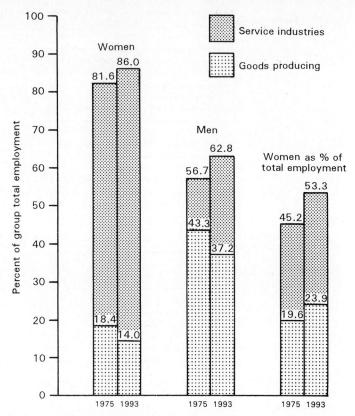

Figure 13.4 Percent distribution of total employment by gender for service and goods producing industries, Canada, 1975 and 1993. (Note that each column sums to 100%).
(*Source:* Compiled by the author from Statistics Canada, 1994, Table 2.7.)

transition to marriage was a fairly normal occurrence for the 1930s group, almost 20 percent of women and 30 percent of men in the 1950s group have not yet been married.

2. As a consequence, marriage after leaving home (or following a separation/divorce) is far more likely for women (33 percent) and men (38 percent) in the 1950s cohort than for those in the 1930s group (Figure 13.1).

3. Although the probability of separation/divorce increases with age for both women and men in the 1930s cohort, so that by age 55 19 percent of both women and men in the group have been divorced once, higher rates of separation/divorce occur at a much younger age for the 1950s cohort. Although this is undoubtedly related to the greater legal ease of divorce in Canada in recent decades, it can also be related to the greater number of two-earner households which makes separation/divorce more feasible for both parties. Particularly noticeable is the much higher probability of separation/divorce for women at age 35 (18 percent) than men (12 percent). It would seem that women with jobs and independent access to financial income are more

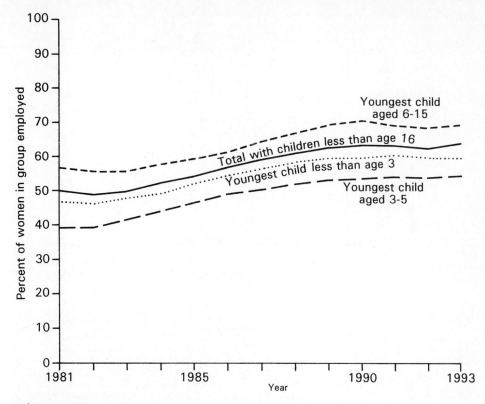

Figure 13.5 The changing percent of mothers in waged employment, Canada, 1981–1993.

(*Source:* Compiled by author from data in Statistics Canada, 1994, Table 6.4.)

likely to terminate a distasteful marriage than those who are more depen-dent on a husband's income (Bergmann, 1986).

Changes in the statistics between the younger cohort and the older cohort demonstrate the changes that are occurring in marital status, partially due to the increase in independence of women in the urban environment.

Rajulton and Ravanera (1995) suggest that the traditional, or typical, se-quence of transitions—leaving home, first marriage, first birth, last birth, first launching, widowhood—is being experienced by a smaller proportion of the population. Other sequences—such as leaving home, cohabitation, first child, first marriage, second child, or leaving home, cohabitation, death of compan-ion, or leaving home, cohabitation, first marriage, first child, first separation/di-vorce, second cohabitation, second child, second marriage—are becoming more common. These varieties of families and households, all of which involve home-workplace commuting for invariably two adults in the unit, imply much greater complexity of living arrangements, transportation needs, and types of residential units. The increasing variety of types of life-course events in recent decades thus imply significant changes in the ways that people live or may wish to live in cities.

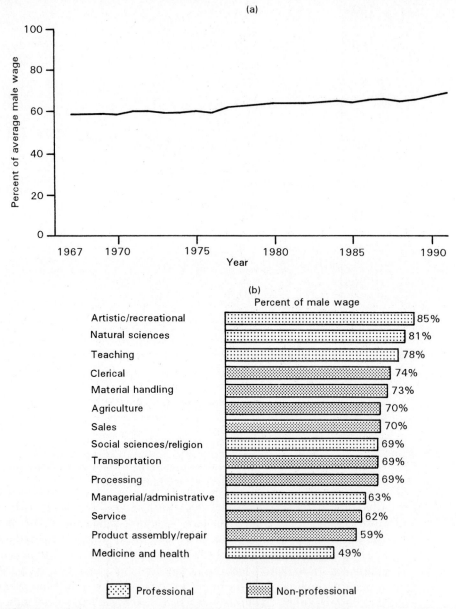

Figure 13.6 Earnings of women in waged employment on a full-time full-year basis as compared with those of men similarly employed, Canada, (a) aggregate comparison; and (b) comparison by occupation, 1991.
(*Source:* Redrawn from Ghalam, 1993, p. 5.)

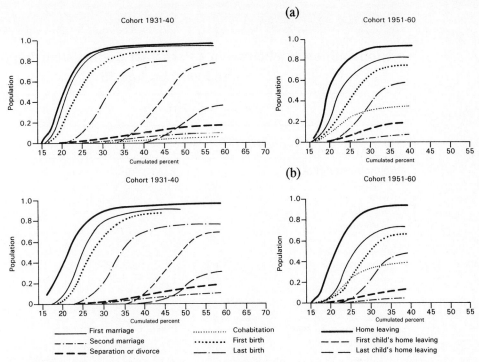

(a)

Cohort 1931-40

Cohort 1951-60

(b)

Cohort 1931-40

Cohort 1951-60

———— First marriage	············· Cohabitation	———— Home leaving
—·—·—· Second marriage	········ First birth	— — — First child's home leaving
— — — Separation or divorce	—·— Last birth	— — Last child's home leaving

Figure 13.7 Cumulative percentages of (a) women and (b) men born in 1931–1940, and 1951–1960, experiencing certain life events.
(*Source:* Extracted and redrawn from Rajulton and Ravanera, 1995, Figures 4.7a and 4.7b, pp. 140–141.)

WOMEN IN MENS' CITIES

Although these changes in female participation rates in the paid labor force and greater ranges of possibilities for self-determination in the life-course have enhanced the position of women, they nevertheless have occurred within the spatial framework of urban forms and structures that have been constructed in a male-oriented society. The internal structure of North American cities has developed in large part based on a sexual division of labor (Holcomb, 1984; MacKenzie, 1989). Theories relating to gender differences and social space must take into account why the sexual division of labor was predominant, and why some of the inequities that it generated, such as the wage gap, are persistent. There are two approaches that may be used to address these questions: a structural approach that indicates how the sexual division of labor may have developed, and a class approach that suggests why elements of dominance-subordination may persist. It should be emphasized, however, that these are only two of a number of theories in an area of research in the social sciences that has developed rapidly (Garber and Turner, 1995).

The structural approach argues that different gender roles developed along with the changing nature of work as society transformed from a preindustrial to

Table 13.1

Probabilities of a Person in a Particular Cohort Incurring a Certain Life Event by Age 35

	COHORT	
Life Event	1931–1940	1951–1960
First marriage (FM)		
Women	.92	.81
Men	.88	.71
First child (FB)		
Women	.86	.73
Men	.80	.63
Cohabitation (FC)		
Women	.02	.33
Men	.04	.38
Separation/Divorce (FS)		
Women	.07	.18
Men	.05	.12
Second Marriage (SM)		
Women	.04	.07
Men	.04	.05

industrial, and then into a mature industrial state (Cooke, 1984). The approach envisages a preindustrial agricultural and craft-based society in which the family was the productive unit and all family members, particularly the spouses, had well-recognized roles. There was a sexual division of labor in the household, but generally women often worked alongside men. In businesses, husbands and wives had fairly equal rights and responsibilities (Arnold and Faulkener, 1985). But, during the eighteenth century, as capital for financing trade and small-scale industry became more important, the control that men exerted over capital, through patrilineal inheritance or marriage, allowed them to gain a dominant position. This dominant position became enshrined in craft guild practices, laws relating to the ownership of property and the regulation of the banking and financial industries, and the political system (Fletcher, 1995). Thus, as men used their control of property to finance and manage industrial capitalism, they, in effect, moved out of the home into the worlds of thought and action, leaving women to assume the unwaged responsibilities of the domestic sphere. As housework and the raising of children became the prime responsibilities of women, their virtual exclusion from education, industry, and business led to de-skilling in nondomestic activities.

In the early stages of industrialization, women and children were used as low-cost, relatively unskilled labor, but they were gradually displaced as production became more labor efficient. This efficiency was generated through the application of new technologies and management techniques derived from engineering. The patriarchal structure of society based on the ownership of capital was, therefore, enhanced through the control by men of technology and management/administration. As women were, in the nineteenth century, largely excluded from formal education and apprenticeship arrangements, they became even more confined to the domestic sphere and to relatively unskilled jobs outside the home. Furthermore, the structure of industry itself became extraordinarily hierarchical, with the widespread adoption of scientific management in the early years of the twentieth century. This system, based on the absolute control by managers of all information and production processes related to a business, and, at its most extreme, the rejection of worker input and initiative with respect to the operation of the enterprise, confirmed the domination of tightly knit male management networks in the economic sphere (Taylor, 1947).

When women did enter the labor force, they were seen by management as occupying less-skilled jobs (even if the jobs, in fact, demanded sophisticated skills), and their wages were regarded as supplemental to those of men. As a consequence, they were paid comparatively less than men for the work they did. With the transformation to a service society, and the increase in the number of jobs in the service sector, certain types of employment (particularly the supportive kind) again became labeled "female," and other kinds "male," and the male-female wage rate differentials based on occupation continued. Only since the 1970s has there been a concerted attempt to eliminate wage differentials between males and females in occupations that require the same general level of skill and training, through pay equity and various types of comparable worth legislation. In spite of these efforts, wage differentials between men and women still exist.

The persistence of elements of gender-based dominance-subordination relationships has also been explained by class formation theory. Previously, in Chapter 11, we made reference to the notion of classes arising out of the struggle between the owners and controllers of capital and labor. In the gender context, the concept of class control can be applied to the relations between men and women, as men to a large extent have controlled the income and women have been dependent economically on them. Since the 1970s, as women have become more prevalent in the workforce and have assumed more "male" job positions, gender-based tensions have developed. Men, who used to be in a privileged employment position, are seeing their traditional role usurped by women. In this sense, men can be viewed as a privileged class trying to hold on to the appurtenances derived from power and control during a time of rapid social and economic change (McDowell, 1993a; 1993b). The mirrored glass facades of corporate office buildings perhaps symbolize this tension, for though the buildings appear to be integrated with their environment, their transparency reveals them to be bastions of male privilege. Wilson (1991) argues that these types of exclusionary situations in cities are the product of male insecurities.

From Sex Roles to Urban Structure

Given that the urban structure in North America has arisen primarily out of capitalist economic activity, and that most of the development that has occurred has taken place in the context of a sexual division of labor, sex-role stereotyping, female occupational segregation, and fairly persistent wage gaps for those women employed outside the home, it is not surprising that this type of segregation is also reflected in the internal structure of cities, and house types and design. This is seen most clearly in the structure of the classic industrial or monocentric city, in the highly suburbanized nature of polycentric urban areas, and in the apparent recent realization of architects and planners of the need for a wider variety of house types and internal living arrangements to cater to a highly diversified society (Weisman, 1992).

Home-Work in Monocentric Urban Areas   The classic monocentric city, with the bulk of paid employment located in or around the center of the urban area, reflects the inherent assumption of the power of capital, the vital importance of male wage labor, and the supportive role of the domestic sphere in the structure of the economy (Hanson and Pratt, 1995). This is illustrated in Figures 7.4, 7.5, and 7.6, wherein the residential areas are defined in relation to the places of employment. The sexual division of labor, which is crystallized in the spatial separation of home from employment in these diagrams, becomes more entrenched and occupation/income related as the industrial city decentralizes. This is illustrated in Figure 13.8, which contrasts the early and late industrial spatial arrangements of home and work on the "poor" and "rich" sides of a monocentric urban area.

1. In the early industrial period, home and work for both the poor and rich households were quite close to each other and the places of employment (Figure 13.8a). The factory's siren, and the time clock, controlled the rhythm of the day in homes in both the rich and poor sectors. Meals and social activities were geared entirely to the cycle of the male working day.
   a) Those women who were in the wage economy, (primarily young single and older married women from the poor side of town, who were generally employed in unskilled factory jobs or service activities), abided by this male-centered routine. Women, therefore, were obliged to seek employment as close to the home as possible in order to minimize travel time and maintain the domestic routine. Married women with young children who were in waged employment relied on older children, close family, and perhaps neighbors for child care.
   b) The physical infrastructure related to housing on both the poor and rich sides of the town was fairly primitive, but the homes in the richer sector had more space around them, and may well have had piped water and inside "water closets." The homes and tenements in the poorer sector had little space, and external water and sewage facilities. Population densities were high, diseases were recurrent, dirt was ever present, and domestic life was a struggle.

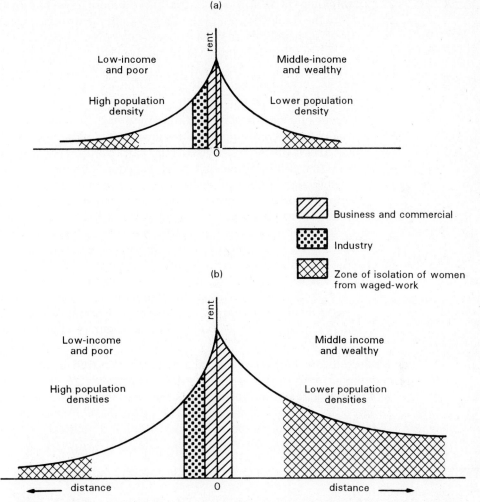

Figure 13.8 The increasing spatial isolation of women from waged employment opportunities for the (a) early, and (b) late industrialized periods in monocentric cities.

2. Decentralization in the later industrial era affected primarily homes in the wealthier sector of the city (Figure 13.8b).

   a) A better physical infrastructure, particularly in the form of transport improvements, such as the streetcar, came first to the wealthier parts of the monocentric city. These improvements permitted greater physical separation between home and work in wealthier households, with public transport catering almost entirely to the needs of male commuters. Home and work in higher-income families became even more clearly delineated, as did the isolation of women in these households from employment opportunities outside the home. Women wishing to pursue waged employment in a monocentric environment now had to travel farther from the domestic sphere than they might have wished given the domestic responsibilities that have been customarily allotted to them. They were, therefore, effectively isolated from the world of waged work except those positions available locally. Volunteer work, often in charitable and service organizations, became an outlet for many women with business and administrative inclinations.

b) Poorer households generally remained in high-density locations close to male employment opportunities located in and around the downtown areas and transport terminals. But, with the gradual decentralization of wealthier families, one source of waged employment for poorer women outside the home—domestic service—became less accessible. Women employed in these activities now had to take public transport to their places of employment. Thus began the phenomena of the *reverse commute,* which in United States cities has racial overtones, with public transit bringing wealthier (usually white) males into the central business district and taking poorer (mainly African-American) women out to the suburbs in the morning, and a reversal of this pattern in the late afternoon and evenings (Johnston-Anumonwo, 1995). In these types of longer-distance employment situations, working women became even more dependent on elder children, close family members, and neighbors for child care.

Monocentric urban areas are thus clearly structured in response to the employment needs of the male labor force. Furthermore, the nature of the city as it has developed not only reflects the gendered division of labor, it also reinforces the patriarchal hierarchy. The power base of men became located in the board rooms, lunch rooms, private clubs, and political offices of the downtown areas from which women were effectively isolated and in which they were not welcome.

Gender Differences in Worktrip Lengths One spatial outcome that could be posited from the domestic division of labor and the wage gap is that women would have shorter work trip lengths than men (Hanson and Johnston, 1985). In fact, Figure 13.9 indicates that in Worcester, Massachusetts, women's worktrip lengths, measured in travel time, tend to be less than men's. The division-of-labor explanation for this difference is rooted in the argument that, since women have incurred greater domestic responsibilities than men, they are more likely to undertake waged work closer to home in order to accommodate exigencies in the domestic sphere. The wage gap argument for the shorter worktrip length is that, since women generally receive less pay than men, they are less likely to indulge in longer and more costly commutes. The outcome for commuting behavior, however, could be influenced by a mix of factors such as marital status, occupation, race, and income level of the persons involved. For example, the division-of-labor argument should show a greater effect on the

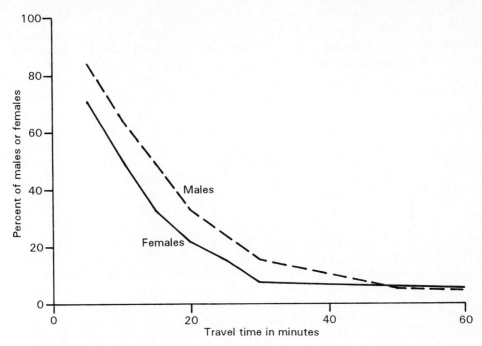

Figure 13.9 Women's work trip lengths tend to be shorter than men's, Worcester, Massachusetts.
(*Source:* Redrawn from Hanson and Pratt, 1995, Figure 4.1, p. 96.)

commuting behavior of married women with children, and the wage gap argument should show that lower paid female workers are more constrained by commuting costs than higher-paid women.

Because of the highly nuanced nature of the issue, and the particular labor market characteristics of various case studies, the evidence with respect to gender differences in worktrip lengths is somewhat contradictory (England, 1993). In general, it seems that the hypothesis applies most directly to white lower- and middle-income women who comprise the majority of those in waged employment in North American cities (Hanson and Pratt, 1994). For example, in a case study of women and men's commuting patterns in Baltimore, Johnston-Anumonwo (1992) demonstrates that when all other factors, such as race, income, and number of children, are controlled, women in two-worker households have shorter trip lengths than men. The nature of the issue is, however, becoming more complex in multicentered cities where employment locations are more highly dispersed.

Home-Work in Polycentric Urban Areas   It can be argued that the massive increase in suburbanization after 1945 in North America demonstrated this home-work–female-male division of labor even more clearly (Fava, 1980). Post-war suburbanization involved a much larger proportion of the population than before because homes were more affordable and greater numbers of people moved into the middle-income range. The houses that were built also catered for needs of many middle- and lower-income nuclear families for greater space, easier maintenance, and privacy (Wagner, 1984). The prominence of model kitchens and garage parking in the suburban home advertisements of the period emphasize the assumptions embedded in the construction of these suburban tracts: They appear to have been built for a "commuting husband and a domestic wife" with child rearing and care for the waged male being their central purpose (Rose, 1993, p. 123; MacKenzie, 1988).

One of the major outcomes of this suburbanization and sprawl was disinvestment in public transport systems in many North American urban areas. Suburban homes are premised on universal private automobile ownership, which by the 1950s had become the second major financial expense (after the home) for most households. In general, the majority of suburban homes possessed one automobile in the 1950s and two in the 1970s. Given the absence or rudimentary nature of public transport beyond the old inner cities, automobile access is a prerequisite for employment in polynucleated urban areas, particularly for jobs in locations other than the central business district (which in some of the older metropolises remains the focus of commuter railroad and rapid transit systems). Auto availability is, therefore, vital for women seeking paid employment who are located outside of older, central cities. Women located within central cities who do not have automobiles are therefore restricted to employment opportunities that are served by public transport. For example, in a sample of women in two downtown neighborhoods in Grand Rapids (Michigan) there are significant differences between racial groups in access to private cars: "Whereas 98% of Anglo-American women . . . had access to a car, 85% of these for their own use, only 75% of African-American women had access to a car, and of these only 72% had them for their personal use" (Peake, 1995, p. 422). Given the state of public transport in the city, taxis are often used by African-American women for mall and supermarket shopping and the journey to work.

Although the decentralization of employment opportunities associated with post–1945 suburbanization would have seemed to increase female job opportunities, women have been faced with the difficulty of getting to them. This is particularly true for minority women who, because of their spatially constrained inner-city residential locations, are having to undertake long and time-consuming trips to suburban areas for different types of service jobs. For example, African-American and Hispanic women in the New York City metropolitan area commute as far as their male counterparts, and both female and male minority workers spend more time traveling to work than white women and men (McLafferty and Preston, 1991). With the growth of employment opportunities in edge cities, minority women of color are likely to experience much longer commuting trips than white women (Johnston-Anumonwo, 1995).

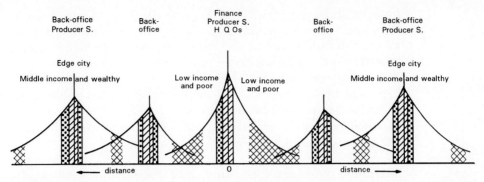

Figure 13.10 The increasing spatial isolation of women from high-order headquarter office and financial services employment, but decreasing isolation from lower-order back-office and other business and service activities in polycentric cities.

The Spatial-Entrapment-of-Women Hypothesis Hanson and Pratt (1991; 1995) have argued that the gendered division of labor, which leads to a need for many women to seek employment close to their residence, and occupational segregation have given rise to subtle forms of labor market segmentation by gender within complex polynucleated cities. The suburbanization of most of the population of metropolitan areas has created reservoirs of female labor in decentralized locations. Many of these women, particularly those who are married with young children, along with those in the inner city, can be viewed as *spatially entrapped* because they place a high premium on waged work close to their residence so that they may continue to accommodate household responsibilities.

By locating in decentralized nucleations and edge cities, business activities can take advantage of less costly and more modern facilities and also tap into the local pools of female labor where the workers are willing to accept lower wages for work closer to home. Thus, businesses downtown tap into central city female labor pools, which include high proportions of single women and persons of color. In suburban locations, by contrast, they tap into pools of primarily white, married women. The types of work and occupations that are available in these decentralized nucleations and edge cities consist of a myriad back-office and customer service (by telephone and computer) activities (Nelson, 1986). This form of labor market segmentation by gender serves not only to perpetuate the wage gap, because many women are prepared to trade off jobs closer to home for lower wages, it also perpetuates occupational segregation, particularly in households of middle income or less.

As headquarter and regional offices, particularly those involving senior management and finance, remain located predominantly in downtown areas (see Chapter 10), and often require high levels of mobility and excessive, irregular hours, many women are excluded from these centers of business power (Figure 13.10). This entrapment of women is exacerbated by differential access between males and females, and between women in different racial groups, to transportation, particularly private cars (Rutherford and Wekerle, 1988). Furthermore, many two-worker households with, or planning, a family may select a suburban residential location close to the wife's job, or expected employment, in order to minimize her travel time and thus accommodate her gendered role in the domestic sphere (Tkocz and Kristensen, 1994). Women in two-earner higher-income households may not be as entrapped as those in middle- and lower-income households because of their greater ability to purchase private automobiles and acceptable substitutes for some domestic activities (England, 1993; Hanson and Pratt, 1994).

Migration, Residential Location, and Patriarchy   Residential and business location decisions can, therefore, reinforce patriarchy in three ways. First, with respect to household labor migration itself, which in the case of male/female-headed households usually involves a wife, husband, and employer, the migration invariably requires a change in location of the male's job with the female as a secondary migrant. This secondary migration "serves to legitimate the sex-typing of occupations by reinforcing both employers' and fellow male workers' perceptions of women as 'uncommitted'" (Halfacree, 1995, p. 173). Thus, even though the move may increase the probability of employment among married women, because the couple will search for a location that maximizes the wife's employability (Cooke and Bailey, 1996), the unintended consequence of women becoming secondary migrants is reinforcement of their domestic role and subordinate position in the workplace.

Secondly, residential location decisions that may be taken to accommodate a gendered division of labor also have the unintended consequence of reinforcing patriarchy. Tkocz and Kristensen (1994) observe that, even in a polynucleated metropolis with men working in central and many peripheral locations, they incur longer worktrips than women primarily because a two-worker household generally adjusts to the wife's job in order to maximize her time at home. This type of residential location decision reinforces patriarchy by restricting women to a local labor market, which implies perpetuation of occupational segregation and the wage gap, while allowing men a wider labor market than women in which to compete. It also assumes that the gender division of labor is natural and immutable.

Third, business location decisions, and the structure of polynucleated metropolises themselves, also reinforce patriarchy. The tendency for businesses to locate back-office and other supportive activities in decentralized nucleations and edge cities, while maintaining headquarter offices in downtown locations, provides a spatial manifestation of the hierarchical nature of business and political/governmental enterprises, which implies that employment in these power-based activities requires patriarchal types of commitments and behaviors. Tkocz and Kristensen (1994) note that women's commuting patterns are more dependent on city structure than the commuting pattern of men. In one sense, the various decentralized locations of employment provide intervening employment opportunities between the suburbs and more centralized headquarter office and producer service employment locations (Figure 13.10).

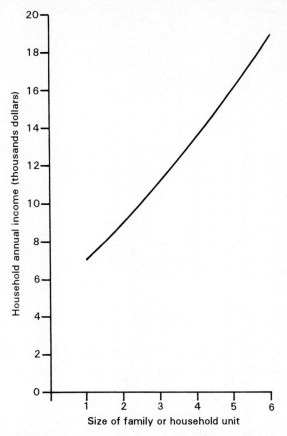

Figure 13.11 Thresholds used in the United States census to define numbers of persons in poverty.
(*Source:* Bureau of the Census, *Poverty in the United States,* 1992, Table A-3.)

Urban Aspects of the Feminization of Poverty

The gender division of labor, women's responsibilities for bearing children, and the perpetuation of the wage gap (particularly for noncollege or university educated women) have also led to a high incidence of poverty among women and children, with a concentration in the central cities of metropolitan areas (Sidel, 1986). Definitions of poverty in the United States and Canada are usually based on market (or private) income, that is, pre-tax income from wages, salaries, investment income, and income from self-employment. Thresholds are then set based on estimates of the levels of income required for subsistence support of families and households of various sizes. Figure 13.11 shows how the estimates increase with size of family or household unit for the United States in 1992. Based on threshold estimates of this type, which are obviously quite crude, particularly when applied in different regions where costs of living vary significantly, and do not represent final income, roughly similar proportions of the population of Canada (15.4 percent) and the United States (14.5 percent) in 1992 may be classified as living in poverty.

Table 13.2

The Feminization of Poverty and Its Concentration in Central Cities of Metropolitan Areas (MSAs), 1992

| | UNITED STATES | | METROPOLITAN AREAS | | | |
| | | | CENTRAL CITIES | | NON-CENTRAL CITIES | |
GROUP	NUMBER	PERCENT	NUMBER	CONCENTRATION	NUMBER	CONCENTRATION
Total Population	255,082		78,381	30.7	124,154	48.7
Population in Poverty	36,880	14.5	15,644	42.4	11,728	31.8
Poverty by Gender						
Female	21,180	57.4	9,037	42.7	6,807	32.1
Male	15,700	42.6	6,607	42.1	4,921	31.3
*Women in Poverty by Race**						
White	14,030	66.2	4,899	34.9	5,157	36.8
African American	6,225	29.4	3,702	59.5	1,358	21.8
*Women in Poverty by Household Type***				percent		percent
Married Couple Families	6,566	31.0	2,332	25.8	2,280	33.5
Lone Female Parent	9,150	43.2	4,618	51.1	2,675	39.3
Unrelated Individuals	4,998	23.6	1,916	21.2	1,688	24.8

(*Source:* Estimates calculated by author from *Poverty in the United States, 1992*, Washington, D.C.: Bureau of the Census, Table 6. Female/male splits calculated on 50/50 basis for ungendered data.)

*Hispanic origin may be in either racial group, or in the residual.

**Excludes residual "other subfamilies."

The data for the United States indicate that poverty is greater among women (57.4 percent) than among men (Table 13.2), with almost exactly the same female-male proportions in both central city and non-central city components of metropolitan areas (MSAs). Both women and men in poverty are, however, concentrated in central cities, with 42.4 percent of all persons in poverty located in central cities as compared with 30.7 percent of the population as a whole. In contrast, in non-central city areas within MSAs, which contain 48.7 percent of the nation's population, 31.8 percent of persons are in poverty. Thus, women are more likely to exist in poverty than men, and women in poverty are as likely as men to be concentrated in central city locations but in larger numbers (Table 13.2). There are, however, distinct differences in the location of women in poverty by race. Whereas the white population has the larger number in poverty, though less than might be expected on the basis of its share of United States population as a whole, 60 percent of all African-American women in poverty are located in the central cities of MSAs as compared with only 35 percent of white women in poverty.

The information in the section subheaded "women in poverty by household type" in Table 13.2 indicates the special nature of the female poverty condition and its connection with the increasing complexity of North American families and households discussed previously. Evidently, 43.2 percent of women in poverty in the United States reside in lone-female parent households (either as parents or children). In central cities more than one-half (51.1 percent) of women reside in such households. More than one-fifth (23.6 percent) of women in poverty live as unrelated individuals, frequently alone. A large number of these are elderly where the poverty condition has developed following the death of a husband. The frequent absence of significant spousal pension benefit rights following the death of a husband (recall that women live, on average, seven or eight years longer than men) often severely reduces the standard of living of the surviving wife.

The spatial concentration of women in poverty is illustrated with respect to lone-female parent families with children in the case of the Wichita MSA in Table 13.3. The information in Table 13.3 emphasizes that poverty in the metropolis is concentrated in the central city of Wichita, and that more than one-half of all families in poverty in the area are headed by lone-parent females. These lone-female-parent-headed families, in turn, are even more concentrated in the central city than poor families in general. There are only twenty-one census tracts, containing 15.2 percent of all families in the MSA, that have seventy-five or more families with children headed by lone female parents and these tracts contain one-half of all the families in poverty in the metropolitan area (Figure 13.12). The location of these clusters of families is coincident with neighborhoods that have a high proportion of public housing—reflective of a process of *residualization*, in which public housing that was once inhabited by low-income African-American and white families has become (through downward filtering) occupied largely by lone-parent families in poverty (Wilson, 1991, p. 148). The data pertaining to the Wichita MSA, therefore, emphasize the highly localized, perhaps even contained, nature of this particular important element of the feminization-of-poverty issue.

Table 13.3

The Concentration of Lone-Female-Parent-Headed Families in Poverty in the Wichita MSA, 1990

	WICHITA MSA	SEDGWICK CO.		WICHITA CITY	
	NUMBER	NUMBER	PERCENT	NUMBER	PERCENT
Population	485,270	403,662	83.2	304,011	62.6
Households	187,099	157,023	83.9	123,682	66.1
Families	131,435	108,670	82.7	80,875	61.5
21 CTs				19,946	15.2
Families in poverty	10,304	8,995	87.3	7,674	74.5
21 CTs				4,076	39.6
Lone-female-parent families	5,269	4,842	91.9	4,331	82.2
21 CTs				2,608	49.5

(*Source:* Extracted from *Population and Housing Characteristics for Census Tracts*, Washington D.C.: Bureau of the Census, Table 19.)

Given this highly metropolitan and central city nature of the location of female poverty, public policies addressing the feminization of poverty have an enormous impact on cities. Jones and Kodras (1990) note that the Aid to Families with Dependent Children (AFDC), which has been a long-standing controversial package of support for lone-female-parent-headed families, has done what it set out to do at its inception in the 1930s; that is, feed children in poor families who may otherwise have gone hungry. It was a policy predicated on the widespread existence of two-parent families. However, with the changing trends in the nature and structure of families and households in North America, policies need to be implemented in the form of programs that provide a safety net in times of temporary distress, but change the dependency role that many poor women have assumed in society. This requires a continuous dismantling of the structures that support the gender division of labor. These are summarized in the concluding section.

IMPROVING URBAN LIVING

The sprawling, polynucleated nature of North American urban areas does not make it easy for anyone, especially women, to lead fulfilling lives that involve families, waged employment if desired, and social activities. This is in part because, in the short-hand of the classical residential location theorists, household incomes (Y) are no longer keeping pace with increases in aggregate housing costs (H), transport costs (T), and costs for the wide array of all other goods and services (G) sought by so many families and individuals in the North American consumer-driven society. Economic restructuring has meant that for many

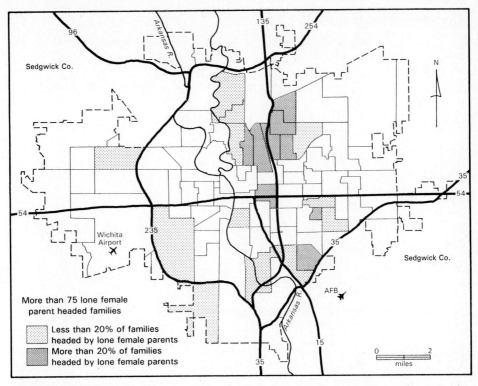

Figure 13.12 The concentration of lone-female headed families in Wichita, Kansas.

families incomes have, at best, remained fairly static in constant dollar terms since the 1970s, while housing costs have fluctuated widely with changes in interest rates and the speculative housing boom/bust of the 1980s and 1990s, and transport costs have increased steadily with rising costs of automobile maintenance and insurance.

Thus, the experience of the 1980s and 1990s is that wage earners in households have to increasingly devote more hours per week to the waged labor market just to meet household expenses. This is illustrated with information relating to wife-and-husband families in Canada in Figure 13.13. Average family incomes (in constant dollars) increased in accordance with the increase in dual-wage-earner families up to 1980, but after that time average family incomes have remained relatively unchanged even though the proportion of dual-wage-earner families has continued to increase to, perhaps, an upper limit of 60 percent of all families in the buoyant labor market of the mid–1980s (Pratt and Hanson, 1991). The information suggests that the increased participation of women in wife-and-husband families in the labor force is occurring as a necessity to meet rising household expenses or as part of the general thrust to diffuse the gender division of labor and its manifest consequences.

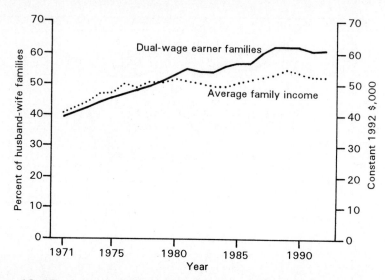

***Figure* 13.13**   Change in family income and proportion of dual-wage earner families in husband-wife families in Canada, 1971–1992.
(*Source:* Redrawn from Statistics Canada, Catalogue 13-208.)

Symmetrical Families

It has been suggested that one effect of the greater involvement of women in the labor force on the domestic sphere would be the emergence of many more symmetrical families (Young and Willmot, 1975). The asymmetrical wife-home/husband-waged-work family may be transforming into one in which household roles are increasingly shared, with both partners being involved to greater or lesser degrees in spheres traditionally defined as either women's work (housekeeping, cooking, child care) or men's work (generating wage income). There is, however, a great deal of skepticism in many circles concerning the extent to which the symmetrical family is emerging (Zinn and Eitzen, 1990). The evidence from daily activity and domestic survey studies suggests that, though there is some movement towards symmetry, many women are at work out of economic necessity and yet still carry the burden of work in the home.

Women's Time Crunch   One of the consequences of the increased participation of women in the labor force in highly suburbanized, often polycentric metropolitan areas is that there is little time within the day for many people, particularly middle- and lower-income women, to accomplish much of what they feel obliged to accomplish within a day (or week). With increasing involvement of women in the labor force, greater complexity of families, and more time being spent on work trips, many women are experiencing severe time-space constraints (Friberg, 1993). As gender-generated inequities are becoming better understood and in some areas addressed, time inequities are becoming important (Dyck, 1989). When a great deal of time has to be spent on commuting and on

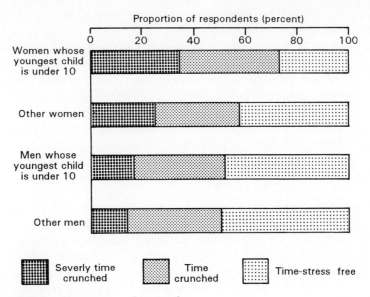

Figure 13.14 Time-stress by gender.
(*Source:* Redrawn from Statistics Canada, Catalogue 11–008 E, No. 31, 1992.)

work-related activities, there is little time for much else. And if there is anything that families require it is time!

The data for the diagram in Figure 13.14 are based on a sample survey of wife-husband families in Canada. The percentages relate to the proportion of respondents who regarded themselves as "severely time crunched," "time crunched," and "time-stress free" during an average day. Women generally regard themselves as more time crunched than men, and mothers whose youngest children are under the age of 10 view themselves as the most time crunched, with less than one-quarter of respondents in this category regarding themselves as having any significant amount of time in a day which is stress free. Men with or without a youngest child under the age of 10 regard themselves as the least time stressed in a day. This undoubtedly reflects the persistence of women's involvement in the domestic sphere.

Shared Domestic Work The continuance of the primary role of women in the domestic sphere is indicated in Peake's (1995) study of a sample of Anglo-American and African American households in two low-income neighborhoods in downtown Grand Rapids (Michigan). In survey households that had a male present (in 87 percent of the cases the male was a husband) little more than 40 percent of the households responded that the male "regularly gives help." This, of course, suggests that almost 60 percent of men in such households do not regularly give help. An analysis of the division of responsibilities indicates that the domestic tasks that men are most likely to perform are repairs, maintenance, and yardwork; the domestic tasks they are least likely to be involved with are house cleaning, shopping, and child care. Thus, even in those households in which men regularly give help, there is a tendency for a women's work–men's work division of labor to persist.

There are, however, significant differences between Anglo-American and African-American households in the involvement of males in domestic activities. Whereas nearly one-half of Anglo-American males regularly give help, barely 30 percent of African-American males do so and nearly one-half rarely help or provide no help whatsoever. Thus it is quite common for African-American women to be living in households in which they are solely responsible for domestic work, even though they may be devoting considerable time during a week to paid work outside the home. The domestic tasks with which African-American men may be involved are also restricted to maintenance and yard work, whereas their Anglo-American counterparts more frequently get involved with a wider variety of domestic activities.

Role Reversals?   The increased participation of women in the labor force, diffused levels of occupational segregation, and the reduction of the wage gap particularly for college- or university-educated younger women, is generating more family situations in which wives are earning more than husbands, or the wife is the sole wage earner. Figure 13.15 indicates that the proportion of wife-husband families in which the wife earns more than the husband, or is the sole wage earner, has increased dramatically in Canada since 1967. Almost 20 percent of husband-wife families in 1993 had the wife as the chief provider of waged income. While part of the reason for this change may have to do with the aging of the population and the retiring—whether it be induced-early, voluntary-early, or statutory—of husbands, much is related to the improved wage-earning capacity of many women in professional, business, and administrative occupations.

Although this may be viewed as a shift in the patriarchal nature of Canadian society, there are, nevertheless, at least two observations that may be made with respect to Figure 13.15. First, whereas women are faring better in the workforce, 80 percent of families are still in a situation where the husband is the chief wage earner. Thus, the general income condition that underpins the gender division of labor to a large degree persists. Therefore, household mobility and residential location decisions continue for the most part to be made within the context of a patriarchal structure of society. Second, in those situations in which the wife is the primary source of waged income, and the husband is either earning less than the wife or is not in the labor force at all, women usually continue to maintain the domestic responsibility of managing the household and organizing or undertaking many of the routine chores. These responsibilities invariably have to be undertaken in highly sprawled urban environments in which home, work, services, and supportive institutions are widely dispersed. A major domestic responsibility with these, and other two-parent and lone-parent families, is child care.

Child Care and Urban Environments

A key to the dissipation of the array of inequities that arise from the gender division of labor is child care. In an egalitarian society, it is understood that both women and men are equally responsible for the raising and nurturing of chil-

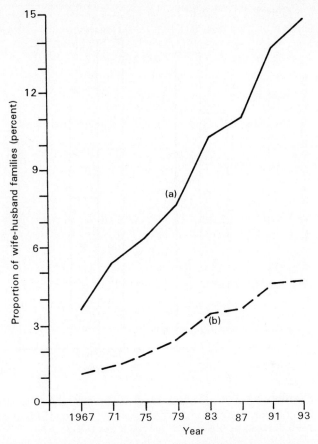

Figure 13.15   Change in the proportion of wife-husband families in which (a) the wife earns more than the husband, or (b) the wife is the sole wage earner.
(*Source:* Compiled by the author from Statistics Canada, 1994, *Women in the Labour Force.*)

dren, and no one gender should bare a disproportionately large share of the parental function nor the private costs. In this context, private costs relate to the *current* time and monetary costs involved with establishing and maintaining a home and the care and nurture of family members, and the *future* costs incurred as a consequence of income or opportunities foregone in the exercise of these responsibilities. Furthermore, the raising of children is a vital social function because the benefits are to a considerable degree realized by society at large. It is for this reason that society has accepted the provision of primary and secondary education for school-age children as a public cost. The raising, care, and education of children has, therefore, been accepted in many ways as a public as well as a private interest.

Increased diversity in the structure of families, and greater participation of women (as a result of financial need or personal choice) in the waged labor force, has created a great need for child care arrangements other than those provided directly by parents in family homes. In the United States in 1991, 51

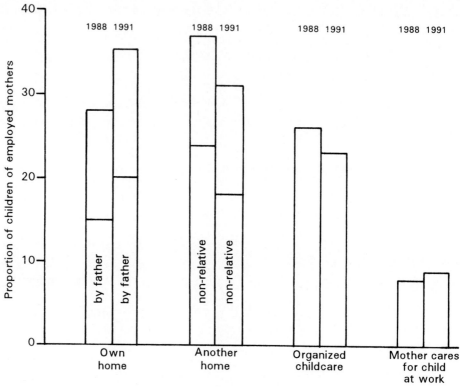

Figure 13.16 Primary child care arrangements used by employed mothers for children under 5 years old, United States, 1988 and 1991.
(*Source:* Illustration prepared from Bureau of the Census, 1993, Table 610, p. 384.)

percent of all children under the age of 5, and a much higher percentage of those between ages 5 and 10, had a mother in waged employment (see Figure 13.5 for Canadian data based on percent of women-with-families employed). Thus, the reality of North American family life in the 1990s is that there is a great need for a range of alternative, nonfamily child care arrangements for young children, and for preadolescents—the so-called "latch-key kids." Child care "is thus neither 'public' nor 'private' but a set of needs and responses located at the intersection of public and private life" (MacKenzie and Truelove, 1993, p. 328).

The range of primary child care arrangements used by mothers of children under the age of 5 in waged employment in the United States is indicated in Figure 13.16. The information outlines not only the major forms of arrangements, but also suggests the sensitivity of these arrangements to buoyancy in the labor market; for example, 1988 was at the end of a boom in employment, and 1991 close to the end of a recession. Although most children of employed mothers are cared for in settings other than the child's home, a greater number are being cared for in situations not involving financial outlays during the recessionary period when a father may be unemployed or on reduced shifts. The "own home" category includes arrangements involving paid sitters, nannies, "au pairs," and maids, some of whom may be illegal female immigrants employed under exploitive working conditions of long hours and low hourly pay. The worst of such cases involve some women living with families in suburbs who are isolated and without transportation and whose work hours appear to be unlimited. As a general rule, their contact with other women in like situations is in local parks where the children are taken to play. The Spanish term for this form of *occupational spatial entrapment* is *encerrado* (locked up). The "another home" category includes care provided by relatives and informal or unlicensed family day care usually offered by nonrelatives, while the "organized child care" category includes licensed family day care and profit and nonprofit day care centers.

Even though child care increasingly lies at the intersection of public and private life, its costs are considered to be the responsibility of parents, whereas the funding of schools is considered to be the responsibility of society as a whole. Although the demand for organized child care has increased dramatically since 1970, as is indicated by the high rate of increase in the number of supervised day care spaces as shown in Figure 13.17, the funding of day care has become of increasing social and political concern. The public aspect of the responsibility is recognized through income tax credits, and a variety of forms of subsidized day care spaces for low-income families. While the income tax credit approach provides a significant benefit for middle- and high-income families, it does little to help low-income families, particularly lone-female-headed families. For low-income families, subsidized day care spaces are vital if mothers are to achieve and maintain independence (Rose, 1993). Typically, it is subsidized spaces that come under threat during times of deficit reduction in public finance.

The Location of Child Care Facilities   When a married working couple with two young children's *affordable* arrangements for child care involve "preschool child care from 7:30 to 9:00 A.M., kindergarten from 9:00 to 11:30 A.M., and formal child care from 11:30 A.M. to 5:30 P.M. for one child and a licensed caretaker from 7:30 A.M. to 5:30 P.M. for a younger child" (Peake, 1995, p. 430), it is evident that the location of child care facilities *vis-a-vis* home and work is crucial to the maintenance of such a complicated schedule. In a study of the distances traveled from homes to day care centers in metro Toronto, Truelove (1989) demonstrates that although 40 percent of children traveled 1 kilometer or less, the average distance traveled by all children was 2.9 kilometers (1.8 miles). Children located in suburban areas had to travel slightly farther, and those in the central city slightly less.

Thus the location of child care facilities has to be examined not in terms of the optimal provision of a set of facilities for a given spatial distribution of potential users, such as location-allocation models, but in terms of the typical daily travel paths of the parent(s) in each family concerned. In dual-worker and lone-parent worker households with young children, an optimal location for child care services would be either in the immediate neighborhood of the home, or in the immediate vicinity of the workplace. An intermediate location

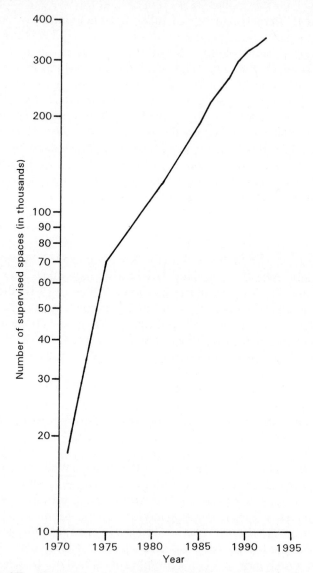

Figure 13.17 The increase in supervised daycare spaces (semi-log scale) in Canada, 1970 to 1992.
(*Source:* Compiled with data from Statistics Canada, 1994, Table 6.10.)

involves extra time and an interrupted or longer worktrip. The choice of neighborhood or workplace would, ideally, be related to the particular needs of the child for flexible connections with parents, relatives, older children, neighborhood friends, and so forth, during the day.

Unfortunately, this type of local neighborhood-workplace choice is not available to the vast majority of parents. Only 3 percent of families with children younger than age 6 using child care in Canada in 1990 had workplace day care. Most of these facilities were available only for public service employees (Statistics Canada, 1994, Table 6.11). During World War II, when womens' labor was required in factories and service activities, workplace day care was quite common. The reluctance of employers in large organizations to institute workplace child care facilities can be interpreted as a manifestation of patriarchy. Given this situation, neighborhood child care facilities are essential, and are probably best associated with the location of primary schools and community centers. However, as many suburban day care facilities are private, profit-seeking organizations, association with publicly funded schools or community centers is often not appropriate unless rentable space is available. It seems abundantly clear, in examining this situation, that child care still awaits its clear recognition, in terms of national policy and appropriately funded programs, as both a private *and* a public concern.

Women-Friendly Urban Areas

Metropolitan areas, because of the intensity of economic and cultural life and social change, often seem to be intimidating places. Disorder, and this does *not* include criminal behaviors, is, however, what urban areas in democratic societies are all about, for disorder leads to new connections (Sennett, 1970). Owing to the concentration of people in urban areas, where different ideas can be met and confronted and a whole range of experiences explored, they are places in which the existing order can be questioned. Wilson (1991) argues that cities offer greater opportunities for personal freedom and the exercise of diversity than small towns and rural areas, which can be quite restricting. Furthermore, whereas urban areas are places for adventure, they can also be places of reassurance because they contain an array of resources that can nurture growth and foster experience. Given the ways in which the gender division of labor has been used to try to confine women to the domestic sphere, urban areas, with their characteristic of disorder, can provide particularly important liberating environments for women.

Nevertheless, the bulk of the North American population appear to want to hive themselves off to suburban enclaves and, at the extreme, a small proportion to gated communities (Bell, 1991). A major reason for this is undoubtedly the fact that urban areas can be dangerous and seductive places, particularly for women and children (and also for men). Women's safety, whether the spatial context be the home, neighborhood, or city-at-large, has to be viewed as a public concern that governments, particularly local governments, must address (Strike, 1995; Rodgers, 1994; MacLeod, 1989/1990). Public policies and recommendations should be formulated with a view to changing the social conditions (the gender division of labor, poverty, patriarchy, unemployment) that lead to violence and exclusionary behaviors against women.

"It is estimated that between one-third and one-half of all sexual assaults take place in urban public space" (England, 1994, p. 633). "Take Back the Night" marches, which take place annually in many cities throughout North America, emphasize that women, far more than men, have legitimate concern about walking alone and using public transit after dark. Many women are concerned about entering parking garages, particularly those that are not well lit, and driving in areas where a vehicle breakdown could lead to an assault. As most assaults in public spaces are crimes of opportunity, various safety features and planning designs can be used to reduce these opportunities, such as safety zones in public transit waiting facilities, eliminating dark corners and unexpected enclaves from public buildings, and providing well-lit streets and adequate community policing (Wekerle, 1992). Neighborhood groups are also beginning to be more active in keeping the outside accessible, in the face of declining or stagnating municipal budgets, by equipping local police with bicycles, patrolling neighborhoods, public housing projects and school yards, and providing security escort services at particularly threatening times of the day (Belluck, 1996).

CONCLUSION

The central message of this chapter is that the internal macro-spatial structure of cities reflects and reinforces the gender division of labor, spatial entrapment, the geographic concentration of the feminization of poverty, and restrictions arising from safety concerns. The diffusion of gender divisions will, therefore, in time have a powerful effect on urban form and structure. Policies for its diffusion might include:

1. Continuing to address the wage gap situation by
   a. Implementing national standards for pay equity legislation with respect to comparable worth, that is, comparable pay for work of comparable value.
   b. Reducing occupational segregation by not only ensuring access of women to appropriate training and education, but also reducing entrenched societal barriers and behaviors that perpetuate notions of distinctions between "women's and men's work."

2. Reducing occupational inequities by
   a. Providing more geared-to-income federal/state/provincial/local subsidized day care for women and mothers who choose to undertake further training and education, and who decide to enter the paid workforce outside the home.
   b. Eliminating distinctions between full- and part-time work with respect to employee and employer compulsory long-term disability and unemployment insurance, and pension contributions.

3. Recognizing the changing nature of the North American family by implementing some sort of "social wage" for mothers and children, perhaps through such means as a negative household/family income tax, which would permit women a degree of choice with respect to labor market participation.

4. Encouraging family stability in a time of severe economic restructuring by establishing a series of minimum family incomes based on family size, again possibly implementable through a form of negative family income tax.

5. Pursuing taxation and planning policies that discourage sprawl and encourage greater densities, thus recompacting cities to provide an environment more conducive to the connection of women and families with the diversity of economic and cultural life in metropolitan areas (see Chapters 15 and 16).

This notion of the reconnection of women and families with the diversity of economic and cultural life in cities is vital for the development and liveability of North American cities. Far too much life in metropolitan areas is constrained by the influence of dominant-subordinate relationships between men and women—which in this chapter is crystallized in the sexual *division* of labor, occupational *segregation* by gender, the female-male *wage gap,* married women as *secondary migrants,* the spatial *entrapment* of women, and the spatial *containment* of the feminization of poverty.

Part **3**

Improving Livability in Urban North America

Chapter

14

Intra- and Interurban Transportation

*C*ommunication and transportation have been vital components in the development of urban hierarchies and the internal spatial structure of metropolises in North America because they provide means by which the various activities that make up cities are interconnected (Ausebel and Herman, 1988). They are even more vital in an age of just-in-time and flexible production systems, transnational corporate decision making, and highly integrated global financial markets, when there is added pressure on transport and electronic communication systems to be fast, dependable, and efficient (Schoenberger, 1995; Kellerman, 1993). These interconnections are composed of an intricate pattern of physical flows, electronic pulses, and face-to-face interactions.

Although an increasing share of communication and information flows occur via the hidden channels of wires, cables, transmitters, and satellites, the bulk of the flows that impinge directly on peoples' lives occur in the physical environment of intra- and interurban transport networks involving roads, rails, and airline terminals. Transportation does not, therefore, just serve the urban environment, it is an integral part of it. Indeed, the physical structure, size, spread, and character of urban areas are dependent on the nature and quality of their transport systems, just as the employment and lifestyles of the residents are dependent in large part on the opportunities made possible by the availability of various means of transport and communication.

In this chapter we will examine more closely the ways in which road transport has come to dominate intra- and interurban transport in North America, and the effect that this is having on intraurban transport choice. We have noted previously that the problems of movement and access to employment and services for certain individuals in urban areas have tended to increase. In part, this increase is related to the change in urban spatial structure from monocentric to polycentric or dispersed. To this change in spatial structure must be added the social and racial polarization of many metropolises, particularly those in the United States, with poorer individuals and families concentrated in inner cities where needs are not equally matched by resources and people are increasingly disadvantaged in terms of movement opportunities. At the same time, there is greater variability in movement patterns than ever before because of the increasingly polynucleated structure of many urban areas, and the more complex configuration of needs and interests of individuals and households.

INTERURBAN TRANSPORTATION

The story of interurban transportation in the latter half of the twentieth century concerns the rapidity with which road transport has come to dominate freight haulage and the transport of people. By the 1980s, most of the interstate highway system in the United States, and limited-access highways in Canada, had been constructed, and highway construction within urban areas virtually ended. A few links and extensions remained to be completed, particularly in the environs of some of the more rapidly expanding metropolises in the South and Southwest, but the network was essentially in place. Continent-wide, this highway system effectively permitted the truck to replace the railroad for all but the long-haul transport of bulk-goods and raw and partially processed materials. In a similar fashion, the highway had replaced the railroad for the transport of people for nearly all short- and medium-distance travel and aircraft for longer-distance travel.

Figure 14.1 shows how the interstate highway system has been built, largely since passage of the Interstate Highway Act of 1956, to connect all metropolitan areas in the United States. This system links at several points into the highway network serving the Windsor, Toronto, Ottawa, Montreal, Quebec City corridor in central Canada, to Winnipeg on the Prairies, and Vancouver and the Calgary-Edmonton corridor in the west. The level of transport interconnection between the two countries expressed by this network was recognized in the North America Free Trade Agreement through clauses allowing United States and Canadian registered vehicles to operate freely in each others' political territory up to a distance of 500 miles from the border. The construction of a North American highway network, and the freeing up of cross-border truck transport, is thus permitting even greater integration of the continent's economy (Cole, 1990).

Interurban Transport: Modal Shares   The change between 1950 and 1990 in interurban freight activity, measured in total weight (tons) transported and revenue received, for rail, truck, pipeline, water, and air transport in Figure 14.2 emphasizes the increasing importance of trucks, pipelines, and aircraft in the United States transport system. The share of total tonnage (Figure 14.2a) car-

——— Interstate highway

········ Canadian limited access

Figure 14.1 The interstate highway system in the United States linking all metropolitan areas.
(*Source:* Redrawn from BTS, 1994, Figure 2.2, p.18.)

ried by truck and pipeline transport has increased considerably, from 34 percent of tonnage in 1950 to 58 percent in 1990, whereas that carried by rail and water has declined from 66 percent to 42 percent. Truck transport has, in effect, assumed the transport of most higher value-added fresh and processed foods and manufactured products, leaving rail for carrying primarily low value-added coal, unprocessed grains, and partially processed chemicals. Pipelines are involved almost exclusively with the transport of petroleum and natural gas. Waterborne traffic is mostly involved with the transport of petroleum and petroleum products, coal and coke. While the amount of freight carried by airlines has increased considerably from almost nothing in 1950, the 1990 tonnage, involving perishables, mail, and technical goods, is still miniscule though of extremely high value.

The change in the amount of freight revenue generated by these various modes indicates the difference in the value of goods carried by the modes and the overwhelming importance of roads and trucks for the transport of manufactured and processed products. Almost as soon as trucks began replacing horse-drawn vehicles in the 1920s, they generated the greatest share of freight revenues. Thus, by 1950, intra- and interurban truck transport accounted for 65 percent of all revenues received by the transport industry, and rail 15 percent (Figure 14.2b). The share generated by the trucking industry increased rapidly between 1950 and 1990 to 79 percent, with rail declining to 9 percent. Interestingly, truck transport within metropolitan areas accounts for nearly one-third

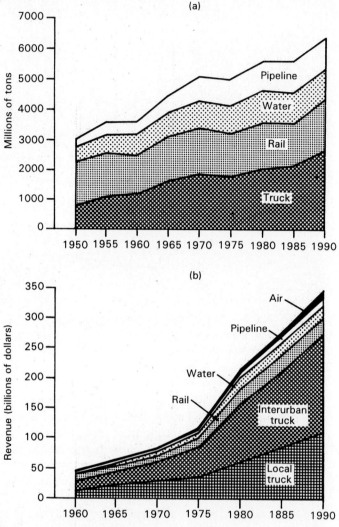

Figure 14.2 Change in (a) interurban freight activity, and (b) revenues received by different transport modes, United States, 1950–1990.
(*Source:* Redrawn from BTS, 1994, Figure 3.13, p. 59.)

of all freight revenues generated in the United States. Air transport now accounts for 4 percent of total freight revenues, exceeding that received from pipelines.

Air Transport The public service area, however, in which the airline industry is a significant force is with the interurban transport of people. Although passenger rail service connects all the major metropolitan areas of the United

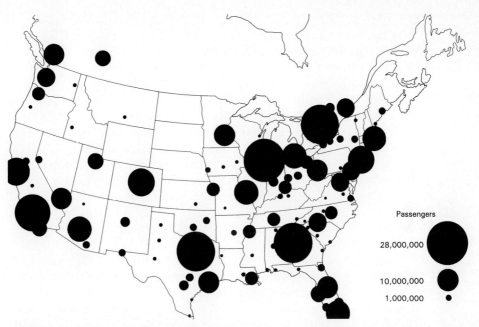

Figure 14.3 Enplanements at the major commercial airports in North America (based on BTS, 1994, Figure 4.3, p. 77; with data for the five leading Canadian airports added by the author).

States (Amtrak) and Canada (Via Rail), service is infrequent outside of the Northeast and Midwest of the United States where Chicago and New York are the main foci, and the Windsor-Quebec City corridor in central Canada. Private automobiles, buses, and aircraft have largely replaced rail for the movement of people, whereas for most interurban business trips air travel is the mode of choice. Air travel is also an important carrier for long-distance vacation trips, though the interstate system has fostered a large volume of road travel during summer periods when families visit relatives, friends, and prime vacation sites.

Figure 14.3 indicates the location of the largest commercial airports in North America. The numbers refer to *enplaned* passengers, which is not necessarily the same as actual passenger destinations, which are the basis of the analysis in Figure 4.8. In some cases, the extensive regional market is served by more than one facility. For example, New York is served by La Guardia, JFK, and Newark. The reason why the volumes are extremely large for certain airports is not simply that they are serving large local markets, but that they are also focal points in a pronounced *hub and spoke network structure.* This structure has been devised by the carriers as an efficient way of linking passengers from small and medium-sized urban areas with collecting points. From these collecting points, or hubs, passengers connect with carriers serving the continent and other parts of the globe. Each major airline has a particular hub as its home base—such as United Airlines in Chicago, American Airlines in Dallas, Air Canada in Toronto, and Delta Air Lines in Atlanta—and these hubs serve as collecting and transfer points for freight as well as passengers.

Travelers from subcenters within the regions of particular hubs invariably have to journey to the hub if they wish to connect to a region served by another major center. For a number of decades, Chicago was the major hub for most of North America, but, with the growth of the Sun Belt, Dallas-Fort Worth and Atlanta have emerged as competitive centers. New York, Los Angeles-Burbank-Long Beach, San Francisco-Oakland, Miami-Fort Lauderdale, Toronto, and Vancouver are hubs, prime destinations, and also serve as major transfer points for international travel. Interestingly, Denver, as the headquarters of a regional carrier (Frontier Airlines), has emerged as a hub linking the mountain states to the major centers, much in the same way as Minneapolis (Northwest Airlines) developed earlier as the link between the upper Midwest and the continental system.

In an age of intense interurban competition for high value-added economic activities, the impact of a hub and spoke system on differential urban growth can be quite considerable. Just as certain large metropolises have become world cities because of the role that they play in the world financial markets, either as transactional centers or as regional conduits for capital, so have certain metropolitan centers with large airports become more attractive locations than others for the conduct of business in organizations that have, or wish to develop, a national or international reach. There is a strong imperative for such organizations to locate in the vicinity of metropolises with airports that serve as hubs within the continental system, or that have a variety of direct international links. As a senior vice president of a large real estate corporation is reported to have commented with respect to the increasing demand for suburban office space: "The advantage of being in the suburbs is that you're closer to customers, closer to where you're employees live, and closer to the airport" (Zehr, 1995).

INTRAURBAN ACTIVITY PATTERNS

The various activities that individuals may be engaged in during a day, week, or year may be classified in a number of ways. An important first distinction based on where activities take place is between domestic and community activities. *Domestic activities*, such as sleeping, housework, eating, watching television, and pursuing hobbies, are engaged in at home and are independent of transportation. *Community activities* are engaged in outside the home, such as shopping, visiting friends, or going to the movies. Another distinction can be made between *obligatory activities*, such as sleeping, shopping for groceries, and working, and *discretionary activities*, such as eating out, cruising malls, or watching

Table 14.1

Average Weekly Time in Travel for Persons Age 16 and Older Engaged in Community-Oriented Obligatory and Discretionary Activities

ACTIVITY TYPE	TIME SPENT TRAVELING (HRS)	
Obligatory		
To and from work	1.76	
Work-related business	.16	
Shopping	.85	
School/church	.36	
Doctor/dentist	.08	
		3.21
Discretionary		
Family or personal business	1.30	
Vacation	.12	
Visiting friends/relatives	.70	
Pleasure driving	.05	
Other social/recreational	1.04	
Other	.05	
		3.26
		6.47

(*Source:* Data retabulated from BTS (1994), Table 1.1, p. 2.)

television. The average time people spend per week in community-oriented obligatory or discretionary activities is indicated in Table 14.1. In a period when time is becoming an increasingly precious commodity (recall Figure 13.14) people are spending on average one hour per day in travel, though journeys to work, which average 42 minutes per round trip on work days, are evidently frequently combined with other activity-related trips.

In terms of transport planning and maintenance, where the focus is (or should be) on making obligatory travel as hassle free as possible, it is important to note that engagement in these activities varies in time as well as space. Obligatory travel, especially that related to work and school, is invariably governed by regulated hours, particularly starting times. Office activities and shift work are still structured around a general 7:00 A.M. to 5:00 P.M. work period. On the other hand, in North America other obligatory activities are frequently available into the evening, and grocery shopping up to 24 hours in most days of the week. Discretionary activities are usually engaged in during evenings or at weekends. In consequence, in North America, as different from urban areas in some other parts of the world (for example, western Europe) where the work day is more highly regulated for all types of employment, there are only a few hours in the night when metropolises are not "on the go." Generally, regulated

Table 14.2

Average Number of Daily Trips, Miles Traveled, and Trip Lengths per Person in United States Metropolitan Areas, by Sex and Trip Purpose

TRIP PURPOSE	MALE	FEMALE	BOTH
Average Daily Person Trips			
Earning a living	0.77	0.57	0.66
Family and personal business	1.12	1.42	1.28
Civic, educational, and religious	0.34	0.36	0.35
Social and recreational	0.78	0.74	0.76
Other	0.02	0.02	0.02
Total	3.03	3.12	3.08
Average Daily Person Miles of Travel			
Earning a living	10.5	5.12	7.69
Family and personal business	8.62	9.23	8.93
Civic, educational, and religious	1.79	1.89	1.84
Social and recreational	10.4	9.38	9.86
Other	0.25	0.20	0.22
Total	31.56	25.83	28.56
Average Person Trip Length (Miles)			
Earning a living	13.91	9.18	11.8
Family and personal business	7.75	6.63	7.11
Civic, educational, and religious	5.39	5.38	5.39
Social and recreational	13.45	12.93	13.19
Other	11.59	9.13	10.30
Total	10.54	8.47	9.45

(*Source:* BTS (1994), Table 3.1, p. 55.)

work hours for many jobs do, however, give rise to distinct rhythms in activity patterns and movement within urban areas, most noticeably the morning and evening rush hours associated with work trips.

Movement Characteristics

Since most urban residents participate in community activities every day, the aggregate number of daily person trips made within urban areas is staggering. Both men and women average a little over three trips per day, with trips related to earning a living, and family and personal business activities, predominating (Table 14.2). Although men and women incur roughly similar miles of travel and average trip length for most activities, there is a big difference between

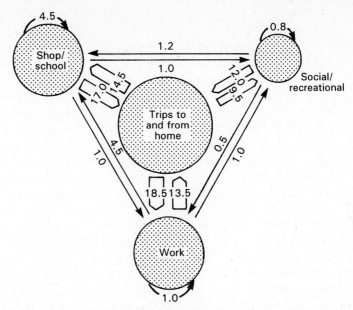

Figure 14.4 The predominance of home-based person trips within urban areas (numbers in percent of total trips, with size of circles indicating the relative share of total trips home-, work-, school/shop-, and social/recreational-based).

men and women in miles traveled and trip lengths associated with earning a living (as noted in Chapter 13). Person miles of travel and average trip length are related to urban population size—the deconcentrated nature of larger metropolises generates greater distances and average trip lengths (Barber, 1995, p. 95). Because of the time-space variations in activity patterns, the trips people make involve many different origins and destinations, some concentrated and others dispersed, depending on the trip purpose; they occur at different times during the day; and they may, in some of the larger metropolises, be made using different modes of transport.

The origins and destinations of activity-oriented trips listed in Table 14.1 are summarized more directly for a typical large metropolis in Figure 14.4. Assuming a population of 1 million and an average of 3.08 trips per person per day (Table 14.1), there would be about 3 million person trips totaling 28.5 million miles within such a metropolis daily. Typically, 32 percent of all journeys are from home to work and back; a second major component, about 31 percent, are the trips focusing on schools and shops; and another 21 percent involve social and recreational trips. Complementing the home-based pattern are the journeys made from one out-of-home activity to another. With decreasing household size, these multiple-destination trips, such as from work to shops and then home, are assuming a larger share of total trips than previously, increasing from about 10 percent in the 1960s to 15 percent in the 1990s. The small proportion of work-work and school-work trips are included to reflect the growing importance of second jobs and student employment in urban economies.

By far the largest proportion of person trips for all purposes in the highly suburbanized urban areas of North America are made by private automobile. Not only is travel by car more comfortable, convenient, and flexible for most urban residents, the automobile is also regarded as safer. The information in Figure 14.5 suggests that about 86 percent of all person trips in metropolitan areas in the United States are undertaken by automobile. Between 1980 and 1990, the proportion of trips involving persons driving alone increased quite dramatically while the number of people using public or other means of transportation decreased. Walking outside of malls—except in the cases of school-age children who reside too close to schools to be bused, adolescents converging on shopping centers, or dog owners being exercised by their canine friends—is decreasing. In many suburban areas, sidewalks are simply not provided. Even in the few metropolitan areas that have good public transport systems, most person trips are made by car. For example, in metro Toronto (1991 population 2,275,000), which is well served by an integrated system of rapid transit, commuter railroad, streetcars, and buses (Cervero, 1986), 73 percent of all person trips are made by car (METROT, 1993).

Typical variations in the pattern of person trips starting at different times of the day in large metropolitan areas are shown in Figure 14.6a. Although the work-related peak concentrations during the morning and evening rush hours stand out, high volumes of traffic occur throughout most of the day. There has been a tendency for the relative magnitude of these peaks, compared with the daytime hourly average, to decline about 10 percent between the 1960s and 1990s. This may be due to the relative decline in manufacturing employment, which has quite rigid shift schedules, and the increase in service employment, a large proportion of which tends to involve work in the evenings as many of the jobs are directly customer oriented (BTS, 1994, pp. 51–52). Also, employers in many companies, particularly those located in downtown areas or in other large nucleations and edge cities, have attempted to reduce employee stress created by peak travel by organizing flextime schedules or earlier or later starting to spread out the peak.

One of the important features of trips associated with earning a living is that they have been increasing in length as urban areas have decentralized. Recent increases in work trip lengths for homeowners and renters (not the entire working population) summarized in Table 14.3 suggest that people on the average traveled 50 to 60 percent farther to work in 1989 than they did in 1975. Homeowners, who, it may be assumed, are predominantly suburbanites, travel farther than renters who, it is similarly assumed, reside predominantly in central cities. On the other hand, the average speeds at which homeowners and renters traveled to work also increased 40 to 60 percent during the same time period, because many suburban workers use limited-access highways to travel to suburban jobs. Thus, travel times to work for both groups have not changed, though, as has been noted in Figure 7.12, roadway congestion is increasing considerably on suburban highways and arterials in the largest metropolises. These changes are important because longer trip lengths, increasing travel speeds,

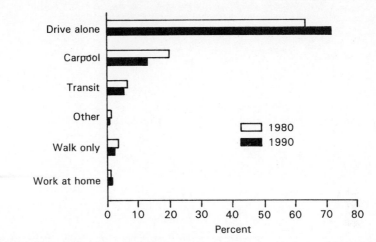

Figure 14.5 The dominance of the automobile in metropolitan travel.
(*Source:* BTS, 1994, Figure 3.5, p. 53.)

and increased congestion add up to increased energy use and exacerbate various types of pollution.

Although trips associated with earning a living tend to peak in the morning and early evenings, other activity-related trips occur during the off-peak times to generate metropolises that are busy with travelers throughout the day. Personal business trips concentrate between 9:00 A.M. and 7:00 P.M., education-related trips in the early morning and later afternoon, and social/recreational trips in the evening hours (Figure 14.6b). Even though the actual number of work-related trips increased between the 1960s and 1990s, there has been a decline in the proportion of total person trips that are work related from around 40 percent to 30 percent (BTS, 1994, p. 52). Personal- and social-related travel has been growing much more rapidly than work-oriented journeys, suggesting that urban congestion is becoming less work related and more social and shopping trip related. This, in turn, implies that the actions of individuals, rather than the designs of employers and planners, will have to be relied on to spread the peak.

Modeling Movements Related to Activity Patterns

With rapid population growth, suburbanization, increasing car ownership, and federal involvement in highway systems and various housing programs, transportation planning emerged as a major area of concern in the United States and Canada in the 1950s and 1960s (Pas, 1995). Region-wide, data-based, goal-oriented, forecast-centered studies were undertaken in many metropolitan areas including Detroit, Chicago, Washington, D.C., Pittsburgh, Philadelphia, and Toronto. The objectives of the studies were to understand better the traffic-generating process, forecast probable future demands, and suggest a variety of integrated transport systems or plans that would cater for these traffic forecasts. Given that the studies were based on forecasting on the basis of current trends,

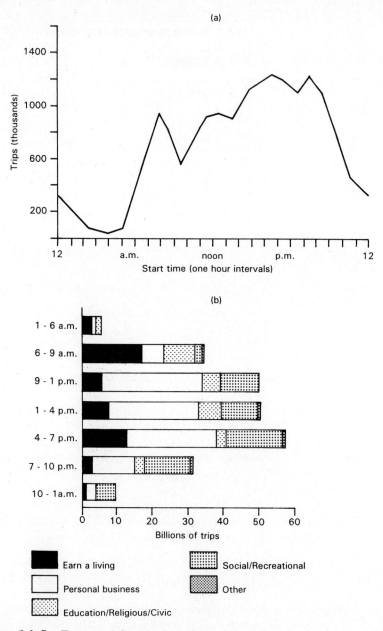

Figure 14.6 Temporal fluctuations in activity patterns for (a) a typical large metropolitan area, and (b) by type of activity.
(*Source:* from BTS, 1994, Figure 3.3, p. 51.)

Table 14.3

One-Way Trip Lengths (in Miles), Speed (Miles per Hour) of Travel and Travel Times (in Minutes) to Work in United States Metropolitan Areas

	HOMEOWNERS		RENTERS	
	1975	**1989**	**1975**	**1989**
Trip length (miles)	7.7	12.0	4.9	8.0
Speed (mph)	22	31	15	25
Travel time (minutes)	19	19	21	21

(*Source:* Data extracted from BTS, 1994, Figure 4.18, p. 87.)

they inevitably emphasized the design of highway systems that would cater for predicted large increases in automobile travel.

Nevertheless, the design of these studies, and their computer-based implementation, pioneered in many ways in the studies for Detroit (DMARTS, 1955) and Chicago (CATS, 1959), has entered the lexicon of transport planning and various elements are reemerging in GIS applications related to transport and diffusion. The data produced in these and other studies also provided the basis for many early studies exploring the connections between land use and activity patterns, such as those relating household location to trip making (Kain, 1962), the structure of the urban retail system and shopping travel behavior (Berry, 1963), and the effect on journey-to-work patterns of manufacturing decentralization (Taaffe et al., 1963). Although research contributions such as these were stimulated by the availability of data generated by the transport studies, the land use focus of the databases invariably restricted the scope of studies to the analysis of economic behavior.

Transport System Modeling   The travel and forecasting models were based in general on four sequential submodels focusing on trip generation, trip distribution, modal choice, and route choice, which together comprise an *urban transport model system* (UTMS; Hutchinson, 1991). An urban region is divided into a number of traffic zones, or cells, in which information concerning the number of trips originating and ending in each zone (hence the name *origin/destination studies*), present land use, and current socioeconomic characteristics of the population, is tabulated in the context of this place-oriented database.

1. The *trip generation model* seeks to identify the factors influencing the number of trips generated by and attracted to each individual zone. These factors for each zone are usually assumed to be the:

- Various types (or proportions) of land use
- Intensity or density of the various land uses
- Socioeconomic characteristics of the persons or uses of the land

For example, a traffic zone with a high proportion of its land in high-rise apartments would have greater traffic generation than one that contains mostly single-family residences. Similarly, a zone involving a shopping center would also attract a large number of trips. Given that the predominance of home-based trips in the 1950s and 1960s was even greater than that depicted for the 1990s in Figure 14.1, most attention was paid to residential trip generation. The models were calibrated using the estimated cross-sectional information.

2. The *trip distribution model* uses the empirical origin destination data, that is, the flow of all trips between all zones, to develop a general mathematical statement which best represents the aggregate pattern of flows. This model is then used for forecasting purposes. The general statement is usually expressed in terms of:

- The trip-generating characteristics of origin zones
- The trip-attracting characteristics of receiving zones
- A measure of the friction of distance between all pairs of zones

As indicated in Chapter 2, the type of general mathematical statement often used to represent such flows is based on the gravity model, a simple representation being:

$$T_{ij} = O_i\, D_j\, /\, F_{ij}$$

where, T_{ij} = number of trips between zone i and zone j

O_i = a measure of the trip-generating potential (e.g., population, number of residences, etc.) of origin zone i

D_j = a measure of the attractiveness (e.g., employment, shopping space, etc.) of destination zone j

F_{ij} = a measure of the friction of distance (e.g., highway mileage, traffic lights, etc.) between zones i and j

The complexity of the basic descriptive model and subsequent forecasting models invariably relates to the various procedures used for estimating and predicting changes in O_i, D_j, and F_{ij}.

3. The *modal choice model* again uses the empirical origin destination data to describe and forecast the variety of means, or modes, of travel used between each origin and each destination. The possible range of travel modes includes: trucks, trains, automobiles, rapid transit (heavy-rail), commuter railroads, buses, streetcars and light-rail transit (LRT), bicycles, and walking. In general, the modal choice model relates:

a) The proportion of trips using various modes between each origin and each destination to the range and intensity of different land uses and socioeconomic characteristics (such as incomes) in the origins; and

b) The relative availability and utility of alternative modes of travel between the origin and the destination to persons and activities in the origin.

For example, if public transport, of any kind, is not available between an origin and destination, the modal choice for person trips is extremely limited. If a variety of modes are available, the choice of modes may well depend on household income. The forecasting model in effect estimates future modal choice in the light of predicted changes in availability of different transport modes, land use, and socioeconomic characteristics for the entire set of traffic zones.

4. The objective of *route choice models* is to find, for the projected flows, the optimal number of trips for each route from each origin to each destination for each mode that it is possible to use. The optimum allocation pattern is usually defined as that which minimizes aggregate travel times for all forecast trips subject to traffic volumes on particular routes being equal to or less than estimated capacity constraints for all routes. An optimal outcome would, therefore, be a pattern of route volumes at different times of the day that not only minimizes aggregate travel times, but also minimizes congestion. These types of route assignment models are exceedingly complex, usually requiring mathematical programming techniques of the type introduced by the location-allocation model mentioned in Chapter 10 with respect to public facility location. During the 1950s and 1960s, these types of models were difficult to resolve due to conceptual and computer limitations, but with advances in both mathematical programming theory and computing capacity they have become more tractable (Sheffi, 1985).

The Impact of Transport System Modeling   In economic terms, it may be argued that the transport systems developed in the planning framework of the urban transport model system (UTMS) have been reasonably effective (Hassell, 1980). That is, even in rapidly expanding urban areas, the transport system has been efficient and flexible enough to permit households to locate in a manner that has contained user travel times (Levinson and Kumar, 1994). Furthermore, the transport systems that were built in the context of UTMS planning were front-end capital loaded during a period (1955 to 1970) when capital was available at reasonable interest rates. Just as important, in terms of local and state/provincial operating expenditures, the systems that were constructed required relatively low maintenance and operating costs, though many bridges are needing to be replaced in the 1990s and roads resurfaced. Also, the new urban transport systems had a considerable multiplier effect on further construction and development as the networks brought more land within the range of possible urban use (Giuliano, 1989). Thus, in purely economic terms, there is a case to be made that good advantage was taken of the availability of federal transport development funds.

On the other hand, the UTMS planning system paid little or no attention to environmental and social impacts. So the structure of the models led inexorably to the planning and incremental construction of urban transport systems based primarily on one mode of transportation—the automobile. In consequence, low-density suburban areas dependent on automobiles and highways were encouraged to sprawl over the countryside, public transport systems were restricted to serve primarily higher-density central cities and the quality of their service allowed to languish, older inner-city neighborhoods were often decimated by highways that were constructed to connect central business areas to suburbs, and urban areas became highly polluted from the noise and emissions emanating from the rapidly increasing volume of trucks and automobiles.

Furthermore, in many cases highways could not be built big enough to cater for the volume of traffic, which usually greatly exceeded most forecasts. Similarly, congestion costs on routes leading to downtown areas added further incentives fostering the decentralization of people and economic activities. The immediate reaction in a number of urban areas was public revolt to halt or slow down the pace of highway construction (Gakenheimer, 1976; Nowlan and Nowlan, 1970). Regardless, if the construction of highways thrusting into the heart of cities became less feasible, beltways in suburban areas around cities were still acceptable, with the result that the sectoral highways decanted the inner city's economy, and the radial highways contained the eviscerated remains. The longer-term reaction in the face of inexorable suburbanization, saturation of car ownership, and spreading suburban gridlock (Cervero, 1986) has been increasing standards regulating emission controls, federal or state/provincial requirements that transport planning take into account socioeconomic and environmental impacts of proposed transport developments, and increasing federal and other public investments for upgrading and expanding existing public transport systems and building new ones to serve extended urban regions (Smerk, 1991; Weiner, 1992).

URBAN SPATIAL STRUCTURE, THE JOURNEY TO WORK, AND MODAL CHOICE

The major feature in the activity patterns of most urban residents is the time spent earning a living and the journey to and from work that is involved (Table 14.1). Although shop/school and social/recreational trips are large in number, trips related to work remain the largest single component in the daily pattern of movement for most households. With the average worker spending 42 minutes per day commuting (Table 14.3), it is evident that traveling to and from work takes significant time out of increasingly busy days. Moreover, commuting is also tiring, and for many people the nervous tension involved with work trips is difficult to withstand. People have developed a variety of coping mechanisms ranging from radios and sound systems in automobiles to walkmans and reading in public transit. The widespread use of cell phones during work trips reflects the pressure that people feel to use time as efficiently as possible.

Transportation in Monocentric and Polycentric Urban Structures

In the monocentric situation, with most economic activities and hence employment concentrated centrally and population and land costs decreasing from the core, public transportation has certain advantages over private automobile transportation. High population densities not only imply relatively short work trip lengths, they also provide sufficient intensity of demand to support a variety of different integrated transport modes that can convey increasing increments of passengers to workplaces (Gordon et al., 1989; Giuliano and Small, 1993). Thus, buses and streetcars can feed passengers into rapid transit facilities which are located on the main radial arteries leading into the center of the city. As the monocentric city spreads through population growth and increasing household demand for space (associated with rising real household incomes), work trips lengthen and travel times increase.

Eventually, the cost and inconvenience, in both time and money, of public transport work trips for some people at a certain distance from the center of the city may increase to a level at which the public-private transport trade-off favors auto work trips despite high downtown parking costs. The level of these downtown parking fees can, therefore, influence the range and size of the public transport market (Shoup and Wilson, 1992). As employment deconcentrates, perhaps in response to the spreading population, auto work trips become more common because public transport is at a cost disadvantage when it attempts to connect low-population-density areas with dispersed places of employment (which also usually provide free parking). In fact, the costs of automobile ownership—which include licensing, insurance, gasoline, maintenance, financing, and asset depreciation—are rarely factored meaningfully into the public-private transport work trip trade-off because the benefits of auto ownership extend to the whole array of obligatory and discretionary activities indicated in Table 14.1 with which an individual or household may be involved.

The Public-Private Transportation Trade-Off in a Polycentric Environment Polynucleated or dispersed urban environments are characterized by: a multiplicity of employment locations, some of which may be quite concentrated; generally low population densities, which may decrease in a negative exponential manner with respect to a particular local center of employment or commerce; and a more complex and fragmented distribution of households by income. In such an situation, low population densities "could mean either shorter or longer trips depending on whether workers choose homes around employment subcenters . . . or whether cross-commuting across metropolitan areas is common" (Gordon et al., 1989, p. 140). Figure 7.9 demonstrates that in the polycentric/dispersed environment of most North American cities the volume of crosstown commuting exceeds those of work trips within central cities, and between central cities and suburbs. Work travel times are further influenced by the dual-worker nature of many households, with, as has been discussed in Chapter 13, a number of households choosing residential locations that minimize travel time for one of the partners.

The automobile is the only means of travel suited to low-density, dispersed environments, because public transport requires high densities, in terms of both population and concentrated employment, to generate traffic volumes sufficient to warrant the costs and visible subsidies. Low-density environments with dispersed employment centers thus compel workers to drive alone to work because there are no other transport alternatives—that is, there is no modal choice. Hence, the inexorable tendency for trips in polycentric and dispersed environments to drive alone (Figure 14.5). In this context, it is also interesting to recall that while work trip lengths have increased in length by more than 50 percent, mean travel times have remained about the same (Table 14.3). This is entirely consistent with the nature of low-density environments, where automobile travel can be at quite high speeds and employment opportunities are dispersed. Traffic congestion at peak times around suburban employment centers may eventually cause average speeds to decrease and travel times to increase. In Vancouver, for example, average trip lengths are increasing, road congestion is causing trip speeds to decrease, and travel times are consequently increasing (Howard, 1995).

Spreading suburban congestion at peak travel periods in the largest metropolitan areas (Figure 7.12) is regarded by Cervero (1989) as the consequence of transferring the national industrial era practice of segregating land uses through zoning, which was a response to the concentration of employment and people in high-density monocentric cities, to the late industrial era when people and employment in most metropolitan areas have become highly dispersed. He argues that the transference of this land-segregating practice from the monocentric industrial city to the current polycentric era is unnecessary because this is an age of less noxious high-technology and service industry employment. The zoning of many of these activities into industrial parks and commercial nucleations in suburban areas induces congestion on surrounding roads at peak travel periods because of the necessity for most people to drive alone to work.

In light of this auto-dependent situation, Cervero (1989) reluctantly recommends greater flexibility in zoning practices for activities providing employment in suburban areas to avoid the congestion arising from high employment concentrations. Ideally, an urban environment would be preferred that would retain modal choice, reduce dependency on the energy-consuming, costly, and pollution creating automobile, and make suburban employment opportunities more accessible to entire metropolitan populations. This, however, requires a level of metropolitan spatial restructuring that would be difficult to implement given the highway-oriented form that polycentrism has taken in North America, and the degree to which the automobile and truck have come to dominate. While public transport alternatives exist in parts of many urban areas, particularly in central cities, ridership in metropolises other than New York, Chicago, San Francisco, Montreal, and Toronto, is quite low compared with the population sizes of the urban areas involved (Figure 14.7).

Outer- and Inner-Urban-Area Contrasts in Modal Choice

The experience of metro Toronto and the extended Toronto CMA casts some light on the modal choice issue. For the purpose of estimating the volume of morning work trips and total daily trips for all activities carried by different transport modes, the Toronto CMA has been divided into four parts by three concentric cordon lines along which traffic surveys are periodically undertaken (Figure 14.8 and Table 14.4). The City of Toronto comprises part of the area located within the inner suburban and central cordon lines. The trip information is based on cordon counts rather than an origin-destination survey based on a large number of traffic zones. Cordon counts are far less expensive, though less

Figure 14.7   Public transport use (all modes) in the 33 metropolises with over 1 million population. These metropolises account for 90 percent of all North American public transport use.
(*Source:* Redrawn from Fielding, 1995, Figure 12.1, p. 289.)

thorough, than origin-destination surveys, but can be used to obtain sample data to address specific issues.

The modal share information is obtained from one-day counts of traffic crossing a number of points along the cordon lines: one cordon is defined as the Metro boundary beyond which are located exurbs and post–1975 suburbs; another cordon roughly separates post–1950 suburbs from the middle of the metro area; and a third cordon generally delimits the pre-1950 and central commercial area (METROT, 1993). The fourfold spatial division that results thus embraces an inner urban area involving the middle Metro area and the central core containing about 1 million people, and an extensive suburban area divided by the Metro boundary that is highly dispersed and polycentric in form involving about 2.9 million people. The extent of decentralization of employment is indicated to a degree in Table 9.3, which suggests that this suburban area is the location of more than two-thirds of the manufacturing employment in the CMA. On the other hand, the central core contains a majority of the office space in the CMA and is the location of about 400,000 jobs, 60 percent of which are full-time in offices (Nowlan and Stewart, 1991).

Change in Modal Diversity and Modal Share   The map of limited-access highways and rapid transit routes superimposed on the pattern of urban development in the Toronto region (Figure 14.8) emphasizes the polycentric-monocentric dichotomy that underlies the structure of transport systems within the

Table 14.4

The Toronto CMA: Cordon Trip Counts (1993) and Population Estimates (1991) for the Outer Suburban, Inner Suburban, Middle Metro, and Central Core

AREA, OR CORDON LINE	RESIDENTIAL POPULATION (IN THOUSANDS)	PERSON TRIPS			
		MORNING INBOUND PEAK		TWO-WAY DAYTIME	
		(IN THOUSANDS)			
		1983	1993	1983	1993
Outside metro	1,620				
Metro boundary		203	288	1,097	1,657
Inside metro	2,273				
Inner suburbs	1,256				
Inner suburban		300	317	1,572	1,801
Middle metro	877				
Central area		306	303	1,339	1,386
Central core	140				
Total (CMA population)	3,893				

(*Source:* Statistics Canada (1994); and METROT (1993).)

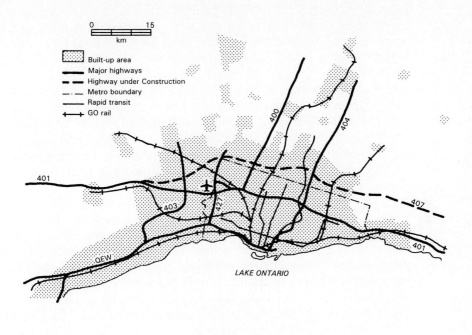

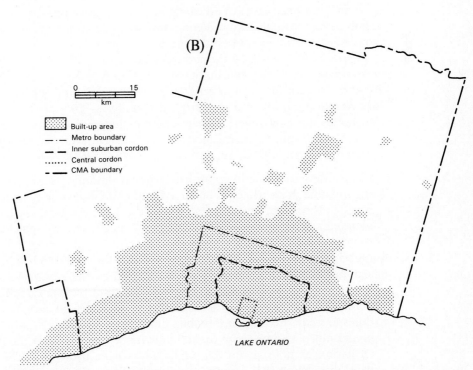

Figure 14.8 (A) Major limited-access highways, rapid transit, and commuter railroad routes superimposed on the built up area of the Toronto CMA, 1991, with (B) trip survey cordon lines.

(*Source:* Compiled by the author from information on several maps in GTA, 1995.)

urban area. On the one hand, the public transit systems, which are represented by the rapid transit lines within the Metro area and the commuter railroad network reaching far into the suburbs, focus on the central core, while the highway network generally follows an extensive grid pattern and is much less centrally focused. Highway 401, which was constructed in the early 1960s as a belt highway for vehicles to circumnavigate the periphery of the city, is now a much congested twelve-lane mid-metropolis throughway, and a new circumferential route farther north (Highway 407) opened in 1997. If the pace of urban expansion continues, another parallel circumferential route can be imagined farther north in the 2030s. Extensive nucleations of business and commercial activity are arrayed along Highway 401, particularly at the intersections with the north-south highways.

The highway system, constructed with federal/provincial funds, thus promotes polycentrism, whereas the public transport system, constructed largely with provincial/local funds and operated with the assistance of subsidies from the province and Metro, reflects monocentrism. Interestingly, all levels of government have attempted to reinforce monocentrism and the use of public transport in off-peak hours in a number of ways, such as supporting (via partnerships) the downtown location of stadia for major league sports teams, and the construction or renovation of theaters in the central core. The trip volume and modal share information recorded at the three cordons suggests, however, that even with these types of efforts to maintain a presence for public transport, the polycentric spatial structure generated by highways, and used by trucks and automobiles, has an overwhelming impact for most of the urban area.

Figure 14.9 shows how the volumes of trips (see Table 14.4 for aggregate trip volumes) and the modal share change at each of the three cordons between 1983 and 1993, for inbound morning rush-hour trips and daytime two-way trips. The evening period is excluded from the analysis. The modes are defined as: drive alone; auto passenger; transit, which includes buses, streetcars, and rapid transit; and GO rail, a provincially supported commuter railroad system which improved and extended its service between 1983 and 1993.

1. The greatest increase in volume of two-way daytime person trips and morning inbound peak trips has occurred at the *Metro boundary cordon* which bifurcates the suburban area. Although the built-up area beyond the Metro boundary is reasonably well served by bus transport, particularly those routes connecting to GO-train and rapid transit connections, crosstown commuting predominates. Thus, although GO-train ridership has increased, the share of morning inbound worktrips carried by the commuter railroad and transit has increased only a little, and for total two-way daytime trips the public transport share has decreased. The pie charts illustrate clearly that in this suburban area the automobile dominates, carrying 83 percent of all inbound morning work trips and 90 percent of two-way daytime person trips. Of particular note, given the public advertising promoting ride-share work trips, drive-alone is the mode that exhibits the greatest volume of increase, a trend entirely consistent with the dispersed employment structure of polynucleated urban environments.

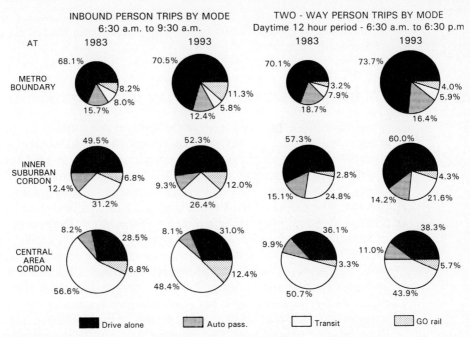

INBOUND PERSON TRIPS BY MODE
6:30 a.m. to 9:30 a.m.

TWO - WAY PERSON TRIPS BY MODE
Daytime 12 hour period - 6:30 a.m. to 6:30 p.m

Figure 14.9   Modal shares of morning in-bound rush-hour person trips, and twelve-hour daytime two-way trips, Metro-Toronto.
(*Source:* Extracted and recompiled from METROT, 1993.)

2. The dominance of the automobile begins to wane at the *inner suburban cordon* where population densities increase and alternative transport modes, particularly buses and rapid transit, are more available. However, the share of trips carried both in the morning peak and the daytime period by transit has decreased, while the share conveyed by automobiles has increased. Nevertheless, at this perceptual junction between the "suburbs" and the "central city," public transport is sufficiently available to be a significant carrier, accounting for 38.4 percent of morning peak inbound trips and about 26 percent of two-way daytime travel.

3. The advantage of public transit is seen most clearly in the modal distribution of trips crossing the *central area cordon*. More than 60 percent of morning peak travel enters the central core on public transit or commuter railroad, and more than 50 percent of all daytime two-way trips also use public transport. Unfortunately, as the share of trips undertaken by automobile has increased and the total volume of daytime traffic entering and leaving the central core has remained about the same since 1983, the market for public transportation in the central city appears to have been decreasing. The economic health of public transport clearly depends on the strength of employment in the central area, and the "population turnaround" that may be occurring in middle Metro and the central core (Bourne, 1991; Nowlan, 1995).

The polycentric nature of modern metropolism is thus making it difficult for all forms of public transportation to survive, leaving little modal "choice" for much of the urban population.

RAPID TRANSIT APPROACHES TO METROPOLITAN SPATIAL RESTRUCTURING

Since 1980 there has been a great increase in the number of heavy-rail rapid transit systems in large metropolises around the world. Heavy-rail systems get their power from a third rail and must occupy grade-separated or fenced rights-of-way. They are popular primarily because of their tremendous carrying capacity, which is ideally suited to high-population-density metropolises such as London, which opened the first underground line in 1863, and Hong Kong and Singapore, which opened their systems in the 1980s. Depending on the length of the trains, the speed capabilities of the system, and the minimum distances (or headway) required between each train for safety purposes, rapid transit can carry ten times (or more) the passenger volume per hour of a four-lane highway with traffic moving freely. The New York rapid transit system, for example, can carry 80,000 people per track-mile per hour, and the Hong Kong system 115,000. In 1980, there were 58 systems in the world that had all, or part, of their rails underground. In 1990 there were 81 systems, some of which incorporate light-rail transit (LRT) for part of their length. LRT is the streetcar reincarnated. They are cheaper to build than heavy-rail because they can draw their power from an overhead wire, and because of their slower speed and short length (usually consisting of only one or two carriages, they can be located at ground level and in the middle of streets). Their passenger carrying capacity is a great deal less than rapid transit.

There have been two phases in the development of rapid transit, powered by electric motors, in North America. The first phase, which commenced in Boston in 1897 with a 2.4-kilometer line to the central business district, lasted from the end of the nineteenth century to the mid–1920s. During this period, lengthy systems were also built in New York and Chicago, and smaller systems elsewhere. New York now has the world's largest network consisting of 27 lines and 416 kilometers of track of which 232 kilometers are underground. These systems reflected the monocentric nature of the classic industrial/commercial monocentric city with their lines focusing on their downtowns. Indeed, in the case of Chicago, the elevated rapid transit lines that converged to circle the central business district provided the adjective describing and defining the downtown area—the Loop.

The second phase, commencing in the 1950s, began with the construction of new systems in Montreal (61 kilometers in length) and Toronto (53 kilometers) as attempts to reinforce access to their downtowns in the face of the first wave of post–World War II suburbanization. These systems demonstrated that if the lines were located along major radials, and coordinated with fairly powerful regional planning controls (for example, limiting the number of suburban regional shopping malls), they could have a major impact on land use development and hence the spatial structure of the expanding metropolitan area. Local surface transit lines feed into the rapid transit stations, and a number of the stations have become the focus for major business and commercial developments with "mini-downtown" clusters of high-rise buildings. Thus, *as long as public investments in rapid transit systems are coordinated with other powerful regional planning arrangements* they can help contain the proliferation of urban sprawl, reduce traffic congestion along some arteries, and support downtown revitalization efforts. Furthermore, by fostering land and property development they help increase the property tax base.

Experience in the United States with New Rapid Transit Systems

Growing concern over sprawling metropolitan development, increasing pollution, extravagant energy use, decaying downtowns, congested circumferential and radial highways, and a desire to influence urban form in metropolitan areas fragmented by many governmental jurisdictions, has led to the construction of a number of new urban rapid transit systems in the United States with federal and local funds (Smerk, 1991). In essence, if federally supported massive funding of highway construction, federally insured and hence promoted low-cost low-density suburban housing tracts, and federally encouraged closure of many commuter rail lines have helped create the conditions leading to these concerns, it is hoped that federally encouraged regional rapid transit systems may ameliorate some of the problems. During this second phase, which has lasted to the 1990s, new rapid transit systems have been constructed in a number of urban regions such as San Francisco, Washington, D.C., Philadelphia, Atlanta, and Miami.

The BART Experience The 115-kilometer (72-mile) Bay Area rapid transit system, extending over an area with about 3 million people, opened in 1972 as the first large-scale regional system to be constructed in the United States for almost fifty years (Figure 14.10). It had three major objectives. The first was to provide a peak-hour, congestion-free, long-distance commuting facility from the suburbs to downtown San Francisco and promote growth in its central business district. Thus, in this most physically divided and economically polynucleated of environments, the BART system was designed to buck the trend. The second was to promote and coordinate regional growth by providing foci for development around the rapid transit stations that were placed on average about 4 kilometers (2.5 miles) apart. The third was to be attractive enough to compete with the automobile by improving on what were thought (in 1962 at the planning stage) to be the car's most attractive features: speed, comfort, style, and downtown delivery without parking problems. By 1976, when the system was essentially complete, the construction costs had reached $1.6 billion (in 1976 prices). Several extensions have been made in the mid–1990s (see Figure 14.10), most noticeably to Colma which connects with San Francisco airport, costing an additional $2.6 billion (in 1995 prices).

Early, and somewhat premature, criticisms of the system were directed mainly at the technical difficulties associated with operating such a massive

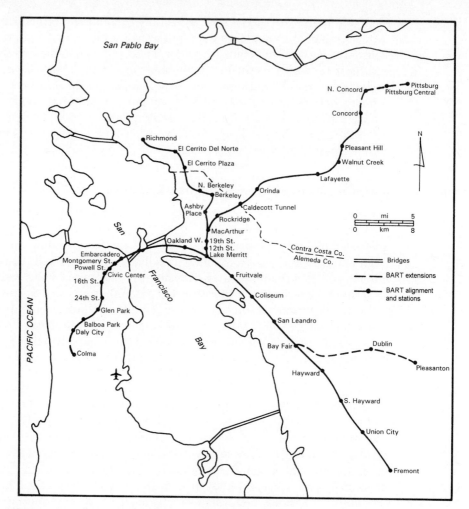

Figure 14.10 The BART rapid transit network.

system, and the initial failure to meet traffic forecasts (Webber, 1976). The technical difficulties have, on the whole, proved to be start-up problems. Ridership, however, did not reach system forecasts (250,000 workday passenger trips) until 1990 (BART,1993). This was largely because the designers of the system emphasized the wrong things. Stress was placed on pleasing passengers once they became part of the system—fast trips, plush trains, handsome stations—rather than providing ease of access to the stations with coordinated feeder buses, easy transfers, and adjacent parking. Most of the stations on the East Bay and in Contra Costa County now provide free parking ("park-and-ride"). Furthermore, the importance of providing modal choice, particularly for the journey to work, was emphasized on October 17, 1989, when a severe earthquake collapsed a section of the Bay Bridge, part of I-280 (the Nimitz Freeway) into San Francisco, and stretches of the Embarcadero Freeway. BART was able to maintain service and cater for the increased cross-Bay demand by carrying a peak of 350,000 passengers on a workday during the crisis.

BART does not, however, appear to have done much to alleviate congestion on the Bay Area highways, or modify land use patterns through the stations serving as foci for commercial and economic development. Information in BTS (1994, p. 92) suggests that the San Francisco PMSA has the third-most recurring levels of highway delay, after the New York PMSA and Los Angeles-Long Beach PMSA, among United States metropolitan areas. The system appears mainly to serve the needs of that part of the Bay Area economy and population that remains within a monocentric paradigm—which is a small part. Consequently, while some development and small increases in property prices and rents have occurred adjacent to certain stations, most of the new development associated with increases in levels of access associated with the system have occurred in the San Francisco central business district (Gatzlaff and Smith, 1993).

The WMATA Experience   The Washington Metropolitan Area Transit Authority (WMATA) operates an extensive integrated regionwide rapid transit and bus system, called Metrorail and Metrobus, over an area with a population of about 3.2 million. The integrated transit region includes: District of Columbia; Montgomery and Prince George's counties in Maryland; Arlington and Fairfax counties and the cities of Alexandria, Fairfax, and Falls Church in Virginia (Figure 14.11). WMATA was created by an interstate compact in 1967, with strong federal leadership, to operate a balanced regional transport system with much the same objectives as those outlined above for BART. Since that time, the environmental objective has gained a higher profile through such measures as the federal Clean Air Act Amendments of 1990, which require that every metropolitan area in the United States adopt a plan to control air pollution and improve air quality. WMATA began building Metrorail in 1969, acquired the area's bus systems in 1973, and commenced operating the first phase of the rapid transit system in 1976. By the time the Metrorail system is completed in 2001, the lines will extend for 165 kilometers (103 miles), and the system will have cost nearly $10 billion provided primarily through federal/local capital grants.

In many ways the system has been quite effective because: a large share of the region's employment remains located in concentrated nucleations on downtown Washington, D.C., and in such places as the Pentagon; important travel facilities, such as Washington National Airport and Union Station are part of the network; major tourist attractions are in the downtown area; and considerable development has occurred at some suburban stations. On an average weekday, nearly 1 million person trips, split roughly 50/50 between Metrorail and Metrobus, occur on the system. If these were made by automobile, there would be at least 150,000 more vehicles on the roads during the morning and evening rush hours with the attendant increased congestion, demand for new highways, and pollution. Yet, in the region served by WMATA commuting to and from work by public transport accounts for only 16 percent of total work trips, and ridership has leveled off since 1989 (WMATA, 1994). On the other

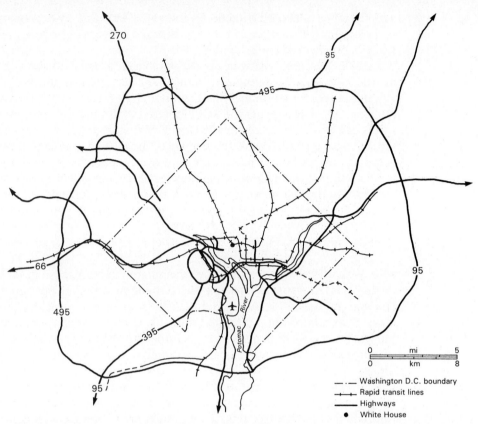

Figure 14.11 Highways and rapid transit lines in the area served by the Washington Metropolitan Area Transit Authority.

hand, 38 percent of all work trips made by residents of the District of Columbia are by public transport.

The increasing dominance of the automobile for journey-to-work trips is related partly to the polynucleation of the urban environment with the decentralization of employment opportunities to edge cities and other places adjacent to and beyond the beltway (see the discussion in Chapter 7). Furthermore, the use of the automobile for commuting trips downtown is exacerbated by a whole array of contradictory policies that, on the one hand, have led to massive investments of public funds in Metrorail and, on the other hand, encourage auto use for the journey to work. For example, 74 percent of employee parking spaces at federal facilities in downtown Washington, D.C., are provided free of charge, 22 percent are discounted, and only 4 percent are priced at market rates (WMATA, 1994). At other nonfederal places of employment less than one-half of parking prices are at market rates. Apart from reducing the financial attractiveness of public transport for the journey to work, benefits such as these reinforce a prevailing view in the United States that public transport is an inferior good for the use of specific populations.

The LRT Revival   Apart from its lower capital cost, the major advantage of light rail, as far as its incorporation into the existing spatial structure of urban areas is concerned, is its flexibility and greater frequency of stopping. Like rapid transit in Washington, D.C., LRTs can be located on old railroad rights-of-way—the San Diego LRT (the Tijuana Trolley) runs on an old railroad line to San Ysidro on the Mexican border—and on existing streets at street level. In congested areas they can be placed below or above ground. As they are slower, with few carriages, they can operate in accordance with existing gridded street patterns, making right-angle turns when necessary. Several cities that had old streetcar systems, such as Pittsburgh, Boston, Cleveland, and Newark, have renovated sections of them in recent years, placing much of the new track on separate rights-of-way (Black, 1993). San Francisco has integrated a renovated old streetcar line on Market Street with its cable-car system and BART. Generally, the new LRT systems (Table 14.5) connect selected suburbs to traditional central business districts, though a line is being considered that would link edge cities around part of the Washington beltway.

The major issue arising with many of the new LRTs, as with the new rapid transit systems, is that their ridership has not met trip forecasts and hence operating subsidies have been far higher than predicted (Moore, 1994). In order to boost possibilities for obtaining capital cost funding, there has been a tendency for proponents of public transport to present optimistic forecasts of the auto/transit modal share trade-off subsequent to the implementation of new systems (Pickrell, 1992). One of the most optimistic forecasts was presented with respect to the new Miami heavy-rail system where initial ridership was 10 percent of that forecast (Gatzlaff and Smith, 1993; Black, 1993). Table 14.5 indicates that there is great variation among the new LRT systems in construction costs per mile, and ridership per mile and per million dollars of capital expenditure. The BART and WMATA experiences would, however, suggest that it takes ten to twenty years for new public transport systems to have a positive impact on ridership.

Subsidizing Public and Private Transport

Public transport not only receives subsidies with respect to capital costs, the various systems also receive large annual subsidies to supplement operating costs. For example, the Toronto Transit Commission covers about 70 percent of its operating costs from rail, bus, and streetcar fares; WMATA manages to cover 54 percent of its operating costs (1994: $615.9 million) from rail and bus fares; and BART covers about 42 percent of annual operating costs from revenues. Figure 14.12 indicates the aggregate level of operating cost subsidies provided by the governments in the United States for public transport. The graph emphasizes that as long as a high level of monocentrism remained in North American

Table 14.5

Recent LRT Project Basic Data and Performances

URBAN AREA	YEAR OPENED	LENGTH (MILES)	COST 1993 DOLLARS (MILLIONS)	AVERAGE WEEKDAY RIDERS		COST PER MILE (MILLION $)	RIDERS PER MILE	RIDERS PER MILLION DOLLARS
				NUMBER	DATE			
Edmonton	1978	6.5	185	25,000	Late 1990	25.0	3846	154
Calgary	1981	17.1	500	111,000	Dec 1991	24.4	6491	266
San Diego	1981	33.2	292	53,000	Aug 1991	8.8	1596	182
Buffalo	1985	6.4	536	31,000	Late 1990	83.8	4844	58
Vancouver	1986	15.2	1,033	110,000	Jan 1991	48.9	7237	148
Portland	1986	15.1	240	24,400	June 1991	15.9	1616	101
Sacramento	1987	18.3	176	23,500	Oct 1990	9.6	1284	134
San Jose	1987	20.3	500	21,600	July 1991	24.6	1064	43
Los Angeles	1990	21.5	877	34,800	Oct 1991	40.8	1619	40
Baltimore	1993	22.5	468?					
St. Louis	1993	18.0	288?					
Dallas	1996	18.5	?					

(*Source:* Compiled from Black (1993), Table 1, p. 151; Table 2, p. 156; and text.)

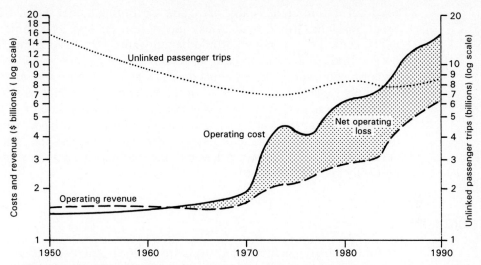

Figure 14.12   The change in net operating loss for public transport companies/authorities in the United States, 1950–1990.
(*Source:* Redrawn and modified from Fielding, 1995, Figure 12.3, p. 298.)

urban spatial structure, revenues covered costs. After 1963, however, when suburbanization and dispersion became prevalent and governmental subsidies became more available to stem the tide of deteriorating systems, costs have exceeded revenues. In 1990, revenues in general covered only about 40 percent of the operating costs of public transport systems.

The level of governmental subsidies for capital and operating costs are, therefore, huge. In the United States, capital costs for new systems constructed since the 1960s have been financed primarily with federal funds, while operating cost subsidies are invariably provided by the local jurisdictions involved. These capital grants and operating cost subsidies are highly visible. So people often think that public transport is the only transport good that is subsidized. Road transport is perhaps even more heavily subsidized—at least $2,300 per car per year (MacKenzie et al., 1992)—despite the license fees, gasoline taxes, and any other user fees that are paid. Thus, taxpayers provide large subsidies for both public and private transportation so that the ever-increasing volume of trips will have to be catered for in some manner. The ways that are chosen will undoubtedly involve large expenditure of public funds in a mix of both public and private transportation.

Given these high levels of taxpayer–funded subsidies for both public and private transportation, questions frequently arise concerning who benefits most from these publicly funded transport investments. Hodge (1988) argues that the issue of fiscal equity in public transport systems can be analyzed in a variety of ways depending on the accounting assumptions that are made. For example, in his analysis of equity issues related to a proposed new regionwide fixed-rail system in the Seattle area, one accounting model suggests that as suburban residents would tend to use the system far less than central city residents, central city residents would benefit more per capita from public transport subsidies than suburban residents. Although one might say that, if higher-income suburban residents do not wish to use a regionwide system that they help subsidize through taxes, it is *their* choice, it should be noted that the estimated large auto subsidies disproportionately favor residents of low-density suburban areas who have more cars per household and have longer average daily trip lengths. The issue of fiscal equity also enters the line location debate because those neighborhoods closest to the stations on a new system would theoretically benefit the most.

CONCLUSION

The increasingly polynucleated structure of North American urban areas favors the use of modes of transport that frequent highways and airways, and those means of communication that can be networked or are wireless. The range of choice of surface means of transport, particularly for trips associated with earning a living are, however, being unduly restricted by a whole series of contradictory public policies with respect to land use planning, economic development, housing, transport development, highway construction, the environment, and employment-associated fringe benefits. The automobile has become the preferred mode of travel because it provides flexibility and comfort, plays into an ethos of freedom and independence, and provides illusions of control and safety. For many it has "proven to be an incredibly effective and liberating technology" (Morrill, 1991). It is a technology that has facilitated the polynucleation of most North American urban areas, a trend in urban form that appears to be irreversible.

Public transport systems that concentrate huge taxpayer–funded investments on fixed-rail transit appear, like mythical King Canute, to be attempting to halt the tide of ever-increasing dispersion. Given the low-density nature of most urban areas and the multiplicity of decentralized workplaces and shopping nucleations, and a need to provide some modal choice, particularly for various disadvantaged segments of society, a far more flexible demand-response system of surface transportation with small- and medium-size nonpolluting buses, using preferred traffic lanes in rush hours, may well be the most efficient capital and operating cost alternative (Morrill, 1991; Pickrell, 1992). There are, however, well-funded lobbies in both the United States and Canada promoting fixed-rail transit systems, whereas the lobbies supporting bus transport are far more diffuse (Fielding, 1995, p. 296). If, however, a metropolis has economic activity and employment concentrated in certain specific areas sufficient to warrant a public transport system that includes rapid transit or LRT, strong regional land use planning and controls, along with policies and practices that promote ridership, should be implemented in parallel with the construction of such systems to achieve associated developmental and environmental benefits and reduce future operating subsidy requirements.

Chapter

15

Managing Urban Areas in a Global Economy

*L*ocal government issues impinge directly on everyone's daily life because most people in North America reside in urban areas. It is one of the main tasks of urban governments to ensure that the services and opportunities over which it has jurisdiction are available to *all* the inhabitants of an urban area. This task of ensuring democracy of access has to be tackled in an increasingly difficult and complex environment resulting not only from the intense fragmentation of local governments arising from suburbanization and polycentricism, but also from enormous changes in the nature and cohesion of family structures, the declining influence of persons (such as police or school teachers) who are supposed to be able to exercise prescribed levels of authority in certain matters and environments, and high levels of violence and abuse particularly associated with the epidemic of drug use in the last thirty years and the widespread availability of guns.

Furthermore, urban areas have become the latest arena of intense competition that is associated with the globalization of various markets since 1970 (Goldberg, 1995). The first arena to become globally competitive were financial markets which were transformed and integrated worldwide by deregulation and new information-communications technologies. One outcome of this has been the emergence of "world cities." New arenas of intense competition followed in such markets as those for manufactured goods—leading to worldwide industrial restructuring—and producer services. One of the latest markets to be globally integrated is the property market, for information concerning property is now transmitted with a facility similar to that for securities and other financial instruments. What this means for local governments is that they have to be in a position to respond to, and deal with, pressures arising from a property market in which choices for the location of capital investments are made in a global context. For example, the property development potential of Phoenix can now be compared as easily with places in Europe, Southeast Asia, and Latin America, as it can with other metropolises in North America.

Thus, local governments are not only in a position of having to attend to issues relating to democracy of access, they also have to compete much more than previously on a world stage as attractive locations for real estate, business, and industrial opportunities—and, hence, jobs. Local governments also have to be able to broker, intervene, monitor, and generally care for the public interest in such developmental situations in a highly sophisticated manner, particularly with the respect to the negotiation of complex property development, financing, and partnership arrangements. It is, therefore, essential that local jurisdictions know how to position themselves nationally and internationally, be aware of their particular locational and structural strengths and weaknesses, and do what they can within their own jurisdictional powers to enhance the ways in which local objectives may be realized. This implies that community-oriented procedures should be in place to provide public input to these types of goal-setting exercises through local political processes (Baker, 1995).

THE CHANGING ROLE OF LOCAL GOVERNMENT

Local governments in North America occasionally go through periods of change and reform and it seems—with current debates over issues like the privatization of some public services, central city/suburban revenue sharing, the role of extended city-states in a more highly competitive world, and the form and function of sustainable cities—that this is such a time (Osborne and Gaebler, 1992; Pierce et al., 1993). The periodicity of these changes occurs primarily because local government, even though it has a direct and immediate influence on people's lives, is not well understood and generally attracts little voter attention. Thus, a local community often continues with a particular form of representation and style of government for some time before pressures build sufficient to generate change. This section emphasizes two essential features of local government. One is that it is the product of the interaction between local politics and the larger aspatial forces that influence urban development (DiGaetano, 1994). The second arises from the first and is that the concerns of local governments in North America have tended to change most dramatically around the onset of each new long wave of economic development.

Local Government at the End of the Nineteenth Century

At the end of the nineteenth century, local governments began to change their increasingly out-of-date preindustrial role as regulators of business and commerce and keepers of the peace, particularly those regulations associated with

Table 15.1

Industrialization, Urban Growth, and the Expansion of City Government Workforces in Cities with 50,000 or More People, 1870–1900, United States

YEAR	MEAN PERCENTAGE MANUFACTURING OF URBAN WORKFORCE	PERCENTAGE OF U.S. POPULATION LIVING IN CITIES 50,000 +	NUMBER OF CITY WORKERS PER 1000 POPULATION	NUMBER OF CITIES
1870	39.5	12.2	3.83	24
1880	44.3	14.1	4.25	34
1890	40.4	18.4	4.62	57
1900	39.4	22.1	5.18	77

(*Source:* DiGaetano (1994), Table 4, p. 198.)

property rights as they related to offensive trades and activities, to providers of services to high-population-density industrial cities. From about 1870 onwards, rapid urbanization, changes in technology, and the growth of the economy, generated a need for greater local government involvement in the provision or regulation of: physical infrastructure, such as paved roads, street lighting, water lines, sanitary sewers, storm drainage systems, and coal/gas and electricity networks; public protection and safety through the provision of municipal fire protection and police; public transportation on public roads as it evolved from horse-drawn vehicles to electric streetcars; and social infrastructure, such as schools and hospitals (which were also supported with charitable funds). Symbolic of this change was the consolidation of New York City in 1898 from five separate boroughs (Figure 11.9).

This was an age involving a mixture of public, volunteer, and religious-based activities in the social sphere as various parts of society attempted to respond to poor living conditions and bad health arising from overcrowding and low household incomes in many families. The infectious disease tuberculosis, which was difficult to contain in congested environments, was rampant in industrial cities and the major cause of death until the discovery of effective inoculation procedures and its widespread public utilization after 1935. The increasing importance of local government during this period is illustrated in Table 15.1 which shows the expansion of city government workforces (per 1,000 population) in United States urban areas with a population of 50,000 or more at the end of the nineteenth century. Between 1870 and 1900, local government workforces (per 1000 pop.) in large manufacturing cities increased by 35 percent.

The change in importance of local government brought with it changes in the form of political leadership and government that in many jurisdictions echo through to the end of the twentieth century. A major element in this change was the massive volume of immigrants that entered larger urban areas in the United States between 1870 and 1910. This volume of immigration reinforced a polarization in local government politics that had been emerging between, on the one hand, those politicians responding to the needs of the less advantaged and newcomers and, on the other hand, those representing the longer-established population (Banfield and Wilson, 1963). The existing established population tended to be comfortable with a political system that emphasized general principles of governmental organization and management, particularly those that moralized the lives of individuals, promoted personal integrity and reliance, minimized the role of government in people's lives, and placed faith in leadership by the business and professional establishment.

The waves of immigrants had, however, need for a political system that would provide jobs and respond to immediate personal and family needs. Many of the newcomers were also more familiar with Old World hierarchical and authoritarian political systems, one element of which was the patron-client relationship (DiGaetano, 1994). These new immigrant familiarities combined with the existing American system of political parties and elections to produce machine politics, in which the party bosses (or patrons) worked to further the interests of their constituency in return for political loyalty. This resulted in a clash of leadership styles between an immigrant-boss-machine complex which, it was argued, permeated local government with patronage appointments and contracts, and "good-government" reformers who emerged intermittently with campaigns to replace patronage politics with "meritocracy" in public administration. These reform movements, while frequently party based, were usually headed by a person who appeared to be of estimable character and worth and seemed to be above party politics—a "mugwump." It is not at all clear that all cases of patronage politics can be dismissed as uneconomic and wasteful, just as reform movements were often not as high-minded as they claimed (Erie, 1988).

Local Government in the Middle of the Twentieth Century

Permutations of this tension between machine politics and reformers colored the politics of local government through the era of national industrial capitalism, which involved increasing urbanization in both the United States and Canada, large-scale rural-to-urban migration, widespread use of public transport in decentralizing monocentric environments, and rising automobile use particularly for social/recreational trips. These tensions were most evident in the larger metropolitan areas which experienced the greatest population growth and had the widest social, economic, and ethnic diversity. One outcome of this tension after 1900 was the Progressive movement, which advocated not so much change in leadership style in city hall, but change in the structural arrangements of local government. New York City, for example, went through three charter reorganizations during the first half of the twentieth century.

The structural reforms advocated by the Progressive movement were aimed at improving the management of local government and reducing the actual or potential impact of machine party politics at the local level (DiGaetano, 1991). The reforms, therefore, embraced a variety of measures designed to make local government more professional and less partisan. As a result, various elements have been implemented in many local jurisdictions across North America. The reforms that were advocated included: city-manager forms of government in which a nonpartisan professional serves as an executive officer responsible to an elected body of representatives; civil service types of arrangements for local government employees including job grading, competitive hiring, job security, requirements concerning confidentiality, and so forth; banning national parties from direct involvement with candidates in local elections; and nominees for municipal elections who could not be obviously identified by party affiliation. Furthermore, the Progressive movement attempted to promote jurisdiction-wide perspectives over narrower neighborhood or community objectives by advocating a high proportion of at-large positions on municipal councils and a low proportion of positions representing wards (or neighborhoods).

Bases of Proliferation   Throughout the first half of the twentieth century, local governments proliferated and their responsibilities grew. In particular, automobiles and faster public transport facilitated greater levels of deconcentration and suburbanization around central cities. The inhabitants of many of these suburban areas were often pleased to be separated from the older central cities and wished their neighborhoods to be independent (Jackson, 1985, pp. 151–153). Thus, opposition to annexation of suburban neighborhoods by central cities grew apace in the United States after 1900, aided by increasing legal and constitutional acceptance of the principle of home rule which supported arguments from smaller communities for jurisdictional autonomy (Garber and Imbroscio, 1996). Suburbs as separate municipal entities increasingly hemmed in the central cities, and subsequently multiplied in number in the post–World War II suburban period. Concomitant with this escalating number of suburban and exurban municipalities is the allocating of a number of service functions that require large populations to achieve economies of scale (such as sewer and water systems) to county levels of government or special districts.

Special districts had been used in some of the older central cities for a number of decades, such as the New York Metropolitan Police Board (1857) and the Chicago Sanitary District (1889), as a means of fostering the idea of municipal consolidation. But, for the developing small suburban municipalities special districts became a way of achieving some of the advantages of larger scale organization in the provision of a few basic services while at the same time maintaining local jurisdictional independence. Thus, a dichotomy has become more and more evident in the late twentieth century between those central cities that have, at various times, been able to extend their boundaries roughly in accordance with population growth and those that have been unable to expand—a phenomenon identified by Rusk (1993) in terms of "elastic" cities which are able to *capture* suburban growth, and "inelastic" cities which are contained and can therefore only *contribute* to suburban growth.

A major stimulus to the changing power and responsibility of local government was the Great Depression, which shattered established procedures for government action. The economic collapse increased the demand for positive action by all levels of government to alleviate the crisis. These pressures were felt most acutely at the local level, where the inherent weaknesses of the administrative and financial structures of local government rapidly became apparent. Most local governments were in a relatively weak financial position because capital costs incurred in the growth of many public and private infrastructure and service functions had been financed largely by borrowing. Furthermore, many of the private companies providing public services (such as public transport) went bankrupt and either closed down or had to be taken over by municipal governments. In fact, between 1913 and the early 1930s the yearly interest payments alone on loans to local governments in the United States rose from $167 million to a staggering (for the times) $1.5 billion. The roles of the federal government in Washington, D.C., and the provincial governments in Canada, had to be extended in order to alleviate the ever-increasing problems, for the financial powers of most local governments were severely limited.

Local Government in the Second Half of the Twentieth Century

The experience of the Depression years left local governments in the United States and Canada after World War II well aware that in times of economic crisis: (1) services provided by private enterprises may well revert to municipal management, or have to cease operation, and in either case the local government may be left carrying the can; (2) social welfare responsibilities may escalate quickly beyond a level at which they can be funded from local revenues; and (3) that there are enormous dangers in overextending local public infrastructure debt loads beyond those that can be supported by conservative forecasts of the revenue base of the municipality. Hence originates the greater acceptance by federal governments in both countries of their role in the funding of the service society that burgeoned in the second half of the twentieth century. As has been mentioned previously (Chapter 4), the approach in Canada has been more national standard and provincially oriented than in the United States where the assistance to municipalities has been piecemeal and, as has been described with respect to new fixed-rail transit transport, program focused.

Since World War II, both the financial responsibility and the number of local government units in urban areas have been increasing at a rapid rate in response to the great growth of metropolitan areas, developer-led suburbanization, and the overwhelming dominance of road transportation (Keating, 1988). Since 1950, state/provincial and local governments in Canada and the United States have been the fastest-growing sectors of their national economies. To illustrate, whereas the United States' GNP grew threefold between 1960 and 1980, state and local government expenditures increased sixfold in the same period. This burgeoning public expenditure is related to the increase in range of public services for the larger urban population. In this regard it is interesting to note that, though the scope of services provided varies greatly among local government jurisdictions, in 1990 the unweighted number of local government employees in the sixty-one largest municipalities in the United States was 18.3 per 1,000 population (Bureau of the Census, 1993, p. 323).

Furthermore, in response to increasing service demands and periodic pressures for reform, most local governments have experienced bouts of reorganization, and in the case of most school districts, consolidation. New York City, for example, experienced the Mayor Lindsay administrative reorganizations of the late 1960s/early 1970s; the Goodman Commission advocacy in the mid–1970s of the decentralization of some administrative and service activities to fifty-nine community boards, which were also to be advisory with respect to land use and zoning applications within their communities; and the Ravitch-Schwartz Commission attempt at restructuring to achieve greater community empowerment in the late 1980s. Savitch (1994) notes that each of these reforms had unintended consequences: The community boards, for example, which were meant to be advisory with respect to land development, quickly became reviewing bodies whose recommendations were followed most of the time.

Thus, during the second half of the twentieth century, local governments were faced with a number of issues. One set of issues arose from the proliferation of adjacent and overlapping local government units within large contiguous dispersed and increasingly polynucleated urban areas, so well documented and discussed by Wood (1961) with respect to the extensive New York metropolitan area of the 1950s. Another concerned the increasing range of functions that had to be provided by local governments, because apart from the traditional array of regulatory, infrastructure, and public service activities, they had also become *agents of social change* through federally legislated requirements to enforce a range of equity concerns such as those subsumed in fair-housing and public accommodation laws; and *agents of development* for the local jurisdiction. All this had to be achieved on a revenue base that was often inadequate to the needs that had to be met, particularly in some of the older less "elastic" central cities which had been left with generally lower income and more disadvantaged populations. Attempts to address these issues were complicated by the enormous proliferation of local government and the fragile revenue base of municipalities (Mikesell, 1992).

LOCAL GOVERNMENT FRAGMENTATION AND METROPOLITAN FINANCE

At the end of the twentieth century, the range of administrative, service, and social activities undertaken by local governments in North America is immense. In 1991, local government expenditures were 26 percent of all governmental expenditures in the United States, with revenues derived from those generated locally (67 percent) and intergovernmental transfers from the federal and state levels (33 percent). Local government is most involved with the provision of public education, with the proportion of all expenditures at the local level devoted to this purpose (36.4 percent) representing 69.3 percent of total national expenditures on education (Table 15.2). Apart from public education, local governments are the principal providers of police and fire protection, sewerage and sanitation, parks and recreation services, and utilities. They also

Table 15.2

The Range of Local Government Activities in the United States by Magnitude of Expenditure, and Concentration

ACTIVITY	EXPENDITURES		LOCAL GOVERNMENT PERCENT OF NATIONAL TOTAL FOR ACTIVITY
	$ BILLIONS	PERCENT	
Education	$229.2	36.7%	69.3%
Public welfare	26.9	4.3	16.0
Health and hospitals	72.3	11.6	43.4
Highways	26.0	4.2	39.6
Police protection	28.0	4.6	72.0
Fire protection	13.7	2.2	100.0
Corrections	9.6	1.5	32.8
Natural resources	2.9	0.5	5.1
Sewerage and sanitation	29.0	4.7	93.5
Housing etc.	14.9	2.4	44.7
Administration	29.5	4.7	45.6
Parks and recreation	13.2	2.1	74.6
Utilities	71.3	11.4	88.0
Interest on debt	28.8	4.6	11.6
Other*	28.0	4.5	6.8
Total expenditures	$623.3	100.0	

(*Source:* Bureau of the Census (1993), p. 293.)
*Such as insurance.

provide considerable support to local hospitals and health services, and for public housing. Expenditures on administration include persons and facilities involved not only with government administration per se (such as accounting and taxation), but also people in a wide variety of social services, such as social workers, and planning.

The Proliferation of Local Governments

It is estimated that 86,743 different local government units existed in the United States in 1992, an increase of 6 percent, primarily through the proliferation of special purpose districts, over the 81,730 that existed in 1982 (Bureau of the Census, 1994). The most recent enumeration of local governments within metropolitan areas indicates that 29,861 of these local governments were located within the 303 urban areas that were defined as metropolitan at that time (Bureau of the Census, 1985). The major types of local government within these metropolitan areas are indicated in Table 15.3, which does not include the many different agencies that are subordinate to local government. School districts, which are responsible for a major share of local government expendi-

Table 15.3

Types of Government Units in Metropolitan Areas, 1977 and 1982

TYPE OF UNITS	NUMBER		% CHANGE
	1982	1977	1977–1982
School districts	5,692	5,751	–1.0%
County governments	670	670	0.0
Municipalities	7,018	6,923	1.4
Townships (incl. New England towns)	4,756	4,752	0.1
Special districts	11,725	10,640	10.2
Total	29,861	28,736	3.9

(*Source:* Bureau of the Census (1985).)

tures, are invariably independent units of local government run by elected school boards with their own administrative structures. This structure and proliferation is similar to that experienced in Canada, for at a comparable date (1986) there were 4,835 units of local government and special districts existing in the ten provinces.

The extent of fragmentation varies markedly among metropolitan areas. Table 15.4 lists the number of units of general government, special districts, and school districts in the fifteen largest metropolitan areas in the United States. In general, younger metropolitan areas have fewer units and hence simpler spatial patterns of local government than older ones. Although the complexity of the local government structure is to a degree related to city size, there is enormous variation in the complexity of governmental jurisdictions among this group of larger municipalities. The most prolific in this regard is Chicago, with 1,194 local government units, followed by Philadelphia, Pittsburgh and Houston (Perrenod, 1984). But size is only a rough guide, for Los Angeles-Long Beach had only 276 local governments and Baltimore as few as 49.

This profusion of governmental entities in urban regions is made even more complicated by the many kinds of jurisdictions that occur. First, there are the elected governments of the municipalities including villages, incorporated towns, cities, boroughs, townships, and counties. Second, there is a multiplicity of special districts which are *ad hoc* units concerned with the provision of single and, less frequently, dual services such as fire protection, policing, health care, water, sewerage, and housing. These districts are by far the most numerous in metropolitan areas today and have been growing at a rapid rate as the scope of public service provision has enlarged and municipalities have used them to avoid other forms of local government reorganization (Bureau of the Census, 1992). Special districts add to the confusion of urban spatial organization because they typically overlap each other and other governmental jurisdictions, levy charges, and about one-half also have their own property-taxing powers.

Table 15.4

The Number of Local Government Units in the Fifteen Largest Metropolitan Areas in the United States, Ranked by Population Size (in thousands)

METROPOLITAN AREA	1980 POPULATION	NUMBER OF LOCAL GOVERNMENT UNITS
New York	10,803	542
Los Angeles-Long Beach	7,478	276
Chicago	7,103	1,194
Philadelphia	4,708	867
Detroit	4,044	238
Houston	3,385	608
San Francisco-Oakland	3,252	331
Washington, D.C.	2,966	87
Dallas-Ft. Worth	2,886	330
Boston	2,603	189
Pittsburgh	2,310	739
St. Louis	2,302	574
Baltimore	2,174	49
Cleveland	1,899	217
Newark	1,763	218

(*Source:* ACIR (1984), Tables A-1 and A-8. Most recent data at time of writing.)

On the average, there are forty-three special districts in each metropolitan area in the United States (Foster, 1996, p. 289). School districts may be regarded as a particular type of special district responsible for the statutory delivery to the public of primary, junior, and secondary education.

Sources of Local Government Revenue

There is a wide variation in the range of services provided by all types of local government units, particularly municipalities, counties, and townships. This variation is illustrated in Table 15.5 which lists the general revenue per capita (that is, all revenue excluding that realized from utilities) for the twenty-five largest municipalities in the United States. Those with high levels of general revenue per capita provide a wide array of services for their inhabitants, whereas those with lower levels of revenue per capita provide less because many services are provided by special districts, county governments, or school districts. At one extreme is Washington, D.C., which, in a service delivery sense is a state by itself as well as a special creature of the Congress, followed by New York, where the city is responsible for most services including education and some postsecondary institutions. At the other extreme are cities in Texas and Ohio, which have low levels of revenue per capita because many services are provided by other types of local government units.

Table 15.5

The Proportion of City Government General Revenue* Derived from Different Sources, 25 Largest Cities

| CITY | INTERGOVERNMENTAL TRANSFERS | TAXES | | | OTHER TAXES AND FEES** | GENERAL REVENUE PER CAPITA |
		PROPERTY	SALES			
New York City, NY	39.1	21.3	10.2		29.4	4,668
Los Angeles, CA	13.7	18.7	21.3		46.3	1,058
Chicago, IL	26.6	18.8	27.3		27.3	1,130
Houston, TX	4.7	28.7	23.6		43.0	822
Philadelphia, PA	24.5	11.6	1.2		62.7	1,722
San Diego, CA	17.9	13.4	17.7		51.0	984
Detroit, MI	44.8	13.1	2.8		39.3	1,473
Dallas, TX	4.4	35.0	24.2		36.4	812
Phoenix, AZ	33.2	11.9	19.3		35.6	1,019
San Antonio, TX	22.5	20.8	16.2		40.5	581
San Jose, CA	15.6	19.4	23.2		41.8	809
Baltimore, MD	50.0	26.0	2.8		21.2	2,371
Indianapolis, IN	28.7	33.9	2.3		35.1	1,204
San Francisco, CA	31.5	19.5	9.4		39.6	3,487
Jacksonville, FL	21.4	26.2	6.5		45.9	1,113
Columbus, OH	19.3	4.0	0.0		76.7	836
Milwaukee, WI	49.2	26.2	0.9		23.7	899
Washington, DC	33.7	19.9	15.2		31.2	7,289
Boston, MA	47.7	33.2	1.6		17.5	2,974
Seattle, WA	13.1	16.5	21.7		48.7	1,304
Cleveland, OH	29.9	8.6	0.5		61.0	1,130
New Orleans, LA	19.9	18.4	19.7		42.0	1,398
Nashville, TN	22.0	27.9	17.5		32.6	1,936
Denver, CO	18.5	11.5	23.5		46.5	2,097
Austin, TX	8.4	19.3	12.7		59.6	1,149

(*Source:* Bureau of the Census, 1993, p. 314.)
*Note that general revenue *excludes* income from utilities and transit.
**Income taxes, licenses, user fees, insurance trust revenues.

Table 15.6

Source of Revenue for Local Government Activities

	REVENUE		LOCAL GOVERNMENT PERCENT OF NATIONAL
ACTIVITY	$ BILLIONS	PERCENT	TOTAL FOR ACTIVITY
Intergovernmental transfers	$201.9	33.0	57.9%
Property tax	161.7	26.4	96.4
Income tax	10.1	1.7	1.7
Corporate taxes	1.9	0.3	1.6
Sales taxes	22.3	3.6	1.8
Selective taxes*	9.8	1.6	9.5
Education fees**	9.8	1.6	27.2
Hospital fees	22.8	3.7	67.4
Sewerage etc. fees	19.4	3.2	98.8
Parks and recreation fees	2.8	0.5	75.7
Housing rents, etc.	2.7	0.4	49.1
Airport fees	5.1	0.8	89.4
Transport and terminal fees	1.2	0.2	43.7
Other charges	15.2	2.5	29.0
Utility fees	54.3	8.9	89.3
Interest earnings	30.8	5.0	44.2
Other***	40.4	6.6	6.7
Total revenue	$612.2	100.0	

(*Source:* Bureau of the Census (1993), p. 293.)
*Motor, fuel, tobacco, alcohol.
**School lunches and tuition fees.
***Mostly insurance trust revenue.

National, state/provincial, and local governments are keen competitors for the tax dollar. Thus, over the years, a rough division of revenue sources has been accepted among the three tiers of government. The federal governments of both the United States and Canada raise the vast bulk of their revenue from personal income, corporation, and sales or excise taxes. They are able to impose relatively high rates of taxation because the respective national territories are completely under their jurisdiction. Many of the states, and all of the Canadian provinces, also raise revenue through income taxes, sales taxes, and other devices such as lotteries. Smaller geographical units, such as municipalities, cannot take full advantage of all these sources of revenue, either because their use is subscribed, as is income tax with the federal government, or because they are afraid to do so in an economic system in which people, businesses, and capital can move relatively freely from one municipality to another. A disproportionate levy of sales taxes among adjacent municipalities, for example, could well encourage people and businesses to move to, or shop in, those municipalities in which the tax is lower.

Consequently, local governments raise or receive revenue from a variety of sources (Table 15.6), among which the most important are direct transfers from federal state governments, property taxes, and various types of user fees. The property tax has, traditionally, been the main way that local governments have been permitted to raise revenues, but it has been declining in importance since the 1960s. This is indicated in Table 15.7, which provides information on the changing revenue base of *municipalities* between 1965 and 1990. In 1965, the property tax generated almost one-third of municipal revenues, but by 1990 it provided less than one-fifth. For various reasons the property tax is no longer regarded as an acceptable means for raising a large share of local government revenue, and is being replaced by increases in direct charges, which may also be debatable on equity grounds, and transfers of funds from higher levels of government. But, with higher levels of federal debt, intergovernmental transfers are being reduced, leading to reexamination of the revenue-generating capability of existing sources and the possibilities of new types of taxes and user fees.

The Property Tax   The general property tax is a holdover from a bygone era when wealth was reflected in landed property, and has provided the basis of local government support since municipalities became legally established in both Canada and the United States. It is still the most productive source of public revenue in North America after federal income and excise taxes. Although its importance to local governments has declined in recent decades, it still accounts for more than one-quarter of revenue to local government and about one-sixth of income for municipalities in the United States (Tables 15.5 and 15.6). In Canada, the contribution of property taxes to local government revenues has not decreased as precipitously as in the United States. In the 1960s, property taxes in Canada provided about 45 percent of local government revenue; but since the 1970s they have provided around 31 percent, transfers from provincial governments 38 percent, with 31 percent from other sources such as user fees (Bird and Slack, 1993, p. 64).

The relative importance of the property tax varies markedly between local government units. This is illustrated in Table 15.5 which, apart from listing the general revenue per capita for a number of cities also shows the share of general revenue generated by intergovernmental transfers, property and sales taxes, and other taxes and fees. For example, Indianapolis and Boston generate about one-third of their general revenue from the property tax, while Cleveland and Columbus receive less than one-tenth from this source. Most municipalities rely on the property tax for about 15 to 30 percent of their revenue. The property tax accounts for about 40 percent of revenues in Metro Toronto and virtually all of Metro Toronto School Board expenditures (GTA, 1996, p. 74), whereas Calgary receives about 25 percent from this source.

The basis of residential property assessment in most countries of the OECD is *market value assessment* (MVA; Youngman and Malme, 1994, p. 147). MVA estimates the value that the market places on property by using sales, property income, or cost of construction data. Sales data are used for non-income generating residential properties, income data usually for income-generating multi-unit properties, and cost of construction information for new properties. A few jurisdictions in North America use *unit assessment* (UA), which is essentially a charge based on size of the building and the land occupied ignoring, in its purest form, the value of location. UA may be modified to reflect market value by weighting space with estimates of local square foot prices, in which case the result is similar to MVA. The property tax is derived from assessments such as these and the local mill rates for the different types of property (see Chapter 8). In large metropolitan areas, where there may be numerous local government units, mill rates can vary between jurisdictions based on local revenue needs even though there may be one form of property assessment, and businesses and households often relocate to low-rated areas.

The Regressive Nature of the Property Tax The residential property tax has traditionally been regarded as quite regressive—that is, lower-income home-owners generally pay a greater share of their household income in property taxes than higher-income households (Frisken, 1991). In both UA and MVA cases, the property tax burden, expressed as a percentage of income, is invariably higher for low-income households than wealthy households because housing is a larger component of consumer spending for low-income groups. Netzer (1974) estimates that whereas the property tax on owner-occupied, single-family houses averaged 4.9 percent of family income in the United States, it was *twice* as high for the lowest-income homeowners as it was for those in the higher-income groups, for whom the tax averaged less than 3.7 percent. Furthermore, in the UA case households occupying large old houses would incur relatively large property taxes even though the residence may be worth little.

Bird and Slack (1993, pp. 88–94) suggest that the regressivity of the property tax may be overstated, in part because households are quite mobile and seek locations where taxes are commensurate with their incomes, and in part because the evidence is based on grouped data that exhibits a great deal of variation in the incidence of the tax within income groups. Furthermore, the fear of rapid property tax increases due to inflating property values and a perception of out-of-control expenditures in some municipalities and school districts, which fueled much of the "Proposition 13" tax revolt in California in 1978 which set a ceiling on property taxes, and Boston's "Proposition $2^1/_2$" in 1980 which set more stringent limits, are not of such great current concern. Thus, even though there is resistance to the property tax, it is not likely to disappear because local governments need a local revenue source, and there are significant constraints on the money that can be raised by other means.

Non-Property Taxes Faced with rising costs and mounting public needs, local governments have been forced to contain expenditures and turn to sources of revenue other than the property tax. Table 15.7 illustrates that for municipalities local sales and income taxes have been providing an increasing share of revenue since 1965. Municipal payroll taxes have proven to be significant sources of revenue in a few central cities such as New York and Philadelphia. In New York City, receipts from the levy amount to about 10 percent of the total revenue received. As the system is commonly employed, residents are taxed on income regardless of source and nonresidents on that portion of their income earned within the central city levying the tax. The main attraction of the tax for the central cities is that it enables them to exploit a source of extraterri-

Table 15.7

The Sources of Revenue Received by Municipalities in the United States, 1965 to 1990

Source	AMOUNT RECEIVED ($ BILLIONS)		PERCENT		CHANGE IN PERCENTAGE SHARE
	1990	1965	1990	1965	1965–1990
Intergovernmental transfers	$34.2	$2.8	18.1	13.8%	+ 4.3
Local property taxes	35.0	6.5	18.6	32.0	–13.4
Local sales taxes	19.2	1.8	10.2	8.9	+ 1.3
Income taxes and licenses	14.6	1.0	7.7	4.9	+ 2.8
Local direct charges	44.2	3.1	23.5	15.3	+ 8.2
Utilities etc. revenue	33.3	3.9	17.7	19.2	–1.5
Other	7.9	1.2	4.2	5.9	–1.7
	188.4	20.3	100.0	100.0	0.0

(*Source:* Bureau of the Census, 1993, p. 311.)

torial revenue from commuters who work in the central city but reside in the suburbs, and for whom the central city government provides services and facilities that they would otherwise enjoy without contributing toward their cost. However, the central cities have to be careful that the tax does not drive employment out to the suburbs.

During the 1970s, sales taxes, licenses, and similar types of charges (such as hotel room taxes) were gradually substituted for property taxes as important sources of revenue in many local government jurisdictions. Table 15.5 indicates that in some large cities, such as Chicago, Houston, Dallas, San Jose, Seattle, and Denver, the tax generates more than 20 percent of general revenue, while in others, such as Detroit, Baltimore, Milwaukee, and Boston, it is hardly used. As with the municipal payroll tax, it enables municipalities to tap nonresidents for revenue. Like the property tax, however, it is decidedly regressive as poorer households spend a higher proportion of their incomes on food and retail goods locally than higher-income households. Moreover, in certain situations it can penalize local merchants, because in a large metropolitan area consumers may be able to avoid the levy by shopping in adjacent territories.

Krmenec (1991), for example, notes, in a study of municipalities in Illinois, that the use of the tax depends to a large extent on the level of local retail competition. Municipalities surrounded by jurisdictions with many attractive retail outlets were reluctant to substitute sales for property taxes, while those that had little surrounding competition were less reluctant. This competition between adjacent jurisdictions takes on another form with competition between local government units for the newest and latest in retail malls and their associated attractions. Lloyd (1991) documents the competition between municipalities in suburban Los Angeles for upscale malls to effectively appropriate sales, and hence sales taxes, from surrounding jurisdictions. For these reasons, local governments often prefer that sales taxes, as with payroll taxes, be administered on a state/province-wide basis and funds then be transferred back to local jurisdictions as unconditional grants.

Service Charges, or User Fees, for Utilities and Public Services A third major source of revenue stems from the charges levied on the users of utilities and services that are provided by local governments from their own facilities, or those operated by other jurisdictions or privately owned. For utilities and public services taken as a whole the revenues provide about one-fifth of the money received by local governments. The direct-user charges for these utilities do not, in the aggregate, cover the entire cost of their provision. For example, about 75 percent of the local government cost of supplying water, and 40 percent of the outlays for public transportation, are recouped from user charges. Service charges are based on the quantity of service used or the benefits received by urban residents. Although in some ways these charges may be regarded as regressive, they are less so in the case of those utilities in which households can decide just how much or how little of a particular service they want to use, particularly when there might otherwise be significant waste, as with unmetered water. There is, however, a certain minimum level of utility service required by all urban residents, regardless of income. So, in many cases, the consumer is not required to pay for the full cost of the service.

Service charges cannot be applied to all local public services. They are, for example, inappropriate, for example, in the case of *pure public goods;* that is, services that no one can be denied, whether they pay for them or not, and for which the benefits to one person do not in any way preclude enjoyment of the same benefits by others. Drinking water in some jurisdictions has been regarded as such a good, though, as is noted later, this may be changing with privatization. Service charges are also inappropriate for *redistributive activities,* especially in a number of cases where the service provides extremely important social as well as private benefits. Thus, although the costs of financing public libraries could be defrayed in part by user charges, libraries are invariably financed from taxes because they serve an educational and, hence, social function. User charges are also difficult, but not impossible, to apply when cost-benefit relationships are inexactly specified, as in the case of pollution control. Nevertheless, local governments are increasing user charges for utilities and public services because of rising constraints on other sources of revenue.

Intergovernmental Transfers   An increasingly large proportion of local government total revenue is derived from intergovernmental transfers. In the United States, these transfers account on the average for one-third of local government revenue (Table 15.6). Among the largest cities, as high as 50 percent depend on the range of activities undertaken at the local level (Table 15.5). In general, the higher the level of general revenue per capita (which is a surrogate for the range of activities), the greater the contribution of intergovernmental transfers to local general revenue. In Canada, about 38 percent of local government revenues are derived from transfers from provincial governments. There is a trickle-down effect involving the transfer of funds from the federal level to the state/provincial governments to the local level, but this trickle is being reduced to a few drops in federal and provincial budget-cutting exercises.

The principal problem with most transfers is that they are *conditional* rather than unconditional; that is, they are usually for specific purposes. *Unconditional grants* can be used for any appropriate purpose that the local government decides. Most of the transfers are for education, to supplement the revenue generated from, primarily, property taxes. Other specific services, such as transportation, communications, and welfare, also receive subsidies, but these are smaller by comparison. Conditional grants, which, for example, might transfer $1.00 for every $1.00 spent by a local government on a specific objective, often compound the fiscal dilemma of local governments. They may feel forced to distort their local priorities to take advantage of the conditional grant and cost-sharing schemes promoted by the higher levels of government.

ISSUES FACING LOCAL GOVERNMENT

The shortcomings of the various ways local governments raise funds would be less pronounced if taxes and direct charges were applied on a uniform basis throughout a metropolitan area and redistributed to local areas on the basis of some principles of social equity and service need. That they often are not exacerbates the problems arising from the fragmentation of local government. Two main kinds of problems stem from this: (1) fiscal problems related to the existing methods of financing local government activities; and (2) organizational problems arising from the difficulties of coordinating the various local government responsibilities within metropolitan areas. The second group of problems is much a result of the first since local revenue generation encourages the implementation of separate rather than cooperative policies in service provision.

Fiscal Problems

Whereas the federal governments of both Canada and the United States raise about 60 percent of their revenues from progressive income, capital gains, and business taxes, local governments depend more on other forms of taxation. They are, therefore, left with three disadvantages. The first is the regressive nature of many of the taxes and charges on which local governments depend. The second is that many of these types of charges and taxes do not keep up with inflation when it occurs as it did in the 1980s. The third is that a property tax system, in particular, fails to keep pace with a growth economy that is not reflected in increasing property values or rising levels of median household income—which has been the experience of much of the 1990s. For example, well-constructed progressive income, capital gains, and business tax systems could (if they were so designed and implemented) capture greater amounts of revenue in times of increasing wealth polarization.

Fiscal Imbalance   The major problem arising from the fragmentation of local government structure is without doubt the disparities in revenue resources and expenditure needs between different units within urban areas. This disparity leads to significant spatial variation in the level and quality of public services provided from place to place, which are most pronounced between older central cities and newer suburbs, although not all suburban communities enjoy favorable resource bases. In general, central cities have older infrastructure wanting more maintenance and fire protection, high concentrations of people in distress who require economic support and a variety of social and health services, neighborhoods in need of good police protection, and children who require a safer environment and improved schooling, but declining business climates often affect negatively most sources of revenue.

As a consequence, financial transfers from state/provincial governments, including the federal government in the United States, to municipalities are vital for the maintenance of their fiscal balance (Table 15.5). Parker (1995) demonstrates, with respect to total United States federal spending between 1983 and 1992 in forty-one of the nation's large metropolitan areas, that:

1. While central cities *and* suburbs were recipients of federal spending, in 1992 total federal support per capita to the central cities was twice that of the suburbs.

2. Between 1983 and 1992 total federal spending per capita in the central cities increased at a rate greater than that for the suburbs.

3. There was great variation in the change in total federal spending in the suburbs as some experienced per capita increases greater than those in the central cities, and some a great deal less.

This last observation perhaps reflects increasing levels of distress in the 1990s in some of the suburbs.

Despite the constraints on revenue raising, and the ever-increasing needs, local governments exhibit on the whole good fiscal responsibility. In the early 1980s, a number of local governments in the United States were in financial disarray. Debts were mounting, and some metropolises, such as New York City, were in a crisis situation, requesting and receiving considerable additional support from the state and federal governments. Since that time, financial order has been restored in most jurisdictions. This has occurred through more efficient administration as well as severe trimming of many line items, increases in user fees and other direct charges, and the delaying of infrastructure maintenance and repair.

Spillovers Existing arrangements for local government taxation cannot adequately cope with spillovers, the effects of which are accentuated by the proliferation of local government units. Spillovers occur when expenditures or actions in one local government unit affect in some way, positively or negatively, another. For example, just as a high quality of police protection in one jurisdiction is commonly reflected in lower crime rates in neighboring communities, a well-developed school system in one district enhances the general cultural and economic health of the urban complex as a whole. Although the identification of spillovers becomes rather complicated in the multi-jurisdictional environments in which most urban households reside, and to which they contribute taxes and from which they receive benefits, the issue of spillovers is as central to local government concerns as intrametropolitan disparities in service provision. Furthermore, as spillover issues are an expression of interdependencies in spatial organization, they are essentially geographic.

The complex nature of the effect of spillover on local government expenditures has been examined by Park (1994) with respect to expenditures within the diverse array of central city, suburban, and county level jurisdictions in fifty-two of the largest metropolitan areas in the United States. Service-and-expenditure interactions are complex because, as has been noted, central city expenditures are supported heavily with federal aid, county expenditures are influenced by state aid and mandates, whereas suburban municipal expenditures are influenced somewhat less by these extraterritorial sources. Nevertheless, there is a tendency for local government units to spend less on some services, such as welfare, if neighboring jurisdictions spend more; and more on other services, such as solid-waste collection and disposal, and policing, if their neighbors spend more.

The problem of spillover is particularly significant whenever the external benefits of providing services are very large in relation to the internal ones, and the services are financed by local taxes. This is particularly the case for those activities whose costs and benefits spread over a large geographic area, such as with pollution control, flood protection and drainage, recreational facilities, and public transportation. With regionwide benefits, it is unreasonable to expect households in a particular part of a metropolis to bear the financial burden for an urban area as a whole. Given the desire of municipalities to maintain as much independence and autonomy as possible, there has been an inexorable tendency for municipalities to address problems of spillover by creating more special districts, and increasing the use of direct charges to those who actually make use of the service or facility.

Organizational and Representational Problems

The interrelated nature of many urban problems is such that they cannot be handled effectively by a large number of local governments acting in isolation. Although this high degree of political and fiscal fragmentation may be defended on the grounds of local autonomy and continuing the democratic tradition of home rule, it is unsuited to the realities of modern metropolitan life. From the governmental point of view, the metropolitan whole should be more than the sum of its administrative parts. In essence, metropolitanism requires *cooperation rather than competition among its many jurisdictions,* and *competition rather than monopoly in those areas of public service provision that are best provided from numerous locations and might benefit particularly from the rigors of public choice*—such as education and some community social service agencies.

Fiscal Zoning　In fact, far from cooperating, many local governments within a metropolitan area compete with each other for acceptable tax-yielding economic activities. The heavy reliance on the property tax forces local governments to meet increased costs, or to hold outlays to a reasonable level, by using their powers of zoning for fiscal ends. Different types of land use demand different services and at the same time are capable of making different-sized contributions to the public purse. Thus, whereas commercial and industrial uses, and high-income residences, often contribute in taxes two or three times the cost of services they use, residential property of moderate value pays only a fraction of the cost of public services it requires. There is, therefore, enormous competition between local government units for high-revenue low-service-cost land uses, and a tendency to discourage low-revenue high-cost uses through various zoning mechanisms.

Fiscal zoning can take many forms, but its basic purpose is to maximize a local revenue base and minimize local costs through various forms of exclusion and economic one-upmanship. Exclusion occurs when a local government unit institutes large-lot zoning for residential construction that effectively promotes low densities and encourages residential construction for high-income house-holds only. Economic one-upmanship occurs when a new suburban commu-nity attracts through various favorable tax deals the construction of a large re-gional shopping center that takes business away from other commercial areas in the metropolis. For example, the West Edmonton Mall has virtually killed retail activities in downtown Edmonton. Fragmentation thus is antithetical to good planning for it encourages communities to make decisions, implemented in part through fiscal zoning, that are in their immediate interest but which may run counter to the wisest course of development for the entire urban area.

Public Awareness and Control Fragmentation not only promotes the wrong kinds of competition, protects particular interests, and fosters exclusion, it also serves to reduce public awareness and involvement in local government. Paradoxically, whereas home rule and local jurisdictional autonomy is meant to promote democracy, fragmentation appears to be doing exactly the opposite. Voter turnout in most local government and board elections is extraordinarily low. There is little doubt that this is in part due to the public being generally un-aware of the ways in which local government influences their daily lives. This lack of public awareness is quite serious, for the political control of local gov-ernment can then fall into the hands of those who may find it difficult to distin-guish between the public interest and their interest.

Although public sector politicians and employees are constrained by codes, understandings, and, increasingly, legislation relating to conflict of interest, it is the shaping of the environment through policy and infrastructure delivery that creates economic opportunities. It is interesting to note that many local govern-ment councils and special district boards are dominated by financial, insurance, real estate, and associated legal interests, broadly associated with the property industry, for whom such representation has special importance. The vast bulk of the public is not really represented, and is usually quite unconcerned about the situation—until such time as a scandal emerges or an objectionable devel-opment occurs, at which time a desire for involvement may be too late.

TOWARD SOLUTIONS

At the heart, therefore, of problems facing local government finance within ur-ban areas is the fragmentation of local government units and dependency on regressive forms of taxation and intergovernmental transfers. As a conse-quence, attempts at addressing the problems of local government finance have tended to focus on the two complementary aspects of the issue—local govern-ment reorganization and fiscal reform. Analyses of these two aspects empha-size that:

1. Most local government units, particularly those at the municipal and county levels, can be regarded as historical anachronisms, albeit (and perhaps fa-tally) rooted in the ethos of home rule and strong local identification.

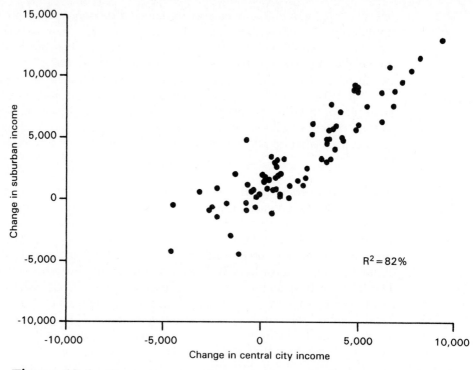

Figure 15.1 The economies of central cities and suburbs are entwined: The close relationship between 1979–1989 change in median household income in United States central cities and suburbs.
(*Source:* Redrawn from Ledebur and Barnes, 1993, Figure 1, p. 4.)

2. Urban regions are *functional economic areas*—that is, each part is interlinked in some ways with the rest of the urban region as exemplified in commuting patterns, which are a principal means for defining metropolitan areas (Ottensman, 1996)—and should be organized on an integrated metropolitan-wide basis. Central cities and suburban communities are thus "all in it together" (Ledebur and Barnes, 1993). This is illustrated in Figure 15.1 in terms of the close relationship between change in suburban and central city median household incomes in the seventy-eight largest metropolitan areas in the United States between 1979 and 1989. Statistical relationships like this (Savitch et al., 1993) do not claim that suburbs depend on their central cities—the highly debatable suburban dependency hypothesis (Hill et al., 1995)—or *vice-versa*, but they do reinforce the functional economic area argument by emphasizing that there are complementary economic and social relationships between central cities and suburbs (Adams et al., 1996).

3. There are considerable *economies of scale* to be achieved in urbanwide provision of some local services, whereas other functions can be and are probably best provided at a more local level to facilitate public choice. For example, on the basis of analyses of economies of scale in the provision and delivery

Table 15.8

An Allocation of Urban Functions and Services Among Three Different Tiers of Local Government According to Economies of Scale in Public Service Provision and Delivery

ACTIVITY	SIZE OF COMMUNITY		
	SMALL	INTERMEDIATE	AREA WIDE
Refuse disposal			X
Refuse collection	X		
Sewage disposal			X
Water treatment and trunk lines			X
Water sales	X		
Airports and ports			X
Bus and truck terminals			X
Railroads and rail yards			X
Mass transit			X
Transport planning			X
Snow removal	X	X	
Street cleaning	X		
Parking lots	X	X	
Cultural facilities			X
Regional parks			X
Community recreational facilities	X		
Planning and land use control			X
Building codes and regulation			X
Building permits and inspection	X	X	
Housing (low-income, elderly)		X	
Housing redevelopment	X	X	
Police	X		X
Fire protection		X	
Ambulance		X	
Health services	X		X
Hospitals		X	X
Welfare	X		X
Social services	X		
Vocational training		X	X
Elementary education	X		
Secondary education	X	X	

(*Source:* Table compiled by the author from a detailed discussion of the economy of scale for urban functions literature in Hallman (1977), pp. 182–194.)

of a wide array of urban services, Hallman (1977) has identified three groups of functions best provided: at a community level, with a population base up to 40,000; at an intermediate level, with a population up to 250,000; and urban regionwide, for populations up to a few million (Table 15.8).

4. In particular, matters related to the representation of an urban area to the nation and the world that are associated with *economic development,* and in-

Table 15.9

Distribution of Types of Municipal Governmental Systems among Canada's Metropolitan Areas

SYSTEM	TERRITORIAL COVERAGE		TOTALS
	COMPREHENSIVE	NONCOMPREHENSIVE	
One-tier	8	3	11
Two-tier	8	4	12
Totals	16	7	23

(*Source:* Sancton (1994), Table 2, p. 16.)

frastructure planning and delivery, require regionwide management (Sancton, 1994). This is because competition between local governments and multiple representation in these activities is counterproductive rather than reinforcing. These are the "highest-order" public services to be provided for an urban area with the largest scale-economies (GTA, 1996).

5. In the case of local government, *structures do determine outcomes.* That is, if the structural and organizational arrangements of local government are congruent with the interrelated nature of urban areas, the social and economic health of the region will be enhanced (Rusk, 1993).

Structural Reforms

The large number of local governments is not of itself bad. The problems arise mostly over the confusion and insularity over what these governments do, not the numbers themselves. Ideally, small jurisdictions should do what they do best, and large jurisdictions should deal with those urban functions that they do best. An implication of Table 15.8 is that some form of two-tier structure of local government in areas with more than 250,000 people would be appropriate, and one-tier government for urban areas with populations less than that. This has been recognized in a number of urban areas in Canada and the United States, though the methods by which two-tier local government has been implemented varies quite significantly.

One- and two-tier systems may be comprehensive or noncomprehensive according to the extent to which a system embraces the entire population of an urban region. Sancton (1994) shows that among Canada's CMAs there are eleven one-tier municipalities that are responsible for all local government functions within their territory that are not assigned to special district bodies (Table 15.9). Eight of these are comprehensive; that is, they embrace more than 70 percent of the population of the relevant urban region, whereas three are noncomprehensive because they encompass less. There are twelve two-tier municipalities, each with a metropolitan-wide government responsible for areawide upper-level functions, and a number of second-tier municipalities responsible for a number of lower-level functions, that are not assigned to special-purpose bodies.

One-Tier Systems   The most prominent examples of one-tier organization are in Canada where metropolitan unification has been brought about in two cases by provincially imposed amalgamation (Winnipeg and Thunder Bay), and for the rest through a series of annexations. The most prominent example of imposed amalgamation is Winnipeg "Unicity" (until Toronto, in 1997) where a one-tier system replaced a two-tier system in 1972 (Brownstone and Plunkett, 1983). One assumption of the amalgamation was that equalized tax levels throughout the metropolitan area would assist the revitalization of the central city. In fact, what happened was that the suburban majority-controlled Unicity council fostered the continued development of suburban infrastructure (Sancton, 1994, pp. 22–28). In consequence, there have been subsequent efforts by the federal and provincial governments in partnership with Unicity to revitalize the central city through such programs as the Winnipeg Core Area Initiative (1981).

The chief reasons for a series of annexations in London (Ontario) in the 1960s and 1970s, which culminated in 1992 in another which tripled the geographic size of the jurisdiction, was the need to cater for population growth, control suburban sprawl, and promote economic development (see Table 4.3 for population data). This series of annexations has maintained London as a one-tier government for a number of decades, and it is, therefore, a good example to assess with respect to the single-tier format. Although unitary government does not seem to have done much to control suburban sprawl, it has provided and delivered services to the areawide population better and more efficiently, that is, at lower per capita costs, than other comparable two-tier jurisdictions (Sancton, 1993, pp. 28–36). These lower costs are reflected in comparatively lower local taxes. Furthermore, the government is regarded as more accountable because the locus of responsibility is clearer.

Two-Tier Systems   The types of two-tier systems that are most common consist of consolidations and federations of municipal units. Consolidation most frequently involves the reallocation of responsibilities for certain functions from all municipalities involved to a preexisting larger jurisdiction, such as a county. A single metropolitan government is created to look after areawide problems, but local provision remains in the hands of existing local governments. Thus, a certain measure of metropolitan control is obtained without creating yet another unit of government. Federation involves local government units agreeing to operate as a collectivity a number of service functions through an areawide organization. The areawide organization may be weak or strong depending on the array of functions, and administrative, coordinating, taxing, and spending powers that are allocated to it. Federations may be more flexible than consolidations because they permit multi-tiered local government to extend in reach as an urban region spreads outwards.

Consolidation   Although school district consolidation to the county level has been common in both Canada and the United States, most cases of municipal consolidation are in the United States. Although it would seem reasonably attractive on paper, such local government restructuring has usually occurred in those cases where the metropolitan area lies almost exclusively in a single county. Examples include: Baton Rouge-East Baton Rouge (1947); Nashville-Davidson County (1962); Jacksonville-Duval County (1967); Columbus-Muscogee County (1970); Lexington-Fayette County (1972); Anchorage-Greater Anchorage Borough (1975); and Indianapolis-Marion County (1969). In only one of these examples—Columbus-Muscogee County—did consolidation initially extend beyond the core county.

The basic problem with city-county consolidation in the United States is that the constitutional arrangements under which municipalities and counties operate emanate from different bases. Consequently, a state usually requires a constitutional amendment (to be voted on by all residents) in order to allow the consolidation. The procedure, as with annexation, can take years and involve so many groups and hearings that the process itself is the greatest impediment. Nevertheless, the evidence available suggests that such mergers may be worthwhile, for not only are economies of scale in the delivery of certain functions realized, but services are extended to cover all parts of the metropolitan area and a more equitable tax system can be developed.

Federation   The idea of federated metropolitan areas has been an appealing one in the United States and Canada since the establishment of the London County Council in England (O'Leary, 1987) in 1888 and 1889, but in the United States various attempts in Boston (1896), Oakland (1921), and Pittsburgh (1929) all failed. An example of a near-federated form of municipal government in the United States is Dade County, Miami, where a two-tier form of government for the incorporated municipalities and counties and a one-tier metropolitan government for the unincorporated portions of the county were approved in 1957. A major reason why federation, or any form of local government reform, is difficult to pursue in the United States arises from the constitutional autonomy of the municipalities and the political requirement for widespread support in each jurisdiction involved. Nevertheless, de facto federation has evolved over a number of years in a number of metropolitan areas such as Nassau County (New York), and Minneapolis-St. Paul (Minnesota).

Despite the political difficulty, Portland (Oregon) area voters, for example, agreed in 1992 to implement a two-tier federation involving three counties and twenty-four municipalities, albeit with limited powers for the upper-level government. Like the Greater Vancouver Regional District (British Columbia) to the north, which was created initially in 1967, the Portland Metro evolved out of the previously established areawide special districts, in this case a unified Portland Metropolitan Area Services District. Its prime function is regional planning, including transport planning, with the Metro Council empowered to coordinate local plans and zoning regulations into a comprehensive framework. Other activities allocated to the metro are: waste collection, recycling, and disposal; environmental monitoring; and operation of a regional zoo and a new convention center (Rusk, 1993, p. 104).

In Canada, the legal responsibility of provincial governments for the establishment and reorganization of municipalities has made local government reform easier to achieve. Commencing with the establishment of federated government in metropolitan Toronto in 1954, a number of forms of federated regional government have been established across Canada embracing large metropolitan areas, such as Montreal and Vancouver, and lower-population-density areas with dispersed municipalities, such as Muskoka (Ontario) (Sancton, 1994). In Metro Toronto, where the federated government has been established the longest, the experience has been encouraging (Sharpe, 1995). Since metropolitan-wide financing is required, and a large number of specific functions are allocated to the Metro to either operate or coordinate with the municipalities, many of the disparities that occur in United States metropolises have been alleviated. In areas of education, public transport, law enforcement, water supply, sewage disposal, waste management, environmental protection, and recreational facilities, many services have improved or their deterioration arrested.

Naturally, the Metro has not been without its critics. Some point especially to problems with respect to welfare, housing, the homeless, urban renewal, and particularly its failure to agree on a common mechanism for property assessment and taxation that would respect the diverse interests of the downtown and suburbs. Furthermore, by the 1990s the Metro is grossly underbounded for newer suburbs now extending far beyond the 1954 Metro boundary (see Figure 14.8). These suburban areas are repeating the cycle of competition through fiscal zoning, preferential taxation, and so forth to attract middle- and upper-income families and new commercial and business activities away from the metro area, a process occurring between the pre–1960 suburbs and the City of Toronto (Sewell, 1993). There are, therefore, severe issues relating to revenue sharing, regional planning, and unified external representation that pertain to an interrelated functional region that now extends well beyond the metro boundary (GTA, 1995). A first stage for addressing these issues was reached with the amalgamation of Metro Toronto in 1997.

Fiscal Reform

Fiscal reforms relate to both revenues and expenditures. On the revenue side, local governments are having to face the issue that, while a large proportion of their income is derived from intergovernmental transfers, higher levels of government are undergoing fiscal retrenchment to contain expenditures and deficits. An element of the strategy of higher levels of government is to pass more responsibilities to lower levels while at the same time constraining financial transfers. As a consequence, although local governments have expanded means of revenue raising they are also looking at revenue sharing as a means of reducing disparities, and privatization and contracting-out as ways of controlling, if not reducing, costs.

Revenue-Sharing The issue of the effect of the suburbanization of the middle-class tax base has been addressed by Rusk (1993) in a detailed analysis of

320 metropolitan areas in the United States. This study argues quite clearly that structure determines outcomes. For example, on the basis of a large quantity of data for central cities and suburbs, he suggests, among a number of findings, that: (1) metropolises in which central cities have been able to expand their boundaries (that is, are "elastic") and are generally less segregated than those that have not been able to expand (that is, are "inelastic"); and, (2) fragmented local government provides an environment more conducive to the formation and perpetuation of highly segregated neighborhoods than metropolitan governments that are more unified (one-tier or two-tier).

One of the most important features of local government reorganization is that urbanwide governments can include redistributive mechanisms for revenue sharing for certain important services, such as public health, water and sewage, policing, and education. The result is that per capita disparities in service levels are not as great within one- or two-tier organizations. Minneapolis-St. Paul provides the only significant example of tax-base sharing in the United States (Rusk, 1993, p. 102–103). Under a state law, the tax-sharing plan applies to 188 municipalities in the area, and involves the pooling of 40 percent of the increase in taxes from commercial-industrial property for redistribution on the basis of per capita property market-value disparities. In 1991, the fiscal-disparities fund amounted to 31 percent of the region's assessed property evaluation, with 157 municipalities net recipients, and 31 net contributors. Interestingly, Rusk (1993) notes that whereas central city Minneapolis was a net recipient in 1980, in 1991 it was a net contributor because of the resurgence of its commercial/industrial base.

Privatization and Contracting-Out Local governments are attempting to control and perhaps reduce expenditures through greater use of competition and more efficient management. In some cases this may involve privatization or contracting-out. Privatization involves the selling or devolution of hitherto publicly provided services to the private sector, with agreements concerning the basic standards of provision and delivery. The competitive element comes in not so much with consumer choice—the lack of choice between cable television companies in particular markets illustrates the problems with replicating complex delivery systems—as with the initial competition between privately owned companies for the purchase of existing systems and, hence, local monopolies. The private company, then, assumes the capital and operating costs, and charges for its service. Contracting-out of management and service delivery may be envisaged for a period of time for those public services with which the local government wishes to retain ownership or rights, such as garbage collection or janitorial services, but does not wish to administer.

Local government units across North America, envisaging mounting bills associated with enlarging and renovating existing public infrastructure relating to sewage and waste treatment, water purification and its delivery, bus transportation, electricity and natural gas networks, and highway construction and its maintenance, are turning to private sector companies, some of which operate globally. By the mid–1990s, the City of Chicago had privatized drug and alcohol treatment, sewer cleaning, and collecting on parking tickets. The City of

Montreal, which is facing a 30 percent budget cut over the next few years and huge investments in upgrading existing facilities, is envisaging privatizing its water system which is currently paid for out of general taxes. A number of international companies, such as Lyonaise des eaux, Generale des eaux, and Bouygues, which virtually control the privatized water systems in France, are interested in buying the system, making the appropriate capital investments, and selling water to Montreal customers on a user fee basis (Corcoran, 1996).

The advantages of these types of privatization are supposed to include more efficient management, lower government spending, tax reductions, lower operating costs, improved service, and reduced waste. The disadvantages may be that: tax reductions will not be sufficient to offset user charges; private companies may be selected on the basis of political objectives rather than on the basis of competitive business plans and experience; regulatory systems need to be put in place to control rates-of-return if more than one provider exists, or prices if there is only one; service may favor the easily served and largest users; and there may be a tendency for private industry to seek and get government subsidies (in the form of offsets, tax deferrals, and so forth) in support of large capital expenditures. Privatization and contracting-out may also be viewed as mechanisms for labor force downsizing and lowering wage rates in a restructuring economy.

CONCLUSION

It is difficult from this review of fragmented local government, variations in the fiscal base, the consequences of fragmentation and the fiscal squeeze, and various approaches to addressing the issues in Canada and the United States, to conclude anything less than that the structure of local government must change as urban areas grow and expand. The form of local government that would appear to be most effective, efficient, and sensitive with respect to local autonomy in extensive urban areas would appear to be some form of two-tier system. To be effective, the system must be flexible enough to be able to expand its jurisdiction as the urban area grows. Planning to cater for population and industrial/commercial growth, to deliver regional infrastructure, and to coordinate outward expansion, are vital functions of the upper level of the two-tier local government. A number of urban areas that have reorganized are facing situations in which their areawide governments are hemmed in by suburbs built since 1980. In such cases, each outer suburban jurisdiction is competing for future growth in its own way. See and Artibise (1991, p. 8), for example, argue that the Greater Vancouver Regional District embraces an area which is far too small with respect to future growth.

If there is more than one level of government, there must be a clear delegation of responsibilities to the different levels. The discussion in this chapter suggests that the functions allocated to an upper tier might include: regional land use and transport planning, and responsibility for the delivery of regional transport infrastructure; public transport; sewer systems and treatment plants that benefit considerably from economies of scale and are also a vital element for controlling and directing urban growth (Figure 10.9); water lines and purification plants; pollution control and cleanups; environmental monitoring and protection; regional parks and conservation areas; waste recycling and disposal facilities; police and public safety; and external representation for economic development. All other aspects appear to be best provided at the local level, including community policing (Table 15.8), to maximize public choice.

Two-tier government must also ensure that, with those services delivered by the lower tier, the interjurisdictional disparities, and tendencies for revenue competition through such means as fiscal zoning, are minimized. This may be addressed using some form of revenue pooling and sharing that exists in most federated two-tier governments, and has been described with respect to Minneapolis-St. Paul. Pooling depends on the implementation of a common system for property assessment throughout the urban region and similar procedures for establishing mill rates. There should also be clear agreement among the participating municipalities on the standards and quality of the service functions to be provided. The principles governing the redistribution mechanism require to be understood and supported by the bulk of the population in the context of the metropolitan whole being viewed as something that is greater than the sum of its parts.

Chapter

16

Future Growth and Sustainable Urban Development

As urban growth has occurred in North America since 1800, it has done so by imposing development fairly swiftly and decisively on the landscape. This development has occurred in large part through private, that is, capitalist, initiative often in partnership with the government. The interest of the government has largely been to foster national objectives, particularly national unity, through such means as promoting the establishment of transport and communications networks, protecting the public interest, and regulating the workings of the private sector. Increasingly, however, society has realized that development has not occurred without substantial costs, and that the costs should be borne, as far as a they are possible to allocate, by the parties incurring them. By the 1960s, there was increasingly widespread concern about the effects of heavy industry, transportation, and chemically intensive agriculture on the pollution of land, water, and air; and the effect of urban growth itself on the consumption of valuable natural resources such as good-quality farmland, wildlife habitats, and recreational spaces.

FROM RESOURCE MANAGEMENT TO SUSTAINABLE DEVELOPMENT

People have been imposing themselves on and polluting the environment for thousands of years. The scale at which this imposition and pollution has occurred, however, has increased through time in association with large-scale urbanization, industrialization, and deindustrialization. Issues concerning the rapid consumption of natural resources and pollution have come to the fore in North America because of society's changing attitudes and perception of it. In some other parts of the world, there is less concern because, for the moment, other issues are higher on the public or political agenda. An important feature of all forms of resource use and pollution is that they are associated with the same types of spillover effects, in this case *environmental spillovers*, as were discussed in the previous chapter with respect to local government fragmentation. That is, the action of a group or individual at one location can affect (positively or negatively) those at other locations and in later times. The essence of pollution and indiscriminate resource use is that it is not reciprocal: Those who pollute or consume irreplaceable resources create negative spillover effects, and others normally have to pay the subsequent cost.

Environmental spillover concerns led, in the 1970s, to a focus on resource management and pollution control in which resources were to be husbanded, and the environment protected, to be used only when the public-private economic payoffs considerably outweighed present and anticipated public costs. The "use" of the environment was, therefore, subject to an economic cost-benefit analysis, in which, for example, polluters pay for environmental transgressions, and impositions on the environment (such as new residential subdivisions, or highways) require environmental impact studies with plans for the mitigation of the most unacceptable effects. This concern about the effect of development on the environment metamorphosed in the late 1980s, following the publication of the report of the Bruntland Commission (1987), to a broader concern with sustainable development.

Sustainable Development

Sustainable development argues that development should occur in a context which "ensures that comparable environmental and economic opportunities are available to members of this and future generations of the world" (Nelson, 1993, p. 14). The concept, therefore, has strong human and environmental premises, being much concerned with *intergenerational equity* with respect to access to resources as well as care for the environment. The environment and the economy are pictured as both sides of the same coin, a healthy environment being regarded as essential for supporting healthy economies. This linkage between the health of the environment and the economy was emphasized in the 1992 Earth Summit where it was emphasized that "sustainable development is not a fixed state of harmony, but rather a process of change in which the exploitation of resources, the direction of investments, the orientation of technological development, and institutional change are made consistent with future as well as present needs" (UNCED, 1992, p. 11).

Although the concept has been subject to much criticism and evaluation, particularly on the grounds of its high level of generality, difficult worldwide application, the overwhelming nature and vagueness of intergenerational responsibility (indefinite, five generations, . . .), and the assumed interrelationship between economic and environmental health, it has at its base an ethic of care and respect that makes sense and resonates among a wide variety of groups and interests (Simon, 1989; Mitchell, 1991). Furthermore, the concept can be used to derive a number of principles for sustainable living which also provide pointers for evaluating progress toward meeting its basic objective. These principles have been summarized by Nelson (1993) as including: respect and care for the community of life; improving the quality of human life; conserving the earth's vitality and diversity; minimizing the depletion of nonrenewable resources; keeping within the earth's carrying capacity; changing personal attitudes and practices; enabling communities to care for their own environments; providing national frameworks for integrating development and conservation; and creating a global alliance (Table 16.1).

North America and Sustainable Living

The role of North America in the movement toward greater sustainable living is vital because the United States and Canada have: populations that are among the wealthiest in the world and the most consumer oriented; land areas that are extensive and occupy large parts of the globe; agricultural, manufacturing, and service economies that have been leaders in technological change and transitions to higher levels of productivity; economies that are highly interrelated and also dependent on other countries in the world; the most automobile-oriented and highly urbanized populations; and are the first countries in which the bulk of the population resides in suburban areas. North America is, therefore, along with the European Community and Japan, in a position to influence the movement to sustainability through domestic actions as well as international leadership.

North American Energy Use and Sustainability   The nature of this domestic and international role is perhaps crystallized with respect to energy consumption, because it is smart energy use in its renewable and nonrenewable forms that is fundamental to development and production and exemplifies a society's wealth and contribution to global sustainability. The basic principle involved concerns that of minimizing the depletion of nonrenewable resources (Table 16.1). Also, as the United States and Canada are both producers and consumers of various types of fuel used for energy production, there are a variety of environmental stresses on the continent that arise from activities associated with the production, conversion, transport, and consumption, of the various sources (MOE, 1991, p. 12.10).

Table 16.1

Principles for Sustainable Living

PRINCIPLES	ELUCIDATION
Respect and Care for the Community of Life	• An ethic based on respect and care for each other and the earth is the foundation of sustainable living. Development ought not to be at the expense of other groups or later generations or threaten the survival of other species. • The benefits and costs of resource use and environmental conservation should be shared fairly among different communities, among people who are poor and those who are affluent, and between our generation and those who come after us.
Improve the Quality of Human Life	• The aim of development is to improve the quality of human life. It should enable people to realize their potential and lead lives of dignity and fulfilment. Economic growth is part of development, but it cannot be a goal in itself; it cannot go on indefinitely. Although people differ in the goals they would set for development, some are virtually universal. These include a long and healthy life, education, access to the resources needed for a decent standard of living, political freedom, guaranteed human rights, and freedom from violence. Development is real only if it makes our lives better in all these respects.

Conserve the Earth's Vitality and Diversity	• Development must be conservation based; it must protect the structure, functions, and diversity of the world's natural systems on which our species depends. To this end we need to:
	  – Conserve life support systems, including process cleansing water and air, and creating and regenerating soils, plants, and animals;
	  – Conserve biodiversity, including all species of plants and animals, genetic stocks and ecosystems;
	  – Ensure that the use of renewable resources is sustainable, including forests, rangelands, cultivated soils, and marine and fresh water ecosystems and fisheries.

Minimize the Depletion of Nonrenewable Resources	• While minerals, oil, gas, and coal cannot be used sustainably, their "life" can be extended by recycling, by use of renewable substitutes, and by other means.
Keep Within the Earth's Carrying Capacity	• Policies that bring human numbers and lifestyles into balance with the Earth's carrying capacity must be complemented by technologies that enhance that capacity by careful management.
Change Personal Attitudes and Practices	• Society must promote values that support the ethic of care and discourage those that are incompatible with a sustainable way of life. Information must be disseminated through formal and informal education so that needed actions are widely understood.
Enable Communities to Care for Their Own Environments	• Communities and local groups provide the easiest channels for people to express their concerns and take ac-

Table 16.1 (continued)

Principles for Sustainable Living

PRINCIPLES	ELUCIDATION
	tion to create securely based sustainable societies. However, such communities need the authority, power and knowledge to act. People who organize themselves to work for sustainability in their own communities can be an effective force whether their community is rich, poor, urban, suburban, or rural.
Provide a National Framework for Integrating Development and Conservation	• All societies need a foundation of information and knowledge, a framework of laws and institutions, and consistent economic and social policies if they are to advance in a rational way. A national program for achieving sustainability should involve all interests and seek to identify and prevent problems before they arise. It must be adaptive, continually redirecting its course in response to experience and to new needs.
Create a Global Alliance	• Global sustainability will depend upon a firm alliance among all countries. But levels of development in the world are unequal, and the lower-income countries must be helped to develop sustainability and to protect their environments. Global and shared resources, especially the atmosphere, oceans, and shared ecosystems, can be managed only on the basis of common purpose and resolve. The ethic of care applies at the international as well as the national and individual levels. No nation is self-sufficient. All stand to gain from worldwide sustainability—and all are threatened if we fail to attain it.

(*Source:* Nelson (1993), p. 17.)

The Main Energy-Producing Fuels   The main sources of energy used in North America are summarized in Figure 16.1. The areas of the circles in the pie charts indicate the amount of fuel used (standardized to BTUs) in the United States for 1970 and 1990, and Canada for 1990. The wedges show that nonrenewable resources provide the main types of fuel used for generating energy, in particular petroleum, natural gas, and coal. Energy generated from nuclear fuels, which have their own horrendous waste storage problems, came significantly on stream after 1970 providing 9.3 percent of the energy used in Canada and 7.6 percent of that used in the United States. Canada is on the whole self-sufficient in fuel production, importing a proportion of its petroleum requirements for eastern Canada from western Europe and the Middle East, but exporting 33 percent more than is imported, along with a large volume of natural gas, to the United States (MOE, 1991, p. 12.12). The United States imports 7.6 percent of its natural gas requirements mainly from Canada, and 45 percent of its crude oil requirements from nearly all the producing areas in the world.

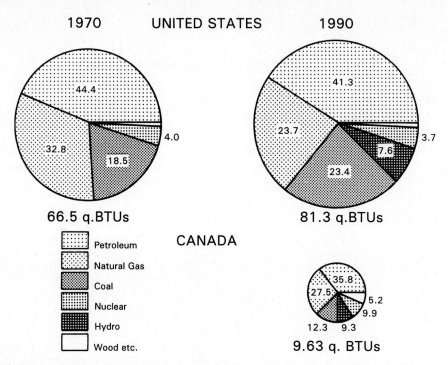

Figure 16.1 Total annual energy consumption in the United States, 1970 and 1990, and Canada, 1990, by major type of fuel used in quadrillion BTUs. (*Source:* Bureau of the Census, 1993, Table 929, p. 573; and MOE, 1991, p. 12.11.)

Changing Energy Consumption The population and economy of North America consume the largest share of the world's energy output (Figure 16.2). This share has, however, decreased considerably from a little over 40 percent in 1960 to under 30 percent in 1990. This decrease in share has not occurred as a consequence of changes in aggregate amount consumed, as is indicated in the consumption-per-capita graph that is also plotted on Figure 16.2. In fact, consumption per capita increased considerably between 1960 and 1970. Therefore, it is tempting to relate this to the enormous deconcentration and suburbanization that occurred in urban areas in North America during this period. The per capita level has drifted down only slightly since 1970, suggesting concerted efforts to constrain per capita energy use. The declining share is, therefore, more indicative of increasing levels of energy consumption in the rest of the world.

The per capita comparisons of total energy use for some of the most developed and leading industrial countries of the world (the "G7") are presented in Figure 16.3. Canada and the United States have by far the highest per capita use of energy of any of the countries in the world, with Canada consuming 23 percent more per capita than the United States. These high comparative numbers are in part related to the great temperature ranges in both countries, and particularly the long cold winters in much of Canada, and the vast expanse of territory integrated by the transport networks. The high per capita use may also be

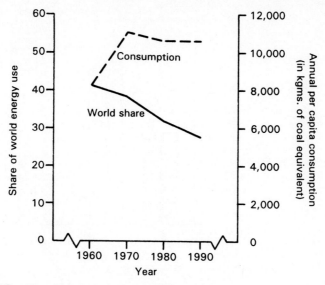

Figure 16.2 The North American share of world energy use and per capita consumption, 1960–1990.
(*Source:* Modified from data in Bureau of the Census, 1993, Table 943, p. 580.)

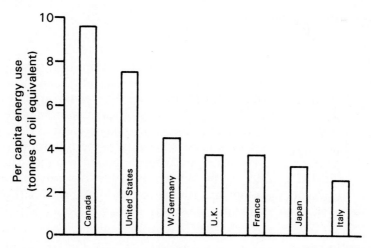

Figure 16.3 Energy consumption per capita: United States and Canada compared with other developed countries.
(*Source:* Diagram from data in OECD, 1991.)

related in part to the more deconcentrated nature of urban form in North America—a matter discussed later in this chapter.

Energy End-Use The pattern of energy consumption among the major end-uses is remarkably similar for both the United States and Canada (Figure 16.4). The largest use is in industry, which consumes about 37 percent of total energy,

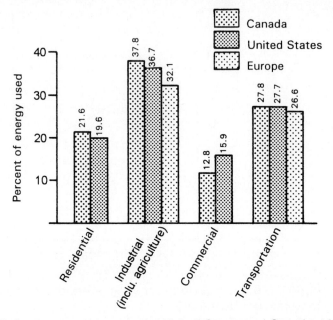

Figure 16.4 Energy end-use in the United States and Canada, 1990.
(*Source:* Compiled by author from data in Bureau of the Census, 1993, p. 76; MOE, 1991, p. 12.12; Aten and Hewings, 1995, p. 353.)

while the second-largest use is transportation consuming about 27 percent. Thus in the drive toward higher levels of user efficiency, it is these two categories along with residential consumption, that require the greatest attention. Given the high levels of per capita energy consumption in North America, the industrial, transportation, and residential end-uses deserve special continuing attention in these respects. The inclusion of western Europe comparative percentage information for industry and transportation (the other categories for Europe are lumped together as "buildings") in Figure 16.4 serves to demonstrate that the transport share remains similar to that for North America even though the per capita total energy use in western Europe is about half (Figure 16.3).

Industry and Sustainability

Given the emphasis in sustainable development on the mutually supportive nature of the environment and development, and that economic change should not evidently compromise the needs of future generations, an *ecosystem* approach is advocated as the most useful way of evaluating sustainable occupancy in particular areas (CF/IRPP, 1990). The boundaries of ecosystems are usually defined by natural features, such as a watershed, for water is a basic prerequisite (along with oxygen) for all forms of life. Furthermore, most impacts find their way into water systems, and often from there into various food chains. Thus ecosystem analysis embraces the whole range of public and private human institutions and activities that have an impact on a defined area. For analytical purposes, impacts may be isolated in terms of types of stress arising from (Regier, 1988):

- Natural processes, such as those arising from the weather and epidemics.

- Additions to the environment arising from human activity, ranging from development-generated sediment, chemicals, and contaminants in soil and water bodies, carbon and nitrogen dioxide in the air, to waste heat discharged to air or water bodies.

- Physical restructuring that alters the environment, including river and stream modifications, wetland drainage, leveling and excavations, shoreline protection, and woodland or forest clearance.

- Removal of renewable and nonrenewable resources arising from rural-to-urban land conversion; transport and power line networks; mineral extraction; and excessive water withdrawal from rivers, lakes, or groundwater supplies.

- The introduction of organisms that are not indigenous to the area (such as the recent introduction of the zebra mussel into the Great Lakes which has proliferated sufficiently to block water inlet pipes) and may adversely shock the more slowly changing local environment to which humans and a myriad other species have become accustomed.

The stresses arising from industrial activity that have accumulated through a number of eras within ecosystems arise preeminently within urban areas (Colten, 1990; Rosen and Tarr, 1994). Some of these impacts have been common to most ecosystems affected by industrialization and urbanization for a long period of time, as is indicated by Hurley (1994) in a discussion of the growth of the oil refining industry in northern New Jersey, Staten Island, and Newtown Creek (on Long Island) at the end of the nineteenth century. The accumulated negative impacts on human health and morbidity rates of the long history of industrial pollution in this area, as well as the positive effects of cleanups and control, have been well documented by Greenburg (1987).

Other impacts relate more specifically to local physiographic and ecological conditions. For example, industrial, residential, and other economic activities in the south Florida area is heavily dependent on groundwater supplies. The aquifers containing this water are being rapidly depleted, leading to local subsidence and lowering of water tables which, along with urban encroachment, is having a negative effect on the Everglades (Craig, 1991). The situation, which is under the control of the South Florida Water Management Water District, probably can be mitigated only through discouragement of high rates of water consumption, perhaps through some combination of pricing and rationing, as alternative methods of provision are too costly or have possibly even greater future environmental and social consequences (Gould, 1995).

The Legacy of Industrial Activity on the Great Lakes Ecosystem   Various phases of the industrial experience are imprinted on the ecosystems of some

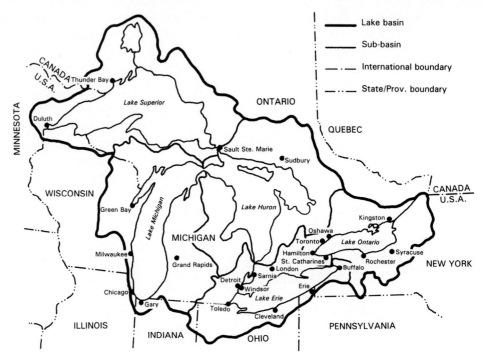

Figure 16.5 Urban areas within the Great Lakes Basin.
(*Source:* Redrawn and modified from CF/IRPP, 1990, Figure 1.1, p. 2.)

regions more than others. The Great Lakes Basin, which is really the upper watershed of the St. Lawrence River, is an ecosystem in which the water, land, and air have been affected by all phases of North American economic development. It is also an area in which various local, provincial/state, national, and transnational programs have been devised to address environmental issues arising from this legacy of development. Probably the best-known program has been initiated by the Great Lakes Water Agreement of 1978 between the United States and Canada "to restore and maintain the chemical, physical, and biological integrity of the waters of the Great Lakes Basin Ecosystem" (IJC, 1988, Article II). This agreement, administered by an International Joint Commission (IJC), is a landmark because it provides one of the few examples of a program jointly conceived and funded by adjacent nation-states to address sustainable development issues relating to a large transnational ecosystem.

The Great Lakes Basin (Figure 16.5) has a 1990–1991 population of 37 million, three-quarters of which is located in the United States side. The vast proportion of the population is around Lakes Erie and Ontario and the southern end of Lake Michigan—the "south lakes" area (Figures 1.1 and 1.11). Three of the ten largest metropolises in North America are located within the ecosystem (Chicago, Detroit, and Toronto), and there is an additional seven metropolises with more than 500,000 people. One of the characteristics of these metropolises is that their growth, at times, has tended to be fairly rapid, such as Buffalo following the opening of the Erie Canal (1825) in the mid-nineteenth century, Chicago during the second half of the nineteenth century, Detroit and Cleveland in the first half of the twentieth century, and Toronto after World War II. These periods of rampant growth have occurred to a large extent at the expense of the environment.

The Early Industrial Era and the Basin Ecosystem   Industrial development in the Great Lakes Basin began during the early industrial capitalism era (1845–1895) following the linking of the area by canal (the Erie Canal and the Welland Canal, 1829, bypassing Niagara Falls), and then the railroad, to New York City and the St. Lawrence River. The opening of the Sault Canal (1855) connected the lower Great Lakes with the iron and copper deposits that had been found on the upper peninsula of Michigan. The integrated waterway provided the means for bringing bulky raw materials together for manufacture at lakeside locations; the invention of the Bessemer steel convertor had provided a way of mass-producing a product far superior to iron; and the continent-wide expansion of the railroad system, growth of machinery manufacturing, and the exigencies of the United States Civil War, provided the demand which stimulated rapid growth of an urban economy based in large part on the manufacture of iron and steel and the smelting of other metals.

The legacy of this unfettered industrial expansion and associated urban growth, is evident in many parts of the Great Lakes Basin today. In particular, the destruction of the forests around the south lakes to provide wood for export, urban housing, fuel, and farmland has proved to be difficult to reverse (when desired) in a number of places. Industrial and municipal wastes were invariably discharged directly into the lakes, and rivers draining to them, leaving bottom sediments contaminated with heavy metals and toxic substances in some locations, such as the Grand Calumet River (Chicago) and Indiana Harbor Canal (Gary) that still persist. Untreated wastes frequently had a devastating effect on the inhabitants of lakeside communities as typhoid, cholera, and other water-related diseases became major causes of death. Also, fisheries in the south lakes were often devastated due to overfishing and loss of habitat due to sedimentation and pollution.

National Industrialization and the Basin Ecosystem   Concern over the impact of industrialization and urbanization in the Great Lakes Basin began to be expressed during the era of national industrial capitalism (1895–1945), which was also a period of continuing rapid growth for most of the urban areas around the south lakes and sporadically on the periphery of the "north lakes"—Lake Huron, Lake Superior, and northern Lake Michigan. By 1920–1921, the population within the basin was about 18 million, heavily concentrated in the largest cities. The concern emanated primarily from issues related to public health. Informed citizens around 1900 noted that "Whereas in the major European cities at this time the annual death rate from typhoid was around 5 per 100,000 population, in the principal U.S. cities in the Great Lakes basin it was four times higher or more" (CF/IRPP, 1990, p. 53). Public health issues became a central concern of the Progressive movement in local government.

Attempts at resolving this environmentally related public health issue, and the impacts of the expanding industrial base which was now including rapidly growing automobile, petroleum refining, and petrochemical industries (particularly focused on Detroit and Cleveland), were essentially piecemeal. For example, as typhoid fever was known to be transmitted by contaminated water, municipalities gradually installed water filtration and chlorination "purification" systems to provide safe water, but did not provide (or require of industry) much by way of waste treatment with respect to outflows, which were causing the contamination in the first place. Few municipalities, as elsewhere in North America (Melosi, 1994), even provided *primary waste treatment*, which merely involves the collection of waste in large tanks and uses settling by gravity for the removal of material. This form of treatment removes no more than 30 percent of the waste in the water and hardly any of the bacteria and chemicals.

During this era there was also an increasing *range* of contaminants being added to the environment. This is illustrated with respect to Lake Erie which, by the 1960s, was choking to death through massive growth of algae stimulated by phosphorous and nitrogen nutrients being discharged into the lake. The sources of these nutrients were the industries located within the Lake Erie sub-basin (Figure 16.5), particularly the petrochemical industries of Cleveland, Detroit, and Sarnia, and the many households that were using greater quantities of detergents and cleansers in labor-saving machines.

The concern over Lake Erie, and realization that binational action was called for, eventually led to the United States–Canada Great Lakes Water Quality Agreement (1972), the precursor of the Great Lakes Water Agreement (1978). This 1972 agreement focused on controlling industrial waste emissions, and funding the provision and upgrading of sewage treatment plants in urban areas bordering the lake to a *secondary waste treatment* level. This level of treatment involves not only removal of wastes by settling and filtering, but also the speeding up of bacterial decay and the reoxygenation of the water by various means. These secondary-level procedures, which may remove up to 90 percent of the waste material, though short of *advanced waste treatment* which includes chemical absorption procedures for the removal of up to 99 percent of waste matter, have proven reasonably successful.

The nonrenewable mining and forest products industries, which, for much of the twentieth century used the land, air, and water as dump sites for untreated wastes and used products, became sectors of particular concern during the national industrial era. For example, the increasing demand for newsprint in the United States after 1900, the banning of the export of pulpwood from Ontario in 1900, and the removal of import duties on newsprint paper by the United States in 1913, stimulated the development of a number of pulp-and-paper-based industrial communities around the edge of the north lakes (Laskin, 1990). These dumped or poured waste solids, and toxic chemical byproducts from the pulp-bleaching process, into the water and on the land. Since about 1970, there have been concerted attempts to mitigate the environmental impact of these plants by restricting the types of chemicals that can be used and improving production processes. In recent years, there has also been an effort to achieve greater sustainability through recycling.

The impact of a myriad chimney stacks over extensive areas, symbolic of this coal-burning and mineral-smelting age when the atmosphere was used as an industrial sewer, is illustrated with respect to nickel and copper smelting in the Sudbury region. This industry expanded rapidly in association with the enormous output of ordnance used in World War I, during which time the surrounding forests were razed for fuel and the land contaminated from sulphur dioxide emissions. The industry continued to grow along with the demand for special steels and wire up to World War II, when there was another armaments-related boom. By the 1960s, the devastation around Sudbury was so great that the area was used as a site-simulator for moon landings (Ripley et al., 1978). Since that time, technological improvements have reduced sulphur dioxide emissions by as much as 80 percent. Concurrently, soil reclamation, along with more resistant varieties of plants, has permitted restoration of grassland, shrub cover, and some woodlands. Also, metal recycling is contributing significantly to sustainable development, with, for example, about 40 percent of the copper produced in Canada coming from recycled material (MOE, 1991, p. 11.22).

Mature Industrialization, Deindustrialization, and the Great Lakes Basin Ecosystem   By the middle of the twentieth century, the Great Lakes Basin had become the principal area for manufacturing in North America. The motor vehicle industry, which along with its numerous associated economic activities has been a prime force in economic development for much of the twentieth century, is focused in the region. After 1945, urban areas around the south lakes also became foci of a growing chemical industry that had been stimulated by research and development into a wide range of synthetic products and pharmaceuticals during World War II. Synthetic products quickly became a vital part of the material basis of manufacturing in the second half of the twentieth century. Thus, the pharmaceutical industry grew rapidly along with increasing per capita expenditures on personal appearance and health care. In consequence, by 1970–1971 the population of the region had almost doubled from its 1920–1921 total reaching 35 million (MOE, 1991, pp. 18.1–31).

Wastes from the chemical industry were deposited in gaseous form into the air, and in liquid form on the land or in the water in what was usually, at the time, conceived as being a relatively safe manner. The problem with many chemical effluents is that it is often difficult to determine their impact on human health, even though their impact on other forms of plant and animal life may appear quite severe. Furthermore, chemical effluents frequently mix through seepage in water and air, creating a *chemical soup* that may have unknown future consequences (CF/IRPP, 1990, pp. 165–169). Although negative impacts on plant and animal life can be regarded as early warnings to humans, it is direct widespread linkages to human health that generate the greatest concern and the most immediate responses (Greenberg, 1987). Chemical deposits may, therefore, be health time-bombs, with unknown impacts on future generations.

To illustrate: Close to the United States side of Niagara Falls is a ditch, remnant of the former Love Canal, about 1,000 yards, or 900 meters, long. Between 1942 and 1953, the Hooker Chemical Company, now absorbed into a larger petrochemical conglomerate, placed 21,800 tons of waste into the ditch. The site was covered and sold to the Niagara Board of Education for $1.00 in 1953 (Levine, 1982). Subsequently, the land became part of a suburban development, and though some signs of impending problems emerged quite soon—land subsidence as the buried barrels of chemicals decayed, rashes and irritated eyes in children following contact with the soil cover, and strong chemical odors—it was not until 1976 that the IJC, tracing sources of chemicals that had induced some mutations and death in a few Lake Ontario fish, revealed the potentially hazardous nature of the chemical cocktail in the covered ditch.

The immediate result was close to panic, as people feared that the buried and leaching chemicals may be extremely carcinogenic and also trigger genetic changes in humans that might, for example, lead to birth defects in subsequent generations. In 1978, the area immediately adjacent to the former canal was declared a federal disaster area. As a result, 239 households relocated and their homes were boarded up. Subsequently, more families have been moved. There has been an expensive cleanup, and a few of the houses have been demolished. Eventually, some families have received financial compensation, but a whole array of lawsuits followed. This has fueled the debate concerning the real severity of the health hazards, and possible intergenerational genetic impacts, and some families have been allowed to return to their homes (Schmitt, 1988; McLafferty, 1992). But, given that the human health impacts remain unclear, who would be willing to continue submitting themselves and their families to these kinds of risks?

The long-term public policy impact of this, and many other mining and industrial hazardous sites in the United States, such as a dump associated with a former chemical factory in Escambia, an African American suburb of Pensacola (Florida), which is leaking carcogenic dioxins into local water supplies and is suspected of being a major cause of above-average cancer rates among local residents, has been the Superfund cleanup program. This program, which is financed in part through a levy on polluting companies and their inheritors had, by 1989, a priority list of 1,163 sites of which 139 were in the Great Lakes Basin. The Province of Ontario has identified 1,748 waste sites in the basin that may be hazardous to humans, and 17 of the most contaminated aquatic sites and a number in residential areas are being cleaned up (PC, 1993, p. 33). There are questions concerning the effectiveness of the various cleanup programs in the two countries, the setting of waste disposal standards that are difficult for industry to keep, and the apparent *underestimation* of the recuperative powers of nature when hazardous emissions are eliminated or greatly reduced (CF/IRPP, 1991, pp. 61–66).

Industrial decline also creates a whole series of issues that impact negatively on sustainable development. The impact of industrial decline in the Great Lakes Basin is crystallized by the population data—between 1970–1971 and 1990–1991 the population within the basin increased by an annual rate one-fifth of that in the previous fifty years. The increase from 35 million to 37 million persons occurred largely as a result of growth in southern Ontario. The impact of price and quality competition from off-shore suppliers from the mid–1970s onwards, who had implemented more efficient quality-oriented production and process technologies in many aspects of manufacturing associated with the automobile industry (see Chapter 5), led to severe industrial restructuring. Thus, at the same time as the complexity of the accumulated legacy of industrial development on the environment and human health was becoming better understood, much of the economic base that had generated growth and urbanization imploded and subsequently re-formed.

The result has not only been a decline in manufacturing employment and a massive Snow Belt–Sun Belt shift in population, but vacation of industrial properties that were constructed for production processes that are now uncompetitive, interindustry linkages that no longer exist, and perhaps products that are no longer required. This abandonment of industrial properties that are incongruent with the needs of a subsequent industrial era is, of course, not unknown in the United Kingdom, continental Europe, and the New England area of the United States (von Tunzelmann, 1995). These large areas of metropolitan industrial rust and decay, which have been identified in Chapter 9 with respect to Detroit as an extensive "zone of deindustrialization" (Figure 9.6), require some kind of redevelopment as land should not be discarded while newer land, perhaps in agricultural use, is being transferred to industrial use elsewhere. Sustainable development implies the recycling of industrial land analogous to the recycling of materials derived from other renewable and nonrenewable resources (Table 16.1).

Many different methods are being used to promote the recycling of industrial land and occasionally the reuse of old industrial buildings. Steel-framed brick-veneered multi-story relics of late-nineteenth-century industrialism are being recycled in a few urban areas for multiple uses, including residential, commercial, and home-work. This is usually only feasible if the sites are located close to transport facilities and downtown employment opportunities, and the loft homesteaders have some confidence in neighborhood security systems. Another method has been to designate zones of deindustrialization as "enterprise zones," which essentially means free from most regulations except those unequivocally averse to the public interest. The theory is that deregulation of this type, perhaps accompanied by a variety of tax holiday and capital subsidies, will speed the transitional process. Experience with this practice in such metropolises as Buffalo suggests that they are not a panacea, and that they usually attract activities that might otherwise have located elsewhere within the metropolitan area.

Transportation and Sustainability

An important factor contributing to sustainable development is transportation which, as has been indicated in Chapter 14, is a good consumed primarily by urban dwellers and within metropolitan areas. Transportation is also a prime consumer of energy (Figure 14.4) and most of this is derived from petroleum. As petroleum is a nonrenewable resource, and its rapid rate of consumption is related primarily to high per capita use in automobiles, sustainable development issues are extremely important in this aspect of North American life. This level of importance is indicated by the data in Table 16.2 which shows that the United States and Canada have the highest motor vehicle ownership rates

Table 16.2

Motor Vehicle Ownership (Commercial and Personal Use) in The United States and Canada Compared with Other OECD Countries

COUNTRY	VEHICLES PER 1,000 POPULATION
United States	761
Canada	622
Australia	578
New Zealand	543
Italy	535
Iceland	531
West Germany[1]	521
Luxembourg	511
France	502
Switzerland	497
Austria	487
Japan	487
Sweden	469
Norway	465
Finland	441
Belgium	438
Netherlands	411
United Kingdom	405
Spain	390
Denmark	372
East Germany[1]	340
Ireland	304
Greece	260
Portugal	233
Turkey	45

[1]Although East and West Germany have been reunited, the vehicle-to-population ratios remain very different and are shown separately.

(*Source:* Organization for Economic Cooperation and Development, OECD Environmental Data 1993.)

(commercial and personal-use vehicles combined) in the world. In this respect, it is not so much a problem that world supplies of petroleum are likely to be exhausted in the near future, but that, increasingly, supplies are having to come from areas where political unrest may periodically interrupt supplies, or the physical environment is becoming exceedingly difficult or too ecologically sensitive to exploit (Brown et al., 1992, p. 29).

Sustainability and Modal Choice   Information produced in Europe comparing the estimated environmental impacts for air, automobile, rail, and bus transport suggests that automobile and air transport are the greatest users of energy, and emit the greatest amount of pollutants, per person per kilometer traveled, as compared to rail or bus (Table 16.3). The automobile, in particular, is the largest user of land of the four transport modes at 120 square meters per

Table 16.3

The Impact of a Number of Transport Modes on Various Aspects of the Environment

ENVIRONMENTAL ASPECT	UNIT OF MEASUREMENT	AIR	AUTOMOBILE	RAIL	BUS
Land use	M^2/person	1.5	120	7	12
Primary energy use	g coal equivalent units/pkm*	365	90	31	27
Carbon dioxide emissions	g/pkm	839.5	200	60	59
Nitrogen dioxide emissions	g/pkm	6.4	0.34	0.08	0.02
Hydrocarbons	g/pkm	1.4	0.15	0.02	0.08
Carbon monoxide emissions	g/pkm	8.1	1.3	0.05	0.15
Air pollution	Polluted air M^3/pkm	95,000	5,900	1,200	3,300
Accident risks	Hours of life lost/1,000 pkm	1.4	11.5	0.4	1

(*Source:* Data compiled by author from Teufel, D. (1989).)
*pkm = passenger kilometer. One passenger traveling 1 kilometer is 1 pkm. Ten passengers traveling 50 km equals 500 pkm.

person; the second-greatest user of energy per person traveling 1 kilometer (pkm) at 90 equivalent grams of coal/pkm; the second-greatest emitter of carbon dioxide (200 g/pkm), nitrogen dioxide (0.34 g/pkm), hydrocarbons (0.15 g/pkm), and carbon monoxide (1.3 g/pkm); the second-largest source of air pollution at 5,900 cubic meters/pkm; and by far the greatest generator of accidents at 11.5 actuarially estimated hours of life lost/1,000 pkm. The chemical pollutants included in the analysis, apart from contributing to the air pollution soup, are also those that are thought to be carcinogenic and increase the likelihood of respiratory diseases.

An interpretation of the environmental performance data in Table 16.3 is that transportation can be provided either at low energy use and low pollution cost per pkm, or high energy use and high pollution cost per pkm. Bus and rail transport represent the modes in the "low" categories, and air and automobile transport the modes in the "high" categories. Recalling the discussion in Chapter 14, and the data in Table 7.2, which emphasize the overwhelming predominance in North America of automobiles for intraurban travel, and airlines and automobiles for interurban travel, it is clear that, in the context of sustainable development, urban areas in North America have the worst of both worlds— high energy use and high pollution. Furthermore, the high energy use–high pollution path also incurs the greatest social costs, as represented by the accident data and implications of the pollutants with respect to human health.

Sustainability and Fuel Taxes The transport system that we have got in North America thus tries to maintain accessibility to employment, shopping, schools, and social/recreational activities, with a mode that contributes among the least to sustainable development. Automobiles are also detrimental to the maintenance of accessibility because of their space requirements and ubiquitous presence which increasingly generate congestion in large metropolitan areas. But this is the "price" that North American society appears to be willing to pay for separation, privacy, convenience, and comfort, provided in the form of private transportation. It can be argued, however, that the price, as has been suggested in Chapter 14 with a reference to road subsidies being of the order of $2,300 per car per year (MacKenzie, et al., 1992), is not high enough.

Taxes on gasoline and road use simply do not reflect the aggregate costs of pollution and the provision of a highly dispersed, low-density, and deconcentrated infrastructure and urban form that are inevitably associated with high rates of automobile ownership and use. Roseland (1992) has estimated that in Canada, if a gasoline tax were used to recover all the estimated social costs of automobile use, the price at the pump would be $5 a liter, or $22.70 per gallon. The sheer magnitude of this estimate indicates why gasoline taxes, and hence prices, in other OECD countries are four to five times what they are in North America, and also why gasoline pricing alone cannot be used to recoup the implied social costs. Because of their dependency on the automobile, North Americans have become incredibly sensitive to even minor short-term fluctuations in gasoline prices. Although improved car-fuel economy in the United States rose from 14 miles per gallon in 1974 to 28 miles per gallon in 1990, greater average distances traveled have increased aggregate gasoline consumption (Brown et al., 1992, p. 29). However, at 28 miles per gallon fuel economy in the United States is still far less than in other developed economies where high gasoline taxes encourage the purchase of vehicles with greater energy efficiency.

Sustainability and Alternative Motor Vehicle Power Systems Some of the problems of universal automobile use, such as its contribution to air pollution and its dependence on a nonrenewable fuel, have been addressed through the 1970 federal Clean Air Act and subsequent emendations, which encouraged a variety of measures directed toward the use of engines with greater fuel economy and emitting fewer pollutants. At the state level, California with its extremely high volume of automobile use in major metropolitan centers, has been a pioneer in regulations designed to spur automobile companies to find more nonpolluting sources of power. And, given the size of the market in the state and the Sun Belt in general, as California goes auto-wise so eventually goes the rest of North America. The emission situation in California is compounded by intense sunlight which reacts with gaseous auto wastes to produce smog containing a variety of chemicals affecting humans and plants. In 1996, California scrapped a regulation requiring that 2 percent of all new vehicles (commercial and private) meet zero-emission standards by 1998, and replaced it with a requirement that 10 percent of new-vehicle sales be zero-emission by 2003.

The change in regulation means that fleets of limited-range battery-powered vehicles will not have to be rushed on to the market to meet the 1998 deadline, and allows more time for the development of clean-air electric vehicles powered by other means such as fuel cells, which are currently used primarily in space craft and submarines. The concept of the fuel cell is older than the internal combustion engine, but is being made more practical with the advance of technology. Whereas batteries require recharging, fuel cells create electricity themselves through an electrochemical process which is fueled with hydrogen. There is no waste, other than water and heat. The technology is developing quite fast, with 1996 automobile prototypes having a range of 150 miles, top speed of 65 miles per hour, and appearing quite stylish. Fuel cell–powered buses and delivery vehicles could well be available at competitive prices within the next ten years. However, while this technological approach addresses some of the problems arising from extensive use of the internal combustion engine, it does not address wider concerns about the costs of automobile-generated and dependent dispersed urban forms.

URBAN FORM AND SUSTAINABILITY

The deconcentrated, low-density, nature of urban form at the end of the twentieth century provides an environment that works in many ways against sustainability by encouraging an excessive use of resources and developments that degrade the environment. "In the streetcar suburbs of the 1880s and 1890s, the average house lot was 3,000 square feet; in the automobile suburbs of the 1920s the average was 5,000 square feet; but in the subdivisions of the 1940s and 1950s the lots were between 4,000 and 8,000 square feet" (Rome, 1994, p. 416). The decline in population densities through time, and with distance from employment nodes in monocentric and polycentric urban areas, in both the United States and Canada, has been discussed in Chapter 8.

The Costs of Dispersion

The costs of dispersion can be demonstrated with respect to the Toronto region which has an average density of 2,000 persons per square mile. This density is comparable with that of other similar-sized metropolitan areas (Gordon et al., 1989, p. 139). The area within the inner suburban and central area cordons (Figure 14.8) accommodates 62 persons per hectare (16,050 p/sq. mi.), between the inner suburban cordon and the metro boundary the average density is about 29 persons per hectare (7,500 p/sq. mi.), and in the *contiguously built-up* areas of the outer suburban zone beyond the Metro boundary the density is about 16 persons per hectare (4,140 p/sq. mi.). Low densities and the "leapfrogging" of suburban developments over lands that for some reason are difficult to develop, lead to an urban form which is extremely expensive to operate.

In a study of the impact of future population growth in the Greater Toronto Area (Figure 10.4), which is larger than the CMA outlined in Figure 14.8, Blais (1995) attempts to put some costs (in 1995 dollars) on this dispersed form of urban development if, as seems likely, it continues. It is estimated that the population of the GTA will increase from 4.5 million in 1994 to 6.9 million in 2021, with 90 percent of the increase occurring in the outer suburbs beyond the metro boundary. The cost estimates are obviously highly aggregated with large possible error terms, and based in part on the use of ratios derived from other

Table 16.4

Comparison of "Costs" of Dispersed and More Compact Suburban Forms (25-year costs, $ billions, 1995)

| | URBAN FORM | | SAVINGS |
COSTS	DISPERSED	COMPACT	DISPERSED–COMPACT
Capital	$54.8*b*	$45.1*b*	$9.7*b*
Operating and maintenance	14.3	11.8	2.5
Environmental*	30.0	17.5	12.5
Total Savings			24.7

(*Source:* GTA (1996), Table 4.1, p. 112, based on Blais (1995).)
*Includes auto-related health, policing, pollution, congestion, and land costs.

metropolitan areas, but they do provide some idea of the magnitudes involved as they relate to a comparison between dispersed and more compact urban forms (Table 16.4).

1. Accepting future population growth projections for the GTA to the year 2021 and continuation of current population and employment densities in the outer suburban area, new *public sector* aggregate 25-year capital requirements would be about $54.8 billion. These capital requirements would be for roads, sewer and water systems, schools, transit, and so forth. On the other hand, if these public sector infrastructure requirements were provided to serve a more compact suburban population, at a density of about 30 persons per hectare, the cost would be about $45.1 billion, a "saving" of $9.7 billion.

2. The aggregate operating and maintenance costs of public sector services in a low-density dispersed environment would be about $14.3 billion, and a more compact situation $11.8, yielding a saving of $2.5 billion.

3. The automobile-related environmental costs—health, policing, pollution, congestion, and land consumption—associated with a dispersed low-density urban form are estimated to amount to $30 billion, and a more compact form $17.5 billion.

The annual saving of the more compact urban form over the dispersed outer suburban form is, therefore, estimated to be $25 billion over 25 years, or $1 billion per year (Table 16.4). To put this $1 billion in perspective, it represents 1.8 percent of the government of Ontario's total expenditures and 10 percent of its deficit in 1995, and the capital savings are 11 percent of its annual capital expenditures.

Recompacting With dispersed suburban developments incurring such high public costs, it might be expected that these costs would be reflected in relatively high initial purchase prices and high continuing property taxes. Information in Chapter 8 with respect to residential values and property taxes suggests that this is not the case. Per unit of area, property prices and taxes tend not to be significantly higher in outer suburban areas, if anything they tend to be lower. People move out to the suburbs in part because they get more space for less money, even though the public capital, operating, and environmental costs of developing low-density areas are much higher than more compact communities. The level of compactness envisaged in Table 16.4 is not at all high—it is basically that of 1950s suburbanism but with contiguous development, that is, no leapfrogging (Blais, 1995). Single-family homes would still predominate, and households could, if they so wished, still have their own front door.

The modern dispersed polycentric city can be viewed as a consequence of a distorted price system in which "the price of using urban fringe space has been set too low—well below the full costs of running pipes, wires, police cars, and fire engines farther than would be necessary if building lots were smaller" (Thompson, 1968, p. 28). The differences in costs of dispersed and more compact suburban forms in Table 16.4 may be interpreted as a corollary to this underpricing. Thus, it can be argued that low-density suburbs are a consequence of widespread suburban infrastructure and service underpricing, which is far more pervasive than the particular case of gasoline. Underpricing of resources, services, and infrastructure invariably leads to overconsumption and reduced care in their use (Barde and Button, 1990, pp. 11–12).

Although some may argue that sprawling polynucleated metropolises are what people like, so don't fiddle with them (Morrill, 1991), there are enormous intergenerational benefits (which is what sustainable development is about) to be achieved by encouraging incremental recompacting. This may be achieved through pricing: gasoline closer to the social costs of automobile use; highway use with tolls to amortize capital outlays and pay for associated maintenance; land with higher site-weighted residential property taxes to reflect infrastructure costs; and, basic services such as sewers and waste treatment plants, water purification and delivery, and solid waste disposal, with user fees to cover the cost of provision (Vojnovic, 1994).

Pricing, Recompacting, and Equity Issues A major question concerns the ways in which more appropriate pricing of urban resources may be achieved without incurring more socioeconomic inequities of access than exist in metropolitan areas today. Recompacting would make public transportation more feasible as densities would be greater, particularly if higher densities and mixed uses were channeled, again through pricing mechanisms, along arteries connecting the major nodes in multinucleated urban areas. Greater provision and use of public transport would in part address some of the equity considerations arising from higher pricing associated with private transport and low-density housing. All individuals and households, however, would benefit from lower taxes which would be achieved through decreases in the need for high public capital expenditures, operating costs, and environmental impacts, associated with low-density dispersion. Greater compactness would facilitate reduced aggregate trip distances traveled—by as much as 20 to 30 percent in some studies

(Paehlke, 1991, p. 12)—thus assisting individuals and households to minimize transport costs.

Recompacting, however, is almost impossible to achieve in metropolitan areas that are highly fragmented and in which the various political jurisdictions are extremely competitive. The coordinated land use, transport, and social planning that is required would be beyond the capability of multijurisdictional urban areas in which there are no effective regionwide higher levels of government. Furthermore, without federal and state/provincial harmonization of taxation and user fee policies it would be impossible for individual metropolitan areas to go too far along the pricing path alone. The notion of a fairly level tax and user fee national playing field would be important for the maintenance of consistent policies. That is why one of the principles for sustainable living emphasizes the need for a state/provincial or federal framework (Table 16.1).

FUTURE URBAN PATTERNS

The pattern of urban areas in North America is now well established. This pattern is dominated by a number of extensive macro-urban regions containing 64 percent of the population of North America, each including a number of metropolitan areas and smaller places (Figure 16.6). These macro-urban regions have been defined on the basis of broad urban fields, overlapping daily urban systems, and minimum population densities of 50 persons per square mile at the county level of aggregation (Yeates, 1980). The data in Table 16.5 provides the populations for each of these late-twentieth-century macro-urban regions in 1920–1921, at the beginning of the automobile era (Table 7.2), and from 1950–1951 when they became identifiable during an era of widespread suburbanization, rapid population growth, buoyant economic expansion, and large-scale interregional migration.

Macro-Urban Regions

The macro-urban regions vary greatly in population and physical size (Table 16.5). The largest, with a 1990 population of more than 46 million, is "Bosnywash" extending from north of Boston, through the urban areas around New York, to south of Washington, D.C. A large part of this area was defined by Gottman in 1961 as a "megalopolis." Megalopolitan urban forms have been described by Doxiadis (1970) as fairly continuous strips of urbanized areas, containing a number of metropolitan nodes, with total populations in excess of 25 million. Because of their size and multi-metropolitan structure, megalopolises are regarded as generating certain issues, such as those involving transportation and the environment, requiring a macro-region perspective. These large macro-urban regions, therefore, present a special planning and organizational challenge because they demand a perspective that is supra-metropolitan. Chapter 15 notes some of the difficulties experienced achieving even a metropolitan-wide governmental perspective.

Table 16.5

Population of the Major Urban Regions in North America, 1920–1921 to 1990–1991, and 1990–1991 Densities

Major Urban Region	1920/21 Number*	1920/21 Percent	1950/51 Number*	1950/51 Percent	1970/71 Number*	1970/71 Percent	1990/91 Number*	1990/91 Percent	Density (p/sq. mi.)
Bosnywash	23.27	20.32	32.62	19.73	43.34	19.27	46.22	16.91	450.6
Lower Great Lakes (United States)	19.99	17.46	28.76	17.39	37.25	16.57	36.81	13.47	286.7
California	3.81	3.33	9.83	5.95	18.62	8.28	28.15	10.30	415.5
The Urban South	6.41	5.60	9.34	5.65	11.99	5.33	14.89	5.45	141.3
Windsor-Quebec City (Canada)	4.60	4.02	7.28	4.40	11.92	5.30	14.68	5.37	217.3
Florida Peninsula	0.64	0.56	2.26	1.37	6.00	2.67	11.07	4.05	307.8
The Gulf Shore	1.97	1.72	3.86	2.33	6.25	2.78	9.02	3.30	211.3
Ohio Valley	3.05	2.66	4.27	2.58	6.00	2.67	6.46	2.36	336.5
The Pacific Northwest (United States/Canada)	1.79	1.56	3.52	2.13	5.40	2.40	7.63	2.79	166.1
Total	65.53	57.23	101.74	61.53	146.77	65.27	174.93	64.00	284.2
United States and Canada	114.50	100.00	165.34	100.00	224.87	100.00	273.36	100.00	

*In millions of persons.

(*Source*: Extracted from Yeates, M. (1980), Tables 2.1 and 2.15; 1990–1991 data compiled by author.)

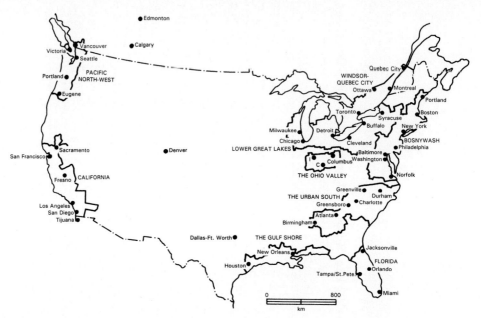

Figure 16.6 The location of macro-urban regions in North America.

Closely allied to Bosnywash in the Middle West is an area around the south lakes and St. Lawrence River. One part of this large area on the United States side of the south lakes, extending farther south than the watershed, is the Lower Great Lakes macro-urban region, with a population of about 37 million. This region extends from Milwaukee, through Chicago, Detroit, Pittsburgh, and Buffalo to Syracuse. The Canadian extension of this region, the Windsor-Quebec City urban axis, has a 1991 population of nearly 15 million, focused primarily around Toronto and Montreal. Though the Lower Great Lakes and Windsor-Quebec City macro-urban regions have not achieved the same aggregate densities as Bosnywash, they have achieved, by 1990–1991, an aggregate population density similar to the average 300 persons per square mile over the entire area of Gottman's (1961) megalopolis. Each macro-urban region includes some farmland, woodland, and other types of open spaces between and peripheral to the higher-density metropolitan areas.

The California macro-urban region, which extends from San Diego in the south, through Los Angeles, Santa Barbara, the San Joaquin Valley, to the San Francisco Bay Area and Sacramento, reached megalopolitan status by 1990. This large region, which extends, in terms of labor supply and *maquiladora* manufacturing, into the Tijuana metropolitan area, has an average population density similar to that of Bosnywash. The region experienced the second-highest annual population growth rate between 1970 and 1990 (after Florida), largely as a result of immigration from Mexico and Central America, Southeast Asia, and elsewhere in the United States (Table 16.6). Recently, following some downsizing in defense-related industries, there has been some outmigration, primarily to other parts of the Southwest. Apart from the area adjacent to San

Table 16.6

Macro-Urban Region Projections for the Twenty-First Century

Macro-Urban Region	1980 FORECAST EVALUATION			1996 FORECAST		
	1990 Actual	2000 Projection	Likely O/U	1970–1990 Annual Growth Rate	2010	Density p./sq. mi.
Bosnywash	46.2	48	C	0.31	49.2	480
Lower Great Lakes	36.8	39	O	-0.06	36.4	285
California	28.2	28	U	2.07	42.6	675
The Urban South	14.9	15	U	1.08	18.5	175
Windsor-Quebec City	14.7	15	U	1.01	17.9	265
Florida Peninsula	11.1	13	C	3.06	20.4	570
The Gulf Shore	9.0	10	C	1.83	13.0	304
Ohio Valley	6.5	7	C	0.34	6.9	360
The Pacific Northwest	7.6	8	C	1.73	10.8	235
United States and Canada	276	291	U	1.03	333.5*	235

Note: O means the 2000 projection probably will overestimate the 2000 actual.
U means the 2000 projection probably will underestimate the 2000 actual.
C means that the 2000 projection will likely prove to be close to the 2000 actual.
*From Bureau of the Census (1993), Table 17; Statistics Canada (1994), Table 15; middle level projections used in both cases.

Francisco Bay, this macro-urban region is largely located in a desert climate, where agribusiness plus population growth have put a severe strain on water supplies as well as waste disposal.

There are three quite large urban regions developing in the south, though none of them has reached megalopolitan status. The highest density of these is the Florida Peninsula (Figure 16.6) which had reached a population of 11 million by 1990 largely as a result of migration from elsewhere in the United States, though there has also been a large Latino and Caribbean influx to the south Florida area around Miami. The growth of service employment, associated with recreational and retirement-based activities and needs, has been phenomenal. The rapid growth of population in the 500-mile-long I-10 Gulf Shore corridor to 9 million is related to petrochemical, high-technology, and defense activities, as well as recreational and retirement-oriented services. The much lower average population density "Urban South" consists of a number of quite separated metropolitan nodes, such as Birmingham, Atlanta, Greenville, Charlotte, Greensboro, and Raleigh-Durham, linked like beads on a string by the I-20/I-85 highways. Many of these urban areas, and the smaller ones surrounding them, are centers of manufacturing concerned with producing a wide variety of electronic, textile, furniture, and other consumer goods (Pollard and Storper, 1996).

The two smallest macro-urban regions epitomize the contrasts in the late North American urban experience. The transnational Pacific Northwest region, or "Cascadia conurbation" (Harvey, 1996), experienced a 1970/71–1990/91 growth rate similar to that of the Gulf Shore. Migrants from the rest of the United States and Canada, and from Southeast Asia, have been attracted by strong employment growth and the equable climate. The Pacific Northwest in the United States has a strong civil aviation, leading technology, software engineering, manufacturing base which has had a high multiplier effect on job growth in consumer and producer services. The Vancouver region has also experienced high employment growth based on a strong Pacific Rim export staples industry. On the other hand, the urban region of the Ohio Valley, linked economically with the employment-downsizing Lower Great Lakes urban region, is undergoing industrial restructuring and has experienced one of the lowest rates of population increase.

Future Population Growth

Figure 16.7 indicates that there have been enormous differences in the rates of growth of population within the areas that are defined as macro-urban regions in the late twentieth century. The areas that have become the Florida and California macro-urban regions have grown at a high compound rate of above 2 percent per annum since 1920. The areas that have become the Urban South, the Windsor-Quebec City corridor, the Gulf Shore, and the Pacific Northwest have increased in population between 1 percent and 2 percent per annum for the entire period. Bosnywash, the Lower Great Lakes, and the Ohio Valley macro-urban regions each grew at a rate of about 1.2 percent per annum between 1920 and 1970, but since 1970 their growth rates have been close to zero. At the met-

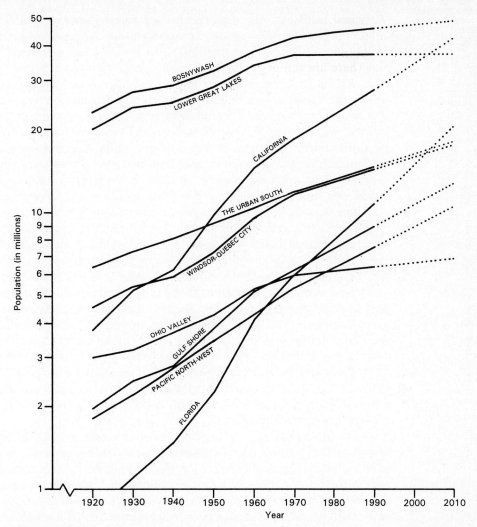

Figure 16.7 Population growth in the macro-urban regions, 1920–1990, and projection of recent trends to the year 2010.

ropolitan level, the variation in growth rates is much greater. For example, the Naples MSA in the Florida urban region grew at a rate of 6.9 percent per annum between 1970 and 1990 to 152,000, while the population of the Buffalo PMSA declined during this same period at a rate of –0.7 percent per annum to 968,532 in 1990. The growth patterns for the macro-urban regions thus reflect these types of intermetropolitan differences, but in a more averaged form.

The Population of the Macro-Urban Regions in the Year 2010 The chief determinant of population growth is net migration, which in turn is related primarily to economic factors in origin and destination areas, and on location

(Mills and Lubuele, 1995). Location is particularly important with respect to immigration from Mexico and Central America—Los Angeles is more likely to be a destination for Mexican immigrants than Buffalo. The economic factors most likely to influence migration are employment and wage rate growth. These are all variables that underly the long-term regional differences in population growth rates that are exhibited in Figure 16.7. The question is, given the discussion relating to the economic and social bases of growth and change in this text, and the ways in which these vary locationally, what kinds of macro-urban region growth patterns might be expected during the next few decades?

One simple way of addressing this question is by projecting the historic experience. This approach can be quite useful in situations in which regional economies, and hence regional populations, follow a fairly stable growth path for a long period of time. Various editions of this book have been published during a twenty-five-year period, and in the third edition (published in 1980) simple projections of macro-urban region populations to the year 2000 were provided on the basis of reasonably consistent growth experiences of the different regions. These projections are listed in Table 16.6 along with the 1990 actual numbers—to modify a famous aphorism, *"projections* are like young chickens, they always come home to roost"* (Robert Southey, *The Curse of Kehama*). It appears that five of the macro-urban region projections may prove to be close to the actual numbers in the year 2000, three will be underestimates because the year-2000 projections have virtually been achieved in these regions by 1990–1991, and one will have proven to be an underestimate because the 1970–1990 population growth rate was close to zero. As a consequence, new projections for the year 2010 are provided in Table 16.6 taking the 1970–1990 growth experiences into account.

The 2010 forecasts for two of the most automobile-addicted parts of North America, the California and Florida macro-urban regions, are particularly sobering. Although recent migration trends suggest that some people are moving from California, immigration and relatively high rates of natural increase should continue to underpin a high growth rate. The average population density, at 675 persons per square mile, will become the highest of any of the urban regions, placing even greater demands on service and transport infrastructure, and pressure on water supplies and waste management. The Florida urban region, with more than 20 million people and the second-highest density of 570 persons per square mile, is likely to be approaching megalopolitan form and size. This area is also experiencing tremendous demands for new infrastructure, which are difficult to meet in a sensitive environment that is already congested. So much so that a $5 billion high-speed passenger rail link—the Florida Overland Express (FOX)—is being considered to connect Miami (and airport) with West Palm Beach, Orlando (and airport), and Tampa (and airport), to lure people off the highways.

Emerging Macro-Urban Regions Furthermore, applications of leading information, communications, and biosynthetic technologies to industrial and service activities may well generate even wider variations in regional growth rates than those experienced thus far, and new macro-urban regions may develop. The 300-mile Dallas-Fort Worth–Austin–San Antonio I-35 corridor experienced an increase in population from 4.1 million in 1970 to about 6.7 million in 1990, a growth rate of almost 2.5 percent per annum. This population increase is consequent to immigration from south of the Rio Grande and elsewhere in the United States attracted by employment in innovation-based manufacturing and the rapidly expanding consumer, producer, and public services. Immigration, in particular, has had a strong stimulative influence on economic performance (Hick and Nivin, 1996; Pollard and Storper, 1996). Economic growth is also related to the establishment of many new high-tech communications and information industries, some of which are relocating to places such as Austin from California. This corridor could well have a population of 11 million by the year 2010.

The growth of a number of smaller urban regions in the High Plains and Mountain States epitomizes a trend that may gain momentum during the next few decades. As many of the newer high-technology industries are far more footloose than traditional manufacturing, requiring primarily venture capital, reliable sources of electric power, an educated workforce, and adjacent university-based research facilities, quite rapid growth is occurring in urban areas that hitherto were regarded as somewhat remote. Three that stand out quite clearly in Figure 1.1 are:

1. Salt Lake City MSA in Utah, where the population reached 1.1 million in 1990 following a growth rate of 2.3 percent per annum between 1970 and 1990. This metropolitan area has the world's second largest cluster of software and computer engineering firms, after California's Silicon Valley located in "Software Valley" along I-15 between the city and Provo.

2. The Edmonton-Calgary corridor (Smith and Johnson, 1978), which experienced a 1971–1991 growth rate of 2.2 percent per annum, bringing its population to 1.6 million in 1991. The economy of this corridor experienced, along with Vancouver, the highest job creation rate in Canada in the early 1990s, based in part on the international demand for leading edge technologies developed in activities associated with the resource sector.

3. The Denver-Colorado Springs-Pueblo I-25 corridor, where the population increased at a rate of 2.0 percent per annum between 1970 and 1990, reaching 2.6 million by 1990. Denver has become the major center for finance and business activities in the Mountain States.

Each of these metropolitan regions is immediately adjacent to excellent outdoor recreational amenities.

CONCLUSION

There is a tendency for people to become somewhat sanguine about the environmental impacts of urban development as scrubbers on smoke stacks, the move to "cleaner" industrial activities, pollution-control equipment on motor vehicles, improved waste disposal facilities, and more careful monitoring, have led to major declines in emissions of chemical, solid, and particle pollutants into the atmosphere, ground, and water. Publicity concerning cleanups has led people to believe that the inherited problems of contaminants and wastes are being resolved, and intergenerational environmental hazards controlled. Unfortunately, only some hazardous sites are being restored, and there is continuous pressure from private industry on governments to reduce the "costs" arising from protective regulations. Also, self-inflicted risks to human health, such as those arising from smoking or unhealthy eating, seem to be more immediately related to quality and length of human life than matters related to urbanization.

Nevertheless, urban development in North America is occurring in a manner that is inimical to sustainable development. Energy use on the continent is the highest in the world, and the production of this energy requires large inputs of nonrenewable fossil fuels some of which are produced domestically while a large share of others are imported. The productivity and lifestyles of North Americans are, therefore, somewhat extravagantly tied to resources that are being depleted and the supplies of which could well be interrupted. The production and consumption of this energy emits wastes that, when concentrated in some ways, can be hazardous to most forms of life. One of the chief uses of this energy is road transport which, since 1950, has facilitated turning North American urban areas inside out, so that they have become deconcentrated, suburbanized, sprawling, and frequently polynucleated.

North American urban form is now virtually dependent on road transport. The concentration of the population into metropolitan areas, and the spread and intermingling of many of these urban areas to form macro-urban regions, some of which are growing quite fast while others are relatively stable, is associated with a geometric increase in vehicle use. This is because motor vehicle ownership and use per capita continues to increase. Thus, even slow-growth urban areas experience an increase in traffic each year, while in faster-growing urban areas, the increase in number of vehicles and trips far outpaces population growth. The path to sustainable living in North America is, therefore, difficult to follow, for sustainable urban development requires implementation of an integrated set of pricing and planning policies directed toward the promotion of urban recompaction rather than a conglomeration of policies that together foster continuing deconcentration.

Bibliography

ABS. 1994. *Homelessness in America*. Special Issue of the *American Behavioral Scientist* 37 (3/4).

Abler, R., and J. S. Adams. 1976. *A Comparative Atlas of America's Great Cities: Twenty Metropolitan Regions*. Minneapolis: University of Minnesota Press.

ACIR. 1984. *Fiscal Disparities: Central Cities and Suburbs, 1981*. Washington, D.C.: Advisory Commission on Intergovernmental Relations.

Adams, C. F., H. B. Fleeter, Y. Kim, M. Freeman, and I. Cho. 1996. "Flight From Blight and Metropolitan Suburbanization Revisited." *Urban Affairs Review* 31 (4): 529–543.

AIEA. 1993. *Arkansas Statistical Abstract*. Little Rock: University of Arkansas, Arkansas Institute for Economic Advancement.

Alba, R. D. 1990. *Ethnic Identity: The Transformation of White America*. New Haven: Yale University Press.

Allen, J. P., and E. Turner. 1996. "Spatial Patterns of Immigrant Assimilation." *The Professional Geographer* 48 (2): 140–155.

Alonso, W. 1964. *Location and Land Use*. Cambridge, MA: Harvard University Press.

Alperovich, G. 1993. "An Explanatory Model of City-Size Distribution: Evidence from Cross-Country Data." *Urban Studies*, 30 (9): 1591–1601.

Anderson, J. E. 1985. "The Changing Structure of a City: Temporal Changes in Cubic Spline Urban Density Patterns." *Journal of Regional Science* 25 (3): 413–425.

Anderson, K. J. 1991. *Vancouver's Chinatown: Racial Discourse in Canada, 1875–1980*. Montreal: McGill-Queen's Press.

Angel, D. P., and J. Engstrom. 1995. "Manufacturing Systems and Technological Change: The U.S. Personal Computer Industry." *Economic Geography* 71 (1): 79–102.

Angel, D. P., and J. Mitchell. 1991. "Intermetropolitan Wage Disparities and Industrial Change." *Economic Geography* 67 (2): 124–135.

Anselin, L. 1988. *Spatial Econometrics: Methods and Models*. Boston: Kluwer Academic Publishers.

Appold, S. J. 1995. "Agglomeration, Interorganizational Networks, and Competitive Performance in the U.S. Metalworking Sector." *Economic Geography* 71 (1): 27–54.

Archer, J. C., and E. R. White. 1985. "A Service Classification of American Metropolitan Areas." *Urban Geography* 6 (2): 122–151.

Archer, W. R., and M. T. Smith. 1993. "Why Do Suburban Offices Cluster?" *Geographical Analysis* 25 (1): 53–64.

Arnold, E., and W. Faulkener. 1985. "The Masculinity of Technology." In W. Faulkener and E. Arnold, eds., *Smothered by Invention: The Masculinity of Technology*. London: Pluto Press, 18–50.

Aten, B. H., and G. J. D. Hewings. 1995. "Transportation and Energy." In S. Hanson, ed., *The Geography of Urban Transportation*. New York: The Guilford Press, 342–358.

Ausebel, J. H., and R. Herman, eds. 1988. *Cities and Their Vital Systems: Infrastructure, Past, Present and Future*. Washington, D.C.: National Academy Press.

Badcock, B. 1993. "A Response to Bourne." *Urban Studies* 30 (1): 191–195.

Baer, W. 1986. "The Shadow Market in Housing." *Scientific American* 255 (5): 29–35.

Baer, W., and C. Williamson. 1988. "The Filtering of Households and Housing Units." *Journal of Planning Literature* 3: 127–152.

Baine, R. P., and A. L. McMurray. 1977. *Toronto: An Urban Study*. Toronto: Clark, Irwin.

Baker, J. R. 1995. "Citizen Participation and Neighborhood Organizations." *Urban Affairs Review* 30 (6): 880–887.

Balakshrinan, T. R. 1982. "Changing Patterns of Ethnic Residential Segregation in the Metropolitan Areas of Canada." *Canadian Review of Sociology and Anthropology* 19: 92–110.

Baldassare, M., and G. Wilson. 1995. "More Trouble in Paradise: Urbanization and the Decline in Suburban Quality-of-Life Ratings." *Urban Affairs Review* 30 (5): 690–708.

Ball, M. 1994. "The 1980s Property Boom." *Environment and Planning A* 26 (5): 671–695.

Banfield, E. C., and J. Q. Wilson. 1963. *City Politics*. New York: Vintage.

Barber, G. 1995. "Aggregate Characteristics of Urban Travel." In S. Hanson, ed., *The Geography of Urban Transportation.* New York: The Guilford Press, 81–99.

Barde, J-P., and K. J. Button. 1990. *Transport Policy and the Environment: Six Case Studies.* London: Earthscan.

Barff, R. A., and P. L. Knight. 1988. "The Role of Federal Military Spending in the Timing of New England Employment Turnaround." *Papers of the Regional Science Association* 65: 151–166.

Barkley, D. L., and S. Hinschberger. 1992. "Industrial Restructuring: Implications for the Decentralization of Manufacturing to Nonmetropolitan Areas." *Economic Development Quarterly* 6 (1): 64–79.

Barlowe, R. 1986. *Land Resource Economics: The Economics of Real Estate.* Englewood Cliffs, NJ: Prentice-Hall.

Barnes, T., and E. Sheppard. 1992. "Is There a Place for the Rational Actor? A Geographical Critique of the Rational Choice Paradigm." *Economic Geography* 68 (1): 1–21.

Barnes, T. J., and R. Hayter. 1994. "Economic Restructuring, Local Development and Resource Towns: Finest Communities in Coastal British Columbia." *Canadian Journal of Regional Science* 17 (3): 289–310.

Barnett, J., H. Lendecker, E. Gunts, and R. Kreloff. 1995. "America's Cities." *Architecture* (April): 55–85.

Barresi, C. 1968. "The Role of the Real Estate Agent in Residential Location." *Sociological Forces* 1: 59–71.

BART. 1993. *BART: History in the Making.* Oakland, CA: Bay Area Rapid Transit District.

Bates, L. J., and R. E. Santerre. 1994. "The Determinants of Restrictive Residential Zoning: Some Empirical Findings." *Journal of Regional Science* 34 (2): 253–263.

Bathelt, H., and A. Hecht. 1990. "Key Technology Industries in the Waterloo Region: Canada's Technology Triangle." *The Canadian Geographer* 34 (3): 225–234.

Baum, A., and D. Burnes. 1993. *A Nation in Denial: The Truth About Homelessness.* Boulder, CO: Westview Press.

Beauregard, R., ed. 1989. *Atop the Urban Hierarchy.* Totowa, NJ: Rowman and Littlefield Publications, Inc.

Beauregard, R. A. 1990. "Trajectories of Neighborhood Change: The Case of Gentrification." *Environment and Planning A* 22: 855–874.

Beauregard, R. A. 1993. *Voices of Decline.* Oxford, UK: Blackwells.

Beauregard, R. A. 1995. "Edge Cities: Peripheralizing the Center." *Urban Geography* 16 (8): 708–721.

Beeson, P. E. 1990. "Sources of the Decline of Manufacturing in Large Metropolitan Areas." *Journal of Urban Economics* 28 (1): 71–86.

Beguin, H. 1992. "Christaller's Central Place Postulates: A Commentary." *Annals of Regional Science* 26 (3): 209–229.

Bell, D. 1976. *The Coming of Post-Industrial Society.* New York: Basic Books.

Bell, D. J. 1991. "Insignificant Others: Lesbian and Gay Geographies." *Area* 23 (4): 323–329.

Belluck, P. 1996. "In Era of Shrinking Budgets, Community Groups Blossom." *The New York Times* (February 25): 1/16.

Bergmann, B. R. 1986. *The Economic Emergence of Women.* New York: Basic.

Berry, B. J. L. 1963. *Commercial Structure and Commercial Blight.* Chicago: University of Chicago Press, Department of Geography Research Paper 85.

Berry, B. J. L. 1967. *Geography of Market Centers and Retail Distribution.* Englewood Cliffs, NJ: Prentice Hall.

Berry B. J. L. 1976. "The Counterurbanization Process: Urban America Since 1970." In B. J. L. Berry, ed., *Urbanization and Counterurbanization.* London: Sage.

Berry, B. J. L. 1981. *Comparative Urbanization: Divergent Paths in the Twentieth Century.* New York: St. Martin s Press.

Berry, B. J. L. 1981. *The Human Consequences of Urbanization.* London: Macmillan.

Berry, B. J. L. 1991a. *Long-Wave Rhythms in Economic Development and Political Behavior.* Baltimore: The Johns Hopkins Press.

Berry, B. J. L. 1991b. "Long Waves in American Urban Evolution." In J. F. Hart, ed., *Our Changing Cities.* Baltimore: Johns Hopkins University Press, 31–50.

Berry, B. J .L., and W. L. Garrison. 1958. "The Functional Bases of the Central Place Hierarchy," *Economic Geography* 34: 145–154.

Berry, B. J. L., and Hak-Min Kim. 1993. "Challenges to the Monocentric Model." *Geographical Analysis* 25 (1): 1–4.

Berry, B. J. L., and J. Parr. 1988. *Market Centers and Retail Location: Theory and Applications.* Englewood Cliffs, NJ: Prentice Hall.

Best, M. 1990. *The New Competition: Institutions of Industrial Restructuring.* Cambridge: Harvard University Press.

Best, P. 1995. "Women, Men, and Work." *Canadian Social Trends* 36: 30–33.

Bettison, D. G. 1975. *The Politics of Canadian Urban Development.* Edmonton: University of Alberta Press.

Bird, R. M., and N. E. Slack. 1993. *Urban Public Finance in Canada.* Toronto: Wiley.

Black, A. 1993. "The Recent Popularity of Light Rail Transit in North America." *Journal of Planning Education and Research* 12 (2): 150–159.

Blackley, P. R., and D. Greytack. 1986. "Comparative Advantage and Industrial Location: An Intrametropolitan Evaluation." *Urban Studies* 23: 221–230.

Blais, P. 1995. *The Economics of Urban Form.* Toronto: Greater Toronto Area Task Force, Background Paper.

Bloomquist, K. M. 1988. "A Comparison of Alternative Methods of Generating Economic Base Multipliers." *Regional Science Perspectives* 18 (1): 58–99.

Bluestone, B., and B. Harrison. 1982. *The Deindustrialization of America: Plant Closings, Community Abandonment, and the Dismantling of Basic Industry.* New York: Basic.

Blumen, O. 1994. "Gender Differences in the Journey to Work." *Urban Geography* 15 (3): 223–245.

Boddy, M. J., and F. Gray. 1979. "Filtering Theory, Housing Policy and the Legitimation of Inequality." *Policy and Politics* 7: 39–54.

Bogardus, E. S. 1926. "Social Distance in the City." In E. W. Burgess, ed., *The Urban Community.* Chicago: University of Chicago Press, 48–54.

Bondi, L. 1990. "Progress in Geography and Gender: Feminism and Difference." *Progress in Human Geography* 14 (3): 438–446.

Borchert, J. R. 1967. "American Metropolitan Evolution." *Geographical Review* 57 (3): 301–332.

Borchert, J. R. 1972. "America's Changing Metropolitan Regions." *Annals of the Association of American Geographers* 62: 352–373.

Borchert, J. R. 1978. "Major Control Points in American Economic Geography." *Annals of the Association of American Geographers* 68: 214–232.

Borchert, J. R. 1983. "Instability in American Metropolitan Growth." *Geographical Review* 73: 127–149.

Borchert, J. R. 1991. "Futures of American Cities." In J. F. Hart, ed., *Our Changing Cities.* Baltimore: The Johns Hopkins Press, 218–250.

Borgers, A., and H. Timmermans. 1993. "Transport Facilities and Residential Choice Behavior: A Model of Multi-Person Choice Processes." *Papers in Regional Science* 72 (1): 45–61.

Bottles, S. L. 1987. *Los Angeles and Utopia: The Making of the Modern City.* Berkeley: University of California Press.

Bourne, L. S. 1981. *The Geography of Housing.* London: Arnold.

Bourne, L. S. 1991. "Recycling Urban Systems and Metropolitan Areas: A Geographical Agenda for the 1990s and Beyond." *Economic Geography* 67 (3): 185–209.

Bourne, L. S. 1993a. "The Myth and Reality of Gentrification: A Commentary on Emerging Urban Forms." *Urban Studies* 30 (1): 183–189.

Bourne, L. S. 1993b. "The Demise of Gentrification? A Commentary and Prospective View." *Urban Geography* 14 (1): 95–107.

Bourne, L. S. 1994. "The Role of Gentrification in the Changing Ecology of Income: Evidence from Canadian Cities and Implications for Further Research." In G. O. Braun, ed., *Managing and Marketing of Urban Development and Urban Life.* Berlin: Reimer, 561–574.

Bourne, L. S. 1995. *Urban Growth and Population Redistribution in North America: A Diverse and Unequal Landscape.* Toronto: Centre for Urban and Community Studies, Major Report 32.

Bourne, L. S., and A. E. Olvet. 1995. *New Urban and Regional Geographies in Canada: 1986–91 and Beyond.* Toronto: Centre for Urban and Community Studies, Major Report 33.

Bradbury, J. H. 1985. "International Movements and Crises in Resource Oriented Companies: The Case of Inco in the Nickel Sector." *Economic Geography* 61 (2): 129–143.

Braid, R. 1984. "The Effects of Government Housing Policies in a Vintage Filtering Model." *Journal of Urban Economics* 16: 272–296.

Brimelow, P. 1995. *Alien Nation.* New York: Random House.

Brown, J., and C. Bennington. 1993. *Racial Redlining: A Study of Racial Discrimination by Banks and Mortgage Companies in the United States.* Washington, D.C.: Essential Information.

Brown, L., and E. Moore. 1970. "The Intra-Urban Migration Process: A Perspective." *Geografiska Annaler* 52B: 1–13.

Brown, L., F. Williams, C. Youngmann, J. Holmes, and K. Walby. 1974. "The Location of Urban Population Service Facilities: A Strategy and Its Application." *Social Science Quarterly* 54 (4): 784–799.

Brown, L. R. et al. 1992. *State of the World, 1992.* New York: W. W. Norton.

Brownstone, M., and T. J. Plunkett. 1983. *Metropolitan Winnipeg: Politics and Reform of Local Government.* Berkeley, CA: University of California Press.

Brunn, S., and T. Leinbach, eds. 1991. *Collapsing Space and Time: Geographic Aspects of Communications and Information.* London: HarperCollins Academic.

Bruntland Commission. 1987. *Our Common Future.* New York: United Nations.

Bryant, C., L. Russworm, and A. G. McLellan. 1982. *The City s Countryside: Land and Its Management in the Rural-Urban Fringe.* London: Longman.

BTS. 1994. *Transportation Statistics: Annual Report, 1994.* Washington, D.C.: Bureau of Transportation Statistics, Department of Transport.

Bureau of Economic Analysis. 1991. *U.S. International Transactions.* Washington, D.C.: U.S. Department of Commerce.

Bureau of the Census. 1975. *Historical Statistics of the United States: Colonial Times to 1970.* Washington, D.C.: U.S. Department of Commerce.

Bureau of the Census. 1985. *Local Government in Metropolitan Areas.* Washington, D.C.: Bureau of the Census, Census of Governments, Volume 5.

Bureau of the Census. 1991. "Metropolitan Area Concepts and Components." *Statistical Abstracts of the United States.* Washington, D.C.: U.S. Department of Commerce, 904–905.

Bureau of the Census. 1992. *Government Units in 1992: Preliminary Report.* Washington, D.C.: Bureau of the Census, Census of Governments.

Bureau of the Census. 1992. *Poverty in the United States: 1992.* Washington, D.C.: U.S. Department of Commerce, P60–185.

Bureau of the Census. 1993. *Statistical Abstract of the United States,* Washington, D.C.: U.S. Department of Commerce.

Bureau of the Census. 1994. *1992 Census of Governments: Government Organization.* Washington, D.C.: Bureau of the Census, Volume 1, No. 1.

Burnell, J. D. 1985. "Industrial Land Use, Externalities, and Residential Location." *Urban Studies* 22: 399–408.

Cameron, G. C. 1973. "Intraurban Location and the New Plant." *Papers of the Regional Science Association* 31: 125–143.

Carey, G. W. 1976. "The New York-New Jersey Metropolitan Region." In J. S. Adams, ed., *Contemporary Metropolitan America, Vol 1.* Cambridge, MA: Ballinger.

Casetti, E. 1984. "Manufacturing Productivity and Snowbelt/Sunbelt Shifts." *Economic Geography* 60 (4): 313–324.

Cashall, J. 1985. "The Form of Micro-spatial Consumer Cognition and Its Congruence with Search Behaviour." *Tijdschrift voor Economische en Sociale Geografie* 76: 345–355.

Castells, M. 1977. *The Urban Question.* London: Edward Arnold.

Castells, M. 1983. *The City and the Grassroots.* Berkeley: University of California Press.

Castells, M. 1989. *The Informational City: Information Technology, Economic Restructuring, and the Urban-Regional Process.* Oxford: Basil Blackwell.

CATS. 1959. *Chicago Area Transportation Study Vol I.*

CCSD. 1987. "Homelessness in Canada: The Report of the National Inquiry." *Social Development Overview* 5 (1): 1–16.

Ceh, S. L. B., and A. Hecht. 1990. "A Spatial Examination of Inventive Activity in Canada: An Urban and Regional Analysis Between 1881 and 1986." *Ontario Geography* 35: 14–24.

Cervero, R. 1986. *Suburban Gridlock.* New Brunswick, NJ: Center for Urban Policy Research.

Cervero, R. 1986. "Urban Transit in Canada: Integration and Innovation at Its Best." *Transportation Quarterly* 40: 293–316.

Cervero, R. 1989. *America's Suburban Centers: The Land Use-Transportation Link.* Boston: Unwin-Hyman.

CF/IRPP. 1990. *Great Lakes, Great Legacy.* Washington, D.C.: The Conservation Foundation. Ottawa: The Institute for Research on Public Policy.

Champion, A. G., ed. 1989. *Counterurbanization: The Changing Face and Nature of Population Deconcentration.* London: Edward Arnold.

Chandler Jr., A. D. 1977. *The Visible Hand: The Managerial Revolution in American Business.* Cambridge, MA: Harvard University Press.

Chang, K. 1989. "Japan's Direct Manufacturing Investment in the United States." *The Professional Geographer* 41 (3): 314–328.

Chapman, K. 1985. "Control of Resources and the Recent Development of the Petrochemical Industry in Alberta." *The Canadian Geographer* 29 (4): 310–326.

Chatterjee, L., and S. Abuttasnath. 1991. "Public Construction Expenditures in the United States: Are There Structural Breaks in the 1921–1987 Period?" *Economic Geography* 67 (1): 42–53.

Checkoway, B. 1984. "Large Builders, Federal Housing Programs and Postwar Suburbanization." In W. K. Tabb and L. Sawers, eds., *Marxism and the Metropolis.* New York: Oxford University Press, 152–173.

Chinitz, B. 1960. *Freight and the Metropolis.* Cambridge, MA: Harvard University Press.

Chressanthis, G. A. 1986. "The Impact of Zoning Changes on House Prices: A Time Series Analysis." *Growth and Change* 17 (3): 49–70.

Christaller, W. 1933. *Die Zentralen Orte in Süddeutschland.* Jena: Gustav Fischer Verlag. Translated by C. W. Baskin. 1966. *Central Places in Southern Germany.* Englewood Cliffs, NJ: Prentice-Hall.

Christerson, B. 1994. "World Trade in Apparel: An Analysis of Trade Flows Using the Gravity Model." *International Regional Science Review* 17 (2): 151–166.

Christopherson, S. 1993. "Market Rules and Territorial Outcomes: The Case of the United States." *International Journal of Urban and Regional Research* 17: 274–288.

Clapp, J. 1980. "The Intrametropolitan Location of Office Activities." *Journal of Regional Science* 20: 387–399.

Clapp, J., H. O. Pollakowski and L. Lynford. 1992. "Intrametropolitan Location and Office Market Dynamics." *Journal of the American Real Estate and Urban Economics Association* 20 (2): 229–257.

Clark, C. 1951. "Urban Population Densities." *Journal of the Royal Statistical Society* A (114): 490–496.

Clark, G. 1989. "Pittsburgh in Transition: Consolidation of Prosperity in an Era of Economic Restructuring." In R. Beauregard, ed., *Economic Restructuring and Political Response.* Newbury Park: Sage Press, 41–69.

Clark, G. 1984. "Who's to Blame for Racial Segregation." *Urban Geography* 5 (3): 193–209.

Clark, W. A. V. 1986. *Human Migration.* Beverly Hills: Sage Publications.

Clark, W. A. V. 1987. "Urban Restructuring from a Demographic Perspective." *Economic Geography* 63: 103–125.

Clark, W. A. V. 1992. "Comparing Cross-sectional and Longitudinal Analyses of Residential Mobility and Migration." *The Professional Geographer Environment and Planning A* 24 (1): 291–302.

Clark, W. A. V., and P. L. Hosking. 1986. *Statistical Methods for Geographers.* New York: Wiley.

Clark, W. A. V., and E. G. Moore. 1982. "Residential Mobility and Public Programs: Current Gaps Between Theory and Practice." *Journal of Social Issues* 38 (3): 35–50.

Clawson, M. 1971. *Suburban Land Conversion in the United States.* Baltimore: Resources for the Future.

Clay, P. L. 1979. *Neighborhood Renewal: Middle-Class Resettlement and Incumbent Upgrading in American Neighborhoods.* Lexington, MA: Lexington Books.

Clemenson, H. 1992. "Are Single-Industry Towns Diversifying?" *Perspectives*, Statistics Canada (Spring): 31–43.

Coffey, W. J. 1994. *The Evolution of Canada's Metropolitan Economies.* Montreal: Institute for Research on Public Policy.

Coffey, W., and M. Polèse. 1987. "Intra-firm Trade in Business Services: Implications for the Location of Office Based Activities." *Papers of the Regional Science Association* 62: 71–80.

Coffey, W. J., and A. S. Bailly. 1991. "Producer Services and Flexible Production." *Growth and Change* 22: 95–117.

Cole, S. 1990. "Indicators of Regional Interaction and the Canada-U.S. Free Trade Agreement." *Canadian Journal of Regional Science* 13 (2/3): 221–246.

Colten, C. E. 1990. "Historical Hazards: The Geography of Relic Industrial Wastes." *The Professsional Geographer* 42 (2): 143–156.

Conway, R. S. 1990. "The Washington Projection and Simulation Model: A Regional Interindustry Econometric Model." *International Regional Science Review* 13 (1/2): 141–166.

Conzen, M. P. 1977. "The Maturing Urban System in the United States, 1840 – 1910." *Annals of the Association of American Geographers* 67 (1): 88–108.

Cook, A. K. 1993. "Growth and Distribution of Producer Services: Metro/Non Metro Differences." *Journal of the Community Development Society* 24 (2): 127–140.

Cook, C. C., and M. J. Bruin. 1993. "Housing and Neighborhood Assessment Criteria Among Black Urban Households." *Urban Affairs Quarterly* 29 (2): 328–339.

Cook, C. C., and M. Lauria. 1995. "Urban Regeneration and Public Housing in New Orleans." *Urban Affairs Review* 30 (4): 538–557.

Cooke, J. J., and A. J. Bailey. 1996. "Family Migration and the Employment of Married Women and Men." *Economic Geography* 72 (1): 38–48.

Cooke, P. 1984. "Region, Class and Gender: A European Comparison." *Progress in Planning* 22 (1): 85–146.

Corcoran, T. 1996. "Privatized Waterworlds" and "Perils of Water Privatization." In *The Globe and Mail.* Toronto: April 26/27.

Cox, K. R. 1986. "Urban Social Movements and Neighborhood Conflicts: Questions of Space." *Urban Geography* 7 (6): 536–546.

Craig, A. K. 1991. "The Physical Environment of South Florida." In T. D. Boswell, ed., *South Florida: The Winds of Change.* Washington, D.C.: Association of American Geographers, 1–16.

Cronon, W. 1991. *Nature's Metropolis: Chicago and the Great West.* New York: W. W. Norton.

Crump, J. R. 1993. "Sectoral Composition and Spatial Distribution of Department of Defense Services Procurement." *Professional Geographer* 45 (3): 286–296.

Curry, M. R. 1991. "Postmodernism, Language, and the Strains of Modernism." *Annals of the Association of American Geographers* 81 (2): 210–228.

Cutler, I. 1982. *Chicago: Metropolis of the Mid-Continent,* 3rd ed. Dubuque, Iowa: Kendall-Hunt.

Cutter, S. 1982. "Residential Satisfaction and the Suburban Homeowners." *Urban Geography* 3 (4): 315–327.

Dagenais, P. 1969. "La métropole du Canada: Montréal ou Toronto." *La révue de géographie de Montreal* 23 (1): 27–38.

Dahmann, D. C., and J. D. Fitzsimmons, eds. 1995. *Metropolitan and Nonmetropolitan Areas.* Washington, D.C.: Bureau of the Census, U.S. Department of Commerce.

Daly, M., I. Gorman, G. Lenjosek, A. MacNevin, and W. Phiriyapreunt. 1993. "The Impact of Regional Investment Decisions on Employment and Productivity: Some Canadian Evidence." *Regional Science and Urban Economics* 23 (4): 559–579.

Daniels, P. W. 1986. "The Geography of Services." *Progress in Human Geography* 10 (3): 436–444.

Daniels, P. W. 1991. "Some Perspectives on the Geography of Services." *Progress in Human Geography* 15 (1): 37–46.

Daniels, P. W. 1991a. "Producer Services and the Development of the Space Economy." In P. W. Daniels and F. Moulaert, eds., op. cit., 135–150.

Daniels, P. W. 1991b. "Internationalization, Telecommunications and Metropolitan Development: The Role of Producer Services." In S. Brunn and T. Leinbach, eds., op. cit., 148–169.

Daniels, P. W. 1993. *Service Industries in the World Economy.* Oxford: Blackwell.

Daniels, P. W., and F. Moulaert, eds. 1991. *The Changing Geography of Advanced Producer Services.* London: Belhaven Press.

Danielson, M. N., and J. Wolpert. 1992. "Rapid Metropolitan Growth and Community Disparities." *Growth and Change* 23 (4): 494–515.

Davies, C. S. 1986. "Life at the Edge: Urban and Industrial Evolution of Texas, Frontier Wilderness-Frontier Space, 1836–1986." *Southwestern Historical Quarterly* 89 (4): 443–554.

Davies S., and M. Yeates. 1991. "Exurbanization as a Component of Migration." *The Canadian Geographer* 35 (2): 177–186.

Davies, W. K. D., and D. P. Donoghue. 1993. "Economic Diversification and Group Stability in an Urban System: The Case of Canada, 1951–1986." *Urban Studies* 30 (7): 1165–1186.

Davies, W. K. D., and R. A. Murdie. 1991. "Consistency and Differential Impact in Urban Social Dimensionality: Intra-Urban Variations in the 24 Metropolitan Areas of Canada." *Urban Geography* 12 (1): 55–79.

Davies, W. K. D., and R. A. Murdie. 1993. "Measuring the Social Ecology of Cities." In L. S. Bourne and D. Ley, eds., *The Changing Social Geography of Canadian Cities.* Montreal: McGill-Queen's University Press, 52–75.

Davis, J. S. 1993. "The Community of Exurban Home Buyers." *Urban Geography* 14 (1): 7–29.

Davis, M. 1990. *City of Quartz: Excavating the Future in Los Angeles.* New York: Verso.

Davis, M. 1992. "Fortress Los Angeles: The Militarization of Urban Space." In M. Sorkin, ed., *Variations on a Theme Park: The New American City and the End of Public Space.* New York: Hill and Wang, 154–180.

Davis, P. 1995. *If You Come This Way: A Journey Through the Lives of the Underclass.* New York: Wiley.

Day, K. M., and S. L. Winer. 1994. "Internal Migration and Public Policy: An Introduction to the Issues and a Review of Empirical Research on Canada." In A. M. Maslove, ed., *Issues in the Taxation of Individuals.* Toronto: University of Toronto Press, 3–61.

Dear, M. 1978. "Planning for Mental Health Care: A Reconsideration of Public Facility Location Theory." *International Regional Science Review* 3 (2): 93–111.

Dear, M., and A. Moos. 1986. "Structuration Theory in Urban Analysis: 2. Empirical Application." *Environment and Planning A* 18: 351–373.

Dear, M. and A. J. Scott. 1981. *Urbanization and Urban Planning in Capitalist Society.* Andover, MA: Methuen.

Dear, M., and M. Taylor. 1982. *Not on Our Street.* London: Pion.

Dear, M., and J. Wolch. 1987. *Landscapes of Despair: From Deinstitutionalization to Homelessness.* Princeton, NJ: Princeton University Press.

Dear, M. J. 1976. "Abandoned Housing." In J. S. Adams, ed., *Urban Policy-Making and Metropolitan Dynamics: A Comparative Geographical Analysis.* Cambridge, MA: Ballinger.

Dear, M. J., and J. Wolch. 1993. "Homelessness." In L. S. Bourne and D. Ley, eds., *The Changing Social Geography of Canadian Cities.* Montreal: McGill-Queen's Press, 298–308.

Deutsche, R. 1990. "Men in Space." *Artforum* (February): 21–23.

Dicken, P. 1992. *Global Shift: The Internationalization of Economic Activity,* 2nd ed. New York: Guilford Press.

DiGaetano, A. 1991. "Urban Political Reform: Did It Kill the Machine?" *Journal of Urban History* 18 (Nov.): 37–67.

DiGaetano, A. 1994. "Urban Governance in a Gilded Age." *Urban Affairs Quarterly* 30 (2): 187–209.

Dingemans, D. 1979. "Redlining and Mortgage Lending in Sacramento." *Annals of the Association of American Geographers* 69: 225–239.

DMARTS. 1955. *Detroit Metropolitan Area Traffic Study, Part I.*

Dobbage, K., and J. Rees. 1993. "Company Perceptions of Comparative Advantage by Region." Chapel Hill: Department of Geography, University of North Carolina.

Domosh, M. 1990. "Shaping the Commercial City: Retail Districts in Nineteenth-Century New York and Boston." *Annals of the Association of American Geographers* 80 (2): 268–284.

Downs, A. 1992. *Stuck in Traffic: Coping with Peak-hour Traffic Congestion.* Washington, D.C.: Brookings Institution.

Doxiadis, C. A. 1970. *Emergence and Growth of an Urban Region.* Detroit: Detroit Edison Co.

Drennan, M. P. 1992. "Gateway Cities: The Metropolitan Sources of U.S. Producer Services Exports." *Urban Studies* 29 (2): 217–235.

Dyck, I. 1989. "Integrating Home and Workplace: Women's Daily Lives in a Canadian Suburb." *The Canadian Geographer* 33, 329–341.

Eaton, B. C., and R. G. Lipsey. 1982. "An Economic Theory of Central Places." *Economic Journal* 92: 56–72.

Edel, M., and E. Sclar. 1975. "The Distributio̶n ... ̶state Value Changes: Metropolitan Boston, ̶870–1970." *Journal of Urban Economics* 2̶

̶ E. Sclar, and D. Lauria. 1984. *Sh̶ ... ̶nership and Social Mobility in Boston's ̶ ̶ation.* New York: Columbia ̶

̶ M. A. Goldberg, and J. M̶ ... ̶ in Canada and the United States: ̶tion of Urban Density Gr̶ ... ̶ 209–217.

England, K. 1993. "Suburban Pink Collar Ghettos: The Spatial Entrapment of Women." *Annals of the Association of American Geographers* 83 (2): 225–242.

England, K. 1994. "From 'Social Justice and the City' to Women-Friendly Cities? Feminist Theory and Politics." *Urban Geography* 15 (7): 628–643.

Erickson, R. A. 1974. "The Regional Impact of Growth Firms: The Case of Boeing." *Land Economics* 50: 127–136.

Erickson, R. A. 1986. "Multinucleation in Metropolitan Economies." *Annals of the Association of American Geographers* 76 (3): 331–346.

Erie, S. P. 1988. *Rainbows End: Irish Americans and the Dilemma of Urban Machine Politics 1840–1885.* Berkeley: University of California Press.

Esparza, A., and A. J. Krmenec. 1994. "Producer Services Trade in City Systems: Evidence from Chicago." *Urban Studies* 31 (1): 29–46.

Ettlinger, N., and B. Clay. 1991. "Spatial Divisions of Corporate Services Occupations in the United States, 1983–88." *Growth and Change* 22 (1): 36–53.

Everitt, J. C., and A. M. Gill. 1993. "The Social Geography of Small Towns." In L. S. Bourne and D. F. Ley, eds., *The Changing Social Geography of Canadian Cities.* Montreal: McGill-Queen's University Press, 252–264.

Ewing, G. O. 1992. "The Bases of Differences Between American and Canadian Cities." *The Canadian Geographer* 36 (3): 266–279.

F M-S. 1995. *Economic Growth in the Mid-South.* Jackson, MS.: Foundation for the Mid-South.

Fabbro, I. 1986. "The Social Geography of Metropolitan Toronto: A Factorial Ecology Approach." Toronto: Department of Geography, York University, unpublished research paper.

Fainstein, S. S. 1992. "Street Scars Undesired." *The Higher* (Sept. 18): 11a.

Fainstein, S. S. 1990. "Economics, Politics, and Development Policy: The Convergence of New York City and London." In J. R. Logan and T. Swanstrom, eds. *Beyond the City Limits: Urban Policy and Economic Restructuring in Comparative Perspective.* Philadelphia, PA: Temple University Press.

Fainstein, S. S., and N. I. Fainstein. 1985. "Economic Restructuring and the Rise of Urban Social Movements." *Urban Affairs Quarterly* 21 (2): 187–206.

Fainstein, S. S., and N. I. Fainstein. 1989. "The Racial Dimension in Urban Political Economy." *Urban Affairs Quarterly* 25 (2): 187–199.

Fainstein, S. S., N. I. Fainstein, R. C. Hill, D. R. Judd, and M. P. Smith. 1983. *Restructuring the City: The Political Economy of Urban Redevelopment.* New York: Longman.

Farley, J. E. 1986. "Segregated City, Segregated Suburbs: To What Extent Are They the Product of Black-White Socio-Economic Differentials." *Urban Geography* 7 (2): 165–171.

Farley, J. E. 1995. "Race Still Matters: The Minimal Role of Income and Housing Cost vs. Causes of Housing Segregation in St. Louis, 1990." *Urban Affairs Review* 31 (2): 244–254.

Farness, D. H. 1989. "Detecting the Economic Base: New Challenges." *International Regional Science Review* 12 (3): 319–328.

Faulkener, W., and E. Arnold, eds. 1985. *Smothered by Invention: Technology in Women's Lives.* London: Pluto Press.

Fava, S. 1980. "Women's Place in the New Suburbia." In G. Wekerle, R. Peterson, and D. Morley, eds., *New Space for Women.* Boulder, CO: Westview Press.

Fernandez, R. M. 1994. "Race, Space, and Job Accessibility: Evidence from a Plant Relocation." *Economic Geography* 70 (4): 390–416.

Ferree, M. 1990. "Beyond Separate Spheres: Feminism and Family Research." *Journal of Marriage and the Family* 52: 866–884.

Feschuk, S. 1995. "Zany Ghermezians Are Back At It." *Globe and Mail*, Toronto, November 9, B4.

Fielding, G. F. 1995. "Transit in American Cities." In S. Hanson, ed., *The Geography of Urban Transportation.* New York: The Guilford Press, 287–304.

Fik, T. J., and G. F. Mulligan. 1990. Spatial Flows and Competing Central Places: Towards A General Theory of Hierarchical Interaction." *Environment and Planning, A* 22 (4): 527–549.

Firey, W. 1945. "Sentiment and Symbolism as Ecological Variables." *American Sociological Review* 10: 140–148.

Fleischmann, A. 1⬛ ⬛⬛nfronting the Rough Edge of Edge City." *Urba⬛ ⬛airs Review* 30 (3): 473–⬛79.

Fletcher, A., 199⬛ ⬛⬛ *ex and Subordination in England.* New H⬛⬛⬛⬛le University Press.

Florida, R., and M. Kenney. 1992. "Restructuring in Place: Japanese Investment, Production Organization, and the Geography of Steel." *Economic Geography D* 68 (2): 146–173.

Florida, R. 1996. "Regional Creative Destruction: Production Organization, Globalization, and the Economic Transformation of the Midwest." *Economic Geography* 72 (3): 314–334.

Foggin, P., and M. Polèse. 1976. *The Social Geography of Montreal.* Toronto: Centre for Urban and Community Studies, University of Toronto Research Paper No. 88.

Fong, E. 1994. "Residential Proximity Among Racial Groups in U.S. and Canadian Neighborhoods." *Urban Affairs Quarterly* 30 (2): 285–297.

Foot, D. K., and D. Stoffman. 1996. *Boom, Bust, and Echo.* Toronto: Macfarlane, Walter, and Ross.

Foster, K. A. 1996. "Specialization in Government: The Uneven Use of Special Districts in Metropolitan Areas." *Urban Affairs Review* 31 (3): 283–313.

Fotheringham, A. S. 1985. "Spatial Competition and Agglomeration in Urban Modeling." *Environment and Planning A* 17 (2): 213–230.

Frey, W. H. 1990. "Metropolitan America: Beyond the Transition." *Population Bulletin* 45 (2): 1–49.

Frey W. H. 1993. "The New Urban Revival in the United States." *Urban Studies* 30 (4/5): 741–774.

Frey, W. H., and A. Speare. 1995. "Metropolitan Areas as Functional Communities." In D. C. Dahmann and J. D. Fitzsimmons, eds., op. cit., 139–190.

Friburg, T. 1993. *Everyday Life: Women's Adaptive Strategies in Time and Space.* Lund, Sweden: Lund Studies in Geography, B, 55.

Friedmann, J. 1973. "The Urban Field as Human Habitat." In S. P. Snow, ed., *The Place of Planning.* Auburn, AL: Auburn University Press.

Friedmann, J. 1986. "The World City Hypothesis." *Development and Change* 17, 1: 69–83.

Friedmann, J., and G. Wolff. 1982. "World City Formation: An Agenda for Research and Action." *International Journal of Urban and Regional Research* 6 (2): 309–343.

Frisken, F. 1991. "Local Constraints on Provincial Initiative in a Dynamic Context: The Case of Property Tax Reform in Ontario." *Canadian Journal of Political Science* 24: 351–378.

Frisken, F., ed. 1994. *The Changing Canadian Metropolis,* Vols. I and II. Berkeley: Institute of Governmental Studies.

Fuchs, L. H. 1990. *The American Kaleidoscope: Race, Ethnicity, and the Civic Culture.* Hanover, NH: Wesleyan University Press.

Fujii, T., and T. A. Hartshorn. 1995. "The Changing Metropolitan Structure of Atlanta, Georgia: Locations of Functions and Regional Structure in a Multinucleated Urban Area." *Urban Geography* 16 (8): 680–707.

Fulton, P. N. 1983. "Public Transportation: Solving the Commuting Problem?" Washington, D.C.: Journey-to-Work and Migration Statistics, Bureau of the Census.

Fulton, W. 1986. "Office in the Dell." *Planning* (July): 13–17.

Furuseth, O. J., and J. T. Pierce. 1983. *Agricultural Land Use in an Urban Society.* Washington, D.C.: Association of American Geographers.

G. and M. 1995. "The Threat to Canada's Great Cities." *Globe and Mail* (September 4): A8.

Gad, G. 1985. "Office Location Dynamics in Toronto: Suburbanization and Central District Specialization." *Urban Geography* 6: 331–51.

Gad, G. 1991. "Office Location." In T. Bunting and P. Filion, eds., *Canadian Cities in Transition.* Toronto: Oxford University Press.

Gakenheimer, R. 1976. *Transportation and Planning as Response to Controversy: The Boston Case.* Cambridge, MA: MIT Press.

Galster, G., and E. W. Hill, eds. 1992. *The Metropolis in Black and White: Place, Power, and Polarization.* New Brunswick, NJ: CUPR Press.

Galster, G., and H. Keeney. 1993. "Subsidized Housing and Racial Change in Yonkers, New York." *Journal of the American Institute of Planners* 59 (2): 172–181.

Galster, G., and J. Rothenberg. 1991. "Filtering in Urban Housing: A Geographical Analysis of a Quality-Segmented Market." *Journal of Planning Education and Research* 11 (1): 37–50.

Gans, H. 1962. "Urbanism and Suburbanism as Ways of Life: A Re-evaluation of Definitions." In A. Rose, ed., *Human Behavior and Social Processes.* Boston: Houghton Mifflin, 625–648.

Gans, H. J. 1995. *The War Against the Poor: The Underclass and Anti-Poverty Policy.* New York: Basic Books.

Garber, J. A. 1993. "Overcoming Ambivalence About American Cities." *Urban Affairs Quarterly* 29 (2): 203–229.

Garber, J. A., and D. L. Imbroscio. 1996. "The Myth of the North American City Reconsidered: Local Constitutional Regimes in Canada and the United States." *Urban Affairs Review* 31 (5): 595–624.

Garber, J. A., and R. S. Turner, eds. 1995. *Gender in Urban Research.* London: Sage.

Garreau, J. 1991. *Edge City: Life on the New Frontier.* New York: Doubleday.

Gatzlaff, D. J., and M. T. Smith. 1993. "The Impact of the Miami Metrorail on the Value of Residences Near Station Locations." *Land Economics* 69 (1): 54–66.

Gaylor, H. 1990. "Changing Aspects of Urban Containment in Canada: The Niagara Case in the 1980s and Beyond." *Urban Geography* 11 (5): 373–393.

Gentilcore, R. L., and G. E. Matthews. 1993. *Historical Atlas of Canada II: The Land Transformed.* Toronto: University of Toronto Press.

Georze, H. (1880) *Progress and Poverty.* New York: Dutton.

Gertler, M. 1992. "Flexibility Revisited: Districts, Nation-States, and the Forces of Production." *Transactions of the Institute of British Geographers* 17: 259–278.

Gertler, M. 1995a. "'Being There: Proximity, Organization, and Culture in the Development and Adoption of Advanced Manufacturing Technologies." *Economic Geography* 71 (1): 1–26.

Gertler, M. 1995b. "Manufacturing Culture: The Spatial Construction of Capital." Toronto: University of Toronto, Department of Geography.

Getis, A. 1983. "Second-Order Analysis of Point Patterns: The Case of Chicago as a Multicenter Urban Region." *The Professional Geographer* 35: 77–107.

Geyer, H. S., and T. Kontuly. 1993. "A Theoretical Foundation for the Concept of Differential Urbanization." *International Regional Science Review* 15 (2): 157–177.

Ghalam, N. Z. 1993. "Women in the Workplace." *Canadian Social Trends* 28: 2–7.

Gibson, L. J., and M. A. Worden. 1981. "Estimating the Economic Base Multiplier: A Test of Alternative Procedures." *Economic Geography* 57: 146–159.

Giddens, A. 1992. "Uprooted Signposts at Century's End." *The Higher* (January 17): 21–22.

Giuliano, G. 1989. "New Directions for Understanding Transportation and Land Use." *Environment and Planning A* 21: 145–159.

Giuliano, G., and K. Small. 1993. "Is the Journey to Work Explained by Urban Structure?" *Urban Studies* 30 (9): 1485–1500.

Glasmeier, A. 1990. "The Role of Merchant Wholesalers in Industrial Agglomeration Formation." *Annals of the Association of American Geographers* 80 (3): 394–417.

Glass, R. (ed.). 1964. *London: Aspects of Change.* London: MacGibbon and Kee.

Gober, P. 1986. "How and Why Phoenix Households Changed: 1970–1980." *Annals of the Association of American Geographers* 76 (4): 536–549.

Gober, P. 1992. "Urban Housing Demography." *Progress in Human Geography* 16 (2): 171–189.

Gober, P. 1993. "Americans on the Move." *Population Bulletin* 48 (3): 1–40.

Goddard, J. B. 1975. *Office Location in Urban and Regional Development.* London: Oxford University Press.

Godfrey, B. J. 1988. *Neighborhoods in Transition: The Making of San Francisco's Ethnic and Non-conformist Communities.* Berkeley: University of California Press.

Goetz, E. G. 1992. "Land Use and Homeless Policy in Los Angeles." *International Journal of Urban and Regional Research* 16 (4): 540–554.

Goetz, E. G. 1995. "A Little Pregnant: The Impact of Rent Control in San Francisco." *Urban Affairs Review* 30 (4): 604–612.

Goheen, P. G. 1990. "The Changing Basis of Inter-Urban Communications in Nineteenth-Century Canada." *Historical Geography* 16 (2): 177–196.

Goheen, P. G. 1993. "The Ritual of the Streets in Mid–19th-century Toronto." *Society and Space* 11 (2): 127–145.

Golant, S. M. 1984. *A Place to Grow Old: The Meaning of Environment in Old Age.* New York: Columbia University Press.

Goldberg, M. 1995. "Urban Futures, Functions, and Forms in a Global Setting." Paper presented at the Urban Regions in a Global Context Conference, Oct. 18–20. Toronto: University of Toronto.

Goldstein, I. J., and G. D. Squires. 1995. "Obfuscating the Reality of Lending Discrimination Through Deceptively Rigorous Statistical Analysis." *Urban Affairs Review* 30 (4): 580–586.

Golledge, C. G., and H. Timmermans. 1988. *Behavioural Modelling in Geography and Planning.* London: Croom Helm.

Golledge, R., and H. Timmermans. 1990. "Applications of Behavioural Research on Spatial Problems I: Cognition." *Progress in Human Geography* 14 (1): 57–99.

Gordon, D. L. A. 1996. "Planning, Design and Managing Change in Urban Waterfront Redevelopment." *Town Planning Review* 67 (3): 261–290.

Gordon, P., A. Kumar, and H. Richardson. 1989. "The Influence of Metropolitan Spatial Structure on Commuting Time." *Journal of Urban Economics* 26 (2): 138–151.

Gordon, P., H. W. Richardson, and H. L. Wong. 1986. "The Distribution of Population and Employment in a Polycentric City: The Case of Los Angeles." *Environment and Planning A* 18, 161–173.

Goss, J. 1993. "The 'Magic of the Mall': An Analysis of Form, Function, and Meaning in the Contemporary Retail Built Economy." *Annals of the Association of American Geographers* 83 (1): 18–47.

Gottman, J. 1961. *Megalopolis.* New York: Twentieth Century Fund.

Gottman, J., and R. H. Harper, eds. 1990. *Since Megalopolis.* Baltimore MD: Johns Hopkins University Press.

Gould, K. 1995. "Just Add Water." *Metropolis* (December): 48ff.

Green, M. B. 1995. "A Geography of U.S. Institutional Investment, 1990." *Urban Geography*, 16 (1): 46–69.

Greenburg, M., ed. 1987. "Health and Risk in Urban-Industrial Society." In M. Greenburg, ed., *Public Health and Environment.* New York: Guilford Press, 3–24.

Grenadier, S. R. 1995. "The Persistence of Real Estate Cycles." *Journal of Real Estate Finance and Economics* 10: 85–120.

Griffith, D. A. 1986. "Central Place Structures Using Constant Elasticity of Substitution Demand Cones." *Economic Geography* 62 (1): 74–84.

Griffith, D. A. 1988. *Advanced Spatial Statistics.* Dordrecht: Kluwer Academic Press.

Gruen, N. J. 1984. "Sociological and Cultural Variables in Housing Theory." *Annals of Regional Science* 18: 1–10.

GTA. 1996. *Report of the Greater Toronto Area Task Force.* Toronto: Queen s Printer for Ontario.

Haag, G., and H. Max. 1995. "Rank-size Distributions of Settlement Systems: A Stable Attractor in Urban Growth." *Papers in Regional Science* 74 (3): 243–258.

Hacker, A. 1992. *Two Nations: Black and White, Separate, Hostile, Unequal.* New York: Scribner's.

Halberstam, D. 1986. *The Reckoning.* New York: William Morrow and Co. Ltd.

Halfacree, K. H. 1995. "Household Migration and the Structuration of Patriarchy: Evidence from the U.S.A." *Progress in Human Geography* 19 (2): 159–182.

Hall, P., ed. 1966. *Von Thünen's Isolated State.* Oxford: Pergamon.

Hall, P., and A. R. Markusen. 1992. "The Pentagon and the Gun Belt." In A. Kirby, ed., *The Pentagon and the Cities.* Newbury Park, CA: Sage Publications, 53–76.

Hall, P., and P. Preston. 1988. *The Carrier Wave: New Information Technology and the Geography of Innovation, 1846–2003.* London: Unwin Hyman.

Hallman, H. W. 1977. *Small and Large Together: Governing the Metropolis.* Beverly Hills: Sage Publications.

Hansen, N. 1990. "Do Producer Services Induce Regional Economic Development?" *Journal of Regional Science* 30 (4): 465–476.

Hansen, N. 1991. "Factories in Danish Fields: How High-Wage, Flexible Production has Succeeded in Peripheral Jutland." *International Regional Science Review* 14 (2): 109–132.

Hansen, N. 1992. "The Location-Allocation versus Functional Integration Debate: An Assessment in Terms of Linkage Effects." *International Regional Science Review* 14 (3): 299–306.

Hanson, S., ed. 1995. *The Geography of Urban Transportation.* New York: Guilford Publications.

Hanson, S., and I. Johnston. 1985. "Gender Differences in Worktrip Length." *Urban Geography* 6: 193–219.

Hanson, S., and G. Pratt. 1991. "Job Search and the Occupational Segregation of Women." *Annals of the Association of American Geographers* 81 (2): 229–253.

Hanson, S., and G. Pratt. 1994. "On Suburban Pink Collar Ghettos: The Spatial Entrapment of Women? by Kim England." *Annals of the Association of American Geographers* 84 (3): 500–504.

Hanson, S., and G. Pratt. 1995. *Gender, Work, and Space.* New York: Routledge.

Harrington, J., A. McPherson, and P. Lombard. 1991. "Interregional Trade in Producer Services: Review and Synthesis." *Growth and Change* 22: 75–94.

Harris, C. D., and E. L. Ullman. 1945. "The Nature of Cities." *Annals of the American Academy of Political Science* 242: 7–17.

Harris, R. 1986. "Homeownership and Class in Modern Canada." *International Journal of Urban and Regional Research* 10 (1): 67–86.

Harris, R. 1991. "Housing." In T. Bunting and P. Filion, eds. *Canadian Cities in Transition.* Toronto: Oxford University Press, 350–378.

Harris, R., and G. J. Pratt. 1993. "The Meaning of Home, Homeownership, and Public Policy." In L. S. Bourne and D. F. Ley, eds. *The Changing Social Geography of Canadian Cities.* Montreal: McGraw-Hill, 281–297.

Harrison, B. 1992. "Industrial Districts: Old Wine in New Bottles." *Regional Studies* 26: 469–483.

Hart J. F., ed. 1991. *Our Changing Cities.* Baltimore: Johns Hopkins University Press.

Hartshorn, T. A., and P. O. Muller. 1989. "Suburban Downtowns and the Transformation of Metropolitan Atlanta's Business Landscape." *Urban Geography* 10 (4): 375–395.

Harvey, D. 1973. *Social Justice and the City.* London: Edward Arnold.

Harvey, D. 1985. *The Urbanization of Capital: Studies in the History and Theory of Capitalist Urbanization.* Baltimore, MD: Johns Hopkins University Press.

Harvey, D. 1989. *The Urban Experience.* Oxford: Basil Blackwell.

Harvey, D. 1990. *The Condition of Postmodernity.* Oxford: Basil Blackwell.

Harvey, D., and L. Chatterjee. 1974. "Absolute Rent and the Structuring of Space by Governmental and Financial Intitutions." *Antipode* 6: 22–36.

Harvey, T. 1996. "Portland: Regional City in a Global Economy." *Urban Geography* 17 (1): 95–114.

Hason, N. 1977. "The Emergence and Development of Zoning Controls in North American Municipalities: A Critical Analysis." Toronto: University of Toronto, Papers on Planning and Development, No. 14.

Hassell, J. S. 1980. "How Effective Has Urban Transport Planning Been?" *Traffic Quarterly* 34: 5–20.

Hawley, A. 1986. *Human Ecology: A Theoretical Essay.* Chicago: University of Chicago Press.

Haynes, K., and A. S. Fotheringham. 1984. *Gravity and Spatial Interaction Models.* Beverly Hills: Sage Publications.

Hayward, D. J., and R. A. Erickson. 1995. "The North American Trade of U.S. States: Comparative Analysis of Industrial Shipment, 1983–1991." *International Regional Science Review* 18 (1): 1–32.

Heikkila, E., D. Dale-Johnson, P. Gordon, J. I. Kim, R. B. Peiser, and H. W. Richardson. 1989. "What Happened to the CBD-Distance Gradient?: Land Values in a Polycentric City." *Environment and Planning A* 21: 221–232.

Henderson, J. 1989. *The Globalisation of High Technology Production: Society, Space and Semiconductors in the Restructuring of the Modern World.* New York: Routledge.

Hewings, G. 1986. *Regional Input-Output Analysis.* Beverly Hills: Sage Publications.

Hick, D. A., and S. R. Nivin. 1996. "Global Credentials, Immigration, and Metro-Regional Economic Performance." *Urban Geography* 17 (1): 23–43.

Hiebert, D. 1993. "Integrating Production and Consumption: Industry, Class, Ethnicity, and the Jews of Toronto." In L. S. Bourne and D. F. Ley, eds., *The Changing Social Geography of Canadian Cities.* Montreal: McGill-Queen's Press, 199–213.

Hill, E. W., H. L. Wolman, and C. Cook Ford. 1995. "Can Suburbs Survive Without their Central Cities: Examining the Suburban Dependence Hypothesis." *Urban Affairs Review* 31 (2): 147–174.

Hinojosa, R. C., and B. W. Pigozzi. 1988. "Economic Base and Input-Output Multipliers: An Empirical Linkage." *Regional Science Perspectives* 18 (2): 3–13.

Hirsch, A. R. 1983. *Making the Second Ghetto: Race and Housing in Chicago 1940–1960.* Cambridge: Cambridge University Press.

Hise, G. 1993. "Home Building and Industrial Decentralization in Los Angeles: The Roots of the Postwar Urban Region." *Journal of Urban History* 19 (2): 95–125.

Hoag, G., and H. Max. 1995. "Rank-size Distributions of Settlement Systems: a Stable Attractor in Urban Growth." *Papers in Regional Science* 74 (3): 243–258.

Hobsbawm, E. 1994. *Age of Extremes: The Short Twentieth Century, 1914–1991.* London: Michael Joseph.

Hobsbawm, E., and T. Ranger, eds. 1983. *The Invention of Tradition.* Cambridge: Cambridge University Press.

Hodge, D. 1988. "Fiscal Equity in Urban Mass Transit Systems: A Geographic Analysis." *Annals of the Association of American Geographers* 78: 288–306.

Hodge, D. 1990. "Geography and the Political Economy of Urban Transportation." *Urban Geography* 11: 87–100.

Hodge, D. 1992. "Urban Congestion: Reshaping Urban Life." *Urban Geography* 13 (6): 577–588.

Hodge, D. G., ed. 1996. "Focus: Defining Spatial Units in Geographic Research." *The Professional Geographer* 48 (3): 298.

Hodge, G. 1965. "The Prediction of Trade Center Viability in the Great Plains." *Papers of the Regional Science Association* 15: 87–118.

Hodgson, J., and J. Oppong. 1989. "Some Efficiency and Equity Effects on Boundaries in Location-Allocation Models." *Geographical Analysis* 21 (2): 167–178.

Holcomb, H. B. 1984. "Women in the City." *Urban Geography* 5 (3): 247–254.

Holcomb, H. B., and R. A. Beauregard. 1981. *Revitalizing Cities.* Washington, D.C.: Association of American Geographers.

Holloway, S. R., and J. O. Wheeler. 1991. "Corporate Headquarters Relocation and Changes in Metropolitan Corporate Dominance, 1980–1987." *Economic Geography* 67 (1): 54–74.

Holmes, J. 1983. "Industrial Reorganization, Capital Restructuring and Locational Change: An Analysis of the Canadian Automobile Industry in the 1960s." *Economic Geography* 59 (3): 251–271.

Holmes, J. 1986. "The Organization and Locational Structure of Production Subcontracting." In A. J. Scott and M. Storper, eds., *Production, Work, Territory.* Boston: Allen and Unwin, 80–106.

Holmes, J. 1992. "The Continental Integration of the North American Automobile Industry: From the Auto Pact to the FTA and Beyond." *Environment and Planning A* 24 (1): 33–48.

Holmes, J., and P. Kumar. 1993. "Labour Movement Strategies in an Era of Free Trade: The Uneven Transformation of Industrial Relations in the North American Automobile Industry." In J. Jenson, R. Mahon, and M. Bienefield, eds., *Production, Space, Indentity: Political Economy Faces the Twenty-first Century.* Toronto: Canadian Scholars Press, 195–224.

Holmes, J., and A. Rusonik. 1991. "The Break-up of the International Labour Union: Uneven Development in North American Auto Industry and the Schism in the UAW." *Environment and Planning, A* 23 (1): 9–35.

Holzer, H. J. 1991. "The Spatial Mismatch Hypothesis: What Has the Evidence Shown? *Urban Studies* 28: 105–122.

Hoover, E. M., and R. Vernon. 1959. *Anatomy of a Metropolis.* New York: Doubleday.

Hopkins, J. S. P. 1991. "West Edmonton Mall as a Centre for Social Interaction." *The Canadian Geographer* 35 (3): 268–279.

Horowitz, D. 1971. "Three Dimensions of Ethnic Politics." *World Politics* 23: 232–244.

Howard, R. 1995. "No Joy in Sight for B.C. Commuters." *Toronto Globe and Mail:* A6.

Howells, C. 1984. "The Location of Research and Development." *Regional Studies* 19: 13–29.

Hoyt, H. 1939. *The Structure and Growth of Residential Neighborhoods in American Cities.* Washington, D.C.: Federal Housing Administration.

Huff, D. L. 1960. "A Topological Model of Consumer Space Preferences." *Papers and Proceedings of the Regional Science Association* 6: 159–173.

Huff, J. O. 1986. "Geographic Regularities in Residential Search Behavior." *Annals of the Association of American Geographers* 76 (2): 208–227.

Hughes, M. A. 1989. "Misspeaking Truth to Power: A Geographical Perspective on the 'Underclass Fallacy'." *Economic Geography* 65 (3): 187–207.

Humphrys, G. 1982. "Power and Industrial Structure." In J. W. House, ed., *The UK Space: Resources, Environment and the Future.* London: Weidenfeld and Nicholson, 282–355.

Hurley, A. 1994. "Ecological Wastelands: Oil Pollution in New York City, 1870–1900." *Journal of Urban History* 20 (3): 340–364.

Hutchinson, B. G. 1991. "Metropolitan Transportation: Patterns and Planning." In T. Bunting and P. Filion, eds., *Canadian Cities in Transition.* Toronto: Oxford University Press, 263–285.

Iacovetta, F. 1992. *Such Hardworking People: Italian Immigrants in Postwar Toronto.* Montreal: McGill-Queen's Press.

Ihlanfeldt, K. R. 1995. "The Importance of the Central City to the Regional and National Economy: A Review of the Arguments and Empirical Evidence." *Cityscape* 1 (2): 125–150.

IJC. 1988. *Revised Great Lakes Water Agreement of 1978: Agreement with Annexes and Terms of Reference, Between the United States and Canada.* Windsor: International Joint Commission.

Illeris, S. 1991. "Location of Services in a Service Society." In P. W. Daniels and F. Moulaert, eds., op. cit. 91–117.

INCO. 1994. *Annual Report.* Toronto/New York.

Ingene, C. A., and A. Ghosh. 1990. "Consumer and Producer Behavior in a Multipurpose Shopping Environment." *Geographical Analysis* 18: 215–226.

Isaac, R. J., and V. C. Armat. 1991. *Madness in the Streets: How Psychiatry and the Law Abandoned the Mentally Ill.* New York: The Free Press.

Jablonsky, T. J. 1993. *Pride in the Jungle: Community and Everyday Life in Back of the Yards, Chicago.* Baltimore: The Johns Hopkins Press.

Jackson, E. L., and D. B. Johnson. 1991. "Geographic Implications of Mega-Malls, with Special Reference to West Edmonton Mall." *The Canadian Geographer* 35 (3): 226–232.

Jackson, J. N., and S. M. Wilson. 1992. *St. Catharines: Canada's Canal City.* St. Catharines, Ontario: St. Catharines Standard Ltd.

Jackson, K. T. 1985. *Crabgrass Frontier: The Suburbanization of the United States.* New York: Oxford University Press.

Jacobs, J. 1989. *Revolving Doors: Sex Segregation and Women's Careers.* Stanford, CA: Stanford University Press.

Jencks, C. 1994. *The Homeless.* Cambridge, MA: Harvard University Press.

Johnson, A. B. 1990. *Out of Bedlam: The Truth About Deinstitutionalization.* New York: Basic Books.

Johnson, D. M. 1991. "Structural Features of West Edmonton Mall." *The Canadian Geographer* 35 (3): 249–261.

Johnson, J. H., and M. L. Oliver. 1991. "Economic Restructuring and Black Male Joblessness in U.S. Metropolitan Areas." *Urban Geography* 12 (6): 542–562.

Johnson, J. H., and C. C. Roseman. 1990. "Recent Black Outmigration from Los Angeles: The Role of Household Dynamics and Kinship Systems." *Annals of the Association of American Geographers* 80 (2): 205–222.

Johnson, M. L. 1985. "Postwar Industrial Development in the Southeast and the Pioneer Role of Labor-Intensive Industry." *Economic Geography* 61 (1): 46–65.

Johnston, R. J. 1983. *The American Urban System: A Geographical Perspective.* New York: St. Martin's.

Johnston-Anumonwo, I. 1992. "The Influence of Household Type on Gender Differences in Work Trip Distance." *Professional Geographer* 44 (2): 161–169.

Johnston-Anumonwo, I. 1995. "Racial Differences in the Community Behavior of Women in Buffalo, 1980–1990." *Urban Geography*, 16 (1): 23–45.

Jones, B. G. and B. D. Lewis. 1990. "The Four Basic Properties of Rank-Size Hierarchical Distributions: Their Characteristics and Interrelationships." *Papers of the Regional Science Association* 68: 83–95.

Jones, J. P., and J. E. Kodras. 1990. "Restructured Regions and Families: The Feminization of Poverty in the U.S." *Annals of the Association of American Geographers* 80 (2): 163–183.

Jones, K. 1991. "Mega-Chaining, Corporate Concentration, and the Mega-Malls." *The Canadian Geographer* 35 (3): 241–249.

Jones, K., W. Evans, and C. Smith. 1994. *New Formats in the Canadian Retail Economy.* Toronto: Ryerson Polytechnic University, Centre for the Study of Commercial Activity.

Jones, K., and J. W. Simmons. 1990. *The Retail Environment.* London: Routledge.

Jones, K., and J. W. Simmons. 1993. *Location, Location, Location: Analysing the Retail Environment.* Toronto: Nelson, Canada.

Jones, R. C. 1991. "Patronage Rates of Supermarket Shopping Centers, San Antonio, Texas." *The Professional Geographer* 43 (3): 345–355.

Jud, G. D., and J. Frew. 1986. "Real Estate Brokers, Housing Prices, and the Demand for Housing." *Urban Studies* 23: 21–31.

Kain, J. F. 1962. *A Multiple Equation Model of Household Locational Behavior and Tripmaking Behavior.* Santa Monica, CA: The Rand Corporation, RM–3086-FF.

Kain, J. F. 1968. "Housing Segregation, Negro Employment, and Metropolitan Decentralization." *Quarterly Journal of Economics* 82 (2): 175–197.

Kaldor, N. 1966. *Causes of the Slow Rate of Economic Growth in the United Kingdom.* Cambridge: Cambridge University Press.

Karl, F. R. 1995. *George Eliot: Voice of a Century.* New York: Norton.

Kasarda, J. 1988. "Economic Restructuring and America's Urban Dilemma." In M. Dogan and J. Kasarda, eds., *The Metropolis Era.* Beverly Hills: Sage Publications.

Kasarda, J. D. 1990. "Structural Factors Affecting the Location and Timing of Underclass Growth." *Urban Geography* 11 (3): 234–264.

Kearns, R. A. 1995. "Medical Geography: Making Space For Difference." *Progress in Human Geography* 19 (2): 251–259.

Keating, A. D. 1988. *Building Chicago: Suburban Developers and the Creation of a Divided Metropolis.* Columbus: Ohio State University Press.

Keinath, W. F. 1985. "The Spatial Composition of the Post-Industrial Society." *Economic Geography* 61: 223–240.

Kellerman, A. 1993. *Telecommunications and Geography.* London: Belhaven Press.

Kelly, K. 1995. "Visible Minorities: A Diverse Group." *Canadian Social Trends* 37: 2–8.

Kemp, K. K., M. F. Goodchild, and R. F. Dodson. 1992. "Teaching GIS in Geography." *Professional Geographer* 44 (2): 181–190.

Kenney, M., and R. Florida. 1992. "The Japanese Transplants: Production Organization and Regional Development." *Journal of the American Planning Association* 58 (1): 21–38.

Kerr, D., D. W. Holdsworth, and G. J. Matthews. 1990. *Historical Atlas of Canada: Vol III, Addressing the Twentieth Century.* Toronto: University of Toronto Press.

Kilgour, D. 1990. *Inside Outer Canada.* Edmonton: Lone Pine Publishing.

King, L. J. 1984. *Central Place Theory.* Beverly Hills: Sage Publications.

Kitchin, J. 1923. "Cycles and Trends in Economic Factors." *Review of Economic Statistics* 5: 10–16.

Klagge, B. 1994. "A New Type of Family Homelessness: Survey Findings, Perceptions, and Local Politics in Madison Wisconsin." *Urban Geography* 15 (2): 168–188.

Knox, P. L. 1982. *Urban Social Geography: An Introduction.* London: Longman.

Knox, P. L. 1991. "The Restless Urban Landscape: Economic and Sociocultural Change and the Transformation of Metropolitan Washington, D.C." *Annals of the Association of American Geographers* 81 (2): 181–209.

Kondratieff, N. D. 1935. "The Long Wave in Economic Life." *Review of Economics and Statistics* 17: 105–115.

Koontz, C. M. 1991. *Market-based Modelling for Public Library Facility Location and Use-Forecasting.* Ann Arbor, MI: University Microfilms International.

Kotlowitz, A. 1991. *There Are No Children Here.* New York: Doubleday.

Kowinski, W. S. 1985. *The Malling of America: An Inside Look at the Great Consumer Paradise.* New York: William Morrow.

Krmenec, A. J. 1991. "Sales Tax as Property Tax Relief? The Shifting Onus of Local Revenue Generation." *The Professional Geographer* 43 (1): 60–67.

Krueger, R. 1984. "The Struggle to Preserve Specialty Cropland in the Rural-Urban Fringe." In M. F. Bunce and M. J. Troughton, eds., *The Pressure of Change in Rural Canada.* Toronto: York University, Department of Geography, Monograph 14: 292–313.

Kuznets, S. 1961. *Capital in the American Economy.* Princeton, NJ: National Bureau of Economic Research.

Lai, D. C. 1988. *Chinatowns: Towns Within Cities in Canada.* Vancouver: UBC Press.

Lal, B. B. 1990. *The Romance of Culture in an Urban Civilization.* London: Routledge.

La Londe, J. B., and R. V. Delaney. 1993. *Trends in Warehousing Costs, Management, and Strategy.* Oak Brook, IL: Warehousing Education and Research Council.

Lamonde, P., and Y. Martineau. 1992. *Désindustrialisation et Restructuration Economique.* Montreal: INRS-Urbanisation.

Lang, M. H. 1989. *Homelessness Amid Affluence.* New York: Praeger.

Laskin, S. L. 1990. "Resource Development on the Shield." In D. Kerr et al., *Historical Atlas of Canada, Volume III,* Toronto: University of Toronto Press, Plate 16.

Lauria, M., and L. Knopp. 1985. "Toward an Analysis of the Role of Gay Communities in the Urban Renaissance." *Urban Geography* 6 (2): 152–169.

Laws, G. 1992. "Emerging Shelter Networks in an Urban Area: Serving the Homeless in Metropolitan Toronto." *Urban Geography* 13 (2): 99–126.

Lea, A. C., and G. L. Menger. 1991. "An Overview of Formal Methods for Retail Site Evaluation and Sales Forecasting Part 3: Location-Allocation Models." *The Operational Geographer* 9 (1): 17–27.

Leacy, F. H. et al. 1983. *Historical Statistics of Canada*. Ottawa: Statistics Canada and Social Science Federation of Canada.

Ledebur, L. C., and W. R. Barnes. 1993. *All in it Together: Cities, Suburbs, and Local Economic Regions*. Washington, D.C.: National League of Cities, Research Report.

Lee, P. 1994. "Housing and Spatial Deprivation: Relocating to the Underclass and the New Urban Poor." *Urban Studies* 31 (7): 1191–1209.

Lees, L. 1994. "Gentrification in London and New York: An Atlantic Gap." *Housing Studies* 9 (2): 199–217.

Lees, L., and L. Bondi. 1995. "De-gentrification and Economic Recession: The Case of New York City." *Urban Geography* 16 (3): 234–253.

Lefebvre, H. 1991. *The Production of Space*. Cambridge, MA: Blackwell.

Leitner, H. 1989. "Urban Geography: The Urban Dimension of Economic, Political, and Social Restructuring." *Progress in Human Geography* 13 (4): 551–565.

Leman, A. B., and I. A. Leman. 1976. *Great Lakes Megalopolis: From Civilization to Ecumenization*. Ottawa: Supply and Services.

Lemann, N. 1991. *The Promised Land: The Great Black Migration and How It Changed America*. New York: Knopf.

Leven, C. L., and M. E. Sykuta. 1994. "The Importance of Race in Home Ownership Mortgage Loan Approvals." *Urban Affairs Quarterly* 29 (3): 479–489.

Levine, A. G. 1982. *Love Canal: Science, Politics and People*. Lexington, MA: Lexington Books.

Levinson, D. M., and A. Kumar. 1994. "The Rational Locator: Why Travel Times Have Remained Stable." *Journal of the American Planning Association* 60 (3): 319–332.

Lewis, P. 1983. "The Galactic Metropolis." In R. Platt and G. Macinko, eds., *Beyond the Urban Fringe*. Minneapolis: University of Minnesota Press, 23–49.

Ley, D. 1990. "Urban Livability in Context." *Urban Geography* 11 (1): 31–35.

Ley, D. 1992. "Gentrification in Recession: Social Change in Six Canadian Inner Cities, 1981–86." *Urban Geography*, 13 (3): 230–256.

Ley, D. 1994a. "Gentrification and the Youth Movement of the 1960s." In G. O. Braun, ed., *Managing and Marketing of Urban Development and Urban Life*. Berlin: Reimer, 575–584.

Ley, D. 1994b. "Gentrification and the Politics of the New Middle Class." *Environment and Planning D* 12: 53–74.

Ley, D., D. Hiebert, and G. Pratt. 1992. "Time to Grow Up? From Urban Village to World City, 1966–1991." In G. Wynn and T. Oke, eds., *Vancouver and Its Region*. Vancouver: University of British Columbia Press.

Ley, D. F., and L. S. Bourne. 1993. "The Social Context and Diversity of Urban Canada." In L. S. Bourne and D. F. Ley, eds. *The Changing Social Geography of Canadian Cities*. Montreal: McGill-Queen's University Press, 3–30.

Lichtenberg, R. M. 1960. *One-Tenth of a Nation*. Cambridge, MA: Harvard University Press.

Lin, J. 1995. "Polarized Development and Urban Change in New York's Chinatown." *Urban Affairs Review* 30 (3): 332–354.

Linge, G. J. R. 1991. "Just-in-Time: More or Less Flexible." *Economic Geography* 67 (4): 316–332.

Linneman, D., and I. F. Megbolugbe. 1992. "Housing Affordability: Myth or Reality?" *Urban Studies* 29 (3/4): 369–392.

Lipset, S. M. 1990. *Continental Divide: The Values and Institutions of the United States and Canada*. New York: Routledge.

Lloyd, M., and K. B. Beesley. 1991. "Perceptions of Change in Downtown Peterborough: Business Perspectives." *The Operational Geographer* 9 (4): 18–21.

Lloyd, R. 1989. "Cognitive Maps: Encoding and Decoding Information." *Annals of the Association of American Geographers* 79: 101–124.

Lloyd, W. J. 1991. "Changing Suburban Retail Patterns in Metropolitan Los Angeles." *The Professional Geographer* 43 (3): 335–344.

Lo, O., and P. Gauthier. 1995. "Housing Affordability Problems Among Renters." *Canadian Social Trends* 36: 14–17.

Logan, J. R., and R. D. Alba. 1993. "Locational Returns to Human Capital: Minority Access to Suburban Community Resources." *Demography* 30 (2): 243–268.

London, B., B. A. Lee, and S. G. Lipton. 1986. "The Determinants of Gentrification in the United States: A City-Level Analysis." *Urban Affairs Quarterly* 27 (3): 369–387.

Long, L. 1988. *Migration and Residential Mobility in the United States.* New York: Russell Sage Foundation.

Long, L. 1991. "Residential Mobility Differences Among Developed Countries." *International Regional Science Review* 14 (2): 133–148.

Long, W. W., and L. I. Nakamura. 1993. "A Model of Redlining." *Journal of Urban Economics* 33 (2): 223–234.

Longino, C. F. 1990. "Geographic Distribution and Migration." In R. H. Binstock and L. K. George, eds., *Handbook of Aging in the Social Sciences.* San Diego, CA: Academic Press, 45–61.

Lorimer, J. 1978. *The Developers.* Toronto: James Lorimer.

Lowry, I. 1960. "Filtering and Housing Standards: A Conceptual Analysis." *Land Economics* 36, 362–370.

Lowry, I. S. 1964. *Model of Metropolis.* Santa Monica, CA: Rand Corporation RM–4035-RC.

Lyons, D. 1995. "Changing Business Opportunities: The Geography of Rapidly Growing Small U.S. Private Firms, 1982–1992." *The Professional Geographer* 47 (4): 388–398.

Mack, R. S. 1993. "Nonmetropolitan Manufacturing in the United States and Product Cycle Theory: A Review of the Literature." *Journal of Planning Literature* 8 (2): 124–139.

MacKenzie, S. 1988. "Building Women, Building Cities: Toward Gender Sensitive Theory in the Environmental Disciplines." In C. Andrew and B. M. Milroy, eds., *Life Spaces: Gender, Household, Employment.* Vancouver: University of British Columbia Press, 13–30.

MacKenzie, S. 1989. "Women in the City." In R. Peet and N. Thrift, eds., *New Models in Geography.* New York: Allen and Unwin, 109–126.

MacKenzie, S., and M. Truelove. 1993. "Changing Access to Public and Private Services: Non-Family Childcare." In L. S. Bourne and D. F. Ley, eds., *The Changing Social Geography of Canadian Cities.* Montreal: McGill-Queen's University Press, 326–342.

MacLeod, L. 1989/1990. "The City for Women: No Safe Place." *Women and Environments* 12: 6–7.

Mager, N. H. 1987. *The Kondratieff Waves.* New York: Praeger.

Maguire, D. J., M. F. Goodchild, and D. W. Rhind. 1991. *Geographical Information Systems, Volume I: Principles; Volume II: Applications.* New York: Wiley.

Maher, C. A. 1974. "Spatial Patterns in Housing Markets: Filtering in Toronto." *The Canadian Geographer* 18: 108–124.

Malecki, E. J. 1981. "Product Cycles, Innovation Cycles and Regional Economic Change." *Technological Forecasting and Social Change* 19: 291–306.

Manson, G. A., and R. E. Groop. 1996. "Ebbs and Flows in Recent U.S. Interstate Migration." *The Professional Geographer* 48 (2): 156–166.

Marchand, B. 1986. *The Emergence of Los Angeles.* London: Pion.

Marden, P. 1992. "The Deconstructionist Tendencies of Postmodern Geographies: A Compelling Logic?" *Progress in Human Geography* 16 (1): 41–57.

Markusen, A. et al. 1991. *The Rise of the Gunbelt.* New York: Oxford University Press.

Markusen, A. 1996. "Sticky Places in Slippery Space: A Typology of Industrial Districts." *Economic Geography* 72 (3): 293–313.

Marr, W. L., and D. G. Paterson. 1980. *Canada: An Economic History.* Toronto: Gage Publishing Ltd.

Marshall, M. 1987. *Long Waves in Regional Development.* London: Macmillan.

Martin, P., and E. Midgley. 1994. "Immigration to the United States: Journey to an Unknown Destination." *Population Bulletin* 49 (2): 1–47.

Martinelli, F. 1991. "A Demand-Oriented Approach to Understanding Producer Services." In P. W. Daniels and F. Moulaert, eds., op. cit. 15–29.

Massey, D. S. 1985. "Ethnic Residential Segregation: A Theoretical and Empirical View." *Sociology and Social Research* 69: 315–350.

Massey, D. S., and N. A. Denton. 1989. "Hypersegregation in U.S. Metropolitan Areas: Black and Hispanic Segregation Along Five Dimensions." *Demography* 26: 373–391.

Matthew, M. R. 1993a. "The Suburbanization of Toronto Offices." *The Canadian Geographer* 37 (4): 293–306.

Matthew, M. R. 1993b. "Towards a General Theory of Suburban Office Morphology in North America." *Progress in Human Geography* 17 (4): 471–489.

McCann, E. 1995. "Neotraditional Developments: The Anatomy of a New Urban Form." *Urban Geography* 16 (3): 210–233.

McCann, L. D., ed. 1982. *Heartland and Hinterland: A Geography of Canada.* Toronto: Prentice-Hall.

McCann, L. D., and P. J. Smith. 1991. "Canada Becomes Urban: Cities and Urbanization in Historical Perspective." In T. Bunting and P. Filion, eds., *Canadian Cities in Transition.* Toronto: Oxford University Press, 69–99.

McCarthy, K. F. 1976. "The Household Life-Cycle and Housing Choices." *Papers of the Regional Science Association.* 37: 55–80.

McCombie, J. 1983. "Kaldor's Laws in Retrospect." *Journal of Post-Keynesian Economics* 5 (3): 414–429.

McDonald, J. 1981. "Spatial Patterns of Business Land Values in Chicago." *Urban Geography* 2 (3): 201–215.

McDonald, J. F. 1989. "Econometric Studies of Urban Population Density: A Survey." *Journal of Urban Economics* 26: 361–385.

McDowell, L. 1993a. "Space, Place, and Gender Relations: Part I." *Progress in Human Geography* 17 (2): 157–179.

McDowell, L. 1993b. "Space, Place, and Gender Relations: Part II." *Progress in Human Geography* 17 (3): 305–318.

McGregor, A., and M. McConnachie. 1995. "Social Exclusion, Urban Regeneration, and Economic Integration." *Urban Studies* 32 (10): 1587–1600.

McHugh, K. E., T. D. Hogan, and S. K. Happel. 1995. "Multiple Residence and Cyclical Migration: A Life Course Perspective." *The Professional Geographer* 47 (3): 251–267.

McInnis, M. 1990. "Economic Growth." In D. Kerr, D. W. Holdsworth, and G. J. Matthews, eds., op. cit., Plate 3.

McKee, B. 1995. "South Bronx." *Architecture* (April): 86–95.

McLafferty, S. 1992. "Health and the Urban Environment." *Urban Geography* 13 (6): 567–576.

McLafferty, S., and V. Preston. 1991. "Gender, Race, and Commuting Among Service Sector Workers." *The Professional Geographer* 43 (1): 1–14.

McMillan, D. P., and J. F. McDonald. 1993. "Could Zoning Have Increased Land Values in Chicago?" *Journal of Urban Economics* 33 (2): 167–188.

Meligrana, J. 1993. "Exercising the Condominium Tenure Option: A Case Study of the Canadian Housing Market." *Environment and Planning A* 25 (7): 961–973.

Melosi, M. V. 1994. "Sanitary Services and Decision Making in Houston, 1876–1945." *Journal of Urban History* 20 (3): 365–406.

Mercer, J., and M. Goldberg. 1986. *The Myth of the North American City.* Vancouver: University of British Columbia Press.

METROT. 1993. *Metro Cordon Count.* Toronto: Metro Planning, Transportation Division, eight pages.

Michalak, W. Z., and K. J. Fairbairn. 1993. "The Location of Producer Services in Edmonton." *The Canadian Geographer* 37 (1): 2–16.

Mikesell, J. 1992. *City Finances, City Futures.* Washington, D.C.: National League of Cities.

Miller, R. E., and P. D. Blair. 1985. *Input-Output Analysis: Foundations and Extensions.* Englewood Cliffs, NJ: Prentice-Hall.

Miller, T., R. L. Tobin, T. L. Friesz. 1992. "Network Facility Location Models in Stackelburg-Nash-Cournot Spatial Competition." *Papers in Regional Science* 71 (3): 277–291.

Mills, E. S. 1967. "An Aggregative Model of Resource Allocation in a Metropolitan Area." *American Economic Review* 57: 197–210.

Mills, E. S. 1995. "Crisis and Recovery in Office Markets." *Journal of Real Estate Finance and Economics* 10: 49–62.

Mills, E. S., and B. W. Hamilton. 1989. *Urban Economics,* 4th ed. Glenview, IL: Scott Foresman.

Mills, E. S., and L. S. Lubuele. 1995. "Projecting Growth of Metropolitan Areas." *Journal of Urban Economics* 37: 344–360.

Miron, J. 1988. *Housing in Postwar Canada: Demographic Change, Household Formation, and Housing Demand.* Montreal: McGill-Queen's University Press.

Miron, J., ed. 1993. *Housing, Home, and Community: Progress in Housing Canadians, 1945–1986.* Montreal: McGill-Queen's University Press.

Mitchell, B., ed. 1991. *Resource Management and Development, Addressing Conflict and Uncertainty.* New York: Oxford University Press.

Mitchell, D. 1995. "The End of Public Space? Peoples Park, Definitions of the Public, and Democracy." *Annals of the Association of American Geographers* 85 (1): 108–133.

Mitchell, D. 1996. "Introduction: Public Space and the City." *Urban Geography* 17 (2): 127–131.

Mitchell, D. 1996. "Political Violence, Order, and the Legal Construction of Public Space: Power and the Public Forum Doctrine." *Urban Geography* 17 (2): 152–178.

MOE. 1991. *The State of Canada's Environment.* Ottawa: Ministry of the Environment.

Mollenkopf, J., and M. Castells, eds. 1991. *Dual City: Restructuring New York.* New York: Russel Sage Foundation.

Monmonnier, M. 1991. *How to Lie With Maps.* Chicago: University of Chicago Press.

Mooman, R. L. 1986. "Have Changes in Localization Economies Been Responsible for Declining Productivity Advantages in Large Cities?" *Journal of Regional Science* 26 (1): 19–32.

Moore, C. L., and M. Jacobsen. 1984. "Minimum Requirements and Regional Economics, 1980." *Economic Geography* 60 (3): 217–224.

Moore, E. G. 1986. "Mobility Intention and Subsequent Relocation." *Urban Geography* 7 (6): 497–514.

Moore, E. G., and M. W. Rosenberg. 1995. "Modelling Migration Flows of Immigrant Groups in Canada." *Environment and Planning A* 27: 699–714.

Moore II, J. E. 1994. "Comment on: The Recent Popularity of Light Rail Transit in North America." *Journal of Planning Education and Research* 13 (1): 50–53.

Morrill, R. L. 1987. "The Structure of Shopping in a Metropolis." *Urban Geography* 8 (2): 97–128.

Morrill, R. L. 1991. "Myths About Metropolis." In J. F. Hart, ed. *Our Changing Cities.* Baltimore: Johns Hopkins University Press, 1–11.

Morrow-Jones, H. A. 1986. "The Geography of Housing: Elderly and Female Households." *Urban Geography* 7 (3): 263–269.

Moser, C. O. N., and L. Peake. 1994. *Seeing the Invisible: Women, Gender and Urban Development.* Toronto: Centre for Urban and Community Studies, University of Toronto, Major Report No. 30.

Moses, L. N., and H. F. Williamson. 1967. "The Location of Economic Activity in Cities." *American Economic Review* 57: 211–222.

MTARTS. 1966. *Metropolitan Toronto and Region Transportation Study.* Toronto: Parliament Building.

Mulligan G. F. 1987. "Consumer Travel Behavior: Extensions of a Multipurpose Shopping Model." *Geographical Analysis* 19: 364–375.

Mulligan, G. F. 1988. "A General Model for Estimating Economic Base Multipliers." *Australian Journal of Regional Studies* 3: 55–71.

Mulligan G. F., and H.-H. Kim. 1991. "Sectoral Level Employment Multipliers in Small Urban Settlements: A Comparison of Five Models." *Urban Geography* 12 (3): 240–259.

Munnell, A., L. E. Browne, J. McEneany, and G. Tootell. 1992. *Mortgage Lending in Boston: Interpreting HMDA Data.* Boston: Federal Reserve Bank of Boston, WP 92–7.

Murdie, R. A. 1969. *Factorial Ecology of Metropolitan Toronto, 1951–1961.* Chicago: University of Chicago Press, Department of Geography, Research Paper No. 116.

Murdie, R. A. 1986. "Residential Mortgage Lending in Metropolitan Toronto: A Case Study of the Resale Market." *The Canadian Geographer* 30 (2): 98–110.

Murdie, R. A. 1988. "Residential Structure of Metropolitan Toronto, 1951–1981." Paper presented at the annual meetings of the Association of American Geographers, Phoenix, AZ.

Murray, C., ed. 1990. *The Emerging British Underclass.* London: Institute for Economic Analysis.

Muth, R. F. 1961. "The Spatial Structure of the Housing Market." *Papers of the Regional Science Association* 7: 207–220.

Muth, R. F. 1969. *Cities and Housing.* Chicago: University of Chicago Press.

Muth, R. F. 1978. "The Allocation of Households to Dwellings." *Journal of Regional Science* 18: 159–178.

Myers, D. 1975. "Housing Allowances, Submarket Relationships, and the Filtering Process." *Urban Affairs Quarterly* 11: 215–240.

Myers, D. 1992. *Analysis With Local Census Data: Portraits of Change.* New York: Academic Press.

NALS. 1981. *U.S. National Agricultural Lands Study: Final Report.* Washington, D.C.: Government Printing Office.

Nader, G. A. 1975. *Cities of Canada* (vols. 1 and 2). Toronto: Macmillan of Canada.

Negrey C., and M. B. Zickel. 1994. "Industrial Shifts and Uneven Development: Patterns of Growth and Decline in U.S. Metropolitan Areas." *Urban Affairs Quarterly* 30 (1): 27–47.

Nelson, H. J. 1983. *The Los Angeles Metropolis.* Dubuque, Iowa: Kendall/Hunt.

Nelson, J. G. 1993. *Toward a Sense of Civics.* Toronto: Council of Ontario Universities.

Nelson K. 1986. "Labor Demand, Labor Supply and the Suburbanization of Low Wage Office Work." In A. J. Scott and M. Storper, eds., *Production, Work, Territory: The Geographical Anatomy of Industrial Capitalism.* Boston: Allen and Unwin.

Netzer, D. 1974. *Economics of Urban Problems.* New York: Basic Books.

Newbold, K. B., S. Birch, and J. Eyles. 1994. "Access to Family Physician Services in Canada: A Tale of Two Provinces." *Canadian Journal of Regional Science* 17 (3): 311–328.

Newling, B. E. 1966. "Urban Growth and Spatial Structure: Mathematical Models and Empirical Evidence." *Geographical Review* 56: 213–225.

NLC. 1993. *A Practical Introduction to Zoning.* Washington, D.C.: National League of Cities.

Norcliffe, G. B. 1975. "A Theory of Manufacturing Places." In *Locational Dynamics of Manufacturing,* edited by L. Collins and D. F. Walker. New York: Wiley.

Norcliffe, G. B., and J. H. Stevens. 1979. "The Heckscher-Ohlin Hypothesis and Structural Divergence in Quebec and Ontario, 1961–1969." *Canadian Geographer* 23 (3): 239–254.

Norton, R. D. 1992. "Agglomeration and Competitiveness: From Marshall to Chinitz." *Urban Studies* 29 (2): 155–170.

Norton, R. D., and J. Rees. 1979. "The Product Cycle and the Spatial Decentralization of American Manufacturing." *Regional Studies* 13: 141–151.

November, S. M., R. G. Cromley, and E. K. Cromley. 1996. "Multi-Objective Analysis of School District Regionalization Alternatives in Connecticut." *The Professional Geographer* 48 (1): 1–13.

Nowlan, D. M. 1994. "The Changing Toronto-Area Economy." Toronto: City of Toronto, Planning Department.

Nowlan, D. M. 1995. "The Effect of Central Area Jobs and Homes on Cordon Crossings: An Update of the Nexus Relationship." Toronto: Metro Toronto Planning Department.

Nowlan, D. M., and N. Nowlan. 1970. *The Bad Trip: The Untold Story of the Spadina Expressway.* Toronto: House of Anansi.

Nowlan, D. M., and G. Stewart. 1991. "Downtown Population Growth and Commuting Trips: Recent Experience in Toronto." *Journal of the American Planning Association* 57 (2): 165–182.

Oakey, R. 1985. "High Technology Industries and Agglomeration Economics." In P. Hall and A. Markusen, eds., *Silicon Landscapes.* Boston: Allen and Unwin, 94–117.

OECD. 1991. *The State of the Environment Report.* Paris: Organization for Economic Cooperation and Development.

O'Hare, W. P. 1992. "America's Minorities–the Demographics of Diversity." *Population Bulletin* 47 (4): 1–47.

Ohls, J. 1975. "Public Policy Toward Low-Income Housing and Filtering in Housing Markets." *Journal of Urban Economics* 2: 144–171.

Ohta, H. 1984. "Agglomeration and Competition." *Regional Science and Urban Economics* 14 (1): 1–18.

Ó hUallacháin, B. 1993. "The Restructuring of the U.S. Steel Industry: Changes in the Location of Production and Employment." *Environment and Planning A* 25: 1339–1359.

Ó hUallacháin, B., and N. Reid. 1991. "The Location and Growth of Business and Professional Services in American Metropolitan Areas, 1976–1986." *Annals of the Association of American Geographers* 81 (2): 254–270.

Ó hUallacháin, B., and N. Reid. 1992. "The Intrametropolitan Location of Services in the United States." *Urban Geography* 13 (4): 334–354.

Ó hUallacháin, B., and N. Reid. 1993. "The Location of Services in the Urban Hierarchy and the Regions of the United States." *Geographical Analysis* 25 (3): 252–267.

Olcott. [1911] 1961. *Blue Book of Chicago Land Values.* Chicago: Olcott.

O'Leary, B. 1987. "Why Was the GLC Abolished?" *International Journal of Urban and Regional Research* 11 (2): 193–217.

Olson, S. H. 1979. "Baltimore Imitates the Spider." *Annals of the Association of American Geographers* 69 (4): 557–574.

Olson, S. H., and A. L. Kobayashi. 1993. "The Emerging Ethnocultural Mosaic." In L. S. Bourne and D. F. Ley, eds., *The Changing Social Geography of Canadian Cities.* Montreal: McGill-Queen's Press, 138–152.

Omi, M., and H. Winant. 1986. *Racial Formation in the United States: From the 1960s to the 1980s.* New York: Routledge.

Orok, B. 1993. "Defining Toronto." *Canadian Social Trends* 28: 22–28.

Osborne, D., and T. Gaebler. 1992. *Reinventing Government: How the Entrepreneurial Spirit Is Transforming the Public Sector.* Reading, MA: Addison-Wesley.

Ottensman, J. R. 1996. "The New Central Cities: Implications of the New Definition of the Metropolitan Area." *Urban Affairs Review* 31 (5): 681–691.

Pacione, M. 1990. "Development Pressure in the Metropolitan Fringe." *Land Development Studies* 7: 69–82.

Paehlke, R. 1991. *The Environmental Effects of Urban Intensification.* Ontario: Ministry of Municipal Affairs, Municipal Planning Policy Branch.

Page, B., and R. Walker. 1991. "From Settlement to Fordism: The Agro-Industrial Revolution in the American Midwest." *Economic Geography* 67 (4): 281–315.

Palen, J. J., and B. London, eds. 1984. *Gentrification, Displacement and Neighborhood Revitalization.* Albany: State University of New York Press.

Palm, R. 1976. "The Role of Real Estate Agents as Information Mediators in Two American Cities." *Geografiska Annaler* 58B: 28–41.

Palm, R. 1985. "Ethnic Segregation of Real Estate Agent Practice in the Urban Housing Market." *Annals of the Association of American Geographers* 75 (1): 58–68.

Park, K. O. 1994. "Expenditure Patterns and Interactions Among Local Governments in Metropolitan Areas." *Urban Affairs Quarterly* 29 (4): 535–564.

Park, R. E., E. W. Burgess, and R. D. McKenzie. 1925. *The City.* Chicago: University of Chicago Press.

Parker, R. A. 1995. "Patterns of Federal Urban Spending: Central Cities and their Suburbs, 1983–1992." *Urban Affairs Review* 31 (2): 184–205.

Parr, J. B., and C. Jones. 1983. "City Size Distributions and Urban Density Functions: Some Interrelationships." *Journal of Regional Science* 23 (3): 283–308.

Pas, E. I. 1995. "The Urban Transportation Planning Process." In S. Hanson, ed., *The Geography of Urban Transportation.* New York: The Guilford Press, 53–77.

Patterson, O. 1991. *Freedom: Freedom in the Making of Western Culture.* New York: Basic Books.

PC. 1993. *Our Environment, Our Health.* Toronto: Premier's Council on Health, Well-being, and Social Justice.

Peake, L. J. 1995. "Towards an Understanding of the Interconnectedness of Women's Lives: The 'Racial' Reproduction of Labor in Low-Income Urban Areas." *Urban Geography* 16 (5): 414–439.

Peet, R., and N. Thrift, eds. 1989. *New Models in Geography: The Political Economy Perspective,* Vols I and II. London: Unwin Hyman.

Perle, E. 1982/3. "Ecology of Urban Social Change: An American Example." *Urban Ecology* 7: 307–324.

Perrenod, V. M. 1984. *Special Districts, Special Purposes: Fringe Governments and Urban Problems in the Houston Area.* College Station, TX: Texas A&M University Press.

Petrakos, G. C. 1992. "Urban Concentration and Agglomeration Economies: Re-examining the Relationship." *Urban Studies* 29 (8): 1219–1229.

Pickrell, D. 1992. "A Desire Named Streetcar: Fantasy and Fact in Rail Transit Planning." *Journal of the American Planning Association* 58: 158–176.

Pierce, N. R., C. W. Johnson, and J. S. Hall. 1993. *Citistates: How Urban America Can Prosper in a Competitive World.* Washington, D.C.: Seven Locks Press.

Pivo, G. 1993. "A Taxonomy of Suburban Office Clusters: The Case of Toronto." *Urban Studies* 30 (1): 31–49.

Platt, R. 1985. "The Farmland Conversion Debate: NALS and Beyond." *The Professional Geographer* 37 (4): 433–442.

Pollard, J., and M. Storper. 1996. "A Tale of Twelve Cities: Metropolitan Employment and Change in Dynamic Industries in the 1980s." *Economic Geography* 72 (1): 1–22.

Pond, B., and M. Yeates. 1993. "Rural-Urban Land Conversion I: Estimating the Direct and Indirect Impacts." *Urban Geography* 14 (4): 323–347.

Pond, B., and M. Yeates. 1994a. "Rural/Urban Land Conversion II: Identifying Land in Transition to Urban Use." *Urban Geography* 15 (1): 25–44.

Pond, B., and M. Yeates. 1994b. "Rural/Urban Land Conversion III: A Technical Note on Leading Indicators of Urban Land Development." *Urban Geography* 15 (3): 207–222.

Popp, H. 1976. "The Residential Location Decision Process: Some Theoretical and Empirical Considerations." *Tijdschrift voor Economische en Social Geografie* 67: 300–306.

Portes, A., and A. Stepick. 1993. *City on the Edge: The Transformation of Miami.* Berkeley: University of California Press.

Potter, R. B. 1982. *The Urban Retailing System.* Aldershot, U.K.: Gower.

Pratt, E. E. 1911. *Industrial Causes of Congestion of Population in New York City.* New York: Columbia University Press.

Pratt, G. and S. Hanson. 1988. "Gender, Class, and Space." *Environment and Planning, D: Society and Space* 6: 15–35.

Pratt, G., and S. Hanson. 1991. "On the Links Between Home and Work: Family Household Strategies in a Buoyant Labor Market." *International Journal of Urban and Regional Research* 15: 55–74.

Pred, A. 1965. "Industrialization, Initial Advantage, and American Metropolitan Growth." *Geographical Review* 55 (2): 158–185.

Pred, A. 1985. "Interpreting Processes: Human Agency and the Becoming of Regional Spatial and Social Structures." *Papers of the Regional Science Association* 57: 7–17.

Prescott, J. R., and D. A. Vandenbroucke. 1992. "Some Comments on Central Place Structure and Rural Planning." *International Regional Science Review* 15 (1): 51, 57.

Pudup, M. B. 1992. "Industrialization after (De)industrialization: A Review Essay." *Urban Geography* 13 (2): 187–200.

Quigley, J. M. 1978. "Housing Markets and Housing Demand: Analytic Approaches." In L. S. Bourne and J. R. Hitchcock, eds., *Urban Housing Markets.* Toronto: University of Toronto Press, 23–44.

Quinzii, M., and J. Thisse. 1990. "On the Optimality of Central Places." *Econometrica* 58: 1101–1109.

Rajulton, F., and Z. R. Ravanera. 1995. "The Family Life Course in Twentieth-Century Canada: Changes, Trends and Interrelationships." In J. Dumas, ed., *Family Over the Life Course.* Ottawa: Statistics Canada, Cat. 91–543E, 115–146.

Ramos, D. 1994. "HUD-dled Masses." *The New Republic* (March 14), 12–16.

Randall, J. E., and G. Viaud. 1994. "A Gender-Sensitive Urban Factorial Ecology: Male, Female, Grouped, and Gendered Social Spaces in Saskatoon." *Urban Geography* 15 (8): 741–777.

Rees, J. 1992. "Regional Development and Policy Under Turbulence." *Progress in Human Geography* 1 (2): 223–231.

Rees, P. 1970. "Concepts of Social Space: Toward an Urban Social Geography." In B. J. L. Berry and F. Horton, eds. *Geographical Perspectives on Urban Systems.* Englewood Cliffs, NJ: Prentice-Hall.

Regier, H. 1988. *Generic Guidelines for the Review of Ecosystem Initiatives in the Great Lakes Basin.* Toronto: University of Toronto, Institute of Environmental Studies.

Reid, C. E. 1977. "Measuring Residential Decentralization of Blacks and Whites." *Urban Studies* 14: 353–357.

Reid, E. P., and M. Yeates. 1991. "Bill 90—An Act to Protect Agricultural Land: An Assessment of Its Success in LaPrairie County, Quebec." *Urban Geography* 12 (4): 295–309.

Relph, E. 1990. *The Toronto Guide: The City, The Metro, the Region.* Toronto: Department of Geography, University of Toronto (Second edition, 1997, Centre for Urban and Community Studies, University of Toronto).

Rice, M. D., and R. K. Semple. 1993. "Spatial Interlocking Directorates in the Canadian Urban System." *Urban Geography* 14 (4): 375–396.

Ring, D., G. Vanderhaeghe, and G. Melnyk. 1993. *The Urban Prairie.* Saskatoon: Fifth House Publishers.

Ripley, E. A., R. E. Redman, and J. Maxwell. 1978. *Environmental Impact of Mining in Canada.* Kingston: Queen's University, Centre for Resource Studies.

Robinson, T. 1995. "Gentrification and Grassroots Resistance in San Francisco s Tenderloin." *Urban Affairs Review* 30 (4): 483–513.

Rodgers, K. 1994. "Wife Assault in Canada." *Canadian Social Trends*, Autumn, 3–8.

Rome, A. W. 1994. "Building on the Land: Toward an Environmental History of Residential Development in American Cities and Suburbs, 1870–1990." *Journal of Urban History* 20 (3): 407–434.

Rose, D. 1993. "Local Child Care Strategies in Montreal, Quebec: The Mediations of State Policies, Class and Ethnicity in the Life Course of Families with Young Children." In C. Katz and J. Monk, (eds.), *Full Circles. Geographies of Women Over the Life Course.* London: Routledge, 188–207.

Rose, G. 1993. *Feminism and Geography: The Limits of Geographical Knowledge.* Cambridge, U.K.: Polity Press.

Rose, H. M. 1971. *The Black Ghetto: A Spatial Behavioral Perspective.* New York: McGraw-Hill.

Rose, H. M. 1976. *Black Suburbanization: Access to Improved Quality of Life or Maintenance of the Status Quo.* Cambridge, MA: Ballinger.

Rose, H. M. 1991. "The Underclass Debate Goes On." *Urban Geography* 12 (6): 491–493.

Rose, H. M., and D. R. Deskins. 1991. "The Link Between Black Teen Pregnancy and Economic Restructuring in Detroit: A Neighbourhood Scale Analysis." *Urban Geography* 12 (6): 494–507.

Roseland, M. 1992. *Towards Sustainable Development.* Ottawa: National Round Table on the Environment and the Economy.

Rosen, C. M., and J. A. Tarr. 1994. "The Importance of an Urban Perspective on Environmental History." *Journal of Urban History* 20 (3): 299–310.

Rossi, P. 1989. *Down and Out in America: The Origins of Homelessness.* Chicago: University of Chicago Press.

Rossi, P. H. [1955] 1980. *Why Families Move,* 2nd ed. Beverly Hills: Sage Publications.

Rossi, P. H., and A. B. Shlay. 1982. "Residential Mobility and Public Policy Issues: Why Families Move Revisisted." *Journal of Social Issues* 38 (3): 21–34.

Rothenberg, J. et al. 1991. *The Maze of Urban Housing Markets: Theory, Evidence, and Policy.* Chicago: University of Chicago Press.

Rowe, S., and J. Wolch. 1990. "Social Networks in Time and Space: Homeless Women in Skid Row, Los Angeles." *Annals of the Association of American Geographers* 80 (2): 184–204.

Ruddick, S. 1995. *Young and Homeless in Hollywood: Mapping Social Identities.* New York: Routledge.

Ruddick, S. 1996. "Constructing Differences in Public Spaces: Race, Class, and Gender as Interlocking Systems." *Urban Geography* 17 (2): 132–151.

Rummel, R. J. 1970. *Applied Factor Analysis.* Evanston IL Northwestern University Press.

Runciman, W. G., ed. 1978. *Weber: Selections in Translation.* Cambridge: Cambridge University Press.

Rushton, G. 1971. "Postulates of Central Place Theory and the Properties of Central Place Systems." *Geographical Analysis* 3: 140–156.

Rusk, D. 1993. *Cities Without Suburbs.* Washington, D.C.: The Woodrow Wilson Center Press.

Russworm, L. H., and C. R. Bryant. 1984. "Changing Population Distribution and Rural/Urban Relationships in Canadian Urban Fields." In M. F. Bunce and M. J. Troughton, eds., *The Pressure of Change in Rural Canada.* Toronto: York University Press, Geographical Monographs, No. 14.

Rutherford, B., and G. Wekerle. 1988. "Captive Rider, Captive Labor: Spatial Constraints and Women's Employment." *Urban Geography* 9: 193–219.

Rybczynski, W. 1995. *City Life: Urban Expectation in a New World.* New York: HarperCollins.

Sancton, A. 1994. *Governing Canada's City-Regions.* Montreal: Institute for Research on Public Policy.

Sands, G. 1982. *Land Office Business: Land and Housing Prices in Rapidly Growing Metropolitan Areas.* Lexington, MA: Heath.

Santiago, A. M., and M. G. Wilder. 1991. "The Impact of Metropolitan Opportunity Structure on the Economic Status of Blacks and Hispanics in Newark." *Urban Geography* 12 (6): 494–507.

Savitch, H. V. 1988. *Post-Industrial Cities: Politics and Planning in New York, London and Paris.* Princeton, NJ: Princeton University Press.

Savitch, H. V. 1994. "Reorganization in Three Cities: Explaining the Disparity Between Intended Actions and Unanticipated Consequences." *Urban Affairs Quarterly* 29 (4): 565–595.

Savitch, H. V., D. Collins, D. Sanders, and J. P. Markham. 1993. "Ties that Bind: Central Cities, Suburbs, and the New Metropolitan Region." *Economic Development Quarterly* 7 (4): 341–357.

Sawers, L., and W. Tabb, eds. 1984. *Sunbelt/Snowbelt: Urban Development and Regional Restructuring.* New York: Oxford University Press.

Saxenian, A. 1984. "The Urban Contradictions of Silicon Valley: Regional Growth and the Restructuring of the Semiconductor Industry." In L. Sawers and W. K. Tabb, eds., *Sunbelt/Snowbelt: Urban Development and Regional Restructuring.* New York: Oxford University Press.

Saxenian, A. 1992. "Contrasting Patterns of Business Organization in Silicon Valley." *Environment and Planning, D* 10 (4): 377–391.

Saxenian, A. 1994. *Regional Advantage: Culture and Competition in Silicon Valley and Route 128.* Cambridge, MA: Harvard University Press.

Schill, M. H., and R. P. Nathan. 1983. *Revitalizing America's Cities: Neighborhood Reinvestment and Displacement.* Albany, NY: SUNY Press.

Schmandt, J., F. Williams, R. H. Wilson, and S. Strover, eds. 1990. *The New Urban Infrastructure.* New York: Praeger.

Schmitt, E. 1988. "Health Chief Says 250 Love Canal Families Can Return." *New York Times* (Sept. 28).

Schneider, M. 1992. "City Limits and the Growth of Suburban Retail Trade, 1982–1987." *Urban Affairs Quarterly* 27 (4): 604–614.

Schoenberger, E. 1985. "Foreign Manufacturing Investment in the United States: Competitive Strategies and International Location." *Economic Geography* 61 (3): 241–259.

Schoenberger, E. 1988a. "From Fordism to Flexible Accumulation: Technology, Competitive Strategies, and International Location." *Environment and Planning D* 6 (3): 245–262.

Schoenberger, E. 1988b. "Multinational Corporations and the New International Division of Labor." *International Regional Science Review* 11 (2): 105–119.

Schoenberger, E. 1995. *The Cultural Crisis of the Firm.* Oxford: Basil Blackwell.

Schoepfle, O. B., and R. L. Church. 1991. "A New Network Representation of a 'Classic' School Districting Problem." *Socio-Economic Planning Sciences* 25 (3): 189–197.

Schwartz, A. 1992. "The Geography of Corporate Services: A Case Study of the New York Urban Region." *Urban Geography* 13 (1): 1–24.

Scott, A. J. 1980. *The Urban Land Nexus and the State.* London: Pion.

Scott, A. J. 1983a. "Industrial Organization and the Logic of Intra-Metropolitan Location: Theoretical Considerations." *Economic Geography* 59 (3): 233–250.

Scott, A. J. 1983b. "Industrial Organization and the Logic of Intra-Metropolitan Location: A Case Study of the Printed Circuits Industry in the Greater Los Angeles Region." *Economic Geography* 59 (4): 343–367.

Scott, A. J. 1984. "Industrial Organization and the Logic of Intra-Metropolitan Location: A Case Study of the Women's Dress Industry in the Greater Los Angeles Region." *Economic Geography* 60 (1): 3–27.

Scott, A. J. 1986. "High Technology Industry and Territorial Development: The Rise of the Orange County Complex, 1955–1984." *Urban Geography* 7 (1): 3–45.

Scott, A. J. 1988a. *New Industrial Spaces.* London: Pion.

Scott, A. J. 1988b. *Metropolis: From Division of Labor to Urban Form.* Berkeley: University of California Press.

Scott, A. J. 1993. "The New Southern Californian Economy: Pathways to Industrial Resurgence." *Economic Development Quarterly* 7 (3): 296–309.

Seeling, M., and A. Artibise. 1991. *From Desolation to Hope: The Pacific Fraser Region in 2010.* Vancouver: Vancouver Board of Trade.

Semple, R. K. 1985. "Toward a Quaternary Place Theory." *Urban Geography* 6: 285–296.

Semple, R. K., and M. B. Green. 1983. "Intraurban Corporate Headquarters Relocation in Canada." *Cahiers de Geographie du Québec* 27: 389–406.

Semple, R. K., and A. G. Phipps. 1982. "The Spatial Evolution of Corporate Headquarters Within an Urban System." *Urban Geography* 3 (3): 258–279.

Sennett, R. 1970. *The Uses of Disorder.* Harmondsworth, U.K.: Penguin.

Sennott, L. I. 1991. "Public Sector Projects in a Mathematical Modelling Course." *Socio-Economic Planning Sciences* 25 (1): 55–56.

Sewell, J. 1993. *The Shape of the City: Toronto Struggles with Modern Planning.* Toronto: University of Toronto Press.

Sewell, J. 1994. *Houses and Homes.* Toronto: Lorimer.

Sharpe, L. J., ed. 1995. *The Government of World Cities: The Future of The Metro Model.* New York: John Wiley and Sons.

Shearer, D. 1982. "How Progressives Won in Santa Monica." *Social Policy* 12 (Winter): 7–14.

Sheffi, Y. 1985. *Urban Transportation Networks: Equilibrium Analysis with Mathematical Programming Methods.* Englewood Cliffs, NJ: Prentice-Hall.

Sherlock, H. 1991. *Cities Are Good for Us.* London: Paladin (HarperCollins).

Shevky, E., and W. Bell. 1955. *Social Area Analysis: Theory, Illustrative Applications and Computational Procedures.* Stanford, CA: Stanford University Press.

Shieh, Y-N. 1992. "Launhardt on von Thünen's Rings: A Note." *Regional Science and Urban Economics* 22 (4): 637–641.

Shilton, L., and J. R. Webb. 1995. "Headquarters Office Employment, and the Wave of Urbanization in the New York Region." *Journal of Real Estate Finance and Economics* 10: 145–159.

Short, J. R., L. M. Benton, W. B. Luce, and J. Walton. 1993. "Reconstructing the Image of an Industrial City." *Annals of the Association of American Geographers* 83 (2): 207–224.

Shoup, D., and R. Wilson. 1992. "Employer-paid Parking: The Problem and Proposed Solutions." *Transportation Quarterly* 46: 169–192.

Sidel, R. 1986. *Women and Children Last: The Plight of Poor Women in Affluent America.* New York: Penguin.

Silver, C., and R. Van Diepen. 1995. "Housing Tenure Trends, 1951–1991." *Canadian Social Trends* 36: 8–13.

Simmons, J. W. 1991. "The Urban System." In T. Bunting and P. Filion, eds. *Canadian Cities in Transition.* Toronto: Oxford University Press, 100–124.

Simmons, J. W. 1995. *International Comparisons of Commercial Structure.* Toronto: Ryerson Polytechnic University, Centre for the Study of Commercial Activity.

Simmons, J. W., et al. 1996. *Commercial Structure of the Greater Toronto Area.* Toronto: Centre for the Study of Commercial Activity, Ryerson Polytechnic University.

Simmons, J. W. 1978. "The Organization of the Urban System." In L. S. Bourne and J. W. Simmons, eds. *Systems of Cities: Readings on Structure, Growth, and Policy.* New York: Oxford University Press.

Simmons, J. W. 1984. "Government and the Canadian Urban System: Income Tax, Transfer Payments, and Employment." *The Canadian Geographer* 28 (1): 18–45.

Simmons, J. W. 1991. "Commercial Structure and Change in Toronto." Toronto: Department of Geography, University of Toronto, Research Paper 182.

Simmons, J. W. 1991. "The Regional Mall in Canada." *The Canadian Geographer* 35 (3): 232–240.

Simmons, J. W. 1994. *Commercial Activities in Canada.* Toronto: Ryerson Polytechnic University, Centre for the Study of Commercial Activity.

Simon, D. 1989. "Sustainable Development: Theoretical Construct or Attainable Goal." *Environmental Conservation* 16 (1): 41–48.

Simon, D. 1993. "The World City Hypothesis: Reflections from the Periphery." London: Department of Geography, Royal Holloway, University of London.

Sinclair, R. 1994. "Industrial Restructuring and Urban Development: An Examination in Metropolitan Detroit." In G. O. Braun, ed., *Managing and Marketing of Urban Development and Urban Life.* Berlin: Reimer, 205–219.

Sinclair, R., and B. Thompson. 1978. *Metropolitan Detroit: An Anatomy of Social Change.* Cambridge, MA: Ballinger.

Sinclair, U. 1906. *The Jungle.* New York: Doubleday.

Sivitanidou, R. 1994. "W&D Facilities and Community Attributes: An Empirical Study." Prepublication paper, University of Southern California, School of Urban and Regional Planning.

Sjoholt, P. 1994. "The Role of Producer Services in Industrial and Regional Development: The Nordic Case." *European Urban and Regional Studies* 1 (2): 115–129.

Small, K. A., and S. Song. 1994. "Population and Employment Densities." *Journal of Urban Economics* 36: 292–313.

Smerk, G. M. 1991. *The Federal Role in Urban Mass Transportation.* Bloomington: Indiana University Press.

Smith, D. 1988. *The Chicago School: A Liberal Critique of Capitalism.* Basinstoke, U.K.: Macmillan Education.

Smith, D. A. 1995. "The New Urban Sociology Meets the Old." *Urban Affairs Review* 30 (3): 432–457.

Smith, D. M. 1994. *Geography and Social Justice.* Oxford: Blackwell.

Smith, G. C. 1985. "Shopping Perceptions of the Inner City Elderly." *Geoforum* 16: 319–331.

Smith, G. C. 1992. "The Cognition of Shopping Centers by the Central Area and Suburban Elderly: An Analysis of Consumer Information Fields and Evaluative Criteria." *Urban Geography* 13 (2): 142–163.

Smith, N. 1979. "Gentrification and Capital: Theory, Practice, and Ideology in Society Hill." *Antipode* 11 (3): 24–35.

Smith, N., and W. Dennis. 1987. "Coalescence and Fragmentation of the Northern Core Region." *Economic Geography* 63 (2): 160–182.

Smith, N., and Schaeffer. 1986. "The Gentrification of Harlem?" *Annals of the Association of American Geographers* 76: 347–365.

Smith, P. J., and D. B. Johnson. 1978. *The Edmonton-Calgary Corridor.* Edmonton: University of Alberta Press.

Smith, P. J., and L. D. McCann. 1981. "Residential Land Use Change in Inner Edmonton." *Annals of the Association of American Geographers* 71 (4): 536–551.

Smith, W. 1964. *Filtering and Neighborhood Change.* Berkeley, CA: University of California, Center for Real Estate and Research in Urban Economics.

Soja, E. W. 1996. *Thirdspace.* Cambridge, MA: Blackwell.

Song, S. 1994. "Modelling Worker Residence Distribution in the Los Angeles Region." *Urban Studies* 31 (9): 1533–1544.

Sorkin, M., ed. 1992. *Variations on a Theme Park: The New American City and the End of Public Space.* New York: Hill and Wang.

South, R. B. 1990. "Transnational 'Maquiladora' Location." *Annals of the Association of American Geographers* 80 (4): 549–570.

Squires, G. D., ed. 1992. *From Redlining to Reinvestment: A Community Response to Urban Disinvestment.* Philadelphia: Temple University Press.

Stabler, J. C. 1987. "Trade Center Evolution in the Great Plains." *Journal of Regional Science* 27 (2): 225–244.

Stafford, H. A., and Q. Wu. 1992. "Manufacturing Plants in Ohio: Spatial Changes, 1978–1987." *Economic Development Quarterly* 6 (3): 273–285.

Statistics Canada. [1979] 1988. *Manufacturing Industries of Canada: Sub-Provincial Areas.* Ottawa: Minister of Supplies and Services, Cat. 31–209.

Statistics Canada. 1988. *Industry Trends 1951–1986.* Ottawa: Supply and Services, Cat. No. 93-152.

Statistics Canada. 1992. *Urban Areas.* Ottawa: Statistics Canada, Cat. 93–305/306.

Statistics Canada. 1993. *Profile of Census Metropolitan Areas and Census Agglomerations, Part B.* Ottawa: Statistics Canada, Cat. 93–338.

Statistics Canada. 1994. *Population Projections for Canada, Provinces, and Territories, 1993–2016.* Ottawa: Statistics Canada, Cat. 95–520.

Statistics Canada. 1994. *System of National Accounts, National Income and Expenditure Accounts: The Annual Estimates.* Ottawa: Supply and Services, Cat. No. 13–201.

Statistics Canada. 1994. *Women in the Labour Force.* Ottawa: Statistics Canada, Cat. No. 75-507E.

Steed, G. P. F. 1973. "Intrametropolitan Manufacturing: Spatial Distribution and Location Dynamics in Greater Vancouver." *The Canadian Geographer* 17: 235–258.

Steed, G. P. F. 1976. "Standardization, Scale, Incubation, and Inertia: Montreal and Toronto Clothing Industries." *The Canadian Geographer* 20: 298–309.

Steinbruner, M., and J. Medoff. 1994. *Jobs and the Gender Gap: The Impact of Structural Change on Worker Pay, 1984–1993.* Washington, D.C.: Center for National Policy.

Stephens, J. D., and B. P. Holly. 1981. "City Systems Behaviour and Corporate Influence: The Headquarters Location of U.S. Industrial Firms, 1955–1975." *Urban Studies* 18: 285–300.

Stern, D. I. 1993. "Historical Path Dependence of the Urban Population Density Gradient." *Annals of Regional Science* 27 (3): 259–283.

St. John, C., M. Edwards, and D. Wenk. 1995. "Racial Differences in Intraurban Residential Mobility." *Urban Affairs Review* 30 (5): 690–708

Stone, J. 1995. "Race, Ethnicity, and the Weberian Legacy." *American Behavioral Scientist* 38 (3): 391–406.

Storper, M. 1992. "The Limits to Globalization: Technology Districts and International Trade." *Economic Geography* 68 (1): 60–93.

Storper, M., and R. Walker. 1989. *The Capitalist Imperative: Territory, Technology, and Industrial Growth.* Oxford: Basil Blackwell.

Stren, R., R. White, and J. Whitney, eds. 1991. *Sustainable Cities: Urbanization and the Environment in International Perspective.* Boulder: Westview Press.

Strike, C. 1995. "Women Assaulted by Strangers." *Canadian Social Trends* (Spring): 2–6.

Sui, D. Z., and J. D. Wheeler. 1993. "The Location of Office Space in the Metropolitan Service Economy of the United States, 1985–1990." *The Professional Geographer* 45 (1): 33–43.

Survey. 1995. "Turn Up the Lights: A Survey of Cities." *The Economist* (July 29): 1–18.

Taaffe, E. J., B. Garner, and M. Yeates. 1963. *The Peripheral Journey to Work.* Evanston, IL: Northwestern University Press.

Taaffe, E. J., S. Krakover, and H. L. Gauthier. 1992. "Interactions Between Spread and Backwash, Population Turnaround and Corridor Effects in the Inter-metropolitan Periphery: A Case Study." *Urban Geography* 13 (6): 505–533.

Taub, R. P., D. G. Taylor, and J. D. Durham. 1984. *Paths of Neighborhood Change: Race and Crime in Urban America.* Chicago: University of Chicago Press.

Taylor, B. C., and P. M. Ong. 1995. "Spatial Mismatch or Automobile Mismatch? An Examination of Race, Residence, and Commuting in U.S. Metropolitan Areas." *Urban Studies* 32 (9): 1453–1473.

Taylor, F. W. 1947. *Scientific Management.* Westport, CT: Greenwood Press.

Teitz, M. 1968. "Toward a Theory of Urban Public Facility Location." *Papers and Proceedings of the Regional Science Association* 21: 35–44.

Teufel, D. 1989. *Die Zukunft des Autoverkehrs.* Heidelberg, Germany: Umwelt-und Prognose Institut, Bericht Nr. 17.

Thirwall, A. P. 1983. "A Plain Man's Guide to Kaldor's Growth Laws." *Journal of Post-Keynesian Economics* 5 (3): 345–358.

Thompson, W. 1968. "The City as a Distorted Price System." *Psychology Today* 2: 28–33.

Tkocz, Z., and G. Kristensen. 1994. "Commuting Distances and Gender: A Spatial Urban Model." *Geographical Analysis* 26 (1): 1–14.

Tomlin, C. D. 1990. *Geographic Information Systems and Cartographic Modeling.* Englewood Cliffs, NJ: Prentice-Hall.

Trachte, K., and R. Ross. 1985. "The Crisis of Detroit and the Emergence of Global Capitalism." *International Journal of Urban and Regional Research* 9 (2): 186–217.

Truelove, M. 1989. "Journey to Day-care Centres in Metropolitan Toronto." *Ohio Geographers* 17: 65–83.

Tuan, Y. F. 1976. "Humanistic Geography." *Annals of the Association of American Geographers* 66: 266–276.

Tucker, W. 1990. *The Excluded Americans: Homelessness and Housing Policies.* New York: Regnery Gateway.

Tulloss, T. K. 1995. "Citizen Participation in Boston's Development Policy." *Urban Affairs Review* 30 (4): 514–537.

Turok, I. 1993. "Inward Investment and Local Linkages: How Deeply Embedded Is Silicon Glen?" *Regional Studies* 27 (5): 401–417.

Tylecote, A. 1992. *The Long Wave in the World Economy.* New York: Routledge.

UNCED. 1992. *Our Common Future Reconvened.* London, U.K.: United Nations Commission on Environment and Development.

Urquhart, M. C., ed. 1965. *Historical Statistics of Canada.* Toronto: Macmillan.

Vance Jr., J. E. 1977. *This Scene of Man.* New York: Harper and Row.

van Duijn, J. J. 1983. *The Long Wave In Economic Life.* London: Allen and Unwin.

Varady, D. P. 1994. "Middle Income Housing Programmes in American Cities." *Urban Studies* 31 (8): 1345–1366.

Veness, A. R. 1994. "Designer Shelters and Models and Makers of Home: New Responses to Homelessness in Urban America." *Urban Geography* 15 (2): 150–167.

Vojnovic, I. 1994. *The Pathway Towards Sustainable Development and Sustainable Urban Forms.* Toronto: Centre for Urban and Community Studies.

von Tunzelmann, G. N. 1995. *Technology and Industrial Progress.* Brookfield, VT: Edward Elgar.

Waddell, P. 1992. "A Multinomial Logit Model of Race and Urban Structure." *Urban Geography* 13 (2): 127–141.

Waddell, P. B., B. J. L. Berry, and I. Hoch. 1993a. "Housing Price Gradients: The Intersection of Space and Built Form." *Geographical Analysis* 25 (1): 5–19.

Waddell, P. B., B. J. L. Berry, and I. Hoch. 1993b. " Residential Property Values in a Multi-nodal Area: New Evidence on the Implicit Price of Location." *Journal of Real Estate Finance and Economics* 7 (2): 117–141.

Waddell, P. B., and V. Shukla. 1993. "Employment Dynamics, Spatial Restructuring and the Business Cycle." *Geographical Analysis* 25 (1): 35–52.

Wagner, P. K. 1984. "Suburban Landscapes for Nuclear Families: The Case of Greenbelt Towns in the United States." *Built Environment* 10: 35–41.

Walby, S. 1990. *Theorizing Patriarchy.* Oxford: Basil Blackwell.

Walker, R. A. 1981. "A Theory of Suburbanization: Capitalism and the Construction of Urban Space in the United States." In M. Dear and A. Scott, eds., *Urbanization and Urban Planning in Capitalist Society.* New York: Methuen, 388–422.

Walker, R., and M. Storper. 1981. "Capital and Industrial Location." *Progress in Human Geography* 5 (4): 473–511.

Ward, C. 1990. *The Child in the City.* London: Bedford Square Press.

Ward, D. 1971. *Cities and Immigrants: A Geography of Change in Nineteenth Century America.* Cambridge, MA: Harvard University Press.

Ward, D. 1989. *Poverty, Ethnicity, and the American City, 1840-1925.* Cambridge: Cambridge University Press.

Ward, D. 1990. "Social Reform, Social Surveys, and the Discovery of the Modern City." *Annals of the Association of American Geographers* 80 (4): 491–503.

Ward, S. K. 1994. "Trends in the Location of Corporate Headquarters, 1969–1989." *Urban Affairs Quarterly* 29 (3): 468–478.

Warf, B. 1989. "Progress Report: Locality Studies." *Urban Geography* 10: 178–185.

Warf, B. 1989. "Telecommunications and the Globalization of Financial Services." *The Professional Geographer* 41 (3): 257–271.

Warf, B., and R. Erickson. 1996. "Introduction: Globalization and the U.S. City System." *Urban Geography* 17 (1): 1–4.

Warner, B. J. 1962. *Streetcar Suburbs: The Process of Growth in Boston.* Cambridge, MA: Harvard University Press.

Warnes, A. M., and R. Ford. 1995. "Housing Aspirations and Migration in Later Life: Developments During the 1980s." *Papers in Regional Science* 74 (4): 361–387.

Warnes, T. 1992. "Migration and the Life Course." In T. Champion and T. Fielding, eds., *Migration Patterns and Processes: Vol I.* London: Bellhaven Press, 175–187.

Waters, M. C. 1990. *Ethnic Options: Choosing Identities in America.* Berkeley, CA: University of California Press.

Weaver, C. L., and R. F. Babcock. 1979. *City Zoning: The Past and Future Frontier.* Chicago: American Planning Association.

Webber, M. J. 1984. *Industrial Location.* Beverly Hills: Sage.

Webber, M. M. 1976. "The BART Experience—What Have We Learned?" *The Public Interest* 45: 79–108.

Weber, A. [1909] 1929. *Theory of the Location of Industries.* Chicago: University of Chicago Press.

Weber, M. 1968. *Economy and Society.* New York: Bedminster. Translation of original work published in 1922, edited by G. Roth and C. Wittich.

Weicher, J., and T. Thibodeau. 1988. "Filtering and Housing Markets: An Empirical Analysis." *Journal of Urban Economics* 23: 21–40.

Weiner, E. 1992. *Urban Transportation Planning in the United States: A Historical Overview.* Washington, D.C.: United States Department of Transportation.

Weisman, L. K. 1992. *Discrimination by Design: A Feminist Critique of the Man-Made Environment.* Urbana, IL: University of Illinois Press.

Wekerle, G., ed. 1992. *A Working Guide for Planning and Designing Urban Environments.* Toronto: City of Toronto Planning and Development Department.

WERC. 1994. *Wercwatch.* Oak Brook, IL: Warehousing Education and Research Council.

West, D. S. 1993. "The Effects of Shopping Center Ownership on Center Composition in Planned and Unplanned Shopping Center Hierarchies." *Papers in Regional Science* 72 (1): 25–43.

Wheaton, W. C. 1983. "Theories of Urban Growth and Metropolitan Spatial Development." In J. Henderson, ed., *Research in Urban Economics.* Greenwich, CT: JAI Press Inc., 3–36.

Wheeler, J. O. 1990. "Corporate Role of New York City in the Metropolitan Hierarchy." *The Geographical Review* 80 (4): 370–381.

Wheeler, J. O., and C. L. Brown. 1985. "The Metropolitan Corporate Hierarchy in the U.S. South, 1960–1980." *Economic Geography* 61 (1): 66–78.

White, R. 1978. "Simulation of Central Place Dynamics and the Rank-size Distribution." *Geographical Analysis* 10: 201–208.

Whitt, J. A., and G. Yago. 1985. "Corporate Strategies and the Decline of Transit in U.S. Cities." *Urban Affairs Quarterly* 21 (Sept.): 37–65.

Wilder, M. G. 1985. "Site and Situation Determinants of Land Use Change: An Empirical Example." *Economic Geography* 61 (4): 332–344.

Wilson, C. 1992. "Restructuring and the Growth of Concentrated Poverty in Detroit." *Urban Affairs Quarterly* 28, 187–205.

Wilson, E. 1991. *The Sphinx in the City: Urban Life, The Control of Disorder, and Women.* London: Virago Press.

Wilson, W. J. 1987. *The Truly Disadvantaged: The Inner City, the Underclass and Public Policy.* Chicago: University of Chicago Press.

Wilson, W. J. 1991. "The Truly Disadvantaged Revisited: A Response to Hochschild and Boxill." *Ethics* 101: 593–609.

Wilson, W. J. 1996. *When Work Disappears: The World of the New Urban Poor.* New York: Alfred A. Knopf.

Wirth, L. 1938. "Urbanism as a Way of Life." *American Journal of Sociology* 44: 3–24.

WMATA. 1994. *Metro at the Crossroads: Fiscal 1994 Budget.* Washington, D.C.: Washington Metropolitan Area Transit Authority.

Wolch, J., M. Dear, and A. Akita. 1988. "Explaining Homelessness." *Journal of the American Planning Association* 5 (4): 443–453.

Wolch, J., A. Rahimian, and P. Koegal. 1993. "Daily and Periodic Mobility Patterns of the Urban Homeless." *Professional Geographer* 45: 159–168.

Wolch, J. R., and R. K. Geiger. 1986. "Urban Restructuring and the Not-for-Profit Sector." *Economic Geography* 62 (1): 3–18.

Wolpert, J., and E. Wolpert. 1976. "The Relocation of Released Mental Hospital Patients into Residential Communities." *Policy Sciences* 7: 31–51.

Wolsink, M. 1994. "Entanglement of Interests and Motives: Assumptions Behind the NIMBY-Theory on Facility Siting." *Urban Studies* 31 (6): 851–866.

Wood, R. C. 1961. *1400 Governments.* Cambridge, MA: Harvard University Press.

Wright, J. D. 1991. *Address Unknown: The Homeless in America.* New York: Aldine de Gruyter.

Yang, C.-H. 1990. "The Optimal Provision of a Central Park in a City." *Journal of Regional Science* 30 (1): 15–36.

Yeates, M. 1965. "Some Factors Affecting the Spatial Distribution of Chicago Land Values." *Economic Geography* 41: 55–70.

Yeates, M. 1974. *An Introduction to Quantitative Analysis in Human Geography.* New York: McGraw-Hill.

Yeates, M. 1975. *Main Street: Windsor to Quebec City.* Toronto: Macmillan of Canada.

Yeates, M. 1980. *North American Urban Patterns.* London: Edward Arnold.

Yeates, M. 1985. "The Core/Periphery Model and Urban Development in Central Canada." *Urban Geography* 6 (2): 101–121.

Yeates, M. 1985a. *Land in Canada's Urban Heartland.* Ottawa: Environment Canada, Lands Directorate.

Yeates, M. 1994. "Urban Canada: Changing Core/Periphery Relations." In L. Bruti Liberati and M. Rubboli, eds., *Canada e Italia Verso il Duemilia: Metropoli a Confronto.* Fasano: Schena, 49–76.

Yeates, M., and P. Lloyd. 1970. *Impact of Industrial Incentives: Southern Georgian Bay Region.* Ottawa: The Queen's Printer.

Yeates, M., J. Piazza, D. Montgomery, M. Robinson, G. Hardman, and S. Simmons. 1996. "The Retail Structure of Non-Metropolitan Urban Centres: Sault Ste. Marie, Sturgeon Falls, Collingwood,

Cobourg, and Port Hope." Toronto: Centre for the Study of Commercial Activity, Ryerson Poly-technic University, Research Report.

Young, M., and P. Willmot. 1975. *The Symmetrical Family.* Middlesex, U.K.: Penguin.

Youngman, J., and J. Malme. 1994. *An International Survey of Taxes on Land and Buildings.* Boston: Kluwer Law and Taxation Publishers.

Zehr, L. 1995. "Suburban Office Markets Post Lower Vacancy Rates." *Globe and Mail* (April 10): B3.

Zinn, B. M., and S. Eitzen. 1990. *Diversity in Families.* New York: Harper and Row.

Zukin, S. 1989. *Loft Living.* New Brunswick, NJ: Rutgers University Press.

Index

Abandoned housing submarket, 391
Abandonment, 394–395
    South Bronx, 395
Accessibility, 523
    automobiles, 524
Accumulated Urban Infrastructure,
    17
Activity patterns, 453
    modelling movements, 455
Affordability, 254
Affordability Issues, 372
African American, 255, 334, 358
    population, 221
After-school care, 87
Agents of development, 483
Agents of social change, 483
Agglomeration economies, 142
    declining productivity, 142
    diseconomies, 142
Agricultural heartlands, 38
Agricultural revolution, 8, 39
Agriculture, 106
Aid to Families with Dependent
    Children (AFDC), 431
Air Transport, 448
Aircraft Manufacturing, 17
Airline passenger outflows, 120
Allocation of land, 255
Amenity rent, 246
Annihilation of space, 41
Art deco style, 198
Asian population, 351
Aspatial processes, 4
    census data, 4
    population, 3
Atlanta, 25, 48, 156, 226, 290, 450,
    532
    political definition of, 25
Atlantic City, 156
Aurora-Elgin, 27
Auto work trips, 461
    parking fees, 461
Automobile companies, 100
Automobile, 213
Automobile assembly and parts, 111
    Auto Pact, 111
Automobile companies, 524
    non-polluting sources of power,
    524
Automobile industry, 15, 79, 144, 284
    automobile ownership, 93
    centers of manufacturing, 144
Automotive Trades Agreement
    ("Auto Pact"), 106

Autonomy-dependence relation-
    ship, 48
Average trip length, 453

"Baby boom", 9, 66, 81, 213
Baby boomers, 413
Baby-bust, 9
Back of the Yards, 201
"Back-office" Activities, 319
Backward and Forward Linkages, 181
Baltimore, 67, 77, 148, 196, 199, 226,
    372, 423, 486, 492
Bank linkages, 46
Barrios, 357
Bases of suburbanization, 481
Basic and non-basic components, 54
    macro-level, 54
    micro-level, 54
Baton Rouge, 502
Battery power vehicles, 524
Bay Area Rapid Transit System, 469
Behavioral processes, 342
Behavioural studies, 23
Belt Railroad Company (BRC), 268
Beltline freeways, 218
Birmingham, 156, 532
Birth/Death/Migration Model, 270
Bloomington, 304
Bonfire of the Vanities, the, 30
Boston, 14, 15, 48, 67, 77, 199, 209,
    215, 223, 239, 251, 489, 492
Bourgeois consolidation, 329
Break-of-bulk points, 40
Bridgeport, 313
British Columbia, 40
Bronx, 203
Brooklyn, 203
Brown-Moore Model, 379
    stress, 379
Bruntland Commission (1987), 13,
    508
Buffalo, 516
Building cycles, 63
Burrough, 138
Business cycle, 75
Business types, 167, 170
Business and professional services,
    184
Business services, 162

Calgary, 3, 29, 44, 102, 111, 129,
    255, 256, 489
California, 3, 29, 40, 131, 255
California census, 3

California macro-urban region,
    530–532
Canadian National Railway, 101
Canadian economy, 97
Canadian Pacific Railway, 101
Canadian Urban growth, 117
Canadian Urban System, 94
    evolution of, 94
Capital, 44, 46
Capital and operating costs, 475
Capital depreciation, 44
Capital investment, 370
Capital/labor relations, 25
Capital-for-labor substitution,
    280–281
Capitalist Mode of Production, 23
Careerism, 347
Causes of urbanization, 38
Causes of abandonment, 391–394
Central Business District (CBD),
    206, 226, 307
Central city/suburban polarization,
    225
Central city, 93, 290, 481, 483,
    population, 251
Central clusters, 273–274
Central densities, 249
Central functions, 166
Central Mortgage and Housing Cor-
    poration (CMHC)
Central Place Hierarchy, 167–177
Central Place Model, 166
    allocation of consumer services,
    166
    location of settlements, 166
Central Place System, 169
Central-city dwellers, 89
Centrifugal/Centripetal Economic
    Forces, 17
Change in Social Space, 342
Changes in market potential, 178
Changes in shopping behavior, 178
Charitable and service organiza-
    tion, 422
Charleston, 39
Charlotte, 532
Chattanooga, 226
Chemical Industry, 519
Chicago, 3, 27, 48, 69, 71, 101, 115,
    121, 148, 156, 174, 180, 186,
    199, 203, 205, 213, 215, 373,
    450, 455, 457, 485, 491, 516
    Gautreaux program, 373
    School of Sociology, 203

Chicago Sanitary District (1889), 481
Chicopee, 40
Child Care, 434, 439
    affordable arrangements, 437
    facilities, 437
    location of, 437
Cincinnati, 69, 71, 88, 156, 199, 226
City of dark/City of light, 224
Civil Rights Act (1968)
Clark Model, 247
    densities, 248
    population "crater", 249
Class and Conflict, 363–364
Classic Deindustrializing Metropo-
    lises, 154
Class formation, 22–24
Classical Political Economy, 23
Classical Residential Location Theo-
    rists, 431
Clean-air electric vehicles, 524
Cleveland, 48, 79, 226, 266, 372, 489,
    517
Cleveland/Akron, 266
"Cold War", 80
Collective agreements, 148
Columbus, 88, 156, 489
Common Law of Nuisances, 258
Community Nucleations, 303
Classical Residential Location Theo-
    rists, 431
Communication, and other public
    utilities, 162
Community Reinvestment Move-
    ment, 388
Community Rehabilitation Agency,
    398
Commuter Traffic, 215
Commuters' Zone, 207
Commuting Behavior, 423
Concentration of Producer Ser-
    vices, 180
Concentric Zone Model, 206
Congestion, 229
Constitutional crises, 121
Consumer behavior and cognition,
    308
Consumer services, 165, 166, 291
    commercial activities, 291
    distribution and exchange of
        goods and services, 165
Consumer society, 66
Contact intensity and variety, 317
Consumption/production link, 190
Continental urban growth, 81
Continuous product innovation, 132
Core/periphery, 79
Core/periphery Urban Structure,
    108, 113, 116

Windsor-Quebec City corridor,
    109
Core region, 108
Core/semi-periphery/periphery dis-
    tinction, 182
Corpus Christi, 156
Cosmopolitanism, 347
Cost-benefit Relationships, 493
Costs of Dispersion, 525
Counterurbanization, 87, 93
Cross-town circumferential com-
    muting, 218, 228
Crow's Nest Pass freight-rate sub-
    sidy, 102
Cutbacks in federal low-income
    housing programs, 372
Cycle of Deprivation, 363

Dallas, 156, 450, 491
Dallas-Fort Worth, 295, 277, 301,
    305, 339, 348,
Daycare, 87
Decentralization, 4, 215, 219, 228,
    232, 421
    and suburban clustering, 320,
    impact of developers and finan-
        cial institutions on, 219
    influence of government on, 216
    of Manufacturing, 281
    of economic activities, 4
Decentralized Clusters, 274, 280
Deconcentration, 4, 132, 195, 196,
    205, 232, 540
Defense expenditures, 15, 93, 150,
    159
Defining social space, 330
Deindustrialization, 159, 508
Demand-side Subsidies, 372
Demographic Changes, 113
Density Gradients, 249
Denver, 226, 277, 450, 492, 540
Depression, 35
Deregulation, 150, 521
    "Open Skies" policy, 150
Des Moines, 156
Designative cognition, 308
Detroit, 48, 52, 79, 101, 148, 223,
    251, 265, 288, 455, 457, 516
Differential rent, 246
    margin of production, 246
Differentiation, 42, 145
    within the urban systems, 42
Diffusion, 42
Direct/Indirect land conversion, 257
    leading/lagging ratios, 258
Direct-user Charges, 492
Discretionary activities, 451
Discrimination, 222

in housing, 222, 254
Diseconomies of size, 58
Distribution of service activities, 190
Dual economies, 359

Early Industrial Era, 199
Early Industrial Capitalism
    (1845–1895), 65, 70, 74
Economic base concept, 53–54
Economic recessions, 58
Economic cycles, 62
Economic Rent, 232–238, 245
Economic Growth, 535
    high-tech communications, 540
Edge cities, 220, 223, 224, 245
Edmonton, 102, 111, 129, 189, 257
Edmonton-Calgary corridor, 540
Education Services, 190
Elderly, 309
    population, 90
Electric Streetcar, 199
El Dorado, 169
Emigration, 96
Employee Accessibility, 318
Employment, 14, 290, 165
    consumer services sector, 290
    Equal Employment Opportuni-
        ties, 87
    in service activities, 290
    locations, 43, 290
    producer services, 290
Enclosed malls, 307
Energy consumption, 13, 509, 512, 519
Energy End-Use, 513
Enplaned passengers, 450
Enterprise zones, 521
Entertainment and recreation, 162
Environmental, 523
    impacts, 540
    performance, 523
    spillovers, 508
Equalization payments (Canada), 150
Equity Issues, 527
    access, 12
    equity Goals, 324
Era of National Industrial Capital-
    ism, 79
Era of Early Industrial Capitalism, 96
Era of Mature Industrial Capitalism,
    104
Erie Canal, 70
Ethnicity and Race, 329, 347
    behavioral models, 348
    human ecology, 348
Eugene-Springfield, 159
Exclusionary Zoning, 261
External Immigration, 9
External economies, 3, 142, 321

Externalities, 259
Exurbanization, 90
Exurbanites, 257
Exurbs, 201

Fabricating plants, 153
Face-to-face contact, 317
Factorial Urban Ecology, 335
Factory system, 8
Fair Housing Act (1966) 220
Families and Households, 412
Familism, 347
Family formation, 212
Family Status Construct, 334
Family stability, 432
Family status, 409
Federal Clean Air Act, 471
Federal Deposit Insurance Corpora-
    tion, 219
Federal Disaster Area, 520
Federal Fair Housing Act (1968), 254
Federally insured loans, 386
Federated Metropolitan Areas, 502
Federation (local government), 501
Female Job Opportunities, 424–425
Feminization of poverty, 426–427
Finance, insurance, and real es-
    tate, 162
Financial centre of the city, 201
Financial Districts and Offices, 313
Financial policies, 219
Financial and Governmental Institu-
    tions, 386
"First world/third world" appear-
    ance, 359
Fiscal imbalance, 494
Fiscal reform, 497, 503
    revenues and expenditures, 503
Fiscal zoning, 496
Fixed capital, 17, 58
Flexible production systems, 276
Florida, 3, 40, 139, 255
    Urban Region, 534
Flow of capital, 42
Formation of social space, 358
Forward linkages, 181
Fourth long wave, 89
Fragmentation, 485, 497
    of Local Government, 494
Fredericton, 95
Friction of distance, 99
Fringe, 206, 291
Frontier Mercantile (to 1845), 65–67
Functional Economic Areas, 498

Gary-Hammond, 27
Gasoline consumption, 524
Gateway Cities, 69, 198

Gender Division of Labor, 431
Gentrification, 252, 404–405
Geography of Retailing, 291
    retail location, 294
Ghettos, 352
Global competition, 160
Globalization of production, 9
Goodman Commission, 483
Government, 11, 12, 81
    and public services, 162, 164
    education, 81
    employment, 164
    health care system, 162
    housing, 81
    provider, 12
    regulator, 12
    services, 189
    social safety net, 81
Gravity Model, 29, 41
    spatial interaction, 458
Great Depression, the, 11, 76, 80,
    100
Great Lakes, 3, 28
    Basin, 516, 519
    Ecosystem, 515
    industrial activity in, 515
Greater Vancouver Regional District
    (BC), 502, 505
Green Belts, 260
Greensboro, 532
Greensboro/Winston-Salem, 265
Greenville, 532
Gross Domestic Product (GDP), 12
    Canada, 12
    United States, 12
Gross Metropolitan Product (GMP),
    14

Halifax, 98, 102, 109, 111
Hamilton, 98, 100, 102
Headquarter Offices, 151, 313
Hierarchical zoning, 260
Hierarchy, 178
    economies, 46
    of Retail Locations, 302
    of urban areas, 46, 52
High technology industries, 150
High Order Centers, 186
Hispanic-American, 358
Holyoke, 40
Home and work, 408
    change in, 409
Home Mortgage Disclosure Act
    (1975), 387
Home-work, 423–424
    in polycentric urban areas, 424
    in monocentric urban areas, 420
Homelessness, 395–398

Homeownership, 368
    rental accommodation, 360
    social class, 359
    wealth, 360
Household size, 10
Housing, 372, 420
    affordability, 383
    condominiums, 368
    macro influences on the demand
        for and supply of, 368
    policies, 216, 373, 374
    rental accommodation, 383
    single-family units, 368
    submarkets, 383, 390
    types, 225
Housing boom/bust, 431
Houston, 17, 156, 223, 256, 485,
    491
Human Ecology, 327
    approach, 327, 328, 342
    Chicago School of Human Ecol-
        ogy, 327
Humanistic approach (in Geogra-
    phy), 21
Huntingdon, 88, 138
Hypersegregation, 356
    legal sanctions supporting, 356

Immigration, 3, 92, 96, 206
Impact zoning, 260
Incipient cores, 120
Incumbent upgrading, 406
Indianapolis, 156, 489
Indigenous peoples of Canada, 121
Indirect impacts, 257
Industrial
    activity, 515
    capitalism, 1895–1945, 98, 99
    decline, 520
    development, 30
    districts, 276
    metropolis, 284
    parks, 277
    properties, 521
    restructuring, 14, 142, 284
    revolution, 8, 39
Industrialization, 508
Industry and sustainability, 514–515
"Inelastic" Cities, 481
Information Technology Indus-
    tries, 15
Information factors, 310
Infrastructure, 17
Inner cities, 4, 17, 113
    historic cores, 17
Inner-city decay, 222
Inner suburban areas, 291, 467
Inner suburban cordon, 467

Innovation, 42, 43
 diffusion (D), 42, 48
 "motors" of development, 42
Input-output, 29, 56–57
Integration, 138, 153
 effect of restructuring on, 154
 fabricated products, 153
 of goods, 139
Inter-industry linkages, 153
Interaction, 42
Intercity migration, 92
Interfirm collaborative action, 141
Intergovernmental transfers, 493
Intermetropolitan changes, 154
Internal migration, 9, 11, 113
Interstate and Defence Highways
 Act (1956), 13, 217, 446
Interurban transportation, 446
Intrametropolitan shifts, 154
Intra-urban, 271
 hierarchy, 301
 manufacturing location, 273
 office location, 312, 313, 315,
 public facility location, 321
Intraurban activity patterns, 450
Invasion-Succession-Dominance,
 328–329
Inventory cycles, 62
Investment, 48, 62
 cycles, 62
 diffusion of, 48

Jersey City, 313
Job demarcation, 148
Jobless economic recovery, 14–15
Joliet, 27
Just-in-time, 276
 input and output requirements,
 276
 systems, 280

Kansas City, 156
Kenney, 139
Kenosha, 27
Kitchener-Waterloo, 257
Kuznets building cycles, 63

Labor costs, 140
 and supplies, 153
 variations in, 140
Labor relations, 147
Labor-market segmentation,
 425
Lake County, 27
Land consumption rates, 255
Land banks, 225
Land consuming, 224
Land use, 232–236

change, 255
 in the Urban Environment, 236
 theory, 233
 zones, 234, 236, 238
 land values, 232
Land value maps, 238
Land-fill site, 18
Laprairie County, 257
Large corporations, 219
Large-scale urbanization, 508
Latch-key kids, 436
Late Industrial Revolution High-
 way Era, 269
Late Twentieth Century Suburbs,
 223, 224, 226
Lawrence, 40
Leakage of money, 541
Leapfrogging, 525
Least-cost theory, 132
Lending, 369
 creditworthiness, 387
 discrimination, 386–388
 minimum loan criteria, 389
 redlining, 387
Levittown, 220
Lexington, 139
Life-course, 344
 considerations, 344, 346, 347
 life-cycle, 344
Limited-access highways, 227
Little Rock, 174
Local Government, 11, 478–504
 changing role of, 478
 competition between, 496
 finance, 497
 fiscal responsibility, 495
 fragmentation, 4, 483, 484, 508
 proliferation of, 484
 public transport, 480
 reorganization, 483, 497, 503, 504
 types, 484
Local hospitals and health services,
 484
Localism, 347
Localization economies, 141, 153
Location of manufacturing, 128, 132
 regional changes in, 128
Location, 186, 325
 of governmental and public services, 186
 of settlements, 59
 rent and taxes, 319
Location/allocation models, 437
Locational factors, 160
Logical positivism, 20
Lone-female parent households, 428
Long wave, 64, 99
Los Angeles, 3, 17, 43, 48, 57, 115,

 174, 180, 185, 204, 216, 220,
 252, 277, 282, 283, 333, 378,
 391
 printed circuits industry, 282
Los Angeles-Long Beach PMSA,
 471
Los Angeles/Burbank/Long Beach,
 450
Los Angeles/Long Beach, 485
Low density suburbs, 527
Low-income groups, 201
Lowell, 40
Lower Great Lakes macro-urban re-
 gion, 3, 523, 530
LRT revival, 473
Lynn, 77

Machine-tool industry, 132
Macro-urban, 533
 population, 533
 regions, 528–534
 wage rate growth, 534
Madison, 40
Major-regional nucleations, 304
Managing urban areas, 477
Manchester, 40
Manhattan, 203, 221, 295, 313, 359
Manufactured goods transportation
 charges, 137
Manufactured goods, 137
Manufacturing, 9, 38, 77, 102, 116,
 123, 165, 266, 271, 278
 and restructuring, 216, 286, 288
 basic activity, 264
 Belt, 124, 131, 138
 economies of scale, 77
 employment, 264, 273, 521
 high-tech production, 116
 importance, 123
 in suburban locations, 271
 industries, 201
 Infrastructure, 266
 nature of, 123
 interurban location of, 264
 pattern of stagnation, 165
 plants, 109, 270
 stage of production processes, 124
 value added, 124
Market value of a property, 259
Market economy, 41
Market value assessment, 262, 489
Mathematical models, 29
Mature Industrial Capitalism, 66, 80
 domestic production, 66
 fourth long wave, 86
Mature industrialization, deindustrialization, 519
Maxi-mills, 139

Megalopolitan urban formations, 81
Megapolis, 86
Memphis, 15, 156
Mercantile, 95
   and staples economies, 94
   Capitalism, 21
   frontier era, 74, 95, 198
   towns and cities, 196
Merchandising technology, 179
Meritocracy, 480
Metalworking sector, 141
Mathematical Location-Allocation
   Programming, 324
Metropolitan areas, 3, 8
   Atlanta Regional Transit Au-
      thority, 25
   counties, 184
   economic and social integration, 3
   spatial restructuring, 467, 468
   stagnation of, 4
Metropolitanism, 81, 203, 496
Miami, 4, 115, 226
Miami/Fort Lauderdale, 450
Michigan, 153
Microcomputer product lines, 275
Middle Industrial Revolution Rail-
   way Era, 267
Middle- and outer suburban rings,
   227
Migration, 42, 43, 44, 48, 206, 221, 426
   differential growth, 44
   from rural areas, 43
   Harlem, 221
   of visible minorities, 221
   secondary, 426
   spatial change, 44
   trends, 534
   White Flight, 221
Milwaukee, 265, 492
   Ghetto, 353
Mini-mills, 139
Minimum requirements approach, 54
Mining and industrial hazardous
   sites, 520
Minneapolis, 15
Minneapolis-St. Paul, 48, 506
Minority groups, 90, 335
Mismatch Hypothesis, 356
Modal Choice Model, 458
Modal Diversity and Modal Share,
   463
Modernity, 22
Modernization of plants, 151
Monitoring social control, 308
Monocentric, 461
   city, 461
   Land Value Models, 238
   model for land values, 239

urban region, 228
Mononucleated cities, 295, 301
Monopoly rent, 246
Montreal, 43, 48, 57, 95, 97, 100,
   101, 102, 109, 111, 114, 117,
   120, 165, 189, 196, 217, 264,
   372, 446, 503, 505
Motor Vehicle Industry, 519
Movement characteristics, 452
Moving assembly line, 145
Moving to Opportunity Program, 373
Multi-national enterprises, 116
Multi-Plant Manufacturing Firms,
   151
Multiculturalism, 23
Multinational corporations, 66
Multiple-Nuclei Model, 209
Multiplier, 54, 57
Multipurpose shopping, 178, 300
Municipal payroll taxes, 490
Municipalities, 489, 492

Napoleonic Wars, 96
Nashua, 40
Nashville, 15
National Affordability Housing Act
   of 1990, 372
National industrial capitalism
   (1895–1945), 65, 76, 77
   the Urban System during the
      Era of, 77
National standards, 106
Natural gas requirements, 511
Neighborhood, 302
   externalities, 326
   gay communities, 347
   nucleations, 302
   shopping areas, 347
Neoclassical Economic Theory, 23,
   375
Net immigration, 114
New England, 131
New Orleans, 69, 71, 156, 198, 199
New Retail Formats, 298
New Stanton, 127
New Rapid Transit Systems, 469, 470
New Urban Sociology, 327
New York, 3, 14, 15, 28, 48, 52, 67,
   71, 77, 101, 115, 120, 121, 131,
   140, 148, 174, 180, 184, 198,
   215, 226, 251, 313, 359, 373,
New York *(continued)*
   405, 450, 468, 471, 479, 480,
   483, 486, 495
   Metropolitan Police Board (1857),
      481
   range of functions, 483
Newark, 199, 222, 313

New York/New Jersey metropolitan
   region, 223
"Night time" maps, 3
Nodal region, 46
Non-farm populations, 90
Non-property taxes, 490
Nondurable goods sector, 125
Norfolk, 77
North American Free Trade Agree-
   ment, 106, 116, 122, 446
North American population, 35
   urbanization, 35
North American Energy Use and
   Sustainability, 509
Not-for-profit sector, 57
Nucleations, 299, 302
   commercial structure of, 302
   minimarts, 302
   street corner developments, 302

Oakland, 251
Obligatory travel, 450
Occupational segregation, 410
Occupational spatial entrapment, 437
Office/business taxes, 319
Omaha, 156
One-Tier Systems, 501
Orange County, 156, 252
Ottawa, 39, 446
Outer Suburbia, 291
Oxford County, 257

Panorama City, 220
PATCO strike, 148
Paternal leave, 87
Patriarachal hierarchy, 422
Patriarchal structural processes, 21
Patriarchy, 426
Patron-client relationship, 480
Patterns of social space, 335, 339
People's Park, 364
Percentage Zoning, 260
Period of concentration, 103
Period of consolidation, 111
Peripheral patterns, 277–278
   greenfield sites, 278
Person miles of travel, 453
Personal services, 162
Personal business trips, 453
Persons per family, 334
Petroleum and natural gas, 106
Phenomenological framework, 21
Philadelphia, 3, 14, 39, 48, 67, 77, 148,
   156, 198, 203, 405, 455, 485,
Philosophical Approaches, 20
Phoenix, 226, 256
Pittsburgh, 79, 148, 161, 199, 455, 485
Planned nucleations, 300
Planned shopping mall, 307

Plant relocation, 357
Polarization of development, 359
Political leadership, 479
Political/Institutional processes, 11
Polycentric, 253
Polycentric Land Value Models, 241, 242
Polycentric urban region, 228
Polycentricity, 3
Polynucleated, 228
  city, 301
  metropolises, 295, 426, 427, 526
  urban areas, 483
Population, 81, 113
  change, 232
  clusters, 4
  decentralization, 203, 251
  decongestion, 229
  densities, 247, 252
  growth, 212, 425, 525, 526, 532
  migration, 15
  Turnaround, 251
Portland, 502, 503
Positivist Approach, 21
Post-Fordist production processes, 285
Post-modernism, 22
Post-war boom, 212
Post-war outmigration, 285
Post-World War II Suburban Period, 481
Poverty, 428
Preindustrial role, 478
Prewar house, 224
Price setting mechanism, 12
Price/distance gradients, 244
Primary waste treatment, 518
Principle of Home Rule, 481
Private ownership of malls, 308
Privately established utility districts, 257
Privatization, 505
  advantages, 505
  and contracting-out, 504
  disadvantages, 505
  of public space, 307
Processing plants, 153
Producer services, 162, 179, 183, 184, 310
  concentration of, 183
  definition of tertiary activities, 179
  employment, 163, 183
  group, 163
  intraurban location of, 310
  trade in, 185
  type of demand, 179
Product life cycles, 278–280

Production, 126
  and consumption, 185, 187
  costs, 139, 140
  facilities, 152
  integration, 126
  workers, 160
Production/Financial System, 8–9
Productive activity, 43
Productive efficiency, 8
Profit maximising locations, 294
Progressive Movement, 480–481
Property Industry, 383, 385
Property tax, 261, 489, 490
  millage rate, 262
  regressive nature, 490
Public housing, 484
Public infrastructure, 504
Public facility location, 321, 324, 325
Public policy context, 326
Public and private transport, 473, 475
Public service provision, 322, 323
Public transit, 12, 214, 215, 224
Public/private partnerships, 225, 401
Publicly induced rent, 246
Puerto Rican, 358
Pure public goods, 493

Quantitative techniques, 29
Quebec City, 3, 39, 57, 69, 95, 97, 165, 360, 446,
Queens, 203

Racial polarization, 4
Racial and residential proximity, 349
Racial separation, 356
Railroad system, 97
Railroads, 267
Raleigh-Durham, 532
Random spread, 277
Rank-size distributions, 74
Rapid transit, 205
Rate of urbanization, 35
Ravitch-Schwartz Commission, 483
Real estate industry, 389
Recompacting, 526–528
  cities, 441
  dispersed suburban developments, 526
Recycling of industrial land, 521
Redevelopment, 400, 401, 403
Redistribution of wealth, 108
Redistributive activities, 493
Regional growth, 37
Regional nucleations, 304
Regional variations, 36
Relative prices of raw materials, 116
Relocation decision, 381–382

Research and development, 81, 151, 152
Research-oriented universities, 156
Residential choice, 378
Residential communities, 346
  Full-care Retirement Homes, 346
  mobile elderly, 346
Residential housing developers, 220
Residential location, 375, 383, 426
  Model, 376, 378
Residential mobility, 346
Residential segregation, 351
Residualisation, 428
Resort cities, 40
Resource management, 508
Restructured manufacturing, 284
Restructuring, 271, 285, 373
Retail activities, 291
Retail nucleations, 300, 305
  hierarchy of, 305
  planned, 300
  retail trade, 162
  Seattle, 305
  unplanned, 300
Retirement age peak, 379
Revenue pooling, 506
Revenue resources, 494
Revenue-sharing, 503
Reverse commuting, 219, 422
Ribbons, 298
  highway-oriented ribbons, 299
  urban- arterial ribbons, 299
Richmond, 203
Rochester, 265
Route Choice Model, 458
Rural land, 257
Rural urban fringe, 228
Rural-urban land conversion research, 257

S-shaped (urbanization) curve, 33–35
Saint John, 98, 102
San Bernardino, 252
San Diego/Los Angeles/San Francisco, 17, 93
San Francisco, 15, 43, 48, 115, 156, 215, 223, 226, 251, 296, 333, 405, 471
San Francisco/Oakland, 180, 450
San Jose, 491
Scales of analysis, 25
Seattle, 17, 43, 121, 148, 209, 251, 305, 491
  metropolitan area, 305
Second long wave, 99
Second order centers, 48
Second ranked outflows, 109
"Second tier'" office location, 320

Second World War, 35
Secondary waste treatment, 518
Sector Model, 208
Sentiment/symbolism, 352
Separation/divorce, 414
Service activities, 116
Service charges, or user fees, 492
Service society, 66, 81, 106
Service employment sector, 163
Sexual division of labor, 441
Shared domestic work, 433
Sherman Anti-Trust Act of 1890, 76
Shift to multi-ethnicity, 10
Shopping malls, 305, 308, 326
Silicon Valley, 43, 141
Single tenant facilities, 311
Single-parent housing, 87
Site selection, 39–40
Slow growth urban areas, 156
Snow belt, 86
Snow Belt/Sun Belt Shifts, 14, 86,
 131, 132
 in population, 11, 521
Social area analysis, 331–333
Social area constructs, 338
Social classes, 201
Social/demographic processes, 9
Social distance, 330
Social fixed overhead capital, 141
Social justice, 388
Social, or class, rent, 246
Social policy development, 122
Social Rank Construct, 334
Social/Recreational Trips, 453
Social spaces, 327, 358
Social wage, 440
Socio-economic trends, 87
South Florida Water Management
 Water District, 515
Spatial Entrapment of Women Hy-
 pothesis, 425
Spatial interaction, 41
Spatial organization, 42, 69
Spatial patterns, 3, 74
Spatial processes, 48
Spatial separation of classes, 199
Spatial spread of cities, 230
Spatial structure, 93
Special District Boards, 497
Specialized areas, 295–298
 automobile rows, 296
 exotic markets, 298
 household-furnishing districts,
  298
 medical districts, 298
 "niche" markets, 295
 theme Malls, 298
Specific zoning, 260

Spillovers, 495, 496
Springfield, 14
St. John, 95, 111
St. John's, 95, 98, 102, 111
Stable Urban Areas in Transition,
 156
Staples economy, 108
Steam engine, 97
Streetcar, 203
Streetcar suburbs, 204
Structural reforms, 500
Structuralist Approaches, 21
Suburban, 196
 congestion, 462
 growth era, 196
 mall, 294
 office parks, 320
 offices, 320
 reinvention of Urban North
  America, 211
 retail shopping malls, 291
 sprawl, 220
 tracts, 225
Suburbanization, 4, 9, 81, 111, 114,
 203, 216, 254, 305
 arcadian environments, 81
Sudbury, 138, 519
Sun belt, 86, 89
Supply-side subsidies, 371
Surplus product, 38
Sustainability, 522, 523, 524
 and Modal Choice, 522
 and private automobiles, 522
 fuel taxes, 523
 transport systems, 523
Sustainable cities, 13
Sustainable development, 13, 508,
 509, 521, 540
Sustainable living, 509
Sustainable urban development,
 507, 540
Symmetrical families, 432

Tangible assets, 44
Techniques for analysis, 26
Technological and process innova-
 tion, 160
Technological control, 153
Technological innovation, 213
Teitz Model, 321
Tertiary activities, 106, 162
Tertiary sector, 81
The BART Experience, 469
Theory of Office Location, 318
Thompson, 138
Thresholds, 167, 171
Thunder Bay, 190
Times Square, 364, 365

Tokyo, 115
Toronto, 15, 43, 48, 52, 54, 97, 100,
 102, 106, 109, 111, 114, 117,
 120, 121, 189, 217, 226, 242,
 251, 264, 295, 301, 342, 360,
 437, 446, 450, 455, 503, 517
Toyota production system, 145
Trade areas, 44
Traditional Shopping Streets, 303
Transactional centers, 450
Transfer zoning, 260
Transnational corporations, 44
Transnational social systems, 349
Transport, costs, 139
 innovations, 199
 network, 41, 100
Transport system modelling, 457,
 459
Transport technology, 139
Transport-oriented settlements, 40
Transportation models, 41
Transportation problem, 324
Transportation, 162, 195, 460, 445,
 521
 airline terminals, 445
 changes, 178
 consumer of energy, 521
 industry, 57
 intra- and interurban, 445
 models, 41
 monocentric and polycentric urban structures, 460
 motor vehicle ownership, 521
 policies, 216
 rails, 445
 roads, 445
 sustainability, 521
Transshipment points, 138
Travel times, 453
Trip Distribution Model, 458
Truck, 215
 impact of, 215
 transport intra- and inter-urban,
  447
Two-parent households, 410
Two-tier local government, 505

U.S. Civil War (1861–1865), 65
U.S. domestic industry, 145
U.S. wholesale prices, 64
Unconditional grants, 492, 493
Underclass debate, 361
Unintended Consequences, 372
Unitary Government, 493
 suburban sprawl, 501
Unplanned Nucleations, 210, 300, 307
Urban agglomeration economies, 280
Urban analysis, 18

Urban areas, 3, 4, 13, 39, 40, 42, 43, 52, 53, 57, 58, 123, 161, 198, 199
  as centers of manufacturing, 123
  as centers of service activities, 161
  as foci of production, 39
  communication, 40
  concentration of surplus product, 40
  and economic strength, 52
  environment, 13
  growth of, 42
  highly educated cadre, 43
  hinterlands, 40
  industrialization, 123
  interaction between, 40
  location of, 39
  manufacturing in, 123
  productivity of, 57
  site, 39
  situation, 39
  spread, 42
Urban consumption, 255, 256
Urban development, 9, 60, 64, 94, 101, 257, 258, 540
  demographic structure, 9
  direct and indirect impact of, 257
  evolution of, 60
  foreign immigration, 9
  internal migration, 9
  major eras, 64
  net natural increase, 9
  waves of, 60
Urban form, 525
Urban freeways, 216
Urban growth, 3, 13–14, 76, 81, 104, 106
  bases of, 104
  and economic linkages, 106
  and ethnic structure, 104
  and immigration, 104
  and Kondratieff long-waves, 77
  rates, 116
  and service activities, 106
  and tertiary activities, 106
  within the periphery, 116, 117
Urban hierarchy, 51, 52
Urban infrastructure, 61, 196
Urban land, 232, 255

Urban living, 431
Urban multiplier, 29, 53
Urban nodes, 41
Urban parks, 12
Urban patterns, 528
Urban phenomena, 18, 25
Urban places, 96
Urban population redistribution, 247
Urban renewal, 371
Urban restructuring, 221
Urban retail structure, 294
Urban revitalization, 367, 399
Urban social space, 342
Urban space, 199
Urban spatial structure, 446, 461
  journey to work, 460
  modal choice, 460
  monocentric, 446
  polycentric, 446
  social and racial polarization, 446
  trips related to work, 460
Urban structure, 226, 420
Urban system, 71, 79, 86, 87, 92, 97, 108
  basic nodes, 97
  changing spatial organization, 108
  development, 67
  diversified economies, 92
  emerging, 71
  Growth, 48
  organization of, 86, 92, 97
  population, 71
  transaction centers, 92
Urban systems, 30
Urban Transport Model System, 457
Urban transport systems, 459
Urbanization, 33, 59, 332
  alienation, 332
  curve S-shaped, 33
  differentiation, 332
  economies, 141
  rural society, 33
  social disorientation, 332
User fees, 489
User-associated externalities, 325
UTMS Planning System, 459

Vacancy rate, 370
Value added, 125, 140

Vancouver, 43, 100, 102, 109, 111, 114, 120, 129, 189, 226, 252, 265, 270, 405, 450, 503, 540
Ventura, 252
Volume of traffic, 460
Volumes of trips, 466

Wage rates, 23, 140
Wage-gap, 440
Washington, D.C., 39, 48, 52, 215, 223, 455, 482, 486
Washington Metropolitan Area Transit Authority (WMATA), 471
Waterfront redevelopment, 403
Weight-loss, 15, 136
White flight, 254
White population, 221
Wholesale and distribution activities, 311
Wholesale trade, 162
Wichita, 3, 333
Windsor, 10, 97, 101, 446
Windsor-Quebec corridor, 217
Winnipeg, 100, 102, 109, 111, 129, 189, 501
Women, 21, 87, 114, 408–437
  homeless, 21
  in the paid labor force, 87
  in men's cities, 417
  sexual division of labor, 417
  socially constructed boundaries, 408
  spatially entrapped, 425
  Time-Crunch, 432
  wage-gap, 417
Women-friendly urban areas, 439
Woodstock, 257
Workplace-residence link, 221
World War II, 519
World cities, 156, 477
"World City" Hypothesis, the, 181

Zoning, 258–261
  and land use change, 261
  and performance standards, 260
  and planning, 260
  hierarchical zoning, 260
  for separation of uses, 259
  specific zoning, 260
  types of, 259